Meromorphic Dynamics

Volume I

This text, the first of two volumes, provides a comprehensive and self-contained introduction to a wide range of fundamental results from ergodic theory and geometric measure theory. Topics covered include finite and infinite abstract ergodic theory, Young towers, measure-theoretic Kolmogorov–Sinai entropy, thermodynamics formalism, geometric function theory, various kinds of conformal measures, conformal graph directed Markov systems and iterated functions systems, semi-local dynamics of analytic functions, and nice sets. Many examples are included, along with detailed explanations of essential concepts and full proofs, in what is sure to be an indispensable reference for both researchers and graduate students.

Janina Kotus is Professor of Mathematics at the Warsaw University of Technology, Poland. Her research focuses on dynamical systems, in particular holomorphic dynamical systems. Together with I.N. Baker and Y. Lü, she laid the foundations for iteration of meromorphic functions.

Mariusz Urbański is Professor of Mathematics at the University of North Texas, USA. His research interests include dynamical systems, ergodic theory, fractal geometry, iteration of rational and meromorphic functions, open dynamical systems, random dynamical systems, iterated function systems, Kleinian groups, Diophantine approximations, topology, and thermodynamic formalism. He is the author of 8 books, 7 monographs, and more than 200 papers.

NEW MATHEMATICAL MONOGRAPHS

Already Published

6 S. Berhanu, P. D. Cordaro and J. Hounie *An Introduction to Involutive Structures*
7 A. Shlapentokh *Hilbert's Tenth Problem*
8 G. Michler *Theory of Finite Simple Groups I*
9 A. Baker and G. Wüstholz *Logarithmic Forms and Diophantine Geometry*
10 P. Kronheimer and T. Mrowka *Monopoles and Three-Manifolds*
11 B. Bekka, P. de la Harpe and A. Valette *Kazhdan's Property (T)*
12 J. Neisendorfer *Algebraic Methods in Unstable Homotopy Theory*
13 M. Grandis *Directed Algebraic Topology*
14 G. Michler *Theory of Finite Simple Groups II*
15 R. Schertz *Complex Multiplication*
16 S. Bloch *Lectures on Algebraic Cycles (2nd Edition)*
17 B. Conrad, O. Gabber and G. Prasad *Pseudo-reductive Groups*
18 T. Downarowicz *Entropy in Dynamical Systems*
19 C. Simpson *Homotopy Theory of Higher Categories*
20 E. Fricain and J. Mashreghi *The Theory of H(b) Spaces I*
21 E. Fricain and J. Mashreghi *The Theory of H(b) Spaces II*
22 J. Goubault-Larrecq *Non-Hausdorff Topology and Domain Theory*
23 J. Śniatycki *Differential Geometry of Singular Spaces and Reduction of Symmetry*
24 E. Riehl *Categorical Homotopy Theory*
25 B. A. Munson and I. Volić *Cubical Homotopy Theory*
26 B. Conrad, O. Gabber and G. Prasad *Pseudo-reductive Groups (2nd Edition)*
27 J. Heinonen, P. Koskela, N. Shanmugalingam and J. T. Tyson *Sobolev Spaces on Metric Measure Spaces*
28 Y.-G. Oh *Symplectic Topology and Floer Homology I*
29 Y.-G. Oh *Symplectic Topology and Floer Homology II*
30 A. Bobrowski *Convergence of One-Parameter Operator Semigroups*
31 K. Costello and O. Gwilliam *Factorization Algebras in Quantum Field Theory I*
32 J.-H. Evertse and K. Gyry *Discriminant Equations in Diophantine Number Theory*
33 G. Friedman *Singular Intersection Homology*
34 S. Schwede *Global Homotopy Theory*
35 M. Dickmann, N. Schwartz and M. Tressl *Spectral Spaces*
36 A. Baernstein II *Symmetrization in Analysis*
37 A. Defant, D. García, M. Maestre and P. Sevilla-Peris *Dirichlet Series and Holomorphic Functions in High Dimensions*
38 N. Th. Varopoulos *Potential Theory and Geometry on Lie Groups*
39 D. Arnal and B. Currey *Representations of Solvable Lie Groups*
40 M. A. Hill, M. J. Hopkins and D. C. Ravenel *Equivariant Stable Homotopy Theory and the Kervaire Invariant Problem*
41 K. Costello and O. Gwilliam *Factorization Algebras in Quantum Field Theory II*
42 S. Kumar *Conformal Blocks, Generalized Theta Functions and the Verlinde Formula*
43 P. F.X. Müller *Hardy Martingales*
44 T. Kaletha and G. Prasad *Bruhat-Tits Theory*
45 J. Schwermer *Reduction Theory and Arithmetic Groups*
46 J. Kotus and M. Urbański *Meromorphic Dynamics I*
47 J. Kotus and M. Urbański *Meromorphic Dynamics II*

Meromorphic Dynamics

Abstract Ergodic Theory, Geometry, Graph Directed Markov Systems, and Conformal Measures

VOLUME I

JANINA KOTUS
Warsaw University of Technology

MARIUSZ URBAŃSKI
University of North Texas

Shaftesbury Road, Cambridge CB2 8EA, United Kingdom

One Liberty Plaza, 20th Floor, New York, NY 10006, USA

477 Williamstown Road, Port Melbourne, VIC 3207, Australia

314–321, 3rd Floor, Plot 3, Splendor Forum, Jasola District Centre,
New Delhi – 110025, India

103 Penang Road, #05–06/07, Visioncrest Commercial, Singapore 238467

Cambridge University Press is part of Cambridge University Press & Assessment,
a department of the University of Cambridge.

We share the University's mission to contribute to society through the pursuit of
education, learning and research at the highest international levels of excellence.

www.cambridge.org
Information on this title: www.cambridge.org/9781009215916

DOI: 10.1017/9781009215930

First published 2023

Printed in the United Kingdom by TJ Books Limited, Padstow Cornwall

A catalogue record for this publication is available from the British Library.

Library of Congress Cataloging-in-Publication data
Names: Kotus, Janina, author. | Urbański, Mariusz. author.
Title: Meromorphic dynamics / Janina Kotus, Warsaw University of
Technology, Mariusz Urbański, University of North Texas.
Description: First edition. | Cambridge, United Kingdom ; New York, NY, USA:
Cambridge University Press, 2023. |
Series: Nmm new mathematical monographs | Includes bibliographical references and index. |
Contents: Volume 1. abstract ergodic theory, geometry, graph directed Markov systems,
and conformal measures – Volume 2. Elliptic functions with an introduction to
the dynamics of meromorphic functions.
Identifiers: LCCN 2022031269 (print) | LCCN 2022031270 (ebook) |
ISBN 9781009216050 (set ; hardback) | ISBN 9781009215916
(volume 1 ; hardback) | ISBN 9781009215978 (volume 2 ; hardback) |
ISBN 9781009215930 (volume 1 ; epub) | ISBN 9781009215985 (volume 2 ; epub)
Subjects: LCSH: Functions, Meromorphic.
Classification: LCC QA331 .K747 2023 (print) | LCC QA331 (ebook) |
DDC 515/.982–dc23/eng20221102
LC record available at https://lccn.loc.gov/2022031269
LC ebook record available at https://lccn.loc.gov/2022031270

ISBN – 2 Volume Set 978-1-009-21605-0 Hardback
ISBN – Volume I 978-1-009-21591-6 Hardback
ISBN – Volume II 978-1-009-21597-8 Hardback

Janina Kotus dedicates this book to the memory of her sister Barbara.
Mariusz Urbański dedicates the book to his family.

Contents

Volume I

PART II COMPLEX ANALYSIS, CONFORMAL MEASURES, AND GRAPH DIRECTED MARKOV SYSTEMS

Volume II

Preface

The ultimate goal of our book is to present a unified approach to the dynamics, ergodic theory, and geometry of elliptic functions mapping the complex plane $\mathbb{C}$ onto the Riemann sphere $\widehat{\mathbb{C}}$. We consider elliptic functions as the most regular class of transcendental meromorphic functions. Poles, infinitely many of them, form an essential feature of such functions, but the set of critical values is finite and an elliptic function is "the same" in all of its fundamental regions. In a sense, this is the class of transcendental meromorphic functions whose resemblance to rational functions on the Riemann sphere is the largest. This similarity is important since the class of rational functions has been, from the dynamical point of view, extensively investigated since the pioneering works of Pierre Fatou [**Fat1**] and Gaston Julia [**Ju**]; also see the excellent historical accounts in [**Al**] and [**AIR**] on the early days of holomorphic dynamics. This similarity can be, and frequently was, a good source of motivation and guidance for us when we were dealing with elliptic functions. On the other hand, the differences are striking in many respects, including topological dynamics, measurable dynamics, fractal geometry of Julia sets, and more. We will touch on them in the course of this Preface. We would just like to stress here that elliptic functions belong to the class of transcendental meromorphic functions and posseses many dynamical and geometric features that are characteristic for this class.

Indeed, the study of iteration of transcendental meromorphic functions, more precisely of transcendental entire functions, began with the pioneering works of Pierre Fatou ([**Fat2**] and [**Fat3**]). Then for about two decades, beginning with paper [**Ba1**], I.N. Baker was actually the sole mathematician dealing with the dynamics of transcendental entire functions. It was Janina Kotus's idea to study the iteration of meromorphic functions despite the existence of poles that, in general, cause the second iterate to not be defined everywhere; more precisely, because it has finite essential singularities, thus, it is not a meromorphic function defined on the complex plane. This is the phenomenon that had been deterring mathematicians from dealing with the iteration of general meromorphic functions. Janina was not afraid and as a result, to our knowledge, the first, and quite systematic, account of the dynamics of general transcendental meromorphic functions was set up in a series of works by I.N. Baker, J. Kotus, and Y. Lü ([**BKL1**]–[**BKL4**]). Since then this subfield of dynamical systems has been flourishing. Of great importance for the development of this subject was the excellent expository article by Walter Bergweiler [**Ber1**]. We would also like to mention an early paper by Alexander Eremenko and Misha Lyubich [**EL**], who introduced and studied class $\mathcal{B}$ of transcendental entire functions. The definition of class $\mathcal{B}$ literally extends to the class of general meromorphic functions and plays an important role in this field too.

The area of transcendental meromorphic dynamical systems is also a beautiful and vast field for investigations of the measurable dynamics they generate and the fractal geometry of their Julia sets and their significant subsets. Here, the early papers by Misha Lyubich [**Ly**] and Mary Rees [**Re**] on measurable dynamics come to the fore. The study of the fractal geometry of Julia sets of meromorphic functions began with two papers by Gwyneth Stallard ([**Sta1**] and [**Sta2**]) and has been continued ever since by her, Krzysztof Barański, Walter Bergweiler, Bogusia Karpińska, Volker Mayer, Phil Rippon, Lasse Rempe-Gillen, Anna Zdunik, the authors of this book, and many more mathematicians, obtaining interesting and sophisticated results; we are not able to list all of them here. We would like, however, to mention the early paper by Anna Zdunik and Mariusz Urbański [**UZ1**], where the concept of conformal measures was adapted for and used in transcendental dynamics, and also Hausdorff and packing measures of (radial) Julia sets were studied in detail.

We would also like to single out a paper by Krzysztof Barański [**Ba**], who initiated the use of thermodynamic formalism in transcendental meromorphic dynamics. More papers using and developing thermodynamic formalism then followed, among them those by Janina Kotus [**KU1**], Anna Zdunik [**UZ2**], and Volker Mayer ([**MyU3**] and [**MyU4**]), all written jointly with Mariusz Urbański. As a source of detailed information on many aspects of the use of this method in meromorphic dynamics, we recommend the expository article by Volker Mayer and Mariusz Urbański [**MyU5**].

As already stated at the beginning of this Preface, we ultimately focus in this book on the dynamics, ergodic theory, and geometry of elliptic functions. To our knowledge, the first dynamical result specific to elliptic functions appeared in Janina Kotus's 1995 paper [**Kot**]. It gave a good lower bound for the Hausdorff dimensions of the Julia setd of elliptic functions. Its was refined, using the theory of countable alphabet conformal iterated function systems, in [**KU3**]. A later paper [**MyU1**] presents a form of thermodynamic formalism for elliptic functions. A quite long series of papers by Jane Hawkins and her collaborators, studying the dynamics and geometry of Weierstrass $\wp$-functions, began in 2002 with [**HK1**]. Our book stems from, has been motivated by, and largely develops our 2004 paper [**KU4**]

We devote the first chapter of the second volume to a rather short and compressed, albeit with proofs, introduction to the topological dynamics of transcendental meromorphic functions. We then move on to elliptic functions, giving first some short, but with proofs, expositions of classical properties of such functions, and then we deal with their measurable dynamics and fractal geometry. We single out several dynamically significant subclasses of elliptic functions, primarily nonrecurrent, compactly nonrecurrent, subexpanding, and parabolic. We devote one long chapter to describing examples of elliptic functions with various properties of their dynamics, Fatou Connected Components, and the geometry of Julia sets.

Our approach to measurable dynamics and the fractal geometry of elliptic functions is founded on the concept of the Sullivan conformal measures. We prove their existence for all (nonconstant) elliptic functions and provide several characterizations, in dynamical terms, of the minimal exponent for which these measures exist. By using the method of conformal iterated function systems with a countable alphabet and essentially reproducing the proof from [**KU3**], we provide a simple lower bound, expressed in terms of orders of poles, of the Hausdorff dimensions of Julia sets of all elliptic functions. It follows from this estimate that such Hausdorff dimensions are always strictly larger than 1. We also provide a closed exact formula for the Hausdorff dimension of the set of points escaping to ∞. We then deal with the class of nonrecurrent elliptic functions $f\colon \mathbb{C} \to \widehat{\mathbb{C}}$ and their subclasses such as compactly nonrecurrent, subexpanding, and parabolic ones. This is the ultimate object of our interests in the book, especially in Volume 2. We would like to add that we do not give separate attention to hyperbolic/expanding elliptic functions; none of them allow critical points or rationally indifferent periodic points in their Julia sets. Indeed, doing this would take up a lot of pages and preparations, and a good account of the thermodynamic formalism of quite general classes of meromorphic functions is given in [**MyU3**] and [**MyU4**], as well as [**MyU2**] and [**MyU5**]. In this book, we focus on elliptic functions that may have critical points and rationally indifferent periodic points in the Julia sets; we do allow them, and our main objective is to deal with and study the various phenomena that they cause.

Our presentation of the theory of nonrecurrent elliptic functions is based on an appropriate version, which we prove, of Mañé's Theorem, which roughly speadking asserts that the connected components of all inverse images of all orders of all sufficiently small sets remain small. Having a Sullivan conformal measure m with a minimal exponent, we prove its uniqueness and atomlessness for compactly nonrecurrent regular elliptic functions.

Next, we prove the existence and uniqueness (up to a nonzero multiplicative factor) of a σ-finite invariant measure μ that is absolutely continuous with respect to the conformal measure m. We prove its ergodicity and conservativity. Restricting our attention to the classes of subexpanding and parabolic functions, in fact to some natural subclasses of them, we prove much more refined stochastic properties of the dynamical system (f, μ). Our approach here stems from and largely develops the methods developed in papers [**ADU**], [**DU4**], [**U3**], and [**U4**]. It is, however, significantly enlarged and enriched, via the powerful tool of nice sets, by the methods of countable alphabet conformal iterated function systems and by graph directed Markov systems as developed and presented in [**MU1**] and [**MU2**]. These in turn are substantially based on the theory of countable alphabet thermodynamic formalism developed in [**MU4**] and [**MU2**]. When dealing with subexpanding functions, especially with the exponential shrinking property, the paper by Przytycki and Rivera-Letelier [**PR**] was also very useful. The finer stochastic properties mentioned above are primarily the exponential decay of correlations, the Central Limit Theorem, and the Law of the Iterated Logarithm in the case of subexpanding elliptic functions. In the case of parabolic elliptic functions for which the invariant measure μ is finite, we prove the Central Limit Theorem. All of these are achieved with the help of the Lai-Sang Young tower techniques from [**LSY3**]. In the case of parabolic elliptic functions for which the invariant measure μ is infinite, we prove an appropriate version of the Darling–Kac Theorem, establishing the strong convergence of weighted Birkhoff averages to Mittag–Leffler distributions.

Last, we would like to mention finer fractal geometry. For both subexpanding and parabolic elliptic functions, we give a complete description and characterization of conformal measures and Hausdorff and packing measures of Julia sets. Because the Hausdorff dimension of the Julia set of an elliptic function is strictly larger than 1, this picture is even simpler than for the subexpanding, parabolic, and nonrecurrent rational functions given in [**DU4**], [**DU5**], and [**U3**].

In order to comprehensively cover the dynamics and geometry of elliptic functions described above, we extensive large preparations. This is primarily done in the first volume of the book, which consists of two parts: Part I, "Ergodic Theory and Geometric Measures" and Part II,"Complex Analysis, Conformal Measures, and Graph Directed Markov Systems." We intend our book to be as self-contained as possible and we use essentially all the major results from Volume 1 in Volume 2 when dealing with dynamics, ergodic theory, and the geometry of elliptic functions.

Our book can thus be treated not only as a fairly comprehensive account of dynamics, ergodic theory, and the fractal geometry of elliptic functions but also as a reference book (with proofs) for many results of geometric measure theory, finite and infinite abstract ergodic theory, Young towers, measure-theoretic Kolmogorov–Sinai entropy, thermodynamic formalism, geometric function theory (in particular the Koebe Distortion Theorems and Riemann–Hurwitz Formulas), various kinds of conformal measures, conformal graph directed Markov systems and iterated function systems, the classical theory of elliptic functions, and the topological dynamics of transcendental meromorphic functions.

The material contained in Volume 1 of this book, after being substantially processed, collects, with virtually all proofs, the results that are essentially known and have been published. However, Chapter 5 contains material on infinite ergodic theory that, to the best or our knowledge, has not been included, with full proofs, in any prior book. Also, Chapter 12, which treats nice sets, is strongly processed and goes, in many respects, far beyond the existing knowledge and use of nice sets and nice families in conformal dynamics. The need for such far-reaching extensions of this method comes from the need for its applications to parabolic elliptic functions.

Most of the material at the end of the second volume of this book is actually new, although we borrow, use, and apply much from the previous results, methods, and techniques. Indeed, Chapter 17, except its last two sections, is entirely new. Also, to the best of our knowledge, Section 19.4 is purely original, providing large classes of a variety of simple examples of various kinds of dynamically elliptic functions. Part VI is, indeed, entirely original.

Acknowledgments

We are very indebted to Jane Hawkins, who gave for us the four images of Julia sets of various Weierstrass elliptic $\wp$-functions that are included in Chapter 19 of Volume 2 of this book. We also thank her very much for fruitful discussions about the dynamics and Julia sets of Weierstrass elliptic $\wp$-functions. Mariusz Urbański would like to thank William Cherry for helpful discussions about normal families and Montel's Theorems. Last, but not least, we want to express our gratitude to the referees of this book, whose numerous valuable remarks, comments, and suggestions prompted us to improve the content of the book and the quality of its exposition. The research of Janina Kotus was supported by the National Science Center, Poland, decision no. DEC-2019/33/B/ST1/00275. The reseach of Mariusz Urbański was supported in part by the NSF grant DMS 1361677.

Introduction

In this introduction, we describe in detail the content of the first volume of the book, simultaneously highlighting its sources and the interconnections between various fragments of the book.

In Part I, "Ergodic Theory and Geometric Measures," we first bring up in Chapter 1, with proofs, some basic and fundamental concepts and theorems from abstract and geometric measure theory. These include, in particular, the three classical covering theorems: $4r$, Besicovitch, and Vitali type. We also include a short section on probability theory: conditional expectations and martingale theorems. We devote a large amount of space to treating Hausdorff and packing measures. In particular, we formulate and prove Frostman Converse Lemmas, which form an indispensable tool for proving that a Hausdorff or packing measure is finite, positive, or infinite. Some of them are frequently called, in particular in the fractal geometry literature, the mass redistribution principle, but these lemmas involve no mass redistribution. We then deal with Hausdorff, packing, and box counting, as well as the dimensions of sets and measures, and provide tools to calculate and estimate them.

In the next four chapters of Part I, we deal with classical ergodic theory, both finite (probability) and, which we would like to emphasize, also infinite.

In Chapter 2, we deal with both finite and infinite invariant measures. We start with quasi-invariant measures and early on, in the second section of this chapter, we introduce the powerful concept of the first return map. This concept, along with that of nice sets (see Chapter 12), will form our most fundamental tool in Part IV in Volume 2 of our book, which is devoted to presenting a refined ergodic theory of elliptic functions. We introduce in Chapter 2 the notions of ergodicity and conservativity (always satisfied for finite invariant measures) and prove the Poincaré Recurrence Theorem, the Birkhoff Ergodic Theorem, and the Hopf Ergodic Theorem, the last pertaining to infinite measures. We also provide a powerful, though perhaps somewhat neglected by the ergodic community, tool for proving the existence of invariant σ-finite measures that are absolutely continuous with respect to given quasi-invariant measures. This stems from the work of Marco Martens [**Mar**].

We then prove, in Chapter 3, the Bogolyubov–Krylov Theorem about the existence of Borel probability invariant measures for continuous dynamical systems acting on compact metrizable topological spaces. We also establish in this chapter the basic properties of invariant and ergodic measures and provide a large variety of examples of such measures.

Chapter 4 is devoted to the stochastic laws for measurable endomorphisms, preserving a probability measure that are finer than the mere Birkhoff Ergodic Theorem. Under appropriate hypotheses, we prove the Law of the Iterated Logarithm. We then describe another powerful method of ergodic theory, namely L. S. Young towers (recently also frequently called Kakutani towers), which she developed in [**LSY2**] and [**LSY3**]; see also [**Go1**] for further progress. With appropriate assumptions imposed on the first return time, the Young construction yields the exponential decay of correlations, the Central Limit Theorem, and the Law of the Iterated Logarithm, which follows too.

In Chapter 5, we deal with refined stochastic laws for dynamical systems preserving an infinite measure. This is primarily the Darling–Kac Theorem. We make use of some recent progress on this theorem and related issues, mainly due to Zweimm̈uller, Thaler, Theresiu, Melbourne, Gouëzel, Bruin, Aaronson, and others, but we do not go into the most recent subtleties and developments of this branch of infinite ergodic theory. We do not need them for our applications to elliptic functions.

In Chapter 6, we provide a classical account of Kolmogorov–Sinai metric entropy for measure-preserving dynamical systems. We prove the Shannon–McMillan–Breiman Theorem and, based on the Abramov formula, define the concept of the Krengel entropy of a conservative system preserving a (possibly infinite) invariant measure.

In Chapter 7, the last chapter of Part I, we collect and prove the basic concepts and theorems of classical thermodynamic formalism. This includes topological pressure, the variational principle, and equilibrium states. We have written this chapter very rigorously, taking care of some inaccuracies which have persisted in some expositions of thermodynamic formalism since the 1970s and 1980s; we have not singled them out explicitly. We do not deal with Gibbs states in this chapter.

We provide many examples in Part I of the book: mainly of invariant measures, ergodic invariant measures, and counterexamples of infinite ergodic theory. We do emphasize once more that we treat the latter with care and detail.

Part II, "Complex Analysis, Conformal Measures, and Graph Directed Markov Systems," devotes the first chapter, Chapter 8, to some selected topics of geometric function theory. Its character is entirely classical, meaning that no dynamics is involved. We deal here at length with Riemann surfaces, normal families, and Montel's Theorem, extremal lengths, and moduli of topological annuli. However, we think that the central theme of this chapter consists of various versions of the Koebe Distortion Theorems. These theorems, which were proved in the early years of the nineteenth century by Koebe and Bieberbach, form a beautiful, elegant, and powerful tool for complex analysis. We prove them carefully and provide their many versions of analytic and geometric character. These theorems also form an absolutely indispensable tool for nonexpanding holomorphic dynamics and their applications occur very frequently throughout the book; most notably when dealing with holomorphic inverse branches, conformal measures, and Hausdorff and packing measures. The version of the Riemann–Hurwitz Formula, appropriate in the context of transcendental meromorphic functions, which we treat at length in the last section of Chapter 8, is a very helpful tool to prove the existence of holomorphic inverse branches and an elegant and probably the best tool to control the topological structure of connected components of inverse images of open connected sets under meromorphic maps, especially to make sure that such connected components are simply connected. Another reason why we devoted a lot of time to the Riemann–Hurwitz Formula is that in the standard monographs on Riemann surfaces this formula is usually formulated only for compact surfaces and its proofs are somewhat sketchy; we do need, as a matter of fact almost exclusively, the noncompact case. Our approach to the Riemann–Hurwitz Formula stems from that of Alan Beardon in [**Bea**], which is designed to deal with rational functions of the Riemann sphere. We modify it to fit to our context of transcendental meromorphic functions.

In Chapter 9, we encounter holomorphic dynamics for the first time in the book. Its settings are somehow technical and it has, on the one hand, a very preparatory character serving the needs of constructing and controlling the Sullivan conformal measures in various subsequent parts of the book; for example, in Chapter 10. On the other hand, this chapter is important and interesting on its own. Indeed, its hypotheses are very general and flexible, and under such weak assumptions it establishes in the context of holomorphic dynamics such important results as Pesin's Theory, Ruelle's Inequality, and Volume Lemmas.

In Chapter 10, we encounter for the first time the beautiful, elegant, and powerful concept of conformal measures, which is due to Patterson (see [**Pat1**] and also [**Pat2**]) in the context of Fuchsian groups and due to Sullivan in the context of all Kleinian groups and rational functions of the Riemann sphere (see [**Su2**]–[**Su7**]). We deal in Chapter 10 with conformal measures being in the settings of the previous chapter, namely Chapter 9. The Sullivan conformal measures and their invariant versions will form the central theme of Volume 2, starting from Chapter 17 onward. In fact, Chapter 10 is the first and essential step for the construction of the Sullivan conformal measures for general elliptic functions that is completed in the last two parts of the second volume of the book. It deals with holomorphic maps f defined on some open neighborhood of a compact f-invariant subset X of a parabolic Riemann surface. We provide a fairly complete account of the Sullivan conformal measures in such a setting. We also introduce in Chapter 10 several dynamically significant concepts and sets, such as radial or conical points and several fractal dimensions defined in dynamical terms. We relate them to exponents of conformal measures,

However, motivated by [**DU1**] and choosing the most natural, at least in our opinion, framework, we do not restrict ourselves to conformal dynamical systems only but present in the first section of Chapter 10 a fairly complete account of the theory of general conformal measures.

Chapter 11 deals with conformal graph directed Markov systems, its special case of iterated function systems, and thermodynamic formalism of countable alphabet subshifts of finite type, frequently also called topological Markov chains. This theory started in the papers [**MU1**] and [**MU4**] and the book [**MU2**]. It was in [**MU1**] and [**MU2**] that the concept of conformal measures due to Patterson and Sullivan was adapted to the realm of conformal graph directed Markov systems and iterated function systems. We present some elements of this theory in Chapter 11, primarily those related to conformal measures and a version of Bowen's Formula for the Hausdorff dimension of limit sets of such systems. In particular, we get an almost cost-free, effective, lower estimate for the Hausdorff dimension of such limit sets. More about conformal graph directed Markov systems can be found in many papers and books, such as [**MU3**]–[**MU5**], [**MPU**], [**MSzU**], [**U5**], [**U6**], [**CTU**], [**CLU1**], [**CLU2**], and [**CU**].

Afterwards, in Parts IV and VI (the second volume of the book), we apply the techniques developed here to get quite a good, explicit estimate from below of Hausdorff dimensions of Julia sets of all elliptic functions and to explore stochastic properties of invariant versions of conformal measures for parabolic and subexpanding elliptic functions. Getting these stochastic properties is possible for us by combining several powerful methods which we have already mentioned. Namely, having proved the existence of nice and pre-nice sets (see Chapter 12 for their definition and construction), it turns out that the holomorphic inverse branches of the first return map that these nice sets generate form a conformal iterated function system. So, the whole theory of conformal graph directed Markov systems applies and we also enhance it by the Young tower techniques developed in [**LSY2**] and [**LSY3**]; see also [**Go1**].

As already discussed in the previous paragraph, in the last chapter of this volume, Chapter 12, "Nice Sets for Analytic Maps," we introduce and thoroughly study the objects related to the powerful concept of nice and pre-nice sets, which will be our indispensable tool in the last part of the second volume of the book, Part VI, "Compactly nonrecurrent Elliptic Functions: Fractal Geometry, Stochastic Properties, and Rigidity," leading, along with the theory of conformal graph directed Markov systems, the first return map method, and the techniques of Young towers, to such stochastic laws as the exponential decay of correlations, the Central Limit Theorem, and the Law of the Iterated Logarithm, which follows for large classes of elliptic functions constituted by subexpanding and parabolic ones. However, the main objective of the discussed chapter is to prove the existence of nice and pre-nice sets.

PART I

Ergodic Theory and Geometric Measures

1

Geometric Measure Theory

1.1 Measures, Integrals, and Measure Spaces

This section has an introductory character. It collects a minimum of knowledge from abstract measure theory needed in subsequent chapters of the book. Most, commonly well known, theorems are brought up without proofs. A full account of measure theory can be found in many books, e.g., [**Coh**], [**Fr**], [**RF**].

Definition 1.1.1 A family $\mathfrak{F}$ of subsets of a set X is said to be a σ-algebra if and only if the following conditions are satisfied:

$$X \in \mathfrak{F}, \tag{1.1}$$

$$A \in \mathfrak{F} \Rightarrow A^c, \in \mathfrak{F}, \tag{1.2}$$

$$\{A_i\}_{i=1}^{\infty} \subseteq \mathfrak{F} \Longrightarrow \bigcup_{i=1}^{\infty} A_i \in \mathfrak{F}. \tag{1.3}$$

It follows from this definition that $\emptyset \in \mathfrak{F}$, i.e., that the σ-algebra $\mathfrak{F}$ is closed under countable intersections and under subtractions of sets. If (1.3) is assumed only for finite subfamilies of $\mathfrak{F}$, then $\mathcal{F}$ is called an algebra. The elements of the σ-algebra $\mathfrak{F}$ are frequently called measurable sets.

Definition 1.1.2 For any family $\mathfrak{F}$ of subsets of X, we denote by $\sigma(\mathfrak{F})$ the least σ-algebra that contains $\mathfrak{F}$, and we call it the σ-algebra generated by $\mathfrak{F}$.

Definition 1.1.3 A function on a σ-algebra $\mathfrak{F}$, $\mu\colon \mathfrak{F} \to [0, +\infty]$, is said to be σ-additive or countably additive if, for any countable subfamily $\{A_i\}_{i=1}^{\infty}$ of $\mathfrak{F}$ consisting of mutually disjoint sets, we have that

$$\mu\left(\bigcup_{i=1}^{\infty} A_i\right) = \sum_{i=1}^{\infty} \mu(A_i). \tag{1.4}$$

We say then that μ is a measure.

If we consider in (1.4) only finite families of sets, we say that μ is additive. The two notions of additivity and of σ-additivity make sense for a σ-algebra as well as for an algebra, provided that, in the case of an algebra, one considers only families $\{A_i\}_{i=1}^{\infty} \subseteq \mathfrak{F}$ such that $\bigcup_{i=1}^{\infty} A_i \in \mathfrak{F}$. The simplest consequences of the definition of measure are the following:

$$\mu(\emptyset) = 0. \tag{1.5}$$

$$\text{If } \ A, B \in \mathfrak{F} \text{ and } \ A \subseteq B, \ \text{ then } \mu(A) \leq \mu(B). \tag{1.6}$$

$$\text{If } A_1 \subseteq A_2 \subseteq \cdots \ \text{ and } \{A_i\}_{i=1}^{\infty} \subseteq \mathfrak{F}, \text{ then } \mu\left(\bigcup_{i=1}^{\infty} A_i\right) = \sup_i \mu(A_i) = \lim_{i\to\infty} \mu(A_i). \tag{1.7}$$

Definition 1.1.4 We say that the triple $(X, \mathfrak{F}, \mu)$ with a σ-algebra $\mathfrak{F}$ and μ, a measure on $\mathfrak{F}$, is a measure space. If $\mu(X) = 1$, the triple $(X, \mathfrak{F}, \mu)$ is called a probability space and μ is a probability measure.

Definition 1.1.5 We say that $\varphi\colon X \to \mathbb{R}$ is a measurable function if $\varphi^{-1}(J) \in \mathfrak{F}$ for every interval $J \subseteq \mathbb{R}$, equivalently for every Borel set $J \subseteq \mathbb{R}$.

Throughout the book, for any set $A \subseteq X$, we denote by $\mathbb{1}_A$ the characteristic function of the set A:

$$\mathbb{1}_A(x) = \begin{cases} 1 & \text{if } x \in A \\ 0 & \text{if } x \notin A. \end{cases}$$

A step function is a linear combination of (finitely many) characteristic functions. It is easy to see that any nonnegative measurable function $\varphi\colon X \to \mathbb{R}$ can be represented as the pointwise limit of a monotone increasing sequence of nonnegative step functions, say

$$\varphi = \lim_{n\to\infty} \varphi_n.$$

The integral of φ against the measure μ is then defined as:

$$\int_X \varphi \, d\mu := \lim_{n\to\infty} \int_X \varphi_n \, d\mu.$$

It is easy to see that this definition is independent of the choice of a sequence $(\varphi)_{n=1}^{\infty}$ of monotone increasing nonnegative step functions. Writing any (not necessarily nonnegative) measurable function $\varphi\colon X \to \mathbb{R}$ in its canonical form

$$\varphi = \varphi_+ - \varphi_-,$$

where

$$\varphi_+ := \max\{\varphi, 0\} \text{ and } \varphi_- := -\min\{\varphi, 0\},$$

we say that the function φ is μ integrable if

$$\int_X \varphi_+ \, d\mu < +\infty \text{ and } \int_X \varphi_- \, d\mu < +\infty.$$

We then define the integral of φ against the measure μ to be

$$\int_X \varphi \, d\mu := \int_X \varphi_+ \, d\mu - \int_X \varphi_- \, d\mu.$$

The integral of φ is also frequently denoted by

$$\mu(\varphi).$$

Since $|\varphi| = \varphi_+ - \varphi_-$, we see that φ is integrable if and only if $|\varphi|$ is, i.e., if $\int_X |\varphi| d\mu < \infty$. We then write $\varphi \in L^1(\mu)$. We now bring up two fundmental properties of integrals – theose that make integrals such powerful and convenient tools.

Theorem 1.1.6 (Lebesgue Monotone Convergence Theorem) *Suppose that* $(\varphi)_{n=1}^{\infty}$ *is a monotone-increasing sequence of integrable, real-valued functions on a probability space* $(X, \mathfrak{F}, \mu)$. *Denote its limit by* φ. *Then*

$$\int_X \varphi \, d\mu = \lim_{n\to\infty} \int_X \varphi_n \, d\mu.$$

In particular, the above limit exists. As a matter of fact, it is enough to assume only that the sequence $(\varphi)_{n=1}^{\infty}$ *is monotone-increasing on a measurable set whose complement is of measure zero.*

Theorem 1.1.7 (Lebesgue Dominated Convergence Theorem) *Suppose that* $(\phi_n)_{n=1}^{\infty}$ *is a sequence of measurable, real-valued functions on a probability space* $(X, \mathfrak{F}, \mu)$, *that* $|\phi_n| \le g$ *for an integrable function* g, *and that the sequence* $(\phi_n)_{n=1}^{\infty}$ *converges* μ*-a.e. to a function* $\varphi\colon X \to \mathbb{R}$. *Then the function* ϕ *is* μ*-integrable and*

$$\int_X \varphi \, d\mu = \lim_{n\to\infty} \int_X \varphi_n \, d\mu.$$

More generally than $L^1(\mu)$, for every $1 \le p < \infty$, we write

$$||\varphi||_p := \left(\int_X |\varphi|^p d\mu\right)^{\frac{1}{p}}$$

and we say that φ belongs to $L^p(\mu) = L^p(X, \mathfrak{F}, \mu)$. If

$$\inf_{\mu(E)=0} \left\{ \sup_{X \setminus E} |\varphi| \right\} < \infty,$$

then we denote the latter expression by $||\varphi||_\infty$, we say that the function φ is essentially bounded, and we write that $\varphi \in L^\infty$. The numbers $||\varphi||_p$, $1 \le p < \infty$, are called L^p-norms of φ. The vector spaces $L^p(X, \mathfrak{F}, \mu)$ become Banach spaces when endowed with respective norms $|| \cdot ||_p$.

Definition 1.1.8 A measure space $(X, \mathfrak{F}, \mu)$ and the measure μ are called

- finite if $\mu(X) < +\infty$,
- probability if $\mu(X) = 1$,
- infinite if $\mu(X) = +\infty$,
- σ-finite if the space X can be expressed as a countable union of measurable sets with finite measure μ.

Given two measures μ and ν on the same measurable space $(X, \mathfrak{F})$, we say that μ is absolutely continuous with respect to ν if, for any set A in $\mathfrak{F}$, $\nu(A) = 0$ entails $\mu(A) = 0$. The famous Radon–Nikodym Theorem gives the following.

Theorem 1.1.9 *Let $(X, \mathfrak{F})$ be a measurable space. Let μ and ν be two σ-finite measures on $(X, \mathfrak{F})$. Then the following statements are equivalent.*

(a) *μ is absolutely continuous with respect to ν ($\nu(A) = 0$ entails $\mu(A) = 0$).*
(b) $\forall_{\varepsilon>0} \exists_{\delta>0} \forall_{A \in \mathfrak{F}} \, [\nu(A) < \delta \Rightarrow \mu(A) < \varepsilon]$.
(c) *There exists a unique (up to sets of measure zero) measurable function $\rho \colon X \to [0, +\infty)$ such that*

$$\mu(A) = \int_A \rho \, d\nu$$

for every $A \in \mathfrak{F}$.

We then write

$$\mu \prec \nu$$

in order to indicate that a measure μ is absolutely continuous with respect to ν. The unique function $\rho \colon X \longrightarrow [0, +\infty)$ appearing in item (c) is denoted by $d\mu/d\nu$ and is called the Radon–Nikodym derivative of μ with respect to ν.

We say that two measures μ and ν on the same measurable space $(X, \mathfrak{F})$ are equivalent if each one is absolutely continuous with respect to the other. To denote this fact, we frequently write

$$\mu \asymp \nu.$$

On the other hand, there is a concept that is somehow opposite to equivalence or even to absolute continuity of measures. Namely, we say that two measures μ and ν on $(X,\mathfrak{F})$ are (mutually) singular if there exists a set $Y \in \mathfrak{F}$ such that

$$\mu(X\backslash Y) = 0 \text{ while } \nu(Y) = 0.$$

We then write that

$$\mu \perp \nu.$$

1.2 Measures on Metric Spaces: (Metric) Outer Measures and Weak* Convergence

In this section, we will show how to construct measures starting with functions of sets that are required to satisfy much weaker conditions than those defining a measure. These are called outer measures. At the end of the section, we also deal with the weak* topology of measures and Riesz Representation Theorem. Again, we refer, for example, to [**Coh**], [**Fr**], [**RF**] for complete accounts.

Definition 1.2.1 An outer measure on a set X is a function μ defined on all subsets of X taking values in $[0,\infty]$ such that

$$\mu(\emptyset) = 0; \tag{1.8}$$

$$\text{if } A \subseteq B, \text{ then } \mu(A) \le \mu(B); \tag{1.9}$$

$$\mu\left(\bigcup_{n=1}^{\infty} A_n\right) \le \sum_{n=1}^{\infty} \mu(A_n) \tag{1.10}$$

for any countable family $\{A_n\}_{n=1}^{\infty}$ of subsets of X.

A subset A of X is called μ-measurable or simply measurable with respect to the outer μ if and only if

$$\mu(B) \ge \mu(A \cap B) + \mu(B\backslash A) \tag{1.11}$$

for all sets $B \subseteq X$. The opposite inequality follows immediately from (1.10). One can immediately check that if $\mu(A) = 0$, then A is μ-measurable.

Theorem 1.2.2 *If μ is an outer measure on X, then the family $\mathfrak{F}$ of all μ-measurable sets is a σ-algebra and restriction of μ to $\mathfrak{F}$ is a measure.*

Proof Obviously, $X \in \mathfrak{F}$. By symmetry (1.11), $A \in \mathfrak{F}$ if and only if $A^c \in \mathcal{F}$. So the conditions (1.1) and (1.2) of the definition of σ-algebra are satisfied.

To check the condition (1.3) that $\mathfrak{F}$ is closed under a countable union, suppose that $A_1, A_2, \dots, \in \mathfrak{F}$ and let $B \subseteq X$ be any set. Applying (1.11) in turn to $A_1, A_2, \dots$, we get, for all $k \geq 1$,

$$\begin{aligned}
\mu(B) &\geq \mu(B \cap A_1) + \mu(B \backslash A_1) \\
&\geq \mu(B \cap A_1) + \mu((B \backslash A_1) \cap A_2) + \mu(B \backslash A_1 \backslash A_2) \\
&\geq \dots \\
&\geq \sum_{j=1}^{k} \mu\left(\left(B \backslash \bigcup_{i=1}^{j-1} A_i\right) \cap A_j\right) + \mu\left(B \backslash \bigcup_{j=1}^{k} A_j\right) \\
&\geq \sum_{j=1}^{k} \mu\left(\left(B \backslash \bigcup_{i=1}^{j-1} A_i\right) \cap A_j\right) + \mu\left(B \backslash \bigcup_{j=1}^{\infty} A_j\right);
\end{aligned}$$

therefore,

$$\mu(B) \geq \sum_{j=1}^{k} \mu\left(\left(B \backslash \bigcup_{i=1}^{j-1} A_i\right) \cap A_j\right) + \mu\left(B \backslash \bigcup_{j=1}^{\infty} A_j\right). \tag{1.12}$$

Since

$$B \cap \bigcup_{j=1}^{\infty} A_j = \bigcup_{j=1}^{\infty} \left(B \backslash \bigcup_{i=1}^{j-1} A_i\right) \cap A_j,$$

using (1.10) we, thus, get

$$\mu(B) \geq \mu\left(\bigcup_{j=1}^{\infty} \left(B \backslash \bigcup_{i=1}^{j-1} A_i\right) \cap A_j\right) + \mu\left(B \backslash \bigcup_{j=1}^{\infty} A_j\right).$$

Hence, condition (1.3) is also satisfied and $\mathfrak{F}$ is a σ-algebra. To see that μ is a measure on $\mathcal{F}$, meaning that condition (1.4) is satisfied, consider mutually disjoint sets $A_1, A_2, \dots, \in \mathfrak{F}$ and apply (1.12) to $B = \bigcup_{j=1}^{\infty} A_j$. We get

$$\mu\left(\bigcup_{j=1}^{\infty} A_j\right) \geq \sum_{j=1}^{\infty} \mu(A_j).$$

Combining this with (1.10), we conclude that μ is a measure on $\mathfrak{F}$. ■

Definition 1.2.3 Let (X, ρ) be a metric space. An outer measure μ on X is said to be a metric outer measure if

$$\mu(A \cup B) = \mu(A) + \mu(B) \tag{1.13}$$

for all positively separated sets $A, B \subseteq X$, i.e, those satisfying the following condition:

$$\rho(A, B) = \inf\{\rho(x, y) : \ x \in A, \ y \in B\} > 0.$$

We assume the convention that $\rho(A, \emptyset) = \rho(A, \emptyset) = \infty$.

Recall that the Borel σ-algebra on X is the σ-algebra generated by open, or equivalently closed, sets. We want to show that if μ is a metric outer measure, then the family of all μ-measurable sets contains this σ-algebra. The proof is based on the following lemma.

Lemma 1.2.4 *Let μ be a metric outer measure on (X, ρ). Let $\{A_n\}_{n=1}^{\infty}$ be an ascending sequence of subsets of X. Denote $A := \bigcup_{n=1}^{\infty} A_n$. If $\rho(A_n, A \backslash A_{n+1}) > 0$ for all $n \geq 1$, then*

$$\mu(A) = \lim_{n \to \infty} \mu(A_n).$$

Proof By (1.9) it is sufficient to show that

$$\mu(A) \leq \lim_{n \to \infty} \mu(A_n). \tag{1.14}$$

If $\lim_{n \to \infty} \mu(A_n) = \infty$, there is nothing to prove. So, suppose that

$$\lim_{n \to \infty} \mu(A_n) = \sup_{n \to \infty} \mu(A_n) < \infty. \tag{1.15}$$

Let $B_1 = A_1$ and $B_n = A_n \backslash A_{n-1}$ for $n \geq 2$. If $n \geq m + 2$, then $B_m \subseteq A_m$ and $B_n \subseteq A \backslash A_{n-1} \subseteq A \backslash A_{m+1}$. Thus, B_m and B_n are positively separated, and applying (1.13) we get, for every $j \geq 1$,

$$\mu\left(\bigcup_{i=1}^{j} B_{2i-1}\right) = \sum_{i=1}^{j} \mu(B_{2i-1}) \quad \text{and} \quad \mu\left(\bigcup_{i=1}^{j} B_{2i}\right) = \sum_{i=1}^{j} \mu(B_{2i}). \tag{1.16}$$

We also have, for every $n \geq 1$, that

$$\begin{aligned} \mu(A) =& \mu\left(\bigcup_{k=n}^{\infty} A_k\right) = \mu\left(A_n \cup \bigcup_{k=n+1}^{\infty} B_k\right) \\ &\leq \mu(A_n) + \sum_{k=n+1}^{\infty} \mu(B_k) \\ &\leq \lim_{l \to \infty} \mu(A_l) + \sum_{k=n+1}^{\infty} \mu(B_k). \end{aligned} \tag{1.17}$$

Since the sets $\bigcup_{i=1}^{j} B_{2i-1}$ and $\bigcup_{i=1}^{j} B_{2i}$ appearing in (1.16) are both contained in A_{2j}, it follows from (1.15) and (1.16) that the series $\sum_{k=1}^{\infty} \mu(B_k)$ converges. Therefore, (1.14) follows immediately from (1.17). The proof is complete. ■

Theorem 1.2.5 *If μ is a metric outer measure on (X, ρ), then all Borel subsets of X are μ-measurable.*

Proof Since the Borel sets form the least σ-algebra containing all closed subsets of X, it follows from Theorem 1.2.2 that it is enough to check (1.11) for every nonempty closed set $A \subseteq X$ and every $B \subseteq X$. For all $n \geq 1$, let $B_n = \{x \in B \backslash A : \rho(x, A) \geq 1/n\}$. Then $\rho(B \cap A, B_n) \geq 1/n$ and by (1.13)

$$\mu(B \cap A) + \mu(B_n) = \mu\big((B \cap A) \cup B_n\big) \leq \mu(B). \tag{1.18}$$

The sequence $\{B_n\}_{n=1}^{\infty}$ is ascending and, since A is closed, $B \backslash A = \bigcup_{n=1}^{\infty} B_n$. In order to apply Lemma 1.2.4, we shall now show that

$$\rho\big(B_n, (B \backslash A) \backslash B_{n+1}\big) > 0$$

for all $n \geq 1$. And, indeed, if $x \in (B \backslash A) \backslash B_{n+1}$, then there exists $z \in A$ with $\rho(x, z) < 1/(n+1)$. Thus, if $y \in B_n$, then

$$\rho(x, y) \geq \rho(y, z) - \rho(x, z) > \frac{1}{n} - \frac{1}{n(n+1)} > 0.$$

Applying now Lemma 1.2.4 with $A_n = B$ shows that $\mu(A \backslash B) = \lim_{n \to \infty} \mu(B_n)$. Thus, (1.11) follows from (1.18). The proof is complete. ■

This theorem, as well as many other reasons disseminated over mathematics, many of which we will encounter in this book, justifies the following definition.

Definition 1.2.6 Any measure on a metric space that is defined on its σ-algebra of Borel sets (or larger) is called a Borel measure.

Let us list the following well-known properties of finite Borel measures.

Theorem 1.2.7 *Any finite Borel measure μ on a metric space X is both outer and inner regular. Outer regularity means that*

$$\mu(A) = \inf\{\mu(G) : G \supseteq A \text{ and } G \text{ is open}\},$$

while inner regularity means that

$$\mu(A) = \sup\{\mu(F) : F \subseteq A \text{ and } F \text{ is closed}\}.$$

In addition, if the space X is completely metrizable, then the closed sets involved in the concept of inner regularity can be replaced by compact ones.

Given a metric space (X, ρ), we denote by $M(X)$ the collection of all Borel probability measures on X. We denote by $C(X)$ the vector space of all real-valued continuous functions on X and by $C_b(X)$ its vector subspace consisting of all bounded elements of $C(X)$. Let us record the following easy theorem.

Theorem 1.2.8 *If (X, ρ) is a metric space, the two measures μ and ν in $M(X)$ are equal if and only if*

$$\nu(g) = \mu(f)$$

for all functions $g \in C_b(X)$.

If X is compact, then $C(X)$ becomes a Banach space if endowed with the supremum metric. Denote by $C^*(X)$ the dual of $C(X)$. Endow $C^*(X)$ with the weak* topology. This means that

$$\text{a net } (F_\lambda)_{\lambda \in \Lambda} \text{ in } C^*(X) \text{ converges to an element } F \in C^*(X)$$

if and only if

$$\text{the net } (F_\lambda(g))_{\lambda \in \Lambda} \text{ converges to } F(g)$$

for every $g \in C(X)$. $M(X)$, the space of all Borel probability measures on X, can then be naturally viewed as a subset of $C^*(X)$: every measure $\mu \in M(X)$ induces the functional

$$C(X) \ni g \longmapsto \mu(g).$$

We will frequently use the following.

Theorem 1.2.9 *Let X be a compact metrizable space. Consider $C^*(X)$ with its weak* topology. Then*

(a) *$M(X)$ is a convex compact subset of $C^*(X)$.*
(b) *$M(X)$ is a metrizable space. In particular, proving continuity or convergence one can restrict oneself to sequences only (as opposed to nets).*
(c) *(Riesz Representation Theorem) Every nonnegative linear functional $F : C(X) \longrightarrow \mathbb{R}$ such that $F(\mathbb{1}) = 1$ is (uniquely) represented by an element in $M(X)$. More precisely, there exists $\mu \in M(X)$ such that*

$$F(g) = \mu(g)$$

for all $g \in C(X)$.

It follows from item (c) of this theorem that the functional F considered therein is bounded. In fact, this is a quite elementary property whose short proof we leave for the reader as an exercise.

Definition 1.2.10 If X is a topological space and μ is a Borel measure on X, then the topological support of μ is defined as the set of all points $x \in X$ such that, for every open set G containing x, $\mu(G) > 0$. It is denoted by $\text{supp}(\mu)$.

The following proposition collects the basic properties of topological supports.

Proposition 1.2.11 *If X is a topological space and μ is a Borel measure on X, then the topological support* $\text{supp}(\mu)$ *of μ is a closed subset of X.*

If, in addition, X is a separable metrizable space, then the following hold.

(1) $\mu(X\backslash\text{supp}(\mu)) = 0$.
(2) *If $F \subseteq X$ is a closed set such that $\mu(X\backslash F) = 0$, then $F \supseteq \text{supp}(\mu)$.*
(3) *If, in addition, μ is a nonzero finite measure, then* $\text{supp}(\mu)$ *is the smallest closed subset of X such that $\mu(F) = \mu(X)$.*

Proof The fact that the topological support $\text{supp}(\mu)$ is closed is immediate from its definition. Let us prove item (1). If $x \in X\backslash\text{supp}(\mu)$, then there exists an open set $G_x \subseteq X$ containing X such that $\mu(G_x) = 0$. Since X is a separable metrizable space, so is $X\backslash\text{supp}(\mu)$. But then $X\backslash\text{supp}(\mu)$ a Lindelöf space. Therefore, the cover G_x, $x \in \text{supp}(\mu)$, of $X\backslash\text{supp}(\mu)$ has a countable subcover. This means that there exists a countable set $D \subseteq X\backslash\text{supp}(\mu)$ such that

$$\bigcup_{x\in D} G_x = X\backslash\text{supp}(\mu).$$

Hence,

$$\mu(X\backslash\text{supp}(\mu)) \le \sum_{x\in D} \mu(G_x) = 0,$$

meaning that item (1) holds.

In order to prove item (2), note that, since $X\backslash F$ is an open set and its measure is equal to zero, it is contained in the complement of $\text{supp}(\mu)$. This means that item (2) holds.

Proving item (3), its hypotheses yield $\mu(X\backslash F) = 0$. Since the set F is also closed, item (2) implies that $F \supseteq \text{supp}(\mu)$, and we are done. ■

We end this section with the following easy fact, which will be frequently used throughout both volumes of the book.

Theorem 1.2.12 *If X is a compact metric space and μ is a finite Borel measure on X with full topological support, then, for every $r > 0$,*

$$M(\mu,r) := \inf\{\mu(B(x,r)) \colon x \in X\} > 0. \tag{1.19}$$

Proof Since the space X is compact, there exists a finite set $F \subseteq X$ such that

$$\bigcup_{y \in F} B(y,r/2) = X.$$

Since $\text{supp}(\mu) = X$, we have that

$$M := \min\{\mu(B(y,r/2)) > 0.$$

Now, if $x \in X$, there exists $y \in F$ such that $x \in B(y,r/2)$. But then $B(x,r) \supseteq B(y,r/2)$, and, therefore,

$$\mu(B(x,r)) \geq \mu(B(y,r/2)) \geq M > 0.$$

The proof of Theorem 1.2.12 is complete. ■

1.3 Covering Theorems: 4*r*, Besicovitch, and Vitali Type; Lebesgue Density Theorem

In this section, we prove first the $4r$ Covering Theorem. Following the arguments of [**MSzU**], we prove it for all metric spaces. If we do not insist on $4r$ but are content with $5r$ (which is virtually always the case), a shorter, less involved proof is possible. This can be found, for example, in [**Heino**]. Then, following [**Mat**], we will prove the Besicovitch Covering Theorem and, as its fairly straightforward consequence, the Vitali-Type Covering theorem. We finally deduce from the latter the Lebesgue Density Points Theorem. All these theorems are classical and can be found in many sources with extended discussions. More applications of covering theorems will appear in further sections of this chapter and throughout the entire book. For every ball $B := B(x,r)$, we put $r(B) = r$ and $c(B) = x$.

Theorem 1.3.1 ($4r$ Covering Theorem). *Suppose that (X,ρ) is a metric space and $\mathcal{B}$ is a family of open balls in X such that* $\sup\{r(B) \colon B \in \mathcal{B}\} < +\infty$. *Then there is a family $\mathcal{B}' \subseteq \mathcal{B}$ consisting of mutually disjoint balls such that $\bigcup_{B \in \mathcal{B}} B \subseteq \bigcup_{B \in \mathcal{B}'} 4B$. In addition, if the metric space X is separable, then $\mathcal{B}'$ is countable.*

Proof Fix an arbitrary $M > 0$. Suppose that there is a family $\mathcal{B}'_M \subseteq \mathcal{B}$ consisting of mutually disjoint balls such that

(a) $r(B) > M$ for all $B \in \mathcal{B}'_M$,
(b) $\bigcup_{B\in\mathcal{B}'_M} 4B \supseteq \bigcup\{B\colon B \in \mathcal{B} \text{ and } r(B) > M\}$.

We shall show that then there exists a family $\mathcal{B}''_M \subseteq \mathcal{B}$ with the following properties:

(c) $\mathcal{B}''_M \subseteq \mathfrak{F} := \{B \in \mathcal{B}\colon 3M/4 < r(B) \leq M\}$,
(d) $\mathcal{B}'_M \cup \mathcal{B}''_M$ consists of mutually disjoint balls,
(e) $\bigcup_{B\in\mathcal{B}'_M\cup\mathcal{B}''_M} 4B \supseteq \bigcup\{B\colon B \in \mathcal{B} \text{ and } r(B) > 3M/4\}$.

Indeed, put

$$\mathcal{B}'''_M = \left\{B \in \mathfrak{F}\colon B \cap \bigcup_{D\in\mathcal{B}'_M} D = \emptyset\right\}. \tag{1.20}$$

Consider $B \in \mathfrak{F}\setminus\mathcal{B}'''_M$. Then there exists $D \in \mathcal{B}'_M$ such that $B \cap D \neq \emptyset$. Hence, $r(B) \leq M < r(D)$ and, in consequence,

$$\rho(c(B),c(D)) < r(B) + r(D) \leq M + r(D) < r(D) + r(D) = 2r(D)$$

and

$$B \subseteq B(c(D), r(B) + 2r(D)) \subseteq B(c(D), 3r(D)) = 3D \subseteq 4D.$$

Therefore,

$$\bigcup_{B\in\mathfrak{F}\setminus\mathcal{B}'''_M} B \subseteq \bigcup_{B\in\mathcal{B}'_M} 4B. \tag{1.21}$$

So, if $\mathcal{B}'''_M = \emptyset$, we are done with the proof by setting $\mathcal{B}''_M = \emptyset$. Otherwise, fix an arbitrary $B_0 \in \mathcal{B}'''_M$ and further, proceeding by transfinite induction, fix some $\mathcal{B}_\alpha \in \mathcal{B}'''_M$ such that

$$c(B_\alpha) \in c\big(\mathcal{B}'''_M\big)\setminus \bigcup_{\gamma<\alpha} \frac{8}{3}B_\gamma$$

for some some ordinal number $\gamma \geq 0$, as long as the difference on the right-hand side above is nonempty. This procedure terminates at some ordinal number λ. First, we claim that the balls $(B_\alpha)_{\alpha<\lambda}$ are mutually disjoint. Indeed, fix $0 \leq \alpha < \beta < \lambda$. Then $c(B_\beta) \notin \frac{8}{3}B_\alpha$. So,

$$\rho\big(c(B_\beta), c(B_\alpha)\big) \geq \frac{8}{3}r(B_\alpha) > \frac{8}{3}\cdot\frac{3}{4}M = 2M$$

and

$$r(B_\beta) + r(B_\alpha) \leq M + M = 2M.$$

Thus, $B_\beta \cap B_\alpha = \emptyset$. Now if $B \in \mathcal{B}'_M$ and $0 \le \alpha < \lambda$, then $B_\alpha \in \mathcal{B}'''_M$ and, by (1.20), $B_\alpha \cap B = \emptyset$. Thus, we proved item (d) with $\mathcal{B}''_M = \{B_\alpha\}_{\alpha<\lambda}$. Item (c) is obvious since $B_\alpha \in \mathcal{B}'''_M \subseteq \mathfrak{F}$ for all $0 \le \alpha < \lambda$. It remains to prove item (e). By the definition of λ, $c(\mathcal{B}'''_M) \subset \bigcup_{\gamma<\lambda} \frac{8}{3} B_\gamma = \bigcup_{B\in\mathcal{B}''_M} \frac{8}{3} B$. Hence, if $x \in B$ and $B \in \mathcal{B}'''_M$, then there exists $D \in \mathcal{B}''_M$ such that $c(B) \in \frac{8}{3} D$. Therefore,

$$\begin{aligned}\rho(x,c(D)) &\le \rho(x,c(B)) + \rho(c(B),c(D)) \le r(B) + \frac{8}{3} r(D)\\ &\le M + \frac{8}{3} r(D) < \frac{4}{3} r(D) + \frac{8}{3} r(D)\\ &= 4r(D).\end{aligned}$$

Thus, $x \in 4D$; consequently, $\bigcup \mathcal{B}'''_M \subseteq \bigcup_{D\in\mathcal{B}''_M} 4D$. Combining this and (1.21), we get that $\bigcup_{B\in\mathfrak{F}} B \subseteq \bigcup_{B\in\mathcal{B}'_M\cup\mathcal{B}''_M} 4B$. This and (b) immediately imply (e). The properties (c), (d), and (e) are established. Now take $S = \sup\{r(B)\colon B \in \mathcal{B}\} + 1 < +\infty$ and define inductively the sequence $(\mathcal{B}'_{(3/4)^n S})_{n=0}^\infty$ by declaring that $\mathcal{B}'_S = \emptyset$ and $\mathcal{B}'_{(3/4)^{n+1} S} = \mathcal{B}'_{(3/4)^n S} \cup \mathcal{B}''_{(3/4)^n S}$. Then

$$\mathcal{B}' = \bigcup_{n=0}^{\infty} \mathcal{B}'_{(3/4)^n S}.$$

It then follows directly from (d) and our inductive definition that $\mathcal{B}'$ consists of mutually disjoint balls. It follows from (e) that $\bigcup_{B\in\mathcal{B}'} 4B \supseteq \bigcup\{B \in \mathcal{B}\colon r(B) > 0\} = \bigcup \mathcal{B}$. The first part of our theorem is, thus, proved. The last part follows immediately from the fact that any family of mutually disjoint open subsets of a separable space is countable. ■

Remark 1.3.2 Assume the same as in Theorem 1.3.1 (no separability of X is required) and suppose that there exists a finite Borel measure μ on X such that $\mu(B) > 0$ for all $B \in \mathcal{B}'$. Then $\mathcal{B}'$ is countable.

We shall now prove the *Besicovitch Covering Theorem*. We consider it to be one of the most powerful geometric tools when dealing with some aspects of fractal sets. We can easily deduce from it two fundamental classical theorems: the *Vitali-Type Covering Theorem* and the *Lebesgue Density Points Theorem.* For the proof of the Besicovitch Covering Theorem, we introduce two concepts. First the following definition.

Definition 1.3.3 Let (X,ρ) be a metric space. A collection $\mathcal{B} = \{B(x_i,r_i)\}_{i=1}^\infty$ of open balls centered at a set $A \subseteq X$, meaning that $x_i \in A$ for all $i \ge 1$, is said to be a packing of A if and only if, for any pair $i \ne j$,

$$\rho(x_i,x_j) \ge r_i + r_j.$$

This property is not in general equivalent to the requirement that all the balls $B(x_i, r_i)$ be mutually disjoint. It is obviously so if X is a Euclidean space. We call the number

$$r(\mathcal{B}) := \sup\{r_i : i \geq 1\}$$

the radius of packing $\mathcal{B}$.

This notion has a far-reaching meaning. It is the key concept to define packing measures and dimensions, which will be done in Section 1.5. The other notion we need is the following.

For any $x \in \mathbb{R}^n$, any $0 < r \leq \infty$, and any $0 < \alpha < \pi$ by $\mathrm{Con}(x,\alpha,r)$, we will denote any solid central cone with vertex x, radius r, and angle α. That is, with the above data, for an arbitrary ray l emanating from x, we denote

$$\begin{aligned}\mathrm{Con}(x,\alpha,r) &= \mathrm{Con}(l,x,\alpha,r)\\ &:= \{y \in \mathbb{R}^n : 0 < |y-x| < r, \angle(y-x,l) \leq \alpha\} \cup \{x\}.\end{aligned}$$

The proof of Theorem 1.3.5 makes substantial use of the following obvious geometric observation.

Observation 1.3.4 Let $n \geq 1$ be an integer. Then there exists $\alpha(n) > 0$ so small that the following holds. If $x \in \mathbb{R}^n$, $0 < r < \infty$, if $z \in B(x,r) \backslash B(x,r/3)$ and if $x \in \mathrm{Con}(z,\alpha(n),\infty)$, then the set $\mathrm{Con}(z,\alpha(n),\infty) \backslash B(x,r/3)$ consists of two connected components: one of z and one of ∞. The one containing z is contained in $B(x,r)$.

Theorem 1.3.5 (Besicovitch Covering Theorem) *Let $n \geq 1$ be an integer. Then there exists an integer constant $b(n) \geq 1$ such that the following holds.*

If A is a bounded subset of $\mathbb{R}^n$, then, for any function $r\colon A \to (0,\infty)$, there exists $\{x_k\}_{k=1}^{\infty}$, a countable subset of A, such that the collection

$$\mathcal{B}(A,r) := \{B(x_k, r(x_k))\colon k \geq 1\}$$

covers A and can be decomposed into $b(n)$ packings of A.

Proof We will construct the sequence $\{x_k : k = 1,2,\ldots\}$ inductively. Let

$$a_0 := \sup\{r(x)\colon x \in A\}.$$

If $a_0 = \infty$, then one can find $x \in A$ with $r(x)$ so large that $B(x, r(x)) \supseteq A$ and the proof is finished.

If $a_0 < \infty$, choose $x_1 \in A$ so that $r(x_1) > a_0/2$. Fix $k \geq 1$. Assume that the points $x_1, x_2, \ldots, x_k$ have already been chosen. If $A \subseteq B(x_1, r(x_1)) \cup \cdots \cup B(x_k, r(x_k))$, then the selection process is finished. Otherwise, put

$$a_k := \sup\left\{r(x)\colon x \in A\backslash\big(B(x_1,r(x_1)) \cup \cdots \cup B(x_k,r(x_k))\big)\right\}$$

and take

$$x_{k+1} \in A\backslash\big(B(x_1,r(x_1)) \cup \cdots \cup B(x_k,r(x_k))\big) \tag{1.22}$$

such that

$$r(x_{k+1}) > a_k/2. \tag{1.23}$$

In order to shorten notation from now on, throughout this proof we will write r_k for $r(x_k)$. By (1.22), we have that $x_l \notin B(x_k, r_k)$ for all pairs k, l with $k < l$. Hence,

$$\|x_k - x_l\| \geq r(x_k). \tag{1.24}$$

It follows from the construction of the sequence (x_k) that

$$r_k > a_{k-1}/2 \geq r_l/2. \tag{1.25}$$

therefore, $r_k/3 + r_l/3 < r_k/3 + 2r_k/3 = r_k$. By combining this and (1.24) we obtain that

$$B(x_k, r_k/3) \cap B(x_l, r_l/3) = \emptyset \tag{1.26}$$

for all pairs k, l with $k \neq l$ since then either $k < l$ or $l < k$.

Now we shall show that the balls $\{B(x_k, r_k)\colon k \geq 1\}$ cover A. Indeed, if the selection process stops after finitely many steps this claim is obvious. Otherwise, it follows from (1.26) that $\lim_{k\to\infty} r_k = 0$ and if $x \notin \bigcup_{k=1}^{\infty} B(x_k, r_k)$ for some $x \in A$, then by construction $r_k > a_{k-1}/2 \geq r(x)/2$ for every $k \geq 1$. The contradiction obtained proves that $\bigcup_{k=1}^{\infty} B(x_k, r_k) \supseteq A$.

The main step of the proof is given by the following.

Claim 1°. For every $z \in \mathbb{R}^n$ and any cone $\mathrm{Con}(z, \alpha(n), \infty)$ ($\alpha(n)$ given by Observation 1.3.4), we have that

$$\#\big\{k \geq 1\colon z \in B(x_k, r_k)\backslash B(x_k, r_k/3) \text{ and } x_k \in \mathrm{Con}(z, \alpha(n), \infty)\big\} \leq (12)^n.$$

Proof Denote the above set by Q. Our task is to estimate its cardinality from above. If $Q = \emptyset$, there is nothing to prove. Otherwise, let $i = \min Q$. If $k \in Q$ and $k \neq i$, then $k > i$ and, therefore, $x_k \notin B(x_i, r_i)$. Therefore, by Observation 1.3.4 applied with $x = x_i$, $r = r_i$, and by the the definition of Q, we get that $\|z - x_k\| \geq 2r_i/3$. Hence,

$$r_k \geq \|z - x_k\| \geq 2r_i/3. \tag{1.27}$$

On the other hand, by (1.25), we have that $r_k < 2r_i$; therefore, $B(x_k, r_k/3) \subseteq B(z, 4r_k/3) \subseteq B(z, 8r_i/3)$. Thus, using (1.26), (1.27) and the fact that the n-dimensional volume of balls in $\mathbb{R}^n$ is proportional to the nth power of radii, we obtain that, $\#Q \leq (8r_i/3)^n/(2r_i/9)^n = 12^n$. The proof of the claim is finished. ■

Clearly, there exists an integer $c(n) \geq 1$ such that, for every $z \in \mathbb{R}^n$, the space $\mathbb{R}^n$ can be covered by at most $c(n)$ cones of the form $\mathrm{Con}(z, \alpha(n), \infty)$. Therefore, it follows from the above claim that, for every $z \in \mathbb{R}^n$,

$$\#\{k \geq 1 \colon z \in B(x_k, r_k) \backslash B(x_k, r_k/3)\} \leq c(n)(12)^n.$$

Thus, applying (1.26),

$$\#\{k \geq 1 \colon z \in B(x_k, r_k) \leq 1 + c(n)(12)^n. \tag{1.28}$$

Since the ball $\overline{B}(0, 3/2)$ is compact, it contains a finite subset P such that

$$\bigcup_{x \in P} B(x, 1/2) \supseteq \overline{B}(0, 3/2).$$

Now, for every $k \geq 1$, consider the composition of the map $\mathbb{R}^n \ni x \longmapsto r_k x \in \mathbb{R}^n$ and the translation determined by the vector from 0 to x_k. Call the image of P under this affine map P_k. Then, $\#P_k = \#P$, $P_k \subseteq \overline{B}(x_k, 3r_k/2)$, and

$$\bigcup_{x \in P_k} B(x, r_k/2) \supseteq \overline{B}(0, 3r_k/2). \tag{1.29}$$

Consider now two integers $1 \leq k < l$ such that

$$B(x_k, r_k) \cap B(x_l, r_l) \neq \emptyset. \tag{1.30}$$

Let $y \in \mathbb{R}^n$ be the only point lying on the interval joining x_l and x_k at the distance $r_k - r_l/2$ from x_k. As $x_l \notin B(x_k, r_k)$, by (1.30) we have that $\|y - x_l\| \leq r_l + r_l/2 = 3r_l/2$ and, therefore, by (1.29) there exists $z \in P_l$ such that $\|z - y\| < r_l/2$. Consequently, $z \in B(x_k, r_l/2 + r_k - r_l/2) = B(x_k, r_k)$. Thus, applying (1.28), with z being the elements of P_l, we obtain the following:

$$\#\big\{1 \leq k \leq l-1 \colon B(x_k, r_k) \cap B(x_l, r_l) \neq \emptyset\big\} \leq \#P(1 + c(n)(1)2^n) \tag{1.31}$$

for every $l \geq 1$. Putting

$$b(n) := \#P(1 + c(n)(12)^n) + 1,$$

this property allows us to decompose the set $\mathbb{N}$ of positive integers into $b(n)$ subsets $\mathbb{N}_1, \mathbb{N}_2, \ldots, \mathbb{N}_{b(n)}$ in the following inductive way. For every $k = 1, 2, \ldots, b(n)$, set $\mathbb{N}_k(b(n)) = \{k\}$ and suppose that, for every

$k = 1, 2, \dots, b(n)$ and some $j \geq b(n)$, the mutually disjoint families $\mathbb{N}_k(j)$ have already been defined so that

$$\mathbb{N}_1(j) \cup \cdots \cup \mathbb{N}_{b(n)}(j) = \{1, 2, \dots, j\}.$$

Then, by (1.31), there exists at least one $1 \leq k \leq b(n)$ such that $B(x_{j+1}, r_{j+1}) \cap B(x_i, r_i) = \emptyset$ for every $i \in \mathbb{N}_k(j)$. We set

$$\mathbb{N}_k(j+1) := \mathbb{N}_k(j) \cup \{j+1\}$$

and

$$\mathbb{N}_l(j+1) = \mathbb{N}_l(j)$$

for all $l \in \{1, 2, \dots, b(n)\} \backslash \{k\}$. Putting now, for every $k = 1, 2, \dots, b(n)$,

$$\mathbb{N}_k := \mathbb{N}_k(b(n)) \cup \mathbb{N}_k(b(n)+1) \cup \cdots,$$

we see from the inductive construction that these sets are mutually disjoint, that they cover $\mathbb{N}$, and that for every $k = 1, 2, \dots, b(n)$ the families of balls $\{B(x_l, r_l) : l \in \mathbb{N}_k\}$ are also mutually disjoint. The proof of the Besicovitch Covering Theorem is finished. ■

We would like to emphasize here that the same statement remains true, if open balls, are replaced by closed ones. It also remains true if, instead of balls, one considers n-dimensional cubes. And, in this latter case, it is even better: namely, the proof based on the same idea is technically considerably easier. There are further, frequently useful, generalizations, especially a theorem of Morse. The reader is advised to consult the book **[Gu]** by Guzman on such topics.

As we have already mentioned, we can easily deduce from the Besicovitch Covering Theorem some other fundamental facts.

Theorem 1.3.6 (Vitali-Type Covering Theorem) *Let μ be a probability Borel measure on $\mathbb{R}^n$, let $A \subset \mathbb{R}^n$ be a Borel set, and let $\mathcal{B}$ be a family of closed balls such that each point of A is the center of arbitrarily small balls of $\mathcal{B}$, i.e.,*

$$\inf\{r : B(x,r) \in \mathcal{B}\} = 0$$

for all $x \in A$. Then there exists a countable (finite or infinite) collection $\mathcal{B}(A)$ of mutually disjoint balls in $\mathcal{B}$ such that

$$\mu\left(A \backslash \bigcup \{B \in \mathcal{B}(A)\}\right) = 0.$$

Proof We assume that A is bounded, leaving the unbounded case to the reader. We may assume that $\mu(A) > 0$. The measure μ restricted to a compact ball $B(0,R)$ such that $A \subset B(0,R/2)$ is Borel, hence regular. Hence, there exists an open set $U \subset \mathbb{R}^n$ containing A such that

$$\mu(U) \le \big(1 + (4b(n))^{-1}\big)\mu(A),$$

where $b(n)$ is as in the Besicovitch Covering Theorem 1.3.5. By that theorem applied for closed balls we can decompose $\mathcal{B}$ in packings $\mathcal{B}_1, \dots, \mathcal{B}_{b(n)}$ of A contained in U, i.e., each $\mathcal{B}_i$ consists of disjoint balls and

$$A \subset \bigcup_{i=1}^{b(n)} \bigcup \mathcal{B}_i \subset U.$$

Then $\mu(A) \le \sum_{i=1}^{b(n)} \mu\big(\bigcup \mathcal{B}_i\big)$; consequently, there exists an i such that

$$\mu(A) \le b(n)\mu\left(\bigcup \mathcal{B}_i\right).$$

Further, for some finite subfamily $\mathcal{B}_i'$ of $\mathcal{B}_i$,

$$\mu(A) \le 2b(n)\mu\left(\bigcup \mathcal{B}_i'\right).$$

Letting $A_1 = A \setminus \big(\bigcup \mathcal{B}_i'\big)$, we get

$$\begin{aligned}
\mu(A_1) &\le \mu\left(U \setminus \bigcup B_i'\right) = \mu(U) - \mu\left(\bigcup B_i'\right) \\
&\le \left(1 + \frac{1}{4}(b(n))^{-1} - \frac{1}{2}(b(n))^{-1}\right)\mu(A) \\
&= u\mu(A)
\end{aligned}$$

with $u := 1 - \frac{1}{4}(b(n))^{-1} < 1$. Now consider A_1 in the role of A before. Since

$$A_1 \subset \mathbb{R}^n \setminus \left(\bigcup \mathcal{B}_i'\right),$$

which is open, we find a packing, playing the role of B_i' contained in

$$\mathbb{R}^n \setminus \left(\bigcup \mathcal{B}_i'\right),$$

so disjoint from $\bigcup \mathcal{B}_i'$. We then get the measure of a noncovered remnant bounded above by $u\mu(A_1) \le u^2\mu(A)$. We can continue, consecutively constructing packings that exhaust the whole set A except at most a set of measure 0. The proof is complete. ■

Now we shall prove two quite straightforward consequences of the Besicovitch Covering Theorem (Theorem 1.3.5), the first one being the celebrated, and to some extent counter-intuitive, Density Points Theorem. It in fact follows from the Vitali-Type Covering Theorem (Theorem 1.3.6), which itself is a consequence of the Besicovitch Covering Theorem.

Theorem 1.3.7 (Lebesgue Density Theorem) *Let μ be a probability Borel measure on $\mathbb{R}^n$, $n \in \mathbb{N}$, and $A \subset \mathbb{R}^n$ be a Borel set. Then the limit*

$$\lim_{r \to 0} \frac{\mu(A \cap B(x,r))}{\mu(B(x,r))}$$

exists and is equal to 1 for μ-almost every point $x \in A$.

Proof First of all, for every Borel set $B \subseteq \mathbb{R}^n$ and every $x \in \mathbb{R}^n$, we obviously have that

$$\lim_{s \nearrow r} \mu(B \cap B(x,s)) = \mu(B \cap B(x,r))$$

and

$$\lim_{s \searrow r} \mu(B \cap B(x,s) \geq \mu(B \cap B(x,r))).$$

Therefore, the function

$$\mathbb{R}^n \ni x \longmapsto \mu(B \cap B(x,r)) \in \mathbb{R}$$

is lower semi-continuous, thus Borel measurable. Hence, the function

$$\mathbb{R}^n \ni x \longmapsto \frac{\mu(A \cap B(x,r))}{\mu(B(x,r))} \in \mathbb{R}$$

is also Borel measurable. Furthermore, since

$$\lim_{\mathbb{Q} \ni s \nearrow r} \mu(B \cap B(x,s)) = \mu(B \cap B(x,r)),$$

it follows that the set of points $x \in \mathbb{R}^n$ for which the limit

$$\lim_{r \to 0} \frac{\mu(A \cap B(x,r))}{\mu(B(x,r))} \tag{1.32}$$

exists is the same as the set of points $x \in \mathbb{R}^n$ for which the limit

$$\lim_{\mathbb{Q} \ni r \to 0} \frac{\mu(A \cap B(x,r))}{\mu(B(x,r))}$$

exists. Since the set $\mathbb{Q}$ of rational numbers is countable, we, thus, conclude that the set of points in X for which in (1.32) exists is Borel measurable.

Seeking contradiction, suppose now that the set of points in A where this limit is either not equal to 1 or does not exist has positive measure μ. Then there exists $0 \leq a < 1$ and Borel $A' \subset A$ of positive measure μ such that, for every $x \in A'$, there exists a sequence $(r_i(x))_{i=1}^{\infty}$ of positive radii converging to 0 such that

$$\frac{\mu\big(A' \cap B(x, r_i(x))\big)}{\mu\big(B(x, r_i(x))\big)} < a$$

for all $i \geq 1$. Given an open set U containing A, let

$$\mathcal{B}_U := \left\{B(x,r_i(x)) \colon x \in A',\ B(x,r_i(x)) \subseteq U\right\}.$$

Then let $\mathcal{B}_U(A')$ be the corresponding collection of balls whose existence is asserted in the Vitali-Type Covering Theorem (Theorem 1.3.6). Then

$$\mu(A') = \sum_{B\in\mathcal{B}_U(A')} \mu(A' \cap B) \le a \sum_{B\in\mathcal{B}_U(A')} \mu(B) \le a\mu(U).$$

Since measure μ is regular, this yields $\mu(A') \le a\mu(A')$. This contradiction finishes the proof. ■

Every point in the set A for which the assertion of the Lebesgue Density Theorem holds will be called a Lebesgue density point of A with respect to the measure μ.

The second consequence of the Besicovitch Covering Theorem (Theorem 1.3.5), which we have mentioned above, is the following technical, but very useful and frequently applied, lemma, which is suitable for proving that one given measure is absolutely continuous with respect to the other. We follow the proof from [**DU2**].

Lemma 1.3.8 *Let μ and ν be Borel probability measures on X, a bounded subset of a Euclidean space $\mathbb{R}^d$, $d \ge 1$. Suppose that there is a constant $M > 0$ and, for every point $x \in Y$, there is a converging to zero sequence $\left(r_j(x)\right)_{i=0}^{\infty}$ of positive radii such that, for all $j \ge 1$ and all $x \in X$,*

$$\mu(B(x,r_j(x)) \le M\nu(B(x,r_j(x)).$$

Then the measure μ is absolutely continuous with respect to ν and the Radon–Nikodym derivative satisfies

$$d\mu/d\nu \le Mb(d),$$

where b(d) is the constant coming from the Besicovitch Covering Theorem, i.e., Theorem 1.3.5.

Proof Consider an arbitrary Borel set $E \subseteq X$ and fix $\varepsilon > 0$. Since $\lim_{j\to\infty} r_j(x) = 0$ and measure ν is regular, for every $x \in E$ there exists a radius $r(x)$ of the form $r_j(x)$ such that

$$\nu\left(\bigcup_{x\in E} B_e(x,r(x)) \setminus E\right).$$

Now, by the Besicovitch Covering Theorem (Theorem 1.3.5) we can choose a countable subcover $\{B(x_i,r_i(x))\}_{i=1}^{\infty}$ from the cover $\{B(x_i,r_i(x))\}_{x\in E}$ of E, of multiplicity bounded above by $b(d)$. Therefore, we obtain that

$$\begin{aligned}\mu(E) &\leq \sum_{i=1}^{\infty}\mu(B(x_i,r_i(x))) \leq M\sum_{i=1}^{\infty}\nu(B(x_i,r_i(x)))\\ &\leq Mb(d)\nu\left(\bigcup_{i=1}^{\infty}B(x_i,r_i(x))\right)\\ &\leq Mb(d)(\varepsilon+\nu(E)).\end{aligned}$$

Letting $\varepsilon \searrow 0$, we, thus, obtain that $\mu(E) \leq Mb(d)\nu(E)$. Therefore, μ is absolutely continuous with respect to ν with the Radon–Nikodym derivative bounded above by $Mb(d)$. ■

1.4 Conditional Expectations and Martingale Theorems

The content of this section belongs to probability theory rather than to classical measure theory. Its culmination (for us), i.e., Theorem 1.4.11, is, however, similar to the Lebesgue Density Theorem, i.e., Theorem 1.3.7, so is natural to place it here. This chapter is about conditional expectations and martingales and is closely modeled on a chapter form Billingsley's book [**Bil2**].

We start with conditional expectations. Let $(X,\mathfrak{F},\mu)$ be a probability space. Let $\mathfrak{D}$ be a sub-σ-algebra of $\mathfrak{F}$. Let

$$\phi\colon X \longrightarrow \mathbb{R}$$

be a measurable function, integrable with respect to the measure μ. We denote by

$$E(\phi|\mathfrak{D}) = E_\mu(\phi|\mathfrak{D})$$

the (conditional) expected value of ϕ with respect to the σ-algebra $\mathfrak{D}$. This is the only function (up to sets of measure zero) that is measurable with respect to the σ-algebra $\mathfrak{D}$ such that

$$\int_D E_\mu(\phi|\mathfrak{D})d\mu = \int_D \phi\, d\mu$$

for every set $D \in \mathfrak{D}$. Its existence for nonnegative integrable functions is a straightforward consequence of the Radon–Nikodym Theorem. In the general case, one sets

$$E_\mu(\phi|\mathfrak{D}) := E_\mu(\phi_+|\mathfrak{D}) - E_\mu(\phi_-|\mathfrak{D}).$$

Uniqueness is obvious.

Conditional expectations exhibit several natural properties. We list the most basic ones below. Their proofs are straightforward and are omitted.

Proposition 1.4.1 *Let $(X, \mathcal{A}, \mu)$ be a probability space, $\mathcal{B}$ and $\mathcal{C}$ denote some sub-σ-algebras of $\mathcal{A}$ and $\varphi \in L^1(X, \mathcal{A}, \mu)$. Then the following hold.*

(a) *If $\varphi \geq 0$ μ-a.e., then*

$$E(\varphi|\mathcal{B}) \geq 0 \quad \mu\text{-a.e.}$$

(b) *If $\varphi_1 \geq \varphi_2$ μ-a.e., then*

$$E(\varphi_1|\mathcal{B}) \geq E(\varphi_2|\mathcal{B}) \quad \mu\text{-a.e.}$$

(c) $\big|E(\varphi|\mathcal{B})\big| \leq E\big(|\varphi| \,\big|\, \mathcal{B}\big)$.

(d) *The functional $E(\cdot|\mathcal{B})$ is linear. In other words, for any $c_1, c_2 \in \mathbb{R}$ and $\varphi_1, \varphi_2 \in L^1(X, \mathcal{A}, \mu)$, we have that*

$$E(c_1\varphi_1 + c_2\varphi_2|\mathcal{B}) = c_1\, E(\varphi_1|\mathcal{B}) + c_2\, E(\varphi_2|\mathcal{B}).$$

(e) *If φ is already $\mathcal{B}$-measurable, then $E(\varphi|\mathcal{B}) = \varphi$. In particular, we have that*

$$E\big(E(\varphi|\mathcal{B})\big|\mathcal{B}\big) = E(\varphi|\mathcal{B}).$$

Also, if $\varphi = c \in \mathbb{R}$ is a constant function, then $E(\varphi|\mathcal{B}) = \varphi = c$.

(f) *If $\mathcal{B} \supseteq \mathcal{C}$, then*

$$E\big(E(\varphi|\mathcal{B})\big|\mathcal{C}\big) = E(\varphi|\mathcal{C}).$$

We will now determine the conditional expectations of an arbitrary integrable function φ with respect to various sub-σ-algebras that are of particular interest and are simple enough.

Example 1.4.2 Let $(X, \mathcal{A}, \mu)$ be a probability space. The family $\mathcal{B}$ of all measurable sets that are either of null or of full measure constitutes a sub-σ-algebra of $\mathcal{A}$. Let $\varphi \in L^1(X, \mathcal{A}, \mu)$. Since $E(\varphi|\mathcal{B})$ is $\mathcal{B}$-measurable,

$$E(\varphi|\mathcal{B})^{-1}(\{t\}) \in \mathcal{B}$$

for each $t \in \mathbb{R}$, meaning that the set $E(\varphi|\mathcal{B})^{-1}(\{t\})$ is either of measure zero or of measure 1. Also bear in mind that

$$X = E(\varphi|\mathcal{B})^{-1}(\mathbb{R}) = \bigcup_{t \in \mathbb{R}} E(\varphi|\mathcal{B})^{-1}(\{t\}).$$

Since the above union consists of mutually disjoint sets of measure zero and 1, it follows that only one of these sets can be of measure 1. In other words, there exists a unique $t \in \mathbb{R}$ such that

$$E(\varphi|\mathcal{B})^{-1}(\{t\}) = A$$

for some $A \in \mathcal{A}$ with $\mu(A) = 1$. Because the function $E(\varphi|\mathcal{B})$ is unique up to a set of measure zero, we may assume without loss of generality that $A = X$. Hence, $E(\varphi|\mathcal{B})$ is a constant function. Therefore,

$$E(\varphi|\mathcal{B}) = \int_X E(\varphi|\mathcal{B})d\mu = \int_X \varphi\, d\mu.$$

Example 1.4.3 Let $(X, \mathcal{A})$ be a measurable space and α be a countable measurable partition of X. The sub-σ-algebra $\sigma(\alpha)$ of $\mathcal{A}$ generated by α is the family of all sets which can be represented as a union of elements of α. When α is finite, so is $\sigma(\alpha)$. When α is countably infinite, $\sigma(\alpha)$ is uncountable; in fact, it is of cardinality continuum. Let μ be a probability measure on $(X, \mathcal{A})$. Let $\varphi \in L^1(X, \mathcal{A}, \mu)$. Since $E(\varphi|\sigma(\alpha))$ is $\mathcal{B}$-measurable,

$$E(\varphi|\mathcal{B})^{-1}(\{t\}) \in \sigma(\alpha)$$

for each $t \in \mathbb{R}$, i.e.,

$$E\big(\varphi|\sigma(\alpha)\big)^{-1}(\{t\}) \in \sigma(\alpha).$$

This means that the set $E\big(\varphi|\sigma(\alpha)\big)^{-1}(\{t\})$ is a union of elements of α. This further means that the conditional expectation function $E(\varphi|\sigma(\alpha))$ is constant on each element of α. Let $A \in \alpha$. If $\mu(A) = 0$, then $E(\varphi|\sigma(\alpha))|_A = 0$. Otherwise,

$$E(\varphi|\sigma(\alpha))|_A = \frac{1}{\mu(A)} \int_A E(\varphi|\sigma(\alpha))d\mu = \frac{1}{\mu(A)} \int_{A_n} \varphi\, d\mu. \tag{1.33}$$

In summary, the conditional expectation $E(\varphi|\mathcal{B})$ of a function φ with respect to a sub-σ-algebra generated by a countable measurable partition is constant on each element of that partition. More precisely, on any given element of the partition, $E(\varphi|\mathcal{B})$ is equal to the mean value of φ on that element. In particular, if α is a trivial partition, i.e., consisting of sets of measure zero and 1 only, then

$$E(\varphi|\sigma(\alpha)) = \int_X \varphi\, d\mu \quad \mu\text{-a.e.} \tag{1.34}$$

The next result is a special case of a theorem originally due to Doob and is commonly called the Martingale Convergence Theorem. In order to discuss it, we first define the martingale itself.

Definition 1.4.4 Let $(X, \mathcal{A}, \mu)$ be a probability space. Let $(\mathcal{A}_n)_{n=1}^{\infty}$ be a sequence of sub-σ-algebras of $\mathcal{A}$. Let also $(\varphi_n : X \longrightarrow \mathbb{R})_{n=1}^{\infty}$ be a sequence of random variables, i.e., a sequence of $\mathcal{A}$-measurable functions. The sequence

$$\big((\varphi_n, \mathcal{A}_n)\big)_{n=1}^{\infty}$$

is called a *martingale* if and only if the following conditions are satisfied:

(a) $(\mathcal{A}_n)_{n=1}^{\infty}$ is an ascending sequence, i.e., $\mathcal{A}_{n+1} \supseteq \mathcal{A}_n$ for all $n \in \mathbb{N}$.
(b) φ_n is $\mathcal{A}_n$-measurable for all $n \in \mathbb{N}$.
(c) $\varphi_n \in L^1(\mu)$ for all $n \in \mathbb{N}$.
(d) $E(\varphi_{n+1}|\mathcal{A}_n) = \varphi_n$ μ-a.e. for all $n \in \mathbb{N}$.

The main, and frequently referred to as the simplest, convergence theorem concerning martingales is this.

Theorem 1.4.5 (Martingale Convergence Theorem) *Let $(X, \mathcal{A}, \mu)$ be a probability space. If $((\varphi_n, \mathcal{A}_n))_{n=1}^{\infty}$ is a martingale such that*

$$\sup\{\|\varphi_n\|_1 : n \in \mathbb{N}\} < +\infty,$$

then there exists $\widehat{\varphi} \in L^1(X, \mathcal{A}, \mu)$ such that

$$\lim_{n\to\infty} \varphi_n(x) = \widehat{\varphi}(x) \quad \textit{for } \mu\textit{-a.e. } x \in X$$

and

$$\|\widehat{\varphi}\|_1 \le \sup\{\|\varphi_n\|_1 : n \in \mathbb{N}\} < +\infty.$$

This is a special case of Theorem 35.5 in Billingsley's book [**Bil2**], proved therein. Its proof is just too long and too involved to be reproduced here. We omit it.

One natural martingale is formed by the conditional expectations of a function with respect to an ascending sequence of sub-σ-algebras.

Proposition 1.4.6 *Let $(X, \mathcal{A}, \mu)$ be a probability space and let $(\mathcal{A}_n)_{n=1}^{\infty}$ be an ascending sequence of sub-σ-algebras of $\mathcal{A}$. For any $\varphi \in L^1(X, \mathcal{A}, \mu)$, the sequence*

$$\big((E(\varphi|\mathcal{A}_n), \mathcal{A}_n)\big)_{n=1}^{\infty}$$

is a martingale.

Proof Indeed, set

$$\varphi_n := E(\varphi|\mathcal{A}_n)$$

for all $n \in \mathbb{N}$. Condition (a) in Definition 1.4.4 is automatically fulfilled. Conditions (b) and (c) follow from the very definition of the conditional expectation function. Regarding condition (d), a straightforward application of Proposition 1.4.1(f) gives

$$E(\varphi_{n+1}|\mathcal{A}_n) = E\big(E(\varphi|\mathcal{A}_{n+1})|\mathcal{A}_n\big) = E(\varphi|\mathcal{A}_n) = \varphi_n$$

μ-a.e. for all $n \in \mathbb{N}$. So $\big((E(\varphi|\mathcal{A}_n), \mathcal{A}_n)\big)_{n=1}^{\infty}$ is a martingale. ■

With the hypotheses of this proposition, by using Proposition 1.4.1(c), we see that

$$\sup_{n\in\mathbb{N}} \|\varphi_n\|_1 = \sup_{n\in\mathbb{N}} \int_X \big|E(\varphi|\mathcal{A}_n)\big| d\mu \le \sup_{n\in\mathbb{N}} \int_X E\big(|\varphi|\big|\mathcal{A}_n\big) d\mu = \int_X |\varphi|\, d\mu < \infty.$$

According to Theorem 1.4.5, there, thus, exists $\widehat{\varphi} \in L^1(X, \mathcal{A}, \mu)$ such that

$$\lim_{n\to\infty} E(\varphi|\mathcal{A}_n)(x) = \widehat{\varphi}(x) \text{ for } \mu\text{-a.e. } x \in X \qquad \text{and} \qquad \|\widehat{\varphi}\|_1 \le \|\varphi\|_1.$$

What is $\widehat{\varphi}$? This is the question we will address now. For this we need the concept of uniform integrability and the convergence theorem that it entails.

Definition 1.4.7 Let $(X, \mathcal{A}, \mu)$ be a measure space. A sequence of measurable functions $(f_n)_{n=1}^{\infty}$ is called uniformly integrable if and only if

$$\lim_{M\to\infty} \sup_{n\in\mathbb{N}} \int_{\{|f_n|\ge M\}} |f_n|\, d\mu = 0.$$

The following theorem is classical in measure theory. It is proved, for example, as Theorem 16.14 in Billingsley's book **[Bil2]**.

Theorem 1.4.8 *Let $(X, \mathcal{A}, \mu)$ be a finite measure space and $(f_n)_{n=1}^{\infty}$ a sequence of measurable functions that converges pointwise μ-a.e. to a function f.*

(a) *If $(f_n)_{n=1}^{\infty}$ is uniformly integrable, then $f_n \in L^1(\mu)$ for all $n \in \mathbb{N}$ and $f \in L^1(\mu)$. Moreover,*

$$\lim_{n\to\infty} \|f_n - f\|_1 = 0 \quad \textit{and} \quad \lim_{n\to\infty} \int_X f_n\, d\mu = \int_X f\, d\mu.$$

(b) *If $f, f_n \in L^1(\mu)$ and $f_n \ge 0$ μ-a.e. for all $n \in \mathbb{N}$, then $\lim_{n\to\infty} \int_X f_n\, d\mu = \int_X f\, d\mu$ implies that $(f_n)_{n=1}^{\infty}$ is uniformly integrable.*

We shall now prove the uniform integrability of the martingale appearing in Proposition 1.4.6.

Lemma 1.4.9 *Let $(X, \mathcal{A}, \mu)$ be a probability space and $(\mathcal{A}_n)_{n=1}^{\infty}$ be a sequence of sub-σ-algebras of $\mathcal{A}$. Then, for every $\varphi \in L^1(X, \mathcal{A}, \mu)$, the sequence $(E(\varphi|\mathcal{A}_n))_{n=1}^{\infty}$ is uniformly integrable.*

Proof Without loss of generality, we may assume that $\varphi \geq 0$. Let $\varepsilon > 0$. Since the measure ν on $(X, \mathcal{A})$ given by

$$\nu(A) := \int_A \varphi \, d\mu$$

is absolutely continuous with respect to μ, it follows from the Radon–Nikodym Theorem (Theorem 1.1.9) that there exists $\delta > 0$ such that

$$A \in \mathcal{A}, \ \mu(A) < \delta \implies \int_A \varphi \, d\mu < \varepsilon. \tag{1.35}$$

Consider any

$$M > \frac{1}{\delta} \int_X \varphi \, d\mu.$$

For each $n \in \mathbb{N}$, let

$$X_n(M) := \big\{ x \in X : E(\varphi|\mathcal{A}_n)(x) \geq M \big\}.$$

Observe that $X_n(M) \in \mathcal{A}_n$ since $E(\varphi|\mathcal{A}_n)$ is $\mathcal{A}_n$-measurable. Therefore, by Tchebyschev's Inequality, we get that

$$\mu(X_n(M)) \leq \frac{1}{M} \int_{X_n(M)} E(\varphi|\mathcal{A}_n) d\mu = \frac{1}{M} \int_{X_n(M)} \varphi \, d\mu \leq \frac{1}{M} \int_X \varphi \, d\mu < \delta$$

for all $n \in \mathbb{N}$. Consequently, by (1.35),

$$\int_{X_n(M)} E(\varphi|\mathcal{A}_n) d\mu = \int_{X_n(M)} \varphi \, d\mu < \varepsilon$$

for all $n \in \mathbb{N}$. Thus,

$$\sup_{n \in \mathbb{N}} \int_{\{E(\varphi|\mathcal{A}_n) \geq M\}} E(\varphi|\mathcal{A}_n) d\mu \leq \varepsilon.$$

Therefore,

$$\lim_{M \to \infty} \sup_{n \in \mathbb{N}} \int_{\{E(\varphi|\mathcal{A}_n) \geq M\}} E(\varphi|\mathcal{A}_n) d\mu = 0,$$

i.e., $(E(\varphi|\mathcal{A}_n))_{n=1}^{\infty}$ is uniformly integrable. ■

Theorem 1.4.10 (Martingale Convergence Theorem for Conditional Expectations) *Let $(X, \mathcal{A}, \mu)$ be a probability space and $\varphi \in L^1(X, \mathcal{A}, \mu)$. Let $(\mathcal{A}_n)_{n=1}^{\infty}$ be an ascending sequence of sub-σ-algebras of $\mathcal{A}$ and*

$$\mathcal{A}_\infty := \sigma\left(\bigcup_{n=1}^{\infty} \mathcal{A}_n \right).$$

Then

$$\lim_{n\to\infty} E(\varphi|\mathcal{A}_n) = E(\varphi|\mathcal{A}_\infty) \;\; \mu\text{-}a.e.\ on\ X$$

and

$$\lim_{n\to\infty} \left\| E(\varphi|\mathcal{A}_n) - E(\varphi|\mathcal{A}_\infty) \right\|_1 = 0.$$

Proof Let

$$\varphi_n := E(\varphi|\mathcal{A}_n).$$

It follows from Proposition 1.4.6 and Lemma 1.4.9 that $((\varphi_n, \mathcal{A}_n))_{n=1}^\infty$ is a uniformly integrable martingale such that

$$\lim_{n\to\infty} \varphi_n = \widehat{\varphi} \;\; \mu\text{-a.e. on } X$$

for some $\widehat{\varphi} \in L^1(X, \mathcal{A}, \mu)$. For all $n \in \mathbb{N}$, the function φ_n is $\mathcal{A}_\infty$-measurable since it is $\mathcal{A}_n$-measurable and $\mathcal{A}_n \subseteq \mathcal{A}_\infty$. Thus, $\widehat{\varphi}$ is $\mathcal{A}_\infty$-measurable, too. Moreover, it follows from Theorem 1.4.8 that

$$\lim_{n\to\infty} \|\varphi_n - \widehat{\varphi}\|_1 = 0 \quad \text{and} \quad \lim_{n\to\infty} \int_A \varphi_n \, d\mu = \int_A \widehat{\varphi} \, d\mu$$

for all $A \in \mathcal{A}$. Therefore, it just remains to show that

$$\widehat{\varphi} = E(\varphi|\mathcal{A}_\infty).$$

Let $k \in \mathbb{N}$ and $A \in \mathcal{A}_k$. If $n \geq k$, then $A \in \mathcal{A}_n \subseteq \mathcal{A}_\infty$ and, thus,

$$\int_A \varphi_n \, d\mu = \int_A E(\varphi|\mathcal{A}_n) d\mu = \int_A \varphi \, d\mu = \int_A E(\varphi|\mathcal{A}_\infty) d\mu.$$

Letting $n \to \infty$, this yields

$$\int_A \widehat{\varphi} \, d\mu = \int_A E(\varphi|\mathcal{A}_\infty) d\mu._k.$$

Since k was arbitrary, this entails

$$\int_B \widehat{\varphi} \, d\mu = \int_B E(\varphi|\mathcal{A}_\infty) d\mu$$

for all $B \in \bigcup_{k=1}^\infty \mathcal{A}_k$. Finally, since $\bigcup_{k=1}^\infty \mathcal{A}_k$ is a π-system generating $\mathcal{A}_\infty$ and both $\widehat{\varphi}$ and $E(\varphi|\mathcal{A}_\infty)$ are $\mathcal{A}_\infty$-measurable, we conclude that

$$\widehat{\varphi} = E(\varphi|\mathcal{A}_\infty) \;\; \mu\text{-a.e. in } X.$$

■

Recall that countable measurable partitions of a measurable space are defined and systematically treated in Section 6.1; they form the key concept for all of Chapter 6. As an immediate consequence of the Martingale Convergence Theorem for Conditional Expectations, i.e., Theorem 1.4.10 and (1.33), we get the following theorem, which is somewhat similar to the Lebesgue Density Theorem, i.e., Theorem 1.3.7 from the previous section.

Theorem 1.4.11 *Let $(X, \mathcal{A}, \mu)$ be a probability space and $(\alpha_n)_{n=1}^{\infty}$ be a sequence of finer and finer (more precisely, α_{n+1} is finer than α_n for all $n \geq 1$) countable measurable partitions of X which generates the σ-algebra $\mathcal{A}$, i.e., $\sigma\left(\bigcup_{n\geq 1} \alpha_n\right) = \mathcal{A}$. Then, for every set $A \in \mathcal{A}$ and for μ-a.e. $x \in A$, we have that*

$$\lim_{n\to\infty} \frac{\mu(A \cap \alpha_n(x))}{\mu(\alpha_n(x))} = 1\!\!1_A(x) = 1.$$

1.5 Hausdorff and Packing Measures: Hausdorff and Packing Dimensions

In this section, we introduce the basic geometric concepts on metric spaces. These are Hausdorff measures, Hausdorff dimensions, packing measures, and packing dimensions. We prove their fundamental properties. While Hausdorff measures and Hausdorff dimensions were introduced quite early, in 1919, by Felix Hausdorff in [**H**], it took several decades more for packing measures and packing dimensions to be defined. It was done in stages in [**Tr**], [**TT**], and [**Su6**]. There are now plenty of books on these concepts; we refer the reader to, for example, [**Fal1**], [**Fal12**], [**Fal13**], [**Mat**], and [**PU2**]. The classical book [**Ro**] by C. A. Rogers is also interesting, and not only for historical reasons, appearing for the first time in 1970. The 1998 edition is particularly interesting because of the comments by Falconer that it contains.

Let $\varphi\colon [0, +\infty) \longrightarrow [0, +\infty)$ be a function with the following properties:

- φ is nondecreasing, meaning that $s \leq t \Rightarrow \varphi(s) \leq \varphi(t)$.
- $\varphi(0) = 0$ and φ is continuous at 0.
- $\varphi((0, +\infty)) \subseteq (0, +\infty)$.

Any function $\varphi\colon [0, +\infty) \longrightarrow [0, +\infty)$ with such properties is referred to in what follows as a gauge function.

Let (X, ρ) be a metric space. For every $\delta > 0$, define

$$\mathrm{H}_{\varphi}^{\delta}(A) := \inf\left\{\sum_{i=1}^{\infty} \varphi(\mathrm{diam}(U_i))\right\}, \tag{1.36}$$

where the infimum is taken over all countable covers $\{U_i\}_{i=1}^{\infty}$ of A of diameter not exceeding δ.

We shall check that, for every $\delta > 0$, $\mathrm{H}_{\varphi}^{\delta}$ is an outer measure. Conditions (1.8) and (1.9) of Definition 1.2.1, defining the concept of outer measures, are obviously satisfied with $\mu = \mathrm{H}_{\varphi}^{\delta}$. To verify (1.10), let $\{A_n\}_{n=1}^{\infty}$ be a countable family of subsets of X. Fix $\varepsilon >$. Then, for every $n \geq 1$, we can find a countable cover $\{U_i^n\}_{i=1}^{\infty}$ of A_n with diameters not exceeding δ such that

$$\sum_{i=1}^{\infty} \varphi(\mathrm{diam}(U_i)) \leq \mathrm{H}_{\varphi}^{\delta}(A_n) + \frac{\varepsilon}{2^n}.$$

Then the family $\{U_i^n : i, n \geq 1\}$ covers $\bigcup_{n=1}^{\infty} A_n$ and

$$\begin{aligned} \mathrm{H}_{\varphi}^{\delta}\left(\bigcup_{n=1}^{\infty} A_n\right) &\leq \sum_{n=1}^{\infty}\sum_{i=1}^{\infty} \varphi(\mathrm{diam}(U_i^n)) \leq \sum_{n=1}^{\infty}\left(\mathrm{H}_{\varphi}^{\delta}(A_n) + \frac{\varepsilon}{2^n}\right) \\ &= \sum_{n=1}^{\infty} \mathrm{H}_{\varphi}^{\delta}(A_n) + \varepsilon. \end{aligned}$$

Thus, letting $\varepsilon \searrow 0$, (1.10) follows, proving that $\mathrm{H}_{\varphi}^{\delta}$ is an outer measure. Define

$$\mathrm{H}_{\varphi}(A) := \sup_{\delta > 0}\left\{\mathrm{H}_{\varphi}^{\delta}(A)\right\} = \lim_{\delta \to 0} \mathrm{H}_{\varphi}^{\delta}(A). \tag{1.37}$$

The limit exists since $\mathrm{H}_{\varphi}^{\delta}(A)$ increases as δ decreases, though it may happen to be infinite. Since all $\mathrm{H}_{\varphi}^{\delta}$ are outer measures. It is therefore immediate that H_{φ} is an outer measure too. Moreover, H_{φ} is a metric measure, since if A and B are two positively separated sets in X, then no set of diameter less than $\rho(A, B)$ can intersect both A and B. Consequently,

$$\mathrm{H}_{\varphi}^{\delta}(A \cup B) = \mathrm{H}_{\varphi}^{\delta}(A) + \mathrm{H}_{\varphi}^{\delta}(B)$$

for all $\delta < \rho(A, B)$. Letting $\delta \searrow 0$, we get the same formula for H_{φ}, which is just (1.13) with $\mu = \mathrm{H}_{\varphi}$. The metric outer measure H_{φ} is called the Hausdorff outer measure associated with the gauge function φ. Its restriction to the σ-algebra of H_{φ}-measurable sets, which by Theorem 1.2.5 includes all the Borel sets, is called the Hausdorff measure associated with the function φ. We should add that even if $E \subseteq X$ is not a Borel set, nor even H_{φ}-measurable, we nevertheless commonly refer to $\mathrm{H}_{\varphi}(E)$ as the Hausdorff measure of E rather than the Hausdorff outer measure of E.

As an immediate consequence of the definition of the Hausdorff measure and the properties of the function φ, we get the following.

Proposition 1.5.1 *For any gauge function φ, the Hausdorff measure H_φ is atomless.*

A particularly important role is played by the gauge functions of the form

$$\varphi_t(r) = r^t$$

for $t > 0$. In this case, the corresponding outer Hausdorff measure is denoted by H^t. So, H^t can be briefly defined as follows.

Definition 1.5.2 Given that $t \geq 0$, the t-dimensional outer Hausdorff measure $\mathrm{H}^t(A)$ of the set A is equal to

$$\mathrm{H}^t(A) = \sup_{\delta>0} \inf \left\{ \sum_{i=1}^{\infty} \mathrm{diam}^t(A_i) \right\},$$

where the infimum is taken over all countable covers $\{A_i\}_{i=1}^{\infty}$ of A by the sets with diameters $\leq \delta$.

Remark 1.5.3 Since $\mathrm{diam}(\overline{A}) = \mathrm{diam}(A)$ for every set $A \subseteq X$, we may, in Definition 1.5.2, restrict ourselves to closed sets A_i only.

Having defined Hausdorff measures, we now move on to define the dual concept, i.e., that of packing measures. As mentioned at the beginning of this section, while Hausdorff measures were introduced quite early, in 1919 by Felix Hausdorff in [**H**], it took several decades more for packing measures to be defined. It was done in stages in [**Tr**], [**TT**], and [**Su6**]. We do it again here now. We recall that, in Definition 1.3.3, we introduced the concept of packing. We will also use it now. For every $A \subseteq X$ and every $\delta > 0$, let

$$\Pi_\varphi^{*\delta}(A) := \sup \left\{ \sum_{i=1}^{\infty} \varphi(\mathrm{diam}(r_i)) \right\}, \tag{1.38}$$

where the supremum is taken over all packings $\{B(x_i, r_i)\}_{i=1}^{\infty}$ of A with radii not exceeding δ. Let

$$\Pi_\varphi^*(A) := \inf_{\delta>0} \left\{ \Pi_\varphi^{*\delta}(A) \right\} = \lim_{\delta\to 0} \Pi_\varphi^{*\delta}(A). \tag{1.39}$$

The limit exists since $\Pi_\varphi^{*\delta}(A)$ decreases as δ decreases. Although the function Π_φ^* satisfies condition (1.9) of outer measures, albeit in contrast to the case of Hausdorff measures, this function does not need to be subadditive, i.e., conditions (1.10) in general fail. In order to obtain an outer measure, we take one more step and we put

$$\Pi_\varphi(A) := \inf\left\{\sum_{i=1}^{\infty} \Pi_\varphi^{*\delta}(A_i)\right\}, \tag{1.40}$$

where the infimum is taken over all countable covers $\{A_i\}_{i=1}^{\infty}$ of A. Analogously, as in the case of Hausdorff measures, one checks, with similar arguments, that Π_φ is already an outer measure. Furthermore, it is a metric outer measure on X. It will be called the outer packing measure, associated with the gauge function φ. Its restriction to the σ-algebra of Π_φ-measurable sets, which by Theorem 1.2.5 includes all Borel sets, will be called the packing measure associated with the gauge function φ.
In the case of gauge functions,

$$\varphi_t(r) = r^t,$$

where $t > 0$, the definition of the outer packing measure takes the following form.

Definition 1.5.4 The t-dimensional outer packing measure $\Pi^t(A)$ of a set $A \subseteq X$ is given by

$$\Pi^t(A) = \inf_{\cup A_i = A}\left\{\sum_i \Pi_*^t(A_i)\right\}$$

(A_i are arbitrary subsets of A), where

$$\Pi_*^t(A) = \sup_{\delta>0}\sup\left\{\sum_{i=1}^{\infty} r_i^t\right\}.$$

Here, the second supremum is taken over all packings $\{B(x_i,r_i)\}_{i=1}^{\infty}$ of the set A by open balls centered at A with radii which do not exceed δ.

From now on, in order to obtain more meaningful geometric consequences, we assume that, for a given gauge function $\phi\colon [0,+\infty) \to [0,+\infty)$, there exists a function $C_\phi\colon (0,\infty) \to (0,\infty)$ such that, for every $a \in (0,\infty)$ and every $t > 0$ sufficiently small (depending on a),

$$C_\phi(a)^{-1}\phi(t) \le \phi(at) \le C_\phi(a)\phi(t). \tag{1.41}$$

We frequently refer to such gauge functions as evenly varying. Since $(at)^r = a^r t^r$, all the gauge functions ϕ of the form $r \mapsto r^t$ satisfy (1.41) with $C_\phi(a) = a^t$.

We shall now establish a simple, but crucial for geometric consequences, relation between Hausdorff and packing measures.

Proposition 1.5.5 *For every set $A \subseteq X$, it holds that* $\mathrm{H}_\varphi(A) \le C_\varphi(2)\Pi_\varphi(A)$.

Proof First, we show that, for every set $A \subseteq X$ and every $\delta > 0$,

$$\mathrm{H}_\varphi^{2\delta}(A) \le C_\varphi(2)\Pi_\varphi^{*\delta}(A). \tag{1.42}$$

Indeed, if there is no finite maximal (in the sense of inclusion) packing of the set A of the form $\{B(x_i,\delta)\}_{i=1}^\infty$, then, for every $k \ge 1$, there is a packing $\{B(x_i,\delta)\}_{i=1}^k$ of A; therefore,

$$\Pi_\varphi^{*\delta}(A) \ge \sum_{i=1}^k \varphi(\delta) = k\varphi(\delta).$$

Since $\varphi(\delta) > 0$, this yields $\Pi_\varphi^{*\delta}(A) = \infty$, and (1.42) holds. Otherwise, let $\{B(x_i,\delta)\}_{i=1}^l$ be a finite maximal packing of A. Then the collection $\{B(x_i,2\delta)\}$ covers A; therefore,

$$\mathrm{H}_\varphi^{2\delta}(A) \le \sum_{i=1}^l \varphi(2\delta) \le C_\varphi(2)l\varphi(\delta) \le C_\varphi(2)\Pi_\varphi^{*\delta}(A).$$

Hence, (1.42) is satisfied. Thus, letting $\delta \searrow 0$, we get that

$$\mathrm{H}_\varphi(A) \le C_\varphi(2)\Pi_\varphi^*(A). \tag{1.43}$$

So, if $\{A_n\}_{n\ge1}$ is a countable cover of A, then

$$\mathrm{H}_\varphi(A) \le \sum_{n=1}^\infty \mathrm{H}_\varphi(A_i) \le C_\varphi(2)\sum_{n=1}^\infty \Pi_\varphi^*(A_i).$$

Hence, applying (1.40), the lemma follows. ■

Definition 1.5.6 HD(A), the Hausdorff dimension of the set A, is defined to be

$$\mathrm{HD}(A) := \inf\{t\colon \mathrm{H}^t(A) = 0\} = \sup\{t\colon \mathrm{H}^t(A) = \infty\}. \tag{1.44}$$

Likewise, PD(A), the packing dimension of the set A, is defined to be

$$\mathrm{PD}(A) := \inf\{t\colon \Pi^t(A) = 0\} = \sup\{t\colon \Pi^t(A) = \infty\}. \tag{1.45}$$

The following theorem is the immediate consequence of the definition of Hausdorff and packing dimensions and the corresponding outer measures.

Theorem 1.5.7 *The Hausdorff and packing dimensions are monotone increasing functions of sets, i.e., if $A \subseteq B$, then*

$$\mathrm{HD}(A) \le \mathrm{HD}(B) \quad \textit{and} \quad \mathrm{PD}(A) \le \mathrm{PD}(B).$$

We shall prove the following theorem, commonly referred to as the σ-stability Hausdorff and packing dimensions.

Theorem 1.5.8 *If* $\{A_n\}_{n=1}^{\infty}$ *is a countable family of subsets of* X*, then*

$$\mathrm{HD}\left(\bigcup_{n=1}^{\infty} A_n\right) = \sup_{n\geq 1}\{\mathrm{HD}(A_n)\}$$

and

$$\mathrm{PD}\left(\bigcup_{n=1}^{\infty} A_n\right) = \sup_{n\geq 1}\{\mathrm{PD}(A_n)\}.$$

Proof We shall prove only the Hausdorff dimension part. The proof for the packing dimension is analogous. Inequality

$$\mathrm{HD}\left(\bigcup_{n=1}^{\infty} A_n\right) \geq \sup_{n\geq 1}\{\mathrm{HD}(A_n)\}$$

is an immediate consequence of Theorem 1.5.7. Thus, if $\sup_n\{\mathrm{HD}(A_n)\} = \infty$, there is nothing to prove. So, suppose that

$$s := \sup_{n\geq 1}\{\mathrm{HD}(A_n)\}$$

is finite and consider an arbitrary $t > s$. Then, in view of (1.44), $\mathrm{H}^t(A_n) = 0$ for every $n \geq 1$; therefore, since H^t is an outer measure,

$$\mathrm{H}^t\left(\bigcup_{n=1}^{\infty} A_n\right) = 0.$$

Hence, by (1.44) again,

$$\mathrm{HD}\left(\bigcup_{n=1}^{\infty} A_n\right) \leq t.$$

The proof is complete. ■

As an immediate consequence of this theorem, Proposition 1.5.1, and (1.44), we obtain the following.

Proposition 1.5.9 *The Hausdorff dimension of any countable set is equal to* 0.

These are the most basic, transparent, and also probably most useful properties of Hausdorff and packing measures and dimensions. We will apply them frequently in both volumes of the book.

1.6 Hausdorff and Packing Measures: Frostman Converse-Type Theorems

In this section, we derive several geometric consequences of Theorem 1.3.1. Their meaning is to tell us when a Hausdorff measure or packing measure is positive, finite, zero, or infinity. We refer to them as the Frostman Converse-Type Theorems. Somewhat strangely, these theorems are frequently called the Mass Redistribution Principle in the fractal literature, as if to indicate that such measures would always have to appear in the process of an iterative construction. At the end of the section, we formulate the Frostman Direct Theorem and compare it with the Frostman Converse-Type Theorems. As already mentioned, the advantage of the latter theorems is that they provide tools to calculate, or at least to estimate, both Hausdorff and packing measures and dimensions. We recall that in this section, as in the entire book, we keep

$$\varphi\colon [0, +\infty) \longrightarrow [0, +\infty),$$

an evenly varying gauge function, i.e., satisfying (1.41). We start with the following.

Theorem 1.6.1 (Frostman Converse-Type Theorem for Generalized Hausdorff Measures) *Let $\varphi\colon [0,+\infty) \longrightarrow [0,+\infty)$ be a continuous evenly varying gauge function. Let (X,ρ) be an arbitrary metric space and μ a Borel probability measure on X. Fix a Borel set $A \subseteq X$. Assume that there exists a constant $c \in (0, +\infty]$ $(1/+\infty = 0)$ such that*

(1)

$$\limsup_{r\to 0} \frac{\mu(B(x,r))}{\varphi(r)} \geq c$$

for all points $x \in A$ except for countably many perhaps. Then the Hausdorff measure H_φ, corresponding to the gauge function φ, satisfies

$$\mathrm{H}_\varphi(E) \leq c^{-1} C_\varphi(8)\mu(E)$$

for every Borel set $E \subseteq A$. In particular,

$$\mathrm{H}_\varphi(A) < +\infty \quad (\mathrm{H}_\varphi(A) = 0 \ \text{if} \ c = +\infty).$$

(2) *If, conversely,*

$$\limsup_{r\to 0} \frac{\mu(B(x,r))}{\varphi(r)} \leq c < +\infty$$

for all $x \in A$, then

$$\mu(E) \leq \mathrm{H}_\varphi(E)$$

for every Borel set $E \subseteq A$. In particular,

$$\mathrm{H}_\varphi(A) > 0$$

whenever $\mu(E) > 0$.

Proof Part (1). Since H_φ of any countable set is equal to 0, we may assume without loss of generality that E does not intersect the exceptional countable set. Fix $\varepsilon > 0$. Then fix $\delta > 0$. Since measure μ is regular, there exists an open set $G \supseteq E$ such that

$$\mu(G) \leq \mu(E) + \varepsilon.$$

Further, for every $x \in E$, there exists $r(x) \in (0, \delta)$ such that $B(x, r(x)) \subseteq G$ and

$$(c^{-1} + \varepsilon)\mu(B(x, r(x))) \geq \varphi(r(x)) > 0.$$

By virtue both of $4r$ Covering Theorem, i.e., Theorem 1.3.1, and of Remark 1.3.2, there exists $\{x_k\}_{k=1}^{\infty}$, a sequence of points in E such that

$$B(x_i, r(x_i)) \cap B(x_j, r(x_j)) = \emptyset \qquad \text{for } i \neq j$$

and

$$\bigcup_{k=1}^{\infty} B(x_k, 4r(x_k)) \supseteq \bigcup_{x \in E} B(x, r(x)) \supseteq E.$$

Hence,

$$\begin{aligned}
\mathrm{H}_\varphi^{2\delta}(E) &\leq \sum_{k=1}^{\infty} \varphi(2 \cdot 4r(x_k)) \leq \sum_{k=1}^{\infty} c_\varphi(8)\varphi(r(x_k)) \\
&\leq c_\varphi(8) \sum_{k=1}^{\infty} (c^{-1} + \varepsilon)\mu(B(x_k, r(x_k))) \\
&= c_\varphi(8)(c^{-1} + \varepsilon)\mu\left(\bigcup_{k=1}^{\infty} B(x_k, r(x_k))\right) \leq c_\varphi(8)(c^{-1} + \varepsilon)\mu(G) \\
&\leq c_\varphi(8)(c^{-1} + \varepsilon)(\mu(E) + \varepsilon).
\end{aligned}$$

So, letting $\delta \searrow 0$, we get

$$\mathrm{H}_\varphi(E) \leq c_\varphi(8)(c^{-1} + \varepsilon)(\mu(E) + \varepsilon)$$

and, since $\varepsilon > 0$ was arbitrary, we finally get

$$\mathrm{H}_\varphi(E) \le c_\varphi(8)c^{-1}\mu(E).$$

This finishes the first part of the proof.
Part (2). Now we deal with the second part of our theorem. Fix an arbitrary $s > c$. Note that, for every $r > 0$, the function

$$X \ni x \longmapsto \frac{\mu(B(x,r))}{\varphi(r)} \quad \text{is Borel measurable.}$$

For every $k \ge 1$, consider the function

$$X \ni x \longmapsto \varphi_k(x) := \sup\left\{\frac{\mu(B(x,r)}{\varphi(r)} : \ r \in \mathbb{Q} \cap (0,1/k]\right\},$$

where $\mathbb{Q}$ denotes the set of rational numbers. This function is Borel measurable as the supremum of countably many measurable functions. Let

$$A_k = A \cap \varphi_k^{-1}((0,s]) \quad \text{for } k \ge 1.$$

All A_k, $k \ge 1$, are then Borel subsets of X. Fix an arbitrary $r \in (0,1/k)$. Then pick $r_j \searrow r$, $r_j \in Q$. Since the function $t \mapsto \mu(B(x,t))$ is nondecreasing and the function φ is continuous, we get, for every $x \in A_k$, that

$$\frac{\mu(B(x,r))}{\varphi(r)} \le \limsup_{j\to\infty} \frac{\mu(B(x,r_j))}{\varphi(r_j)} \le s.$$

Now fix $k \ge 1$. Then fix a Borel set $F \subseteq A_k$. Our first objective is to prove the assertion of Part (2) for the set F. To do this, fix $r < 1/k$ and then $\{F_i\}_1^\infty$, a countable cover of F by subsets of F that are closed relative to F and have diameters less than $r/2$. For every $i \ge 1$, pick $x_i \in F_i$. Then $F_i \subset B(x_i, \mathrm{diam}(F_i))$. Since all the sets F_i, $i \ge 1$, are also Borel in X, we, therefore, have that

$$\sum_{i=1}^\infty \varphi(\mathrm{diam}(F_i)) \ge s^{-1}\sum_{i=1}^\infty \mu(B(x_i,\mathrm{diam}(F_i))) \ge s^{-1}\sum_{i=1}^\infty \mu(F_i) \ge s^{-1}\mu(F).$$

Hence, invoking Remark 1.5.3, we get that

$$\mathrm{H}_\varphi(F) \ge s^{-1}\mu(F). \tag{1.46}$$

Moving on, by our hypothesis, we have that

$$\bigcup_{k=1}^\infty A_k \cap A = A.$$

Define inductively

$$B_1 := A_1 \cap A$$

and

$$B_{k+1} := A_{k+1} \cap \left(A \backslash \bigcup_{j=1}^{k} A_j \cap A \right).$$

Obviously, the family $\{B_k\}_1^\infty$ consists of mutually disjoint sets, with each set B_k contained in A_k, $k \geq 1$, and

$$\bigcup_{k=1}^{\infty} B_k = \bigcup_{k=1}^{\infty} A_k = A.$$

Hence, if E is a Borel subset of A, then applying (1.46) for sets $F = E \cap B_k$, $k \geq 1$, we get

$$\mathrm{H}_\varphi(E) = \bigcup_{k=1}^{\infty} \mathrm{H}_\varphi(E \cap B_k) \geq s^{-1} \sum_{k=1}^{\infty} \mu(E \cap B_k) = s^{-1}\mu(E).$$

Letting $s \searrow c$ then finishes the proof. ■

Now let us prove the corresponding theorem for packing measures.

Theorem 1.6.2 (Frostman Converse-Type Theorem for Generalized Packing Measures) *Let $\varphi\colon [0,\infty) \longrightarrow [0,\infty)$ be a continuous evenly varying gauge function. Let (X,ρ) be an arbitrary metric space and μ be a Borel probability measure on X. Fix a Borel set $A \subset X$ and assume that there exists $c \in (0,+\infty]$* $(1/+\infty = 0)$ *such that*

(1)

$$\liminf_{r \to 0} \frac{\mu(B(x,r))}{\varphi(r)} \leq c \quad \textit{for all } x \in A.$$

Then

$$\mu(E) \leq \Pi_\varphi(E)$$

for every Borel set $E \subseteq A$, where, we recall, Π_φ denotes the packing measure corresponding to the gauge function φ. In particular, if $\mu(E) > 0$, then

$$\Pi_\varphi(E) > 0.$$

(2) *If, conversly,*

$$\liminf_{r\to 0} \frac{\mu(B(x,r))}{\varphi(r)} \geq c \quad \textit{for all } x \in A,$$

then

$$\Pi_\varphi(E) \leq c^{-1}\mu(E)$$

for every Borel set $E \subseteq A$*. In particular, if* $\mu(E) < +\infty$*, then*

$$\Pi_\varphi(E) < +\infty.$$

Proof Part (1). Let $\varepsilon > 0$. Fix an arbitrary subset $F \subseteq A$. Define a decreasing sequence $(G_n)_{n\geq 1}$ of open sets containing F as follows. By our hypothesis, for every $x \in A$, there exists $0 < r_1(x) < 1$ such that

$$\frac{\mu\big(B(x,r_1(x))\big)}{\phi(r_1(x))} \leq c+\varepsilon.$$

Take the family of balls $\big\{B\big(x,\frac{1}{4}r_1(x)\big)\big\}_{x\in F}$. According to the $4r$ Covering Theorem (Theorem 1.3.1), there is a countable set $F_1 \subseteq F$ such that the subfamily $\big\{B\big(x,\frac{1}{4}r_1(x)\big)\big\}_{x\in F_1}$ consists of mutually disjoint balls satisfying

$$F \subseteq \bigcup_{x\in F} B\Big(x,\frac{1}{4}r_1(x)\Big) \subseteq \bigcup_{x\in F_1} B\big(x,r_1(x)\big).$$

Let $G_1 := \bigcup_{x\in F_1} B(x,r_1(x))$. For the inductive step, suppose that G_n has been defined for some $n \geq 1$. By our hypothesis again, for every $x \in A$, there exists some $0 < r_{n+1}(x) < \frac{1}{n+1}$ such that $B(x,r_{n+1}(x)) \subseteq G_n$ and

$$\frac{\mu\big(B(x,r_{n+1}(x))\big)}{\phi(r_{n+1}(x))} \leq c+\varepsilon. \tag{1.47}$$

Consider the family of balls $\big\{B\big(x,\frac{1}{4}r_{n+1}(x)\big)\big\}_{x\in F}$. According to the $4r$ Covering Theorem (Theorem 1.3.1), there exists a countable set $F_{n+1} \subseteq F$ such that the subfamily $\big\{B\big(x,\frac{1}{4}r_{n+1}(x)\big)\big\}_{x\in F_{n+1}}$ consists of mutually disjoint balls satisfying

$$F \subseteq \bigcup_{x\in F} B\left(x,\frac{1}{4}r_{n+1}(x)\right) \subseteq \bigcup_{x\in F_{n+1}} B\big(x,r_{n+1}(x)\big).$$

Let

$$G_{n+1} := \bigcup_{x\in F_{n+1}} B(x,r_{n+1}(x)).$$

It is clear that G_{n+1} is an open set and $F \subseteq G_{n+1} \subseteq G_n$. Moreover, for all pairs $x, y \in F_{n+1} \subseteq F$, we know that

$$d(x,y) \geq \frac{1}{4}\max\{r_{n+1}(x), r_{n+1}(y)\} \geq \frac{1}{8}(r_{n+1}(x) + r_{n+1}(y)).$$

Therefore, the collection $\left\{\left(x, \frac{1}{8}r_{n+1}(x)\right)\right\}_{x\in F_{n+1}}$ forms an $\left(\frac{1}{n+1}\right)$-packing of F. Using (1.47), it follows that

$$\begin{aligned}
\Pi_\phi^{*\frac{1}{n+1}}(F) &\geq \sum_{x\in F_{n+1}} \phi\left(\frac{1}{8}r_{n+1}(x)\right) \geq (C_\phi(8))^{-1} \sum_{x\in F_{n+1}} \phi(r_{n+1}(x)) \\
&\geq (C_\phi(8))^{-1} \sum_{x\in F_{n+1}} \frac{\mu\big(B(x, r_{n+1}(x))\big)}{c+\varepsilon} \\
&= \frac{(C_\phi(8))^{-1}}{c+\varepsilon}\mu\left(\bigcup_{x\in F_{n+1}} B(x, r_{n+1}(x))\right) \\
&\geq \frac{(C_\phi(8))^{-1}}{c+\varepsilon}\mu(G_{n+1}).
\end{aligned}$$

Letting n increase to infinity, we, thus, obtain that

$$\Pi_\phi^*(F) \geq \frac{(C_\phi(8))^{-1}}{c+\varepsilon} \inf_{n\geq 1} \mu(G_n) = \frac{(C_\phi(8))^{-1}}{c+\varepsilon} \lim_{n\to\infty} \mu(G_n) = \frac{(C_\phi(8))^{-1}}{c+\varepsilon}\mu(G_F),$$

where $G_F := \bigcap_{n\geq 1} G_n$ is a G_δ set and, therefore, in particular, is a Borel set. Consequently, for every Borel set $E \subseteq A$, we have that

$$\begin{aligned}
\Pi_\phi(E) &= \inf\left\{\sum_{k=1}^{\infty} \Pi_\phi^*(A_k) \colon \{A_k\}_{k=1}^{\infty} \text{ is a cover of } E\right\} \\
&= \inf\left\{\sum_{k=1}^{\infty} \Pi_\phi^*(A_k) \colon \{A_k\}_{k=1}^{\infty} \text{ is a partition of } E\right\} \\
&\geq \frac{(C_\phi(8))^{-1}}{c+\varepsilon} \inf\left\{\sum_{k=1}^{\infty} \mu(G_{A_k}) \colon \{A_k\}_{k=1}^{\infty} \text{ is a partition of } E\right\} \\
&\geq \frac{(C_\phi(8))^{-1}}{c+\varepsilon} \inf\left\{\mu\left(\bigcup_{k=1}^{\infty} G_{A_k}\right) \colon \{A_k\}_{k=1}^{\infty} \text{ is a partition of } E\right\} \\
&\geq \frac{(C_\phi(8))^{-1}}{c+\varepsilon} \inf\left\{\mu(E) \colon \{A_k\}_{k=1}^{\infty} \text{ is a partition of } E\right\} \\
&= \frac{(C_\phi(8))^{-1}}{c+\varepsilon}\mu(E).
\end{aligned}$$

Since this holds for all $\varepsilon > 0$, we deduce that $\Pi_\phi(E) \geq (C_\phi(8)c)^{-1}\mu(E)$. The proof of Part (1) is complete.
Part (2). The sequence of functions $(\psi_k)_{k=1}^\infty$, where

$$X \ni x \longmapsto \psi_k(x) := \inf\left\{\frac{\mu(B(x,r))}{\phi(r)} : r \in \mathbb{Q} \cap \left(0, \frac{1}{k}\right]\right\},$$

forms an increasing sequence of measurable functions. Let $0 < s < c$. For each $k \geq 1$, let

$$A_k := \psi_k^{-1}([s, +\infty)).$$

As $(\psi_k)_{k=1}^\infty$ is increasing, so is the sequence $(A_k)_{k=1}^\infty$. Moreover, since $s < c$, it follows from our hypothesis that

$$\bigcup_{k=1}^\infty A_k \supseteq A.$$

Furthermore, since $[s, +\infty)$ is a Borel subset of $\mathbb{R}$, the measurability of ψ_k ensures that A_k is a Borel subset of X. Fix $k \geq 1$. Choose some arbitrary $r \in (0, 1/k]$ and pick a sequence $(r_j)_{j\geq 1} \in \mathbb{Q}$ such that r_j increases to r. Since μ is a measure and ϕ is continuous, we deduce that, for all $x \in A_k$,

$$\frac{\mu(B(x,r))}{\phi(r)} = \lim_{j\to\infty} \frac{\mu(B(x,r_j))}{\phi(r_j)} \geq \psi_k(x) \geq s.$$

Thus, if $x \in A_k$, then

$$\inf\left\{\frac{\mu(B(x,r))}{\phi(r)} : r \in (0, 1/k]\right\} \geq s.$$

Now fix any set $F \subseteq A_k$ and any $r \in \left(0, \frac{1}{k}\right]$. Let $\{(x_i, t_i)\}_{i\geq 1}$ be an r-packing of F. Then

$$\sum_{i=1}^\infty \phi(t_i) \leq s^{-1} \sum_{i=1}^\infty \mu(B(x_i, t_i)) = s^{-1}\mu\left(\bigcup_{i=1}^\infty B(x_i, t_i)\right) \leq s^{-1}\mu(F_r),$$

where F_r denotes the open r-neighborhood of F. Taking the supremum over all r-packings yields

$$\Pi_\phi(F) \leq \Pi_\phi^*(F) \leq \Pi_\phi^{*r}(F) \leq s^{-1}\mu(F_r).$$

Thus, we have that $P_\phi(F) \leq s^{-1}\mu(F_r)$ for all $r \in (0, 1/k]$ and each subset $F \subseteq A_k$. Consequently, $\Pi_\phi(F) \leq s^{-1}\mu(F_0) = s^{-1}\mu(\overline{F})$ for all $F \subseteq A_k$. In particular, if C is a closed subset of E, then

$$\Pi_\phi(C \cap A_k) \leq s^{-1}\mu(\overline{C \cap A_k}) \leq s^{-1}\mu(C) \leq s^{-1}\mu(E).$$

As this holds for all integers $k \geq 1$ and closed sets $C \subseteq E \subseteq A = \cup_{k\geq 1} A_k$, we deduce that $\Pi_\phi(C) \leq s^{-1}\mu(E)$. By the regularity of μ, on taking the supremum over all closed sets C contained in E, we conclude that

$$\Pi_\phi(E) \leq s^{-1}\mu(E).$$

Letting s increase to c finishes the proof. ∎

Replacing $\varphi(r)$ by r^t and H_φ by H^t in Theorem 1.6.1 and Π_φ by Π^t in Theorem 1.6.2, we immediately get the following two results.

Theorem 1.6.3 (Frostman Converse-Type Theorem for Hausdorff Measures) *Fix $t > 0$ arbitrary. Let (X, ρ) be a metric space and μ a Borel probability measure on X. Fix a Borel set $A \subseteq X$. Assume that there exists a constant $c \in (0, +\infty]$ $(1/+\infty = 0)$ such that*

(1)

$$\limsup_{r\to 0} \frac{\mu(B(x,r))}{r^t} \geq c$$

for all points $x \in A$ except for countably many perhaps. Then the Hausdorff measure H_t satisfies

$$H_t(E) \leq c^{-1}8^t\mu(E)$$

for every Borel set $E \subseteq A$. In particular,

$$H_t(A) < +\infty \quad (H_t(A) = 0 \text{ if } c = +\infty).$$

(2) *If, conversely,*

$$\limsup_{r\to 0} \frac{\mu(B(x,r))}{r^t} \leq c < +\infty$$

for all $x \in A$, then

$$\mu(E) \leq H_t(E)$$

for every Borel set $E \subseteq A$. In particular,

$$H_t(A) > 0$$

whenever $\mu(E) > 0$.

Theorem 1.6.4 (Frostman Converse-Type Theorem for Packing Measures) *Fix $t > 0$ arbitrary. Let (X, ρ) be a metric space and μ a Borel probability measure on X. Fix a Borel set $A \subset X$ and assume that there exists $c \in (0, +\infty]$ $(1/+\infty = 0)$ such that*

(1)

$$\liminf_{r\to 0} \frac{\mu(B(x,r))}{r^t} \le c \quad \textit{for all } x \in A.$$

Then

$$\mu(E) \le \Pi_t(E)$$

for every Borel set $E \subseteq A$*. In particular, if* $\mu(E) > 0$*, then*

$$\Pi_t(E) > 0.$$

(2) *If conversely,*

$$\liminf_{r\to 0} \frac{\mu(B(x,r))}{r^t} \ge c \quad \textit{for all } x \in A,$$

then

$$\Pi_t(E) \le c^{-1}\mu(E)$$

for every Borel set $E \subseteq A$*. In particular, if* $\mu(E) < +\infty$*, then*

$$\Pi_t(E) < +\infty.$$

In the opposite direction to Frostman Converse Theorems, there is the following well-known theorem.

Theorem 1.6.5 (Frostman Direct Lemma) *Let* X *be either a Borel subset of a Euclidean space* $\mathbb{R}^d$*,* $d \ge 1$*, or an arbitrary compact metric space. If* $t > 0$ *and* $\mathrm{H}_t(X) > 0$*, then there exists a Borel probability measure* μ *on* X *such that*

$$\mu(B(x,r)) \le r^t$$

for every point $x \in X$ *and all radii* $r > 0$*.*

This is a very interesting theorem, although Frostman Converse Theorems seem to be more suitable for estimating and calculating Hausdorff and packing measures and dimensions.

1.7 Hausdorff and Packing Dimensions of Measures

In this section, we define the concepts of dimensions, both Hausdorff and packing, of Borel measures. We then provide tools to calculate and estimate them. We also establish some relations between them. The dimensions of measures play an important role in both fractal geometry and dynamical systems.

We start this section with the following simple but crucial consequence of Theorem 1.6.3.

Theorem 1.7.1 (Volume Lemma for Hausdorff Measures) *Suppose that μ is a Borel probability measure on a metric space (X,ρ) and that A is a bounded Borel subset of $\mathbb{R}^n$. Then*

(a) *If $\mu(A) > 0$ and there exists θ_1 such that, for every $x \in A$,*

$$\liminf_{r\to 0} \frac{\log \mu(B(x,r))}{\log r} \geq \theta_1,$$

then $\mathrm{HD}(A) \geq \theta_1$.

(b) *If there exists θ_2 such that, for every $x \in A$,*

$$\liminf_{r\to 0} \frac{\log \mu(B(x,r))}{\log r} \leq \theta_2,$$

then $\mathrm{HD}(A) \leq \theta_2$.

Proof (a) Take any $0 < \theta < \theta_1$. Then, by assumption,

$$\limsup_{r\to 0} \mu(B(x,r))/r^\theta = 0.$$

It, therefore, follows from Theorem 1.6.3(2) that $\mathrm{H}^\theta(A) = +\infty$. Hence, $\mathrm{HD}(A) \geq \theta$. Consequently, $\mathrm{HD}(A) \geq \theta_1$.

(b) Now take an arbitrary $\theta > \theta_2$. Then, by assumption,

$$\limsup_{r\to 0} \mu(B(x,r))/r^\theta = +\infty.$$

Therefore, applying Theorem 1.6.3(1), we obtain that $\mathrm{H}^\theta(A) = 0$. Thus, $\mathrm{HD}(A) \leq \theta$ and, consequently, $\mathrm{HD}(A) \leq \theta_2$. The proof is finished. ■

Similarly, one proves the following consequence of Theorem 1.6.4.

Theorem 1.7.2 (Volume Lemma for Packing Measures) *Suppose that μ is a Borel probability measure on $\mathbb{R}^n$, $n \geq 1$, and A is a bounded Borel subset of $\mathbb{R}^n$.*

(a) *If $\mu(A) > 0$ and there exists θ_1 such that, for every $x \in A$,*

$$\limsup_{r\to 0} \frac{\log \mu(B(x,r))}{\log r} \geq \theta_1,$$

then $\mathrm{PD}(A) \geq \theta_1$.

(b) *If there exists θ_2 such that, for every $x \in A$,*

$$\limsup_{r\to 0} \frac{\log \mu(B(x,r))}{\log r} \leq \theta_2,$$

then $\mathrm{PD}(A) \leq \theta_2$.

We will now apply Theorem 1.7.1(a) to get quite a general lower bound for the Hausdorff dimension, the one that is a generalization of a result due to McMullen [**McM1**], the proof of which is taken from [**U1**]. Although this result is usually applied in a dynamical context, it really does not require any dynamics to formulate and to prove it.

As always in this section, let (X,ρ) be a metric space and μ be a Borel probability upper Ahlfors measure on X, meaning that there exist constants $h>0$ and $C\ge 1$ such that, for every $x\in X$ and $r>0$,

$$\mu(B(x,r))\le Cr^h. \tag{1.48}$$

We then call h the exponent of μ. For any integer $k\ge 1$, let E_k be a finite collection of compact subsets of X, each element of which has positive measure μ. We denote:

$$K:=\bigcup_{F\in E_1}F. \tag{1.49}$$

We assume the following.

$$\text{If } k\ge 1,\ F,G\in E_k, \text{ and } F\ne G, \text{ then } \mu(F\cap G)=0. \tag{1.50}$$

$$\text{Every set } F\in E_{k+1} \text{ is contained in a unique element } G\in E_k. \tag{1.51}$$

For every integer $k\ge 1$ and every set $F\in E_k$, define

$$\text{density}\left(\bigcup_{D\in E_{k+1}}D,F\right):=\frac{\mu\left(D\cap\bigcup_{D\in E_{k+1}}D\right)}{\mu(F)} \tag{1.52}$$

and assume that

$$\Delta_k:=\inf\left\{\text{density}\left(\bigcup_{D\in E_{k+1}}D,F\right):\ F\in E_k\right\}>0 \tag{1.53}$$

for every $k\ge 1$. Put also

$$d_k:=\sup\left\{\text{diam}(F):\ F\in E_k\right\}.$$

Suppose that $d_k<1$ for every $k\ge 1$ and that

$$\lim_{k\to\infty}d_k=0. \tag{1.54}$$

We then call the collection

$$\{E_k\}_{k=1}^{\infty}$$

a McMullen sequence of sets. Let

$$E_\infty := \bigcap_{k=1}^{\infty} \bigcup_{F \in E_k} F.$$

We shall prove the following generalization of a McMullen result from [**McM1**], the proof of which is taken from [**U1**].

Proposition 1.7.3 *If $\{E_k\}_{k=1}^\infty$ is a McMullen sequence of subsets of a metric space (X,ρ) endowed with a Borel probability upper Ahlfors measure μ having exponent h, then*

$$\mathrm{HD}(E_\infty) \geq h - \limsup_{k\to\infty} \frac{\sum_{j=1}^{k-1} \log \Delta_j}{\log d_k}.$$

Proof We construct inductively a sequence of Borel probability measures $\{\nu_k\}_{k=1}^\infty$ on K as follows.

Put $\nu_1 := \mu$ and define ν_{k+1} by putting, for each Borel set $A \subseteq K$,

$$\nu_{k+1}(A) := \sum_{F \in E_k} \frac{\mu\left(A \cap F \cap \bigcup_{D \in E_{k+1}} D\right)}{\mu\left(F \cap \bigcup_{D \in E_{k+1}} D\right)} \nu_k(F). \tag{1.55}$$

This definition makes sense since, by (1.52) and (1.53), we see that $\mu\left(F \cap \bigcup_{D \in E_{k+1}} D\right) > 0$. By induction, we get for every $k \geq 1$, that

$$\nu\left(\bigcup_{D \in E_k} D\right) = 1, \tag{1.56}$$

and it follows from properties (1.49)–(1.51) that ν_{k+1} is a Borel probability measure indeed. In view of (1.55) and (1.50), we have that $\nu_{k+1}(F) = \nu_k(F)$ for each $F \in E_k$. Hence, using (1.50) and (1.51), we conclude by induction that $\nu_n(F) = \nu_k(F)$ for every $n \geq k$. Since $\lim_{k\to\infty} d_k = 0$, we, therefore, obtain a unique probability measure ν on K (being the weak limit of measures ν_k) such that

$$\nu(F) = \nu_k(F) \tag{1.57}$$

for every $F \in E_k$. Looking now at (1.56) and the definition of the set E, one gets

$$\nu(E_\infty) = 1. \tag{1.58}$$

Making use of (1.55) and (1.57), one easily estimates, for every $F \in E_k$, that

$$\nu(F) \leq \frac{\mu(F)}{\Delta_{k-1}, \ldots, \Delta_1}. \tag{1.59}$$

In view of Theorem 1.7.1, the Volume Lemma for Hausdorff Measures, in order to prove that $\mathrm{HD}(E) \geq \delta$ for some $\delta \geq 0$ it is enough to show that

$$\liminf_{r \to 0} \frac{\log \nu(B(x,r))}{\log r} \geq \delta \tag{1.60}$$

for ν-a.e. $x \in E$.

Now consider $x \in E_\infty$ and $0 < r < \sup_{k \geq 1}(d_k)$ arbitrary. Then there exists an integer $k = k(r) \geq 1$ such that $d_{k+1} \leq r \leq d_k$. Let $\tilde{B}(x,r)$ be the union of all sets in E_{k+1} which meet $B(x,r)$. Then $\tilde{B}(x,r) \subseteq B(x,2r)$ and, using (1.59) and (1.48), we get

$$\begin{aligned} \frac{\log \nu(B(x,r))}{\log r} &\geq \frac{\log \mu(\tilde{B}(x,r)) - \sum_{j=1}^{k-1} \log \Delta_j}{\log r} \\ &\geq \frac{\log C + h \log 2 + h \log r}{\log r} - \frac{\sum_{j=1}^{k-1} \log \Delta_j}{\log d_j}. \end{aligned}$$

Since $\lim_{r \to 0} k(r) = \infty$, we, therefore, obtain that

$$\liminf_{r \to 0} \frac{\log \nu(B(x,r))}{\log r} \geq h - \limsup_{k \to \infty} \frac{\sum_{j=1}^{k-1} \log \Delta_j}{\log d_k}.$$

In view of (1.58) and by applying Theorem 1.7.1(a), this finishes the proof. ■

Now we define the following main concepts of this section.

Definition 1.7.4 Let μ be a Borel measure on a metric space (X, ρ). We write

$$\mathrm{HD}_\star(\mu) := \inf\{\mathrm{HD}(Y) \colon \mu(Y) > 0\} \text{ and } \mathrm{HD}^\star(\mu) = \inf\{\mathrm{HD}(Y) \colon \mu(X \backslash Y) = 0\}.$$

Of course,

$$\mathrm{HD}_\star(\mu) \leq \mathrm{HD}^\star(\mu),$$

and, in the case when $\mathrm{HD}_\star(\mu) = \mathrm{HD}^\star(\mu)$, we call this common value the *Hausdorff dimension* of the measure μ and we denote it by $\mathrm{HD}(\mu)$.

An analogous definition can be formulated for packing dimensions, with respective notation $\mathrm{PD}_\star(\mu)$, $\mathrm{PD}^\star(\mu)$, and $\mathrm{PD}(\mu)$, and the name *packing dimension* of the measure μ.

The next definition introduces concepts that are effective tools to calculate the dimensions introduced above.

Definition 1.7.5 Let μ be a Borel probability measure on a metric space (X,ρ). For every point $x \in X$, we define the *lower and upper pointwise dimension*of the measure μ at the point $x \in X$, respectively, as

$$\underline{d}_\mu(x) := \liminf_{r\to 0} \frac{\log \mu(B(x,r))}{\log r} \quad \text{and} \quad \overline{d}_\mu(x) := \limsup_{r\to 0} \frac{\log \mu(B(x,r))}{\log r}.$$

In the case when both numbers $\underline{d}_\mu(x)$ and $\overline{d}_\mu(x)$ are equal, we denote their common value by $d_\mu(x)$. We then obviously have that

$$d_\mu(x) = \lim_{r\to 0} \frac{\log \mu(B(x,r))}{\log r},$$

and we call $d_\mu(x)$ the pointwise dimension of the measure μ at the point $x \in X$.

The following theorem about Hausdorff and packing dimensions of a Borel measure μ follows easily from Theorems 1.7.1 and 1.7.2.

Theorem 1.7.6 *If μ is a Borel probability measure on a metric space (X,ρ), then*

$$\mathrm{HD}_\star(\mu) = \operatorname{ess\,inf} \underline{d}_\mu, \quad \mathrm{HD}^\star(\mu) = \operatorname{ess\,sup} \underline{d}_\mu$$

and

$$\mathrm{PD}_\star(\mu) = \operatorname{ess\,inf} \overline{d}_\mu, \quad \mathrm{PD}^\star(\mu) = \operatorname{ess\,sup} \overline{d}_\mu.$$

Proof Recall that the μ-essential infimum ess inf of a measurable function ϕ and the μ-essential supremum ess sup of this function are, respectively, defined by

$$\operatorname{ess\,inf}(\phi) := \sup_{\mu(N)=0} \inf_{x\in X\setminus N} \phi(x) \quad \text{and} \quad \operatorname{ess\,sup}(\phi) := \inf_{\mu(N)=0} \sup_{x\in X\setminus N} \phi(x).$$

Put $\phi_* ::= \operatorname{ess\,inf} \phi$. We shall prove that

$$\mu(\phi^{-1}((0,\phi_*))) = 0 \quad \text{and} \quad \mu(\phi^{-1}((0,\theta))) > 0 \tag{1.61}$$

for all $\theta > \phi_*$. Indeed, if we had $\mu(\phi^{-1}((0,\phi_*))) > 0$, then there would exist $\theta < \phi_*$ with $\mu(\phi^{-1}((0,\theta]) > 0$. Hence, for every measurable set $N \subseteq X$ with $\mu(N) = 0$, we would have that $\inf_{X\setminus N} \phi \le \theta$. Thus, ess $\inf \phi \le \theta$, which is a contradiction, and the first part of (1.61) is proved.

For the second part, proceeding also by way of contradiction, assume that there exists $\theta > \phi_*$ with

$$\mu(\phi^{-1}((0,\theta)) = 0.$$

Then for $N := \phi^{-1}((0,\theta))$, we would have that $\inf_{X\setminus N}(\phi) \geq \theta$. Hence, $\operatorname{ess\,inf}\phi \geq \theta$, which is a contradiction, and this finishes the proof of (1.61). This formula, applied to the function $\phi := \underline{d}_\mu$, tells us that, for every Borel set $A \subseteq X$, with $\mu(A) > 0$, there exists a Borel set $A' \subseteq A$ with $\mu(A') = \mu(A) > 0$ such that, for every $x \in A'$, we have that $\underline{d}_\mu(x) \geq d_{\mu*}$. Hence,

$$\mathrm{HD}(A) \geq \mathrm{HD}(A') \geq d_{\mu*}$$

by Theorem 1.7.1(a). Thus,

$$\mathrm{HD}_\star(\mu) \geq d_{\mu*}. \tag{1.62}$$

On the other hand, for every $\theta > \theta_1$, we have that $\mu\big(\{x \in X : \underline{d}_\mu(x) < \theta\}\big) > 0$. Hence, by Theorem 1.7.1(b),

$$\mathrm{HD}(\{x : \underline{d}_\mu(x) < \theta\}) \leq \theta.$$

Therefore, $\mathrm{HD}_\star(\mu) \leq \theta$. So, letting $\theta \searrow \theta_1$, we get

$$\mathrm{HD}_\star(\mu) \leq d_{\mu*}.$$

Along with (1.62), we, thus, conclude that $\mathrm{HD}_\star(\mu) = d_{\mu*}$.

One should proceed similarly to prove that $\mathrm{HD}^\star(\mu) = \operatorname{ess\,sup}\underline{d}_\mu(x)$ and to obtain corresponding results for packing dimensions. For the latter, one should refer to Theorem 1.7.2 instead of Theorem 1.7.1. ■

Definition 1.7.7 A Borel probability measure μ on a metric space (X,ρ) is called dimensional exact if and only if, for μ-a.e. $x \in X$, $d_\mu(x)$, the pointwise dimension of the measure μ at x exists and is μ-a.e. constant.

As an immediate consequence of Theorem 1.7.6, we get the following.

Proposition 1.7.8 *If μ is a Borel probability dimensional exact measure on a metric space (X,ρ), then both* HD(μ) *and* PD(mu) *exist; moreover*

$$\mathrm{HD}(\mu) = \mathrm{PD}(\mu) = d_\mu,$$

where d_μ is the μ-a.e. constant value of the pointwise dimension of μ.

1.8 Box-Counting Dimensions

We shall now examine a slightly different type of dimension; namely, the box-counting dimension. This dimension, as we will shortly see, is not given by means of any outer measure. It behaves worse: it is not σ-stable and a set and its closure have the same box-counting dimension. Its definition is, however, substantially simpler than those of Hausdorff and packing dimensions and is frequently easier to calculate or to estimate; it also frequently agrees with Hausdorff and packing dimensions and is widely used in the physics literature.

Definition 1.8.1 Let $0 < r < 1$ and $A \subseteq X$ be a bounded set. Define $N(A,r)$ to be the minimum number of balls of radius at most r with centers in A needed to cover A. Then the *upper and lower box-counting* (or, more simply, *box*) dimensions of A are, respectively, defined to be

$$\overline{\mathrm{BD}}(A) := \limsup_{r\to 0} \frac{\log N(A,r)}{-\log r}$$

and

$$\underline{\mathrm{BD}}(A) := \liminf_{r\to 0} \frac{\log N(A,r)}{-\log r}.$$

If these two quantities are equal, their common value is called the *box-counting dimension*, or simply the *box dimension*, of A, and we denote it by $\mathrm{BD}(A)$.

As said, the box-counting dimensions do not share all the congenial properties of the Hausdorff dimensions. In particular, they are not σ-stable. To see this, observe that

$$\mathrm{BD}(\mathbb{Q} \cap [0,1]) = 1 \neq 0 = \sup\{\mathrm{BD}(\{q\}) : q \in \mathbb{Q} \cap [0,1]\}.$$

The box-counting dimension is, however, easily seen to be finitely stable; see Proposition 1.8.2.

Proposition 1.8.2 *If (X,ρ) is a metric space and $F_1, F_2, \dots, F_n$ is a finite collection of subsets of X, then*

$$\overline{\mathrm{BD}}\big(F_1 \cup F_2 \cup \cdots \cup F_n\big) = \max\big\{\overline{\mathrm{BD}}(F_1), \overline{\mathrm{BD}}(F_2), \dots, \overline{\mathrm{BD}}(F_n)\big\}$$

and the same formula holds for the lower box-counting dimension.

The terminology "box counting" comes from the fact that in Euclidean spaces we may use boxes from a lattice rather than balls to cover the set under scrutiny. Indeed, let $n \geq 1$, $X = \mathbb{R}^n$, and $\mathcal{L}(r)$ be any lattice in $\mathbb{R}^n$ consisting of cubes (boxes) with edges of length r. For any $A \subseteq X$, define

$$L(A,r) = \mathrm{card}\{C \in \mathcal{L}(r) : C \cap A \neq \emptyset\}.$$

Proposition 1.8.3 *If A is a bounded subset of $\mathbb{R}^n$, then*

$$\overline{\mathrm{BD}}(A) = \limsup_{r\to 0} \frac{\log L(A,r)}{-\log r}$$

and

$$\underline{\mathrm{BD}}(A) = \liminf_{r\to 0} \frac{\log L(A,r)}{-\log r}.$$

Proof Without loss of generality, let $0 < r < 1$. Select points $x_i \in A$ so that

$$A \subseteq \bigcup_{i=1}^{N(A,r)} B(x_i, r).$$

Fix $1 \le i \le N(A,r)$ momentarily. If $C \in \mathcal{L}(r)$ is such that $d(C, x_i) < r$, where d denotes the standard Euclidean metric on $\mathbb{R}^n$, one immediately verifies that $C \subseteq B(x_i, r + r\sqrt{n}) = B(x_i, (1+\sqrt{n})r)$. Thus, for any given $1 \le i \le N(A,r)$, we have that

$$\begin{aligned}\#\{C \in \mathcal{L}(r) : d(C, x_i) < r\} &= \frac{\lambda(B(x_i,(1+\sqrt{n})r))}{\lambda(\text{cube of side } r)} \le \frac{c_n\big[(1+\sqrt{n})r\big]^n}{r^n} \\ &= c_n(1+\sqrt{n})^n,\end{aligned}$$

where λ denotes the Lebesgue measure on $\mathbb{R}^n$ and c_n denotes the volume of the unit ball in $\mathbb{R}^n$. Since every $C \in L(A,r)$ admits at least one number $1 \le i \le N(A,r)$ such that $d(C, x_i) < r$, we deduce that $L(A,r) \le N(A,r)c_n(1+\sqrt{n})^n$. Therefore,

$$\log L(A,r) \le \log\big(c_n(1+\sqrt{n})^n\big) + \log N(A,r).$$

Hence,

$$\frac{\log L(A,r)}{-\log r} \le \frac{\log\big(c_n(1+\sqrt{n})^n\big)}{-\log r} + \frac{\log N(A,r)}{-\log r}.$$

So,

$$\limsup_{r\to 0} \frac{\log L(A,r)}{-\log r} \le \overline{\mathrm{BD}}(A) \text{ and } \liminf_{r\to 0} \frac{\log L(A,r)}{-\log r} \le \underline{\mathrm{BD}}(A).$$

For the opposite inequality, again let $0 < r < 1$ and, for each $C \in L(A,r)$, choose $x_C \in C \cap A$. Then $C \cap A \subseteq B(x_C, r\sqrt{n})$. Thus, the family of balls

$$\big\{B(x_C, r\sqrt{n}) : C \in L(A,r)\big\}$$

covers A. Therefore, $N(A, r\sqrt{n}) \le L(A,r)$. It then follows that

$$\overline{\mathrm{BD}}(A) \le \limsup_{r\to 0} \frac{\log L(A,r)}{-\log r} \text{ and } \underline{\mathrm{BD}}(A)(A) \le \liminf_{r\to 0} \frac{\log L(A,r)}{-\log r}.$$

■

Now we return to our general setting. Let (X,ρ) be again a metric space, $A \subseteq X$, and $r > 0$. Further, define $P(A,r)$ to be the supremum of the cardinalities of all packings of A of the form $\{B(x_i,r)\}_{i=1}^{\infty}$, so

$$P(A,r) := \sup\left\{\#\{B(x_i,r)\}_{i=1}^{\infty}\right\}.$$

Such packings will be called r-packings of A in what follows. We shall prove the following technical, though interesting in itself, fact.

Lemma 1.8.4 *If $A \subseteq X$ and $r > 0$, then $N(A,2r) \le P(A,r) \le N(A,r)$.*

Proof The first inequality certainly holds if $P(A,r) = \infty$. So, assume that this is not the case and let $\{(x_i,r)\}_{i=1}^{k}$ be an r-packing of A that is maximal in the sense of inclusion. Then $\{B(x_i,2r)\}_{i=1}^{k}$ is a cover of A and, consequently, $N(A,2r) \le P(A,r)$. For the second inequality, there is nothing to prove if $N(A,r) = \infty$. So, again, let $\{(x_i,r)\}_{i=1}^{k}$ be a finite r-packing of A and assume that

$$\{B(y_j,r)\}_{j=1}^{\ell}$$

is a finite cover of A with centers in A. Then, for each $1 \le i \le k$, there exists $1 \le j(i) \le \ell$ such that

$$x_i \in B\big(y_{j(i)},r\big).$$

We will show that $k \le \ell$. In order to do this, it is enough to show that the function $i \mapsto j(i)$ is injective. But, for each $1 \le j \le \ell$, the cardinality of the set

$$\left\{\{x_i\}_{i=1}^{k} \cap B(y_j,r)\right\}$$

is at most 1 (otherwise, $\{(x_i,r)\}_{i=1}^{k}$ would not be an r-packing), and so the function $i \mapsto j(i)$ is injective, as required. Thus, $P(A,r) \le N(A,r)$. ■

These inequalities have the following immediate implications.

Corollary 1.8.5 *If X is a metric space and $A \subseteq X$, then*

$$\overline{\mathrm{BD}}(A) = \limsup_{r\to 0} \frac{\log P(A,r)}{-\log r}$$

and

$$\underline{\mathrm{BD}}(A) = \liminf_{r\to 0} \frac{\log P(A,r)}{-\log r}.$$

As an immediate consequence of this corollary and of the second part of Definition 1.8.1, we obtain the following.

Corollary 1.8.6 *If X is a metric space and $A \subseteq X$, then*

$$\mathrm{HD}(A) \leq \underline{\mathrm{BD}}(A) \leq \mathrm{PD}(A) \leq \overline{\mathrm{BD}}(A).$$

We end this section with the following.

Proposition 1.8.7 *Let (X, ρ) be a metric space endowed with a finite Borel measure μ such that*

$$\mu(B(x,r)) \geq Cr^t$$

for some constant $C > 0$, all $x \in X$, and all radii $0 \leq r \leq 1$. Then

$$\overline{\mathrm{BD}}(X) \leq t.$$

If, on the other hand, $\mu(X) > 0$ and

$$\mu(B(x,r)) \leq Cr^t$$

for some constant $C < +\infty$, all $x \in X$, and all radii $0 \leq r \leq 1$, then

$$\underline{\mathrm{BD}}(X) \geq \mathrm{HD}(X) \geq t.$$

Finally, if $\mu(X) > 0$ and

$$C^{-1}r^t \leq \mu(B(x,r)) \leq Cr^t$$

for some constant $C \in [1, +\infty)$, all $x \in X$, and all radii $0 \leq r \leq 1$, then

$$\mathrm{BD}(X) = \mathrm{PD}(X) = \mathrm{HD}(X) = t.$$

Proof We start with the first inequality. Let $\{B(x,r)\}_{i=1}^k$ be an r-packing of X. Then

$$kr^t \leq C^{-1} \sum_{i=1}^{k} \mu(B(x_i,r)) \leq C^{-1}.$$

Hence, $k \leq C^{-1}r^{-t}$. Therefore, $P(X,r) \leq C^{-1}r^{-t}$. Consequently,

$$\log P(X,r) \leq -\log C - t \log r.$$

In conjunction with the first formula of Corollary 1.8.5, this yields

$$\overline{\mathrm{BD}}(X) \leq t.$$

The second assertion of our proposition directly follows from the first inequality of Corollary 1.8.6 and from item (2) of the Frostman Converse-Type Theorem for Hausdorff Measures (Theorem 1.6.3).

The last assertion of our proposition is now an immediate consequence of the two first assertions and Corollary 1.8.6. ∎

2

Invariant Measures: Finite and Infinite

In this chapter, we begin to investigate the class of all measurable dynamical systems that posseses an invariant measure. This means that we are in the field of ergodic theory. We do want to emphasize that most of this chapter pertains to all the systems regardless of whether the reference invariant measure is finite or infinite. With one notable exception though; namely, that of the Birkhoff Ergodic Theorem proved in [**Bir**]. This theorem, formulated in the realm of probability spaces, is important for at least three reasons. First, establishing equality of time averages and space averages, it yields profound theoretical and, one could say, philosophical consequences. Second, it has countless applications throughout various types of dynamical systems; our book is an example of this. And third, it is an indispensable tool to develop ergodic theory with infinite invariant measures. Otherwise, unless explicitly stated (rarely), we do not assume the reference measure to be finite.

Invariant measures, both finite and infinite, abound but we postpone examples to further chapters. Indeed, Section 3.3, contains a fairly large collection of invariant and ergodic measures. Further examples, dealt with in detail in this book, are provided in Section 11.3, particularly Theorem 11.6.6, and especially in Chapter 22 (see Volume 2), which is almost entirely devoted to investigating (ergodic) invariant measures, both finite and infinite, for dynamical systems generated by elliptic meromorphic functions.

2.1 Quasi-invariant Transformations: Ergodicity and Conservativity

In this section, $(X,\mathfrak{F},m)$ is a measure space and a map $T\colon X \longrightarrow X$ is a measurable map with respect to the σ-algebra $\mathfrak{F}$.

Definition 2.1.1 If $T\colon (X,\mathfrak{F}) \longrightarrow (X,\mathfrak{F})$ is a measurable map, then a measure m on $(X,\mathfrak{F})$ is called quasi-invariant with respect to the map T if and only if

$$m \circ T^{-1} \prec m.$$

Equivalently, $m(T^{-1}(A)) = 0$ whenever A is measurable and $m(A) = 0$. We would like to emphasize that, in this section, we do not yet assume the measure m to be T-invariant. This concept will be defined in the next section.

Definition 2.1.2 If m is a quasi-invariant measure for a measurable map $T\colon X \longrightarrow X$, then we say that the map $T\colon X \longrightarrow X$ (or the measure m) is ergodic if and only if for every measurable set $A \subseteq X$, the following implication holds:

$$T^{-1}(A) = A \implies [m(A) = 0 \text{ or } m(X\backslash A) = 0].$$

It is easy to prove the following.

Proposition 2.1.3 *If m is a quasi-invariant measure for a measurable map $T\colon X \longrightarrow X$, then the map $T\colon X \longrightarrow X$ (or the measure m) is ergodic if and only if, for every measurable set $A \subseteq X$,*

$$m(T^{-1}(A) \div A) = 0 \implies [m(A) = 0 \text{ or } m(X\backslash A) = 0].$$

The second important concept in the theory of quasi-invariant measures is conservativity. We introduce it now.

Definition 2.1.4 Let $T\colon (X,\mathfrak{F}) \longrightarrow (X,\mathfrak{F})$ be a measurable map. A measurable set $W \subset X$ is called a wandering set if and only if the sets

$$\{T^{-k}(W)\}_{n=0}^{\infty}$$

are mutually disjoint.

One way of constructing wandering sets is now described.

Lemma 2.1.5 *Let $T\colon (X,\mathfrak{F}) \longrightarrow (X,\mathfrak{F})$ be a measurable map. If $A \in \mathfrak{F}$, the set*

$$W_A := A\backslash \bigcup_{n=1}^{\infty} T^{-n}(A)$$

is wandering with respect to T.

Proof Suppose, for a contradiction, that W_A is not wandering for T, i.e.,

$$T^{-k}(W_A) \cap T^{-l}(W_A) \neq \emptyset$$

for some $0 \le k < l$. This means that

$$T^{-k}\big(W_A \cap T^{-(l-k)}(W_A)\big) \neq \emptyset.$$

This means that, on the one hand,

$$W_A \cap T^{-(l-k)}(W_A) \neq \emptyset.$$

On the other hand, by the very definition of W_A,

$$W_A \cap T^{-(l-k)}(W_A) = \emptyset.$$

This contradiction finishes the proof. ■

Definition 2.1.6 A measurable map $T\colon (X, \mathfrak{F}) \longrightarrow (X, \mathfrak{F})$ is called conservative if and only if there is no wandering set $W \subseteq X$ with $m(W) > 0$.

For every set $B \subseteq X$, let

$$\begin{aligned}\hat{B}_\infty &:= \Big\{x \in X\colon T^n(x) \in B \quad \text{for infinitely many} \ n \ge 1\Big\} \\ &= \left\{x \in X\colon \sum_{n=1}^{\infty} \mathbb{1}_B \circ T^n(x) = +\infty\right\} \\ &= \bigcap_{k=1}^{\infty}\bigcup_{n=k}^{\infty} T^{-n}(B)\end{aligned}$$

and

$$B_\infty := B \cap \hat{B}_\infty.$$

Obviously,

$$T^{-1}(\hat{B}_\infty) = \hat{B}_\infty = T(\hat{B}_\infty) \tag{2.1}$$

and

$$T^{-1}(X\backslash\hat{B}_\infty) = X\backslash\hat{B}_\infty = T(X\backslash\hat{B}_\infty). \tag{2.2}$$

Notice also that if W is wandering, then

$$W \cap \bigcup_{n=1}^{\infty} T^{-n}(W) = \emptyset = W \cap \bigcup_{n=1}^{\infty} T^{n}(W). \tag{2.3}$$

In particular

$$W_\infty = \emptyset.$$

The celebrated Poincaré Recurrence Theorem says that if the measure $m(X)$ is finite, then $m(B\backslash B_\infty) = 0$. We shall now prove its generalization in two respects, assuming only that the measure is merely quasi-invariant and need not be finite. The Poincaré Recurrence Theorem, immediately following on from this generalization, will be stated as Theorem 2.2.5.

Theorem 2.1.7 (Halmos Recurrence Theorem). *Let m be a quasi-invariant measure for a measurable map $T: X \longrightarrow X$. Fix a measurable set $A \subseteq X$. Then*

$$m(B \backslash B_\infty) = 0$$

for all measurable sets $B \subseteq A$ if and only if

$$m(A \cap W) = 0$$

for all measurable wandering sets $W \subseteq X$.

Proof Note that if $m(A) = 0$, then $m(B \backslash B_\infty) = 0$ for all measurable sets $B \subseteq A$ and $m(A \cap W) = 0$ for all measurable wandering sets $W \subseteq X$. Thus, our equivalence is trivially satisfied, and we may assume without loss of generality that $m(A) > 0$.

We shall prove the implication ($\Rightarrow$). Suppose for a contradiction that $m(A \cap W) > 0$ for some measurable wandering set $W \subseteq X$. Then $(A \cap W)_\infty = \emptyset$ and, therefore,

$$m\big(A \cap W \backslash (A \cap W)_\infty\big) > 0.$$

Let us now prove the implication ($\Leftarrow$). Fix a measurable set $B \subseteq A$. For all $n \geq 0$, let

$$B_n := B \cap T^{-n}(B) \cap \bigcup_{k=n+1}^{\infty} T^{-k}(X \backslash B).$$

Seeking contradiction suppose that

$$m(B \backslash B_\infty) > 0.$$

Then there exists $n \geq 0$ such that $m(B_n) > 0$. But, as $B_n \subseteq B \subseteq A$, it follows from our hypotheses and Lemma 2.1.5 that $m(W_{B_n}) = 0$. This means that $m(B_n \backslash B_n^-) = 0$ or, equivalently,

$$m\left(B_n \backslash \bigcup_{k=1}^{\infty} T^{-k}(B_n)\right) = 0.$$

So, (as $m(B_n) > 0$) there exists $x \in B_n \cap \bigcup_{k=1}^{\infty} T^{-k}(B_n)$. Hence,

$$x \in B \cap T^{-k}(B_n) \subseteq B \cap T^{-(n+k)}(B)$$

for some $k \geq 1$. Thus, $x \notin B_n$. This contradiction finishes the proof that $m(B \backslash B_\infty) = 0$. We are done. ■

Taking, in Theorem 2.1.7, $A = X$, we immediately get the following corollary.

Corollary 2.1.8 *A measurable transformation $T\colon (X, \mathfrak{F}) \longrightarrow (X, \mathfrak{F})$ of a measure space $(X, \mathfrak{F})$ possessing a quasi-invariant measure m is conservative with respect to m if and only if*

$$m(B \backslash B_\infty) = 0$$

for all $B \in \mathfrak{F}$.

As an immediate consequence of this corollary and the Borel–Cantelli Lemma, we obtain the following.

Corollary 2.1.9 *If a quasi-invariant transformation $T\colon X \longrightarrow X$ of a measure space $(X, \mathfrak{F}, m)$ is conservative, then*

$$\sum_{n=0}^{\infty} m(T^{-n}(A)) = +\infty$$

for any set $A \in \mathfrak{F}$ with $m(A) > 0$.

We shall now prove the following characterization of ergodicity and conservativity.

Theorem 2.1.10 *A transformation $T\colon X \longrightarrow X$ of a measure space $(X, \mathfrak{F})$ possessing a quasi-invariant measure m is conservative and ergodic with respect to m if and only if, for every set $A \in \mathfrak{F}$ with $m(A) > 0$,*

$$\sum_{n=1}^{\infty} \mathbb{1}_A \circ T^n = +\infty$$

m-a.e. on X. Equivalently, if and only if

$$\mu(X \backslash \hat{A}_\infty) = 0$$

for every $A \in \mathfrak{F}$ such that $\mu(A) > 0$.

Proof Assume first that the map $T\colon X \longrightarrow X$ is conservative and ergodic. Fix $A \in \mathfrak{F}$ such that $\mu(A) > 0$. It then follows from (2.1) that either $m(\hat{A}_\infty) = 0$ or $m\big(X \backslash \hat{A}_\infty\big) = 0$. If $m\big(X \backslash \hat{A}_\infty\big) = 0$, we are done. So, suppose that

$$m(\hat{A}_\infty) = 0.$$

By Corollary 2.1.8, this implies that $m(A) = 0$. This contradiction finishes the proof of our implication.

Now let us prove the converse. We first show that the map $T\colon X \longrightarrow X$ is conservative. Keep $A \in \mathfrak{F}$ with $m(A) > 0$ and assume that $m(X\backslash\hat{A}_\infty) = 0$. Since $A_\infty = A \cap \hat{A}_\infty$, we, thus, get that $m(A\backslash A_\infty) = 0$. It, therefore, follows from Corollary 2.1.8 that $T\colon X \to X$ is conservative.

Now let us turn to ergodicity. Keep $A \in \mathfrak{F}$ and assume that $T^{-1}(A) = A$. Then $\hat{A}_\infty = A$ and $\widehat{(X\backslash A)}_\infty = X\backslash A$. If $m(A) > 0$, then, by assumption $\mu(X\backslash A_\infty) = 0$, we get that $\mu(X\backslash A) = 0$. Likewise, if $m(X\backslash A) > 0$, we get that $\mu(A) = 0$. We are done. ■

With a little stronger hypothesis, we obtain yet another consequence of Theorem 2.1.10.

Definition 2.1.11 A measurable transformation $T\colon (X,\mathfrak{F},m) \to (X,\mathfrak{F},m)$ is called nonsingular if, for every set $F \in \mathfrak{F}$,

$$m(T^{-1}(F)) = 0 \iff m(F) = 0.$$

Of course, every nonsingular map is quasi-invariant. The, just announced, consequence of Theorem 2.1.10 is the following.

Corollary 2.1.12 *If $T\colon X \longrightarrow X$ is a nonsingular transformation of a measure space $(X,\mathfrak{F},m)$, then the following are equivalent.*

(a) *The map $T\colon X \longrightarrow X$ is conservative and ergodic.*
(b) *For every set $A \in \mathfrak{F}$ with $m(A) > 0$,*

$$\sum_{n=1}^{\infty} 1\!\!1_A \circ T^n(x) = +\infty$$

for m-a.e. $x \in X$.

(c)

$$\mu(X\backslash\hat{A}_\infty) = 0$$

for every $A \in \mathfrak{F}$ such that $\mu(A) > 0$.

(d)

$$m\left(X\backslash\bigcup_{n=0}^{\infty} T^{-n}(A)\right) = 0$$

for every set $A \in \mathfrak{F}$ with $m(A) > 0$.

Proof Because of Theorem 2.1.10, our only task is to show that (d) $\Rightarrow$ (b). Indeed, for every $n \geq 1$, let

$$A_n^* := X \setminus \bigcup_{k=0}^{\infty} T^{-k}(T^{-n}(A)).$$

By nonsingularity of T, $m(T^{-n}(A)) > 0$ for every $n \geq 0$. Hence, by (d), $m(A_n^*) = 0$ for every $n \geq 0$. Therefore,

$$m\left(\bigcup_{n=0}^{\infty} A_n^*\right) = 0.$$

But for a point $x \in X$ to be in the complement of this set means that $T^j(x) \in A$ for infinitely many js. This, in turn, means that (b) holds for the point x. ■

Suppose now that $(X, \mathfrak{F}, m)$ is a probability space and that $T: X \longrightarrow X$ is a measurable quasi-invariant mapping. Assume, further, that $T(A) \in \mathfrak{F}$ for every set $A \in \mathfrak{F}$. We call $T: X \longrightarrow X$ weakly metrically exact if and only if

$$\limsup_{n \to \infty} m(B \cap T^n(A)) = m(B) \tag{2.4}$$

for all measurable sets $A, B \subseteq X$ with $m(A) > 0$.

If m is a probability space, then this just means that

$$\limsup_{n \to \infty} m(T^n(A)) = 1 \tag{2.5}$$

for all measurable sets $A \subseteq X$ with $m(A) > 0$.

We shall prove the following theorem.

Theorem 2.1.13 *Each weakly metrically exact mapping $T: (X, \mathfrak{F}) \longrightarrow (X, \mathfrak{F})$ possessing a quasi-invariant probability measure m is conservative and ergodic with respect to m.*

Proof To prove ergodicity, suppose that $m(A) > 0$ and $T^{-1}(A) = A$. Then

$$T^n(A) \subseteq A$$

for all $n \geq 0$. Hence,

$$m(X \setminus A) = \limsup_{n \to \infty} \mu\big((X \setminus A) \cap T^n(A)\big) \leq m(A \cap (X \setminus A)) = 0.$$

So $m(X \setminus A) = 0$ and T is ergodic.

In order to prove conservativity of T, note that if $W \subseteq X$ is a wandering set, then, by (2.3), we have that

$$\bigcup_{n=0}^{\infty} T^n(W) \subseteq X \backslash W.$$

Therefore,

$$m(W) = \limsup_{n\to\infty} m(W \cap T^n(W)) \leq m(W \cap (X \backslash W)) = 0.$$

Hence, $m(W) = 0$. The map $T\colon X \to X$ is, thus, conservative and we are done. ■

Now let $(X, \mathfrak{F}, m)$ be a σ-finite measure space and keep $T\colon X \longrightarrow X$ as an $\mathfrak{F}$-measurable transformation with respect to which the measure m is quasi-invariant. Let $f \in L^1(m)$ be a nonnegative function. Then the measure $(fm) \circ T^{-1}$ is absolutely continuous with repect to m and we put

$$\mathcal{L}_m(f) := \frac{d((fm) \circ T^{-1})}{dm} \tag{2.6}$$

as the Radon–Nikodym derivative of $fm \circ T^{-1}$ with respect to m. For any function $f \in L^1(m)$, we then define

$$\mathcal{L}_m(f) := \mathcal{L}_m(f_+) - \mathcal{L}_m(f_-),$$

where $f = f_+ - f_-$ is the canonical decomposition of f into its positive and negative parts $f_+ = \max\{f, 0\}$ and $f_- = -\min\{f, 0\}$, respectively. Obviously,

$$\mathcal{L}_m\colon L^1(m) \longrightarrow L^1(m)$$

is a linear operator and is characterized by the property that

$$\int_X \mathcal{L}_m(f) g \, dm = \int_X f \cdot g \circ T \, dm \tag{2.7}$$

for all $f \in L^1(m)$ and $g \in L^\infty(m)$. Taking $g = 1\!\!1$, we see that $\mathcal{L}_m\colon L^1(m) \to L^1(m)$ is bounded and $\|\mathcal{L}_m\|_{L^1(m)} = 1$. The bounded linear operator

$$\mathcal{L}_m\colon L^1(m) \longrightarrow L^1(m) \tag{2.8}$$

is called the transfer, or Perron–Frobenius, operator associated with the quasi-invariant measure m. We shall now prove one more characterization of ergodicity and conservativity, this time in terms of the transfer operator $\mathcal{L}_m$.

Theorem 2.1.14 *Let $(X, \mathfrak{F}, \mu)$ be a σ-finite measure space and let $T\colon X \longrightarrow X$ be an $\mathfrak{F}$-measurable transformation with respect to which the measure m is quasi-invariant. Then the transformation $T\colon X \longrightarrow X$ is ergodic and conservative with respect to m if and only if*

$$\sum_{n=0}^{\infty} \mathcal{L}_m^n(f) = +\infty$$

m-a.e. for every nonnegative (a.e.) function $f \in L^1(m)$ with $\int_X f dm > 0$.

Proof ($\Rightarrow$) Proving by contradiction, suppose that there exists a function $f \in L^1(m)$ with the following properties.

(a) $f \geq 0$ m-a.e.
(b) $\int_X f dm > 0$.
(c) $m\left(\left\{x \in X\colon \sum_{n=0}^{\infty} \mathcal{L}_m^n(f)(x) < +\infty\right\}\right) > 0$.

By virtue of (c) there exists a measurable set

$$B \subseteq \left\{x \in X\colon \sum_{n=0}^{\infty} \mathcal{L}_m^n(f)(x) < \infty\right\}$$

such that $m(B) \in (0, \infty)$ and

$$\sup\left\{\sum_{n=0}^{\infty} \mathcal{L}_m^n(f)(x)\colon x \in B\right\} < +\infty.$$

In particular,

$$\int_B \left(\sum_{n=0}^{\infty} \mathcal{L}_m^n(f)\right) dm < +\infty.$$

Also using (2.7), we, therefore, get that

$$\begin{aligned}
\int_X f\left(\sum_{n=1}^{\infty} \mathbb{1}_B \circ T^n\right) dm &= \sum_{n=1}^{\infty} \int_X f \mathbb{1}_B \circ T^n dm = \sum_{n=1}^{\infty} \int_X \mathbb{1}_B \mathcal{L}_m^n(f) dm \\
&= \sum_{n=1}^{\infty} \int_B \mathcal{L}_m^n(f) dm \\
&= \int_B \left(\sum_{n=1}^{\infty} \mathcal{L}_m^n(f)\right) dm \\
&< +\infty. \qquad (2.9)
\end{aligned}$$

Since (b) implies that f is positive on a set of positive measure m, (2.9) and (a) yield that $\sum_{n=1}^{\infty} 1\!\!1_B \circ T^n$ is finite on a set of positive measure m. So, T is not conservative and ergodic by virtue of Theorem 2.1.14. This contradiction finishes the proof of our implication.

Proceeding in the opposite direction, assume that T is not ergodic and conservative. By Theorem 2.1.10, this means that there exist two sets $C, F \in \mathfrak{F}$ such that $m(C) > 0$, $m(F) > 0$, and

$$\sum_{n=0}^{\infty} 1\!\!1_F \circ T^n(x) < \infty$$

for all $x \in C$. Therefore, there exists a measurable subset D of C such that $m(D) \in (0, +\infty)$ and

$$M := \sup \left\{ \sum_{n=0}^{\infty} 1\!\!1_F \circ T^n(x) : \ x \in D \right\} < \infty.$$

Since $1\!\!1_D \geq 0$ and since $1\!\!1_D \in L^1(m)$, using (2.7), we get that

$$\begin{aligned} \int_F \sum_{n=0}^{\infty} \mathcal{L}_m^n(f) dm &= \int_X 1\!\!1_F \sum_{n=0}^{\infty} \mathcal{L}_m^n(1\!\!1_D) dm = \int_X 1\!\!1_D \left(\sum_{n=0}^{\infty} 1\!\!1_F \circ T^n \right) dm \\ &= \int_D \sum_{n=0}^{\infty} 1\!\!1 \circ T^n dm \\ &\leq M < +\infty. \end{aligned}$$

Hence,

$$m \left(\left\{ x \in F : \ \sum_{n=0}^{\infty} \mathcal{L}_m^n(f)(x) < +\infty \right\} \right) = m(F) > 0$$

and we are done by contrapositive. ■

2.2 Invariant Measures: First Return Map (Inducing); Poincaré Recurrence Theorem

In this section, we define and extensively investigate a very powerful tool of ergodic theory, the one constituted by the first return time and map, also known as inducing. This tool is extremely useful in both the ergodic theory of transformations preserving probability measures as well as those preserving infinite measures. In the former case, after inducing, the map obtained is usually somewhat worse in regard to a minor aspect but is much better in regard to a major, more desirable, aspect. In the infinite case, inducing allows us to study systems preserving infinite measures by means of systems preserving probability measures. And this is a big advantage indeed, especially since, for such maps, we have the Birkhoff Ergodic Theorem.

Throughout this section, $(X, \mathfrak{F}, \mu)$ is a measure space and $T\colon (X, \mathfrak{F}) \longrightarrow (X, \mathfrak{F})$ is a measurable conservative map-preserving measure μ, which means that

$$\mu \circ T^{-1} = \mu.$$

This is the central equation of ergodic theory. We then also say that the measure μ is T-invariant. This concept will be explored extensively throughout the current chapter and the entire book. Since, obviously, every measure-preserving map is nonsingular, as an immediate consequence of Corollary 2.1.12, we obtain the following.

Theorem 2.2.1 *If $T\colon X \longrightarrow X$ is a measure-preserving transformation of a measure space $(X, \mathfrak{F}, \mu)$, then the following are equivalent.*

(a) *The map $T\colon X \to X$ is conservative and ergodic.*
(b) *For every set $A \in \mathfrak{F}$ with $\mu(A) > 0$,*

$$\sum_{n=1}^{\infty} 1\!\!1_A \circ T^n = +\infty$$

μ-a.e. on X.

(c)

$$\mu(X \backslash \hat{A}_\infty) = 0$$

for every $A \in \mathfrak{F}$ such that $\mu(A) > 0$.

(d)

$$\mu\left(X \backslash \bigcup_{n=0}^{\infty} T^{-n}(A)\right) = 0$$

for every set $A \in \mathfrak{F}$ with $\mu(A) > 0$.

As an immediate consequence of this theorem, we get the following.

Corollary 2.2.2 *If $T\colon X \longrightarrow X$ is a conservative and ergodic measure-preserving transformation of a measure space $(X, \mathfrak{F}, \mu)$, then the measure space $(X, \mathfrak{F}, \mu)$ is σ-finite if and only if there exists at least one set $F \in \mathfrak{F}$ with $0 < \mu(F) < +\infty$.*

We know from Theorem 2.1.13 that a sufficient condition for ergodicity and conservativity of a quasi-invariant measure is that it is weakly metrically exact. If the quasi-invariant measure is invariant, then we can say a little bit more. Namely, we say that if $(X, \mathfrak{F}, \mu)$ is a measure space, then a measurable map $T: X \longrightarrow X$ preserving measure μ such that $T(A) \in \mathfrak{F}$ for every set $A \in \mathfrak{F}$ is called metrically exact if and only if

$$\lim_{n\to\infty} \mu(B \cap T^n(A)) = \mu(B) \tag{2.10}$$

for all measurable sets $A, B \subseteq X$ with $\mu(A) > 0$.

If μ is a probability space, then this just means that

$$\lim_{n\to\infty} \mu(T^n(A)) = 1 \tag{2.11}$$

for all measurable sets $A \subseteq X$ with $\mu(A) > 0$. Noting that, for invariant measures μ,

$$\mu(T(A)) \geq \mu(A)$$

for every measurable set $A \subseteq X$ such that $\mu(A)$ is also measurable, and applying Theorem 2.1.13, we obtain the following.

Proposition 2.2.3 *If $(X, \mathfrak{F}, \mu)$ is a probability space and $T: X \longrightarrow X$ is a measurable map-preserving measure μ, then T is metrically exact if and only if it is weakly metrically exact. If this holds, the map T is ergodic and conservative.*

We also have the following immediate result.

Proposition 2.2.4 *Let $(X, \mathfrak{F}, m)$ be a probability space and let $T: X \longrightarrow X$ be an $\mathfrak{F}$-measurable transformation with respect to which the measure m is quasi-invariant and weakly metrically exact. Then any T-invariant probability measure on $(X, \mathfrak{F})$ absolutely continuous with respect to m is metrically exact.*

In the context of finite measures, as an immediate consequence of the very definition of conservativity and Corollary 2.1.8, we get the following celebrated theorem.

Theorem 2.2.5 (Poincaré Recurrence Theorem) *If $(X, \mathfrak{F}, \mu)$ is a finite measure space, then every measurable map $T: X \longrightarrow X$ preserving measure μ is conservative. This means that*

$$\mu(F \backslash F_\infty) = 0$$

for every set $F \in \mathfrak{F}$.

This theorem has profound philosophical and physical consequences, particularly because of its apparent contradiction with the Second Law of Thermodynamics, growth of entropy. It is also a frequently used tool in everyday dealing with ergodic theory.

We now define and investigate the main concept of this section, the inducing scheme. So, fix $F \in \mathfrak{F}$ with

$$0 < \mu(F) < +\infty.$$

Consider the function

$$\tau_F \colon F_\infty \longrightarrow \mathbb{N} = \{1, 2, 3, \ldots\},$$

given by the formula

$$\tau_F(x) := \min\{n \geq 1 \colon T^n(x) \in F\},$$

and the map

$$T_F \colon F_\infty \to F_\infty,$$

given by the formula

$$T_F(x) := T^{\tau_F(x)}(x).$$

Since, by Corollary 2.1.8,

$$\mu(F \backslash F_\infty) = 0,$$

we may, somewhat informally, say that the map

$$T_F \colon F \longrightarrow F$$

and all its iterates are defined m-a.e. on F. The number $\tau_F(x) \geq 1$ is called the first return time to the set F, and the map $T_F \colon F \to F$ is called the first return map or the induced map.

Let $\phi \colon X \longrightarrow \mathbb{R}$ be a measurable function. For any $n \geq 1$, we define the nth Birkhoff sum of the function φ by

$$S_n\phi := \phi + \phi \circ T + \cdots + \phi \circ T^{n-1}. \tag{2.12}$$

Given a function $g \colon X \longrightarrow \mathbb{R}$, let $g_F \colon F \longrightarrow \mathbb{R}$ be defined by the formula,

$$g_F(x) := \sum_{j=0}^{\tau_F(x)-1} g \circ T^j(x). \tag{2.13}$$

Let μ_F be the conditional measure on F, i.e.,

$$\mu_F(A) := \frac{\mu(A)}{\mu(F)}$$

for every measurable set $A \subseteq F$. We shall prove the following.

Theorem 2.2.6 *Suppose that $T : X \longrightarrow X$ is a measurable transformation of a measure space $(X, \mathfrak{F}, m)$.*

(a) *Let μ be a T-invariant measure on X. If $F \in \mathfrak{F}$ satisfies*

$$0 < \mu(F) < +\infty \text{ and } \mu(F \backslash F_\infty) = 0,$$

then measure μ_F is T_F-invariant on F.

(b) *Conversely, if ν is a probability T_F-invariant measure on $(F, \mathfrak{F}|_F)$, then there exists a T-invariant measure μ on $(X, \mathfrak{F})$ such that*

$$\mu_F = \nu.$$

In fact, the formula

$$\begin{aligned} \mu(B) &:= \sum_{k=0}^{\infty} \nu\left(F \cap T^{-k}(B) \backslash \bigcup_{j=1}^{k} T^{-j}(F)\right) \\ &= \sum_{k=0}^{\infty} \nu\big(\{x \in F \cap T^{-k}(B) : \tau_F(x) > k\}\big) \end{aligned} \tag{2.14}$$

defines such a T-invariant measure on X.

(c) *Consequently,*

$$\mu\left(X \backslash \bigcup_{k=1}^{\infty} T^{-k}(F)\right) = 0.$$

(d) *In particular, the measure μ is σ-finite.*

Proof We first prove (a). For every $B \in \mathfrak{F}|_F$, we have that

$$\begin{aligned} \mu(T_F^{-1}(B)) &= \sum_{n=1}^{\infty} \mu\big(\tau_F^{-1}(n) \cap T_F^{-1}(B)\big) = \sum_{n=1}^{\infty} \mu\big(\tau_F^{-1}(n) \cap T^{-n}(B)\big) \\ &= \sum_{n=1}^{\infty} \mu\left(F \cap T^{-n}(B) \backslash \bigcup_{k=1}^{n-1} T^{-j}(F)\right) \\ &= \sum_{n=1}^{\infty} \mu\big(F \cap T^{-1}(B_{n-1})\big), \end{aligned} \tag{2.15}$$

where

$$B_0 := B \quad \text{and} \quad B_n := T^{-n}(B) \setminus \bigcup_{k=0}^{n-1} T^{-k}(F)$$

for all $n \geq 1$. Observe that

$$\mu(B_n) \leq \mu(T^{-n}(B)) = \mu(B) \leq \mu(A) < +\infty$$

for every $n \geq 0$. Since

$$T^{-1}(B_{n-1}) = \big(F \cap T^{-1}(B_{n-1})\big) \cup B_n$$

and

$$\big(F \cap T^{-1}(B_{n-1})\big) \cap B_n = \emptyset,$$

we have that

$$\mu\big(F \cap T^{-1}(B_{n-1})\big) = \mu\big(T^{-1}(B_{n-1})\big) - \mu(B_n) \leq \mu(B_{n-1}) - \mu(B_n). \tag{2.16}$$

Therefore, by (2.15),

$$\mu\big(T_F^{-1}(B)\big) = \lim_{n\to\infty} \big(\mu(B) - \mu(B_n)\big) \leq \mu(B). \tag{2.17}$$

Hence, also

$$\mu\big(T_F^{-1}(B)\big) = \mu(F) - \mu\big(T_F^{-1}(F \setminus B)\big) \geq \mu(F) - \mu(F \setminus B) = \mu(B). \tag{2.18}$$

These last two formulas complete the proof of item (a).

Proving item (b), we shall show first that the measure μ given by (2.14) is T-invariant. Indeed,

$$\begin{aligned}
\mu \circ T^{-1}(B) &= \sum_{k=0}^{\infty} \nu\big(\{x \in F \cap T^{-(k+1)}(B) \colon \tau_F(x) > k\}\big) \\
&= \sum_{k=0}^{\infty} \nu\big(\{x \in F \cap T^{-(k+1)}(B) \colon \tau_F(x) > k+1\}\big) \\
&\quad + \sum_{k=0}^{\infty} \nu\big(\{x \in F \cap T^{-(k+1)}(B) \colon \tau_F(x) = k+1\}\big) \\
&= \mu(B) - \nu(B \cap F) + \sum_{k=1}^{\infty} \nu\big(\{x \in F \cap T^{-k}(B) \colon \tau_F(x) = k\}\big) \\
&= \mu(B) - \nu(B \cap F) + \sum_{k=1}^{\infty} \nu\big(\{x \in F \colon \tau_F(x)\, k \text{ and } x \in T_F^{-1}(B)\}\big)
\end{aligned}$$

$$= \mu(B) - \nu(B \cap F) + \sum_{k=1}^{\infty} \nu\big(\{x \in F : \tau_F(x)\ k \text{ and } x \in T_F^{-1}(F \cap B)\}\big)$$
$$= \mu(B) - \nu(B \cap F) + \nu\big(T_F^{-1}(F \cap B)\big)$$
$$= \mu(B).$$

If $B \subseteq F$, then (2.14) reduces to $\mu(B) = \nu(F \cap B) = \nu(B)$. Hence,

$$\mu_F(B) = \frac{\mu(B)}{\mu(F)} = \frac{\nu(B)}{\nu(F)} = \nu(B).$$

Item (b) is, thus, proved.

Proving (c), formula (2.14) gives

$$\mu\left(X \setminus \bigcup_{k=1}^{\infty} T^{-k}(F)\right) = \sum_{n=0}^{\infty} \nu\left(\left\{x \in F \setminus \bigcup_{k=1}^{\infty} T^{-(n+k)}(F) : \tau_F(x) > n\right\}\right)$$
$$= \sum_{n=0}^{\infty} \nu(\emptyset) = 0.$$

Item (c) is thus proved.

Item (d) is now an immediate consequence of Theorem 2.2.1 and Corollary 2.2.2. ■

Remark 2.2.7 It follows from (2.15) and (2.18) that $\lim_{n\to\infty} \mu(B_n) = 0$. In particular, taking $B = F$, we get that

$$\lim_{n\to\infty} \mu\left(T^{-n}(F) \setminus \bigcup_{k=0}^{n-1} T^{-k}(F)\right) = 0.$$

In the second part of Theorem 2.2.6, we have defined the invariant measure μ by the formula (2.14). We shall now show that this choice of the invariant measure μ is actually unique.

Proposition 2.2.8 *If T is a measure-preserving transformation of a measure space $(X, \mathfrak{F}, \mu)$, $F \in \mathfrak{F}$ with $0 < \mu(F) < +\infty$, and*

$$\mu\left(X \setminus \bigcup_{n=1}^{\infty} T^{-n}(F)\right) = 0,$$

then, for every set $B \in \mathfrak{F}$ we have that

$$\mu(B) = \sum_{k=0}^{\infty} \mu\left(F \cap T^{-k}(B) \backslash \bigcup_{j=1}^{k} T^{-j}(F)\right)$$
$$= \sum_{k=0}^{\infty} \mu\big(\{x \in F \cap T^{-k}(B) \colon \tau_F(x) > k\}\big).$$

Proof Let $B \in \mathfrak{F}$. Assume first that $\mu(B) < +\infty$. For every $k \geq 0$, let

$$B_k := T^{-k}(B) \backslash \bigcup_{j=0}^{k} T^{-j}(F).$$

Observe that $\mu(B_k) \leq \mu(T^{-j}(B)) = \mu(B) < +\infty$ for every $j \geq 0$. As μ is T-invariant and $T^{-1}(B_{k-1}) \supseteq B_j$ for all $k \geq 1$, we get, for every $n \geq 0$, that

$$\mu(B\backslash F) - m(B_n) = \sum_{j=1}^{n} \mu(B_{j-1}) - \mu(B_j)$$
$$= \sum_{j=1}^{n} \mu\big(T^{-1}(B_{j-1})\big) - \mu(B_j) = \sum_{j=1}^{n} \mu\big(T^{-1}(B_{j-1})\backslash B_j\big)$$
$$= \sum_{j=1}^{n} \mu\left(\left(T^{-j}(B) \backslash \bigcup_{i=1}^{j} T^{-i}(F)\right) \Bigg\backslash \left(T^{-j}(B) \backslash \bigcup_{i=0}^{j} T^{-i}(F)\right)\right)$$
$$= \sum_{j=1}^{n} \mu\left(F \cap T^{-j}(B) \backslash \bigcup_{i=1}^{j} T^{-i}(F)\right).$$

Then

$$\mu(B) - \mu(B_n) = \mu(B \cap F) + \mu(B\backslash F) - \mu(B_n)$$
$$= \sum_{j=0}^{n} \mu\left(F \cap T^{-j}(B) \Bigg\backslash \bigcup_{k=1}^{j} T^{-k}(F)\right).$$

Letting $n \to \infty$, we see that, in order to complete the proof, it is enough to show that

$$\lim_{n\to\infty} \mu(B_n) = 0.$$

Since $\mu\big(X \backslash \bigcup_{k=0}^{\infty} T^{-k}(F)\big) = 0$, we have that

$$\mu\left(X \backslash \bigcup_{k=0}^{\infty} F^{(k)}\right) = 0, \tag{2.19}$$

where

$$F^{(k)} := T^{-k}(F)\setminus\bigcup_{j=0}^{k-1} T^{-j}(F).$$

The usefulness of the $F^{(k)}$s lies in their mutual disjointness. Indeed, relation (2.19) implies that $\mu\big(B\setminus\bigcup_{k=0}^{\infty} F^{(k)}\big) = 0$. Then

$$\mu(B) = \mu\left(B\cap\bigcup_{k=0}^{\infty} F^{(k)}\right) = \mu\left(\bigcup_{k=0}^{\infty} B\cap F^{(k)}\right) = \sum_{k=0}^{\infty}\mu(B\cap F^{(k)}). \tag{2.20}$$

Fix $\varepsilon > 0$. Since $\mu(B) < \infty$, there exists $k_\varepsilon \in \mathbb{N}$ so large that

$$\sum_{k=k_\varepsilon+1}^{\infty} \mu\big(B\cap F^{(k)}\big) < \frac{\varepsilon}{2}. \tag{2.21}$$

Relation (2.19) also ensures that $\mu\big(B_n\setminus\bigcup_{k=0}^{\infty} F^{(k)}\big) = 0$ for every $n \ge 0$. So, as for B,

$$\mu(B_n) = \sum_{k=0}^{\infty}\mu\big(B_n\cap F^{(k)}\big).$$

But

$$\begin{aligned}
&B_n\cap F^{(k)}\\
&= \left[T^{-n}(B)\Big\backslash\bigcup_{i=0}^{n} T^{-i}(F)\right]\cap\left[T^{-k}(F)\Big\backslash\bigcup_{j=0}^{k-1} T^{-j}(F)\right]\\
&= \big[T^{-n}(B)\cap T^{-k}(A)\big]\Big\backslash\bigcup_{i=0}^{\max\{n,k-1\}} T^{-i}(F)\\
&= \begin{cases} \emptyset & \text{if} \quad k\le n\\ T^{-n}\big(B\cap T^{-(k-n)}(F)\big)\Big\backslash\bigcup_{i=0}^{k-1} T^{-i}(F) & \text{if} \quad k>n \end{cases}\\
&= \begin{cases} \emptyset & \text{if} \quad k\le n\\ \left[T^{-n}\big(B\cap T^{-(k-n)}(F)\big)\Big\backslash\bigcup_{i=n}^{k-1} T^{-i}(F)\right]\Big\backslash\bigcup_{i=0}^{n-1} T^{-i}(F) & \text{if} \quad k>n \end{cases}\\
&= \begin{cases} \emptyset & \text{if} \quad k\le n\\ T^{-n}\big(B\cap F^{(k-n)}\big)\Big\backslash\bigcup_{i=0}^{n-1} T^{-i}(F) & \text{if} \quad k>n. \end{cases}
\end{aligned}$$

Consequently,

$$\mu(B_n) = \sum_{k>n} \mu(B_n \cap F^{(k)}) = \sum_{k=n+1}^{\infty} \mu\left(T^{-n}\big(B \cap F^{(k-n)}\big) \Big\backslash \bigcup_{i=0}^{n-1} T^{-i}(F)\right)$$
$$= \sum_{k=1}^{\infty} \mu\big(B_n^{(k)}\big),$$

where, for each $n, k \geq 0$,

$$B_n^{(k)} := T^{-n}(B \cap F^{(k)}) \backslash \bigcup_{j=0}^{n-1} T^{-j}(F) \subseteq T^{-(n+k)}(F) \backslash \bigcup_{j=0}^{n+k+1} T^{-j}(F) = B^{(n+k)}.$$

Thus, by Remark 2.2.7, we have, for every $k \geq 0$, that

$$\lim_{n\to\infty} \mu\big(B_n^{(k)}\big) = 0. \tag{2.22}$$

Since, for every $n \geq 0$, we have that

$$\mu\big(B_n^{(k)}\big) \leq \mu\big(T^{-n}\big(B^{(k)}\big)\big) = \mu\big(B^{(k)}\big),$$

we therefore get from (2.21), for every $n \geq 0$, that

$$\sum_{k=k_\varepsilon+1}^{\infty} \mu\big(B_n^{(k)}\big) < \frac{\varepsilon}{2}.$$

By virtue of (2.22), there exists $n_\varepsilon \geq 1$ so large that, for all $1 \leq k \leq k_\varepsilon$ and all $n \geq n_\varepsilon$,

$$\mu\big(B_n^{(k)}\big) \leq \frac{\varepsilon}{2k_\varepsilon}.$$

So, for all $n \geq n_\varepsilon$, we have that

$$\mu(B_n) = \mu\left(\bigcup_{k=1}^{\infty} B_n^{(k)}\right) = \sum_{k=1}^{\infty} \mu\big(B_n^{(k)}\big) = \sum_{k=1}^{k_\varepsilon} \mu\big(B_n^{(k)}\big) + \sum_{k=k_\varepsilon+1}^{\infty} \mu\big(B_n^{(k)}\big)$$
$$\leq \sum_{k=0}^{k_\varepsilon} \frac{\varepsilon}{2k_\varepsilon} + \frac{\varepsilon}{2}$$
$$= \frac{\varepsilon}{2} + \frac{\varepsilon}{2} = \varepsilon.$$

The proof is, thus, complete for all $B \in \mathfrak{F}$ with $\mu(B) < +\infty$.

Now let $B \in \mathfrak{F}$ be any set. Since

$$\mu\big(F^{(k)}\big) \leq \mu\big(T^{-k}(F)\big) = \mu(F) < +\infty$$

for all $k \geq 0$, the sets

$$\left(B \cap F^{(k)}\right)_{k=0}^{\infty}$$

are all of finite measure. Thus, the first part of this proof shows that

$$\mu\left(B \cap F^{(k)}\right) = \sum_{n=0}^{\infty} \mu\left(F \cap T^{-n}\left(B \cap F^{(k)}\right) \setminus \bigcup_{j=1}^{n} T^{-j}(F)\right).$$

By (2.20), the mutual disjointness of the sets $F^{(k)}$s, and Theorem 2.2.6(c) along with its formula (2.14) (to have the last equality), we then conclude that

$$\begin{aligned}
\mu(B) &= \sum_{k=0}^{\infty} \mu\left(B \cap F^{(k)}\right) = \sum_{k=0}^{\infty} \sum_{n=0}^{\infty} \mu\left(F \cap T^{-n}\left(B \cap F^{(k)}\right) \setminus \bigcup_{j=1}^{n} T^{-j}(F)\right) \\
&= \sum_{n=0}^{\infty} \sum_{k=0}^{\infty} \mu\left(F \cap T^{-n}\left(B \cap F^{(k)}\right) \setminus \bigcup_{j=1}^{n} T^{-j}(F)\right) \\
&= \sum_{n=0}^{\infty} \mu\left(F \cap T^{-n}(B) \setminus \bigcup_{j=1}^{n} T^{-j}(F)\right).
\end{aligned}$$

The proof is complete. ■

Proposition 2.2.9 *Let $T : (X, \mathfrak{F}) \longrightarrow (X, \mathfrak{F})$ be a measurable map-preserving measure μ on $(X, \mathfrak{F})$. Let $\varphi : X \longrightarrow \mathbb{R}$ be a measurable function. Assume that $F \in \mathfrak{F}$ is such that $0 < \mu(F) < +\infty$ and*

$$\mu\left(X \setminus \bigcup_{k=1}^{\infty} T^{-k}(F)\right) = 0.$$

Then

(a) *$\varphi_F \in L^1(\mu_F)$ whenever $\varphi \in L^1(\mu)$.*
(b) *If $\varphi \geq 0$ or $\varphi \in L^1(\mu)$, then*

$$\int_F \varphi_F \, d\mu_F = \frac{1}{\mu(F)} \int_X \varphi \, d\mu.$$

If, in addition, T is conservative and ergodic, then the above two statements apply to all sets $F \in \mathfrak{F}$ such that $0 < \mu(F) < +\infty$.

Proof Suppose first that $\varphi = 1\!\!1_B$ for some $B \in \mathfrak{F}$ such that $0 < \mu(B) < +\infty$. In view of Proposition 2.2.8, we have that

$$\begin{aligned}
\int_X 1\!\!1_B d\mu = \mu(B) &= \sum_{k=0}^{\infty} \mu(\{x \in T^{-k}(B) \cap F\colon \ \tau_F(x) > k\}) \\
&= \sum_{n=1}^{\infty}\sum_{j=0}^{n-1} \mu(\{x \in F \cap T^{-j}(B)\colon \ \tau_F(x) = n\}) \\
&= \sum_{n=1}^{\infty}\sum_{j=0}^{n-1} \int_{\tau_F^{-1}(n)} 1\!\!1_{T^{-j}}(B) d\mu \\
&= \sum_{n=1}^{\infty}\sum_{j=0}^{n-1} \int_{\tau_F^{-1}(n)} 1\!\!1_B \circ T^j d\mu \\
&= \sum_{n=1}^{\infty} \int_{\tau_F^{-1}(n)} S_n 1\!\!1_B d\mu \\
&= \sum_{n=1}^{\infty} \int_{\tau_F^{-1}(n)} \varphi_F d\mu \\
&= \int_F \varphi_F d\mu = \mu(F) \int_F \varphi_F d\mu_F,
\end{aligned}$$

and we are done in this case.

If $\varphi\colon X \longrightarrow \mathbb{R}$ is a simple measurable function, i.e., $\varphi = \sum_{i=1}^n a_i \varphi^{(i)}$, where all $a_i \in \mathbb{R}$, $1 \le i \le n$, and all $\varphi^{(i)}$, $1 \le i \le n$, are characteristic functions of some measurable sets with positive and finite measures, then

$$\begin{aligned}
\int_X \varphi d\mu = \sum_{i=1}^{n} a_i \int_X \varphi^{(i)} d\mu &= \mu(F) \sum_{i=1}^{n} a_i \int_F \varphi_F^{(i)} d\mu_F \\
&= \mu(F) \int_F \sum_{i=1}^{n} a_i \varphi_F^{(i)} d\mu_F \\
&= \mu(F) \int_F \left(\sum_{i=1}^{n} \varphi^{(i)}\right)_F d\mu_F \\
&= \mu(F) \int_F \varphi_F d\mu_F.
\end{aligned}$$

We, thus, are done in this case as well.

The next case is to consider an arbitrary nonnegative measurable function $\varphi\colon X \longrightarrow [0, +\infty)$. Then φ is a pointwise monotone increasing limit of nonnegative step functions, say $\left(\varphi^{(n)}\right)_{n=1}^{\infty}$. Then, also, the sequence $\left(\varphi_F^{(n)}\right)_{n=1}^{\infty}$ converges pointwise in a monotone increasing way to φ_F. Hence, applying the Lebesgue Monotone Convergence Theorem twice, we then get that

$$\begin{aligned}\int_X \varphi d\mu &= \lim_{n\to\infty}\int_X \varphi^{(n)} d\mu \\ &= \lim_{n\to\infty}\mu(F)\int_F \varphi_F^{(n)} d\mu_F = \mu(F)\int_F \varphi_F d\mu_F.\end{aligned}$$

Since $|\varphi_F| \le |\varphi|_F$, we have, in particular, shown that if $\varphi\colon X \to \mathbb{R}$ is μ-integrable, then $\varphi_F\colon F \to \mathbb{R}$ is μ_F-integrable. Moreover, writing $\varphi = \varphi^+ - \varphi^-$, where $\varphi^+ = \max\{\varphi, 0\}$ and $\varphi^- = \max\{-\varphi, 0\}$, we have that both functions φ^- and φ^- are μ-integrable and

$$\begin{aligned}\int_X \varphi d\mu &= \int_X \varphi^+ d\mu - \int_X \varphi^- d\mu = \mu(F)\int_F \varphi^+ d\mu_F - \mu(F)\int_F \varphi^- d\mu_F \\ &= \mu(F)\int_F (\varphi_F^+ - \varphi_F^-) d\mu_F \\ &= \mu(F)\int_F \varphi_F d\mu_F.\end{aligned}$$

The proof is complete. ■

Observe that if $\varphi \equiv 1$, then $\varphi_F \equiv \tau_F$; therefore, and as an immediate consequence of this proposition, we get the following celebrated result.

Theorem 2.2.10 (Kac's Lemma). *Let $T\colon (X, \mathfrak{F}, \mu) \to (X, \mathfrak{F}, \mu)$ be a measurable map-preserving measure μ. If $F \in \mathfrak{F}$ is such that $0 < \mu(F) < +\infty$ and*

$$\mu\left(X \setminus \bigcup_{k=1}^{\infty} T^{-k}(F)\right) = 0,$$

then

$$\int_F \tau_F d\mu_F = \frac{\mu(X)}{\mu(F)}.$$

In particular,

(a) $\mu(X) < +\infty \iff \int_F \tau_F d\mu < +\infty$.
(b) *If μ is a probability measure, then*

$$\int_F \tau_F d\mu_F = \frac{1}{\mu(F)}.$$

If, in addition, T is conservative and ergodic, then the above two statements apply to all sets $F \in \mathfrak{F}$ such that $0 < \mu(F) < +\infty$.

Item (a) of this theorem tells us that the measure μ is finite if and only if the first return time is integrable. This is a very powerful tool to check whether an invariant measure is finite or infinite. This will be explored in depth in Chapter 22 in Volume 2 in the context of the dynamics of elliptic functions.

Proposition 2.2.11 *Suppose that $T\colon (X,\mathfrak{F},\mu) \longrightarrow (X,\mathfrak{F},\mu)$ is a measure-preserving transformation.*

(a) *If T is ergodic and conservative with respect to μ, then, for every set $F \in \mathfrak{F}$ with $0 < m(F) < +\infty$, we have that $T_F\colon F \longrightarrow F$ is ergodic with respect to μ_F.*
(b) *Conversely, if, for some such set F, the map $T_F\colon F \longrightarrow F$ is ergodic with respect to μ_F and*

$$\mu\left(X\setminus\bigcup_{n=1}^{\infty}T^{-n}(F)\right)=0,$$

then the map $T\colon X \longrightarrow X$ is ergodic with respect to μ.

Proof Suppose that $T\colon X \to X$ is ergodic and conservative. If $A \subseteq F$ is T_F-invariant and $\mu(A) > 0$, then, for μ almost all $x \in F\setminus A$, we have by Theorem 2.1.10 that

$$0=\sum_{n=1}^{\infty}1\!\!1_A\circ T_F^n(x)=\sum_{n=1}^{\infty}1\!\!1_A\circ T^n(x)=+\infty.$$

This contradiction shows that the map $T_F\colon F \to F$ is ergodic.

In order to prove the converse, suppose that $T^{-1}(B) = B$ and $m(B)>0$. Since $\mu\big(X\setminus\bigcup_{n=1}^{\infty}T^{-n}(F)\big) = 0$, there exists $k \geq 1$ such that $\mu(B \cap T^{-k}(F)) > 0$. But then

$$\mu(B\cap F)=\mu(T^{-k}(B\cap F))=\mu(T^{-k}(B)\cap T^{-k}(F))=\mu(B\cap T^{-k}(F))>0.$$

Since m_F is a probability measure and $T_F\colon F \to F$ is ergodic, this implies that

$$1=m_F\left(\bigcup_{n=0}^{\infty}T^{-n}(B\cap F)\right)\le m_F\left(\bigcup_{n=1}^{\infty}T^{-n}(B\cap F)\right).$$

This means that $m\left(F\setminus\bigcup_{n=0}^{\infty}T^{-n}(B\cap F)\right)=0$. As $m\left(X\setminus\bigcup_{n=0}^{\infty}T^{-n}(F)\right)=0$, this implies that

$$m\left(X\setminus\bigcup_{n=0}^{\infty}T^{-n}(B\cap F)\right)=0.$$

Consequently,

$$m(X\backslash B) = m\left(X\backslash\bigcup_{n=1}^{\infty}T^{-n}(B)\right) \leq m\left(X\backslash\bigcup_{n=1}^{\infty}T^{-n}(B\cap F)\right) = 0.$$

We are done. ■

2.3 Ergodic Theorems: Birkhoff, von Neumann, and Hopf

In this section, we deal with the Birkhoff, von Neumann, and Hopf Ergodic Theorems. These are central pillars of abstract ergodic theory. The Birkhoff Ergodic Theorem, proved for the first time in [**Bir**], establishing equalities of time and space averages, has profound theoretical and philosophical consequences. It abounds in applications. The present book is evidence of this. The original proof provided by Birkhoff in [**Bir**] was very long and complicated. Since then, some simplifications have been made. We provide a short simple proof of the Birkhoff Ergodic Theorem, taken from [**KH**]. This theorem concerns measure-preserving dynamical systems acting on probability spaces. As its first, fairly straightforward, consequence, we prove the L^p-von Neumann Ergodic Theorem. Second, as a more involved consequence of the Birkhoff Ergodic Theorem, by utilizing the induced procedure described in the previous section, we prove the Hopf Ergodic Theorem, which holds for general measure-preserving transformations, whose σ-finite invariant measure can be infinite. We would like to remark that the Hopf Ergodic Theorem is not proved in Aaronson's book [**Aa**].

The formulation and the proof of the Birkhoff Ergodic Theorem we provide below utilizes the concept, widely used in probability theory, of an expected value with respect to a sub-σ-algebra, which we introduced and studied in Section 1.4.

Let $(X, \mathfrak{F}, \mu)$ be a probability space and $T: X \longrightarrow X$ be a measurable map preserving the probability measure μ. Let

$$\mathcal{I} := \{A \in \mathfrak{F}: \mu(T^{-1}(A)\triangle A) = 0\}.$$

It is easy to check that $\mathcal{I}$ is a σ-algebra and we call it the $\mathcal{I}$ σ-algebra of T-invariant (mod 0) sets. Let us record the following obvious theorem.

Theorem 2.3.1 *A measurable map $T: X \longrightarrow X$ preserving a probability measure μ is ergodic if and only if $\mathcal{I}$, the σ-algebra of T-invariant (mod 0) subsets of X, is trivial, i.e., it consists of sets of measure 1 and zero only.*

Recall that, for any integer $n \geq 1$, the nth Birkhoff sum of the function ϕ was defined as

$$S_n\phi = \phi + \phi \circ T + \cdots + \phi \circ T^{n-1}.$$

Theorem 2.3.2 (Birkhoff Ergodic Theorem) *If $T: X \longrightarrow X$ is a measure-preserving endomorphism of a probability space $(X, \mathfrak{F}, \mu)$ and if $\phi: X \longrightarrow \mathbb{R}$ is an integrable function, then*

$$\lim_{n\to\infty} \frac{1}{n} S_n\phi(x) = E_\mu(\phi|\mathcal{I}) \quad \text{for} \quad \mu\text{-a.e. } x \in X.$$

If, in addition, T is ergodic, then

$$\lim_{n\to\infty} \frac{1}{n} S_n\phi(x) = \int_X \phi \, d\mu \quad \text{for} \quad \mu\text{-a.e. } x \in X.$$

Proof Let $f \in L^1(\mu)$. For every integer $n \geq 1$, let

$$F_n := \max\left\{\sum_{i=0}^{k-1} f \circ T^i : \; 1 \leq k \leq n\right\}.$$

Of course, the sequence $(F_n)_{n=1}^\infty$ is monotone increasing. Then, for every $x \in X$, we have that

$$F_{n+1}(x) - F_n(T(x)) = f(x) - \min\{0, F_n(T(x))\} \geq f(x).$$

The sequence $(F_{n+1}(x) - F_n \circ T)_{n=1}^\infty$ is monotone decreasing, since $(F_n)_{n=1}^\infty$ is monotone increasing. Define

$$A := \left\{x \in X: \; \sup_{n\geq 1}\left\{\sum_{i=0}^{n} f(T^i(x))\right\} = +\infty\right\}.$$

Note that $A \in \mathcal{I}$. If $x \in A$, then $F_{n+1} - F_n(T(x))$ monotonously decreases to $f(x)$ as $n \to \infty$. The Lebesgue Monotone Convergence Theorem implies then that

$$0 \leq \int_A (F_{n+1} - F_n) d\mu = \int_A (F_{n+1} - F_n \circ T) d\mu \longrightarrow \int_A f \, d\mu. \tag{2.23}$$

Notice that $\frac{1}{n}\sum_{k=0}^{n-1} \phi \circ T^k \leq F_n/n$; so, outside A, we have that

$$\limsup_{n\to\infty} \frac{1}{n} \sum_{k=0}^{n-1} f \circ T^k \leq 0. \tag{2.24}$$

Therefore, if the conditional expectation value $f_{\mathcal{I}}$ of f is negative a.e., i.e., if

$$\int_C f d\mu = \int_C f_{\mathcal{I}} d\mu < 0$$

for all $C \in \mathcal{I}$ with $\mu(C) > 0$, then, as $A \in \mathcal{I}$, (2.23) implies that $\mu(A) = 0$. Hence, (2.24) holds a.e. Now fix $\varepsilon > 0$ and let

$$f := \phi - \phi_{\mathcal{I}} - \varepsilon.$$

Then, obviously,

$$f_{\mathcal{I}} = -\varepsilon < 0.$$

The equality $\phi_{\mathcal{I}} \circ T = \phi_{\mathcal{I}}$ entails

$$\frac{1}{n}\sum_{k=0}^{n-1} f \circ T^k = \left(\frac{1}{n}\sum_{k=0}^{n-1} \phi \circ T^k\right) - \phi_{\mathcal{I}} - \varepsilon.$$

So, (2.24) yields

$$\limsup_{n\to\infty} \frac{1}{n}\sum_{k=0}^{n-1} \phi \circ T^k \le \phi_{\mathcal{I}} + \varepsilon \quad \text{a.e.}$$

Call the sets of points where this inequality holds by $X_+(\varepsilon)$. Replacing ϕ by $-\phi$ gives

$$\liminf_{n\to\infty} \frac{1}{n}\sum_{k=0}^{n-1} \phi \circ T^k \ge \phi_{\mathcal{I}} - \varepsilon \quad \text{a.e.}$$

Call the sets of points where this inequality holds by $X_-(\varepsilon)$. Then, denoting

$$X_* := \bigcap_{n=1}^{\infty} \big(X_+(1/n) \cap X_-(1/n)\big),$$

we have that $\mu(X_*) = 1$ and

$$\lim_{n\to\infty} \frac{1}{n}\sum_{k=0}^{n-1} \phi \circ T^k = \phi_{\mathcal{I}}$$

on X_*. The proof of the first part of our theorem is, thus, complete. The second part, i.e., that concerning ergodic maps, directly follows from the first part along with Theorem 2.3.1 and (1.34). ■

As the first consequence of the Birkhoff Ergodic Theorem we prove the von Neumann Ergodic Theorem, which asserts that the converges in the Birkhoff Ergodic Theorem are not only almost everywhere but also in $L^p(\mu)$ whenever the input function ϕ is in $L^p(\mu)$. It was first proved by von Neumann in [**vN**] and has had many generalizations and extensions since then.

Theorem 2.3.3 (L^p-von Neumann Ergodic Theorem). *If $T: X \longrightarrow X$ is a measurable endomorphism of a probability space $(X, \mathfrak{F}, \mu)$ preserving measure μ and if $\phi: X \longrightarrow \mathbb{R}$ is a measurable function that belongs to $L^p(\mu)$ with some $p \geq 1$, then $E(\phi|\mathcal{I}) \in L^p(\mu)$ and*

$$\lim_{n\to\infty} \frac{1}{n} S_n\phi = E(\phi|\mathcal{I}),$$

where the convergence and equality are understood in the Banach space $L^p(\mu)$. If, in addition, T is ergodic, then

$$\lim_{n\to\infty} \frac{1}{n} S_n\phi(x) = \int \phi \, d\mu$$

and, as above, both convergence and equality are taken in $L^p(\mu)$.

Proof As an auxiliary step, consider first an essentially bounded function $\psi \in L^p(\mu)$. This means that

$$\|\psi\|_\infty := \text{ess sup}\{|\psi(x)| : x \in X\} < +\infty.$$

Then also $||n^{-1} S_n\psi \leq ||\psi||_\infty < +\infty$ for all $n \geq 1$. Hence,

$$\left\| \frac{1}{n} S_n\psi - \psi^* \right\|_\infty \leq ||\psi||_\infty < +\infty. \tag{2.25}$$

But by the Birkhoff Ergodic Theorem (Theorem 2.3.2), the sequence $\left(n^{-1} S_n\psi\right)_1^\infty$ converges almost everywhere to $\psi^* := E(\psi|\mathcal{I})$. Hence, $\psi^* \in L^\infty(\mu) \subseteq L^p(\mu)$ and

$$\lim_{n\to\infty} \left| \frac{1}{n} S_n\psi(x) - \psi^*(x) \right| = 0 \ \mu\text{-a.e.}$$

Therefore, looking at (2.25), the Lebesgue Dominated Convergence Theorem yields

$$\lim_{n\to\infty} \left\| \frac{1}{n} S_n\psi - \psi^* \right\|_p = 0,$$

where, by $|| \cdot ||_p$, we denoted the L^p norm in the Banach space $L^p(\mu)$. Consequently, we got the following.

Claim 1°. $\left(n^{-1} S_n\psi\right)_1^\infty$ is a Cauchy sequnce in $L^p(\mu)$.

Passing to the general case, consider an arbitrary function $\phi \in L^p(\mu)$. We aim to prove the following.

Claim 2°. $\left(n^{-1} S_n\phi\right)_1^\infty$ is a Cauchy sequence in $L^p(\mu)$.

Proof In order to prove this claim, fix $\varepsilon > 0$ arbitrary. There then exists a function $\phi_\varepsilon \in L^\infty(\mu)$ such that

$$\|\phi - \phi_\varepsilon\|_p < \varepsilon/4. \tag{2.26}$$

In consequence,

$$0 \leq \left\| \frac{1}{n} S_n\phi - \frac{1}{n} S_n\phi_\varepsilon \right\|_p \leq \varepsilon/4 \tag{2.27}$$

for all $n \geq 1$. By virtue of Claim 1 there exists $N \geq 1$ such that if $k, l \geq N$, then

$$\left\| \frac{1}{l} S_l\phi_\varepsilon - \frac{1}{k} S_k\phi_\varepsilon \right\|_p < \varepsilon/2.$$

From this and (2.27), we get, for such k and l, that

$$\begin{aligned}
\left\| \frac{1}{l} S_l\phi - \frac{1}{k} S_k\phi \right\|_p &\leq \left\| \frac{1}{l} S_l\phi - \frac{1}{l} S_l\phi_\varepsilon \right\|_p + \left\| \frac{1}{l} S_l\phi_\varepsilon - \frac{1}{k} S_k\phi_\varepsilon \right\|_p \\
&\quad + \left\| \frac{1}{k} S_k\phi_\varepsilon - \frac{1}{k} S_k\phi \right\|_p \\
&\leq \frac{\varepsilon}{4} + \frac{\varepsilon}{2} + \frac{\varepsilon}{4} \\
&= \varepsilon.
\end{aligned}$$

Claim 2° is proved. ■

Let $\hat{\phi}$ be the limit of the sequence in

$$\left(n^{-1} S_n\phi\right)_1^\infty$$

in $L^p(\mu)$; the latter is a Banach space, so complete. Since convergence in the L^p norm entails convergence in measure and since any sequence convergent in measure contains a subsequence converging almost everywhere, invoking the Birkhoff Ergodic Theorem (Theorem 2.3.2), we conclude that $\hat{\phi} = E(\phi|\mathcal{I})$. The last assertion of our theorem is now also an immediate consequence of Theorem 2.3.2. ■

Of particular significance is the case of $p = 2$. Then $L^p(\mu)$ is a Hilbert space with the inner product

$$(\phi, \psi) = \int_X \phi\bar{\psi}\, d\mu.$$

For every $\phi \in L^2(\mu)$, define

$$U_T(\phi) := \phi \circ T. \tag{2.28}$$

Then

$$\int_X |U_T(\phi)|^2 \, d\mu = \int_X |\phi|^2 \circ T \, d\mu = \int_X |\phi| \, d\mu < +\infty.$$

So, $U_T(\phi) \in L^2(\mu)$ and (2.27) thus defines a bounded, with norm ≤ 1, linear oprator from $L^2(\mu)$ into itself. It is called the Koopman operator associated with the measure-preserving map $T\colon X \to X$. In fact, the same calculation gives the following.

Theorem 2.3.4 *$U_T(\phi)\colon L^2(\mu) \to L^2(\mu)$ is a unitary operator, meaning that*

$$(U_T g, U_T h) = (g,h)$$

for all $g,h \in L^2(\mu)$.

Denote by $U_T^*\colon L^2(\mu) \to L^2(\mu)$ the operator conjugate to U_T. Let

$$L^2_{inv}(\mu) := \{\phi \in L^2(\mu)\colon U_T(\phi) = \phi\} \text{ and } L^2_{*inv}(\mu) := \{\phi \in L^2(\mu)\colon U_T^*(\phi) = \phi\}$$

be the respective spaces of all U_T-invariant and U_T^*-invariant elements of $L^2(\mu)$. Obviously, both $L^2_{inv}(\mu)$ and $L^2_{*inv}(\mu)$ are closed vector subspaces of $L^2(\mu)$. We shall prove the following.

Lemma 2.3.5 *We have that*

$$L^2_{inv}(\mu) = L^2_{*inv}(\mu).$$

Proof Let $g \in L^2_{inv}(\mu)$. Then $U_T g = g$ and, since by Theorem 2.3.4, $U_T^* U_T = \mathrm{Id}$, we, therefore, get that $U_T^* g = U_T^* U_T g = g$, meaning that $g \in L^2_{*inv}(\mu)$. The inclusion

$$L^2_{inv}(\mu) \subseteq L^2_{*inv}(\mu)$$

is proved. Conversely, if $g \in L^2_{*inv}(\mu)$, then

$$\begin{aligned}
\|U_T g - g\|_2^2 &= (U_T g - g, U_T g - g) \\
&= \|U_T g\|_2^2 - (g, U_T g) - (U_T g, g) + \|g\|_2^2 \\
&= \|g\|_2^2 - \left(U_T^* g, g\right) - \left(g, U_T^* g\right) + \|g\|_2^2 \\
&= 2\|g\|_2^2 - 2(g,g) = 2\|g\|_2^2 - 2\|g\|_2^2 \\
&= 0. \qquad (2.29)
\end{aligned}$$

Hence, $U_T g = g$, meaning that $g \in L^2_{inv}(\mu)$ and the inclusion

$$L^2_{*inv}(\mu) \subseteq L^2_{inv}(\mu)$$

is also proved. We are done. ■

We will need to know what the orthogonal complement of $L^2_{inv}(\mu)$ in $L^2(\mu)$ is. In fact, let G_T be the closed vector subspace of $L^2(\mu)$ generated by all coboundaries, i.e., by vectors of the form

$$u - u \circ T, \ \ u \in L^2(\mu).$$

Of course, all coboundaries form a vector space, so G_T is just the closure of all coboundaries.

The orthogonal complement of $L^2_{inv}(\mu)$ is described by the following.

Lemma 2.3.6 *We have that*

$$L^2(\mu) = L^2_{inv}(\mu) \bigoplus G_T,$$

i.e., the Hilbert space $L^2(\mu)$ can be represented as the orthogoal sum of its closed vector subspaces $L^2_{inv}(\mu)$ and G_T.

Proof Fix a coboundary $\psi = u - u \circ T$. Then, for every $\phi \in L^2_{inv}(\mu)$, we have that

$$\begin{aligned}(\psi,\phi) &= (u - U_T u,\phi) = (u,\phi) - (U_T u,\phi) \\ &= (u,\phi) - (U_T u,\phi) \\ &= (u,\phi) - (U_T u,U_T\phi) = (u,\phi) - (u,\phi) \\ &= 0.\end{aligned} \tag{2.30}$$

This means that $\phi \perp \psi$, whence

$$L^2_{inv}(\mu) \subseteq G_U^\perp. \tag{2.31}$$

Now fix $\phi \in G_U^\perp$. Then, for every $\psi \in\in L^2(\mu)$, we have that $(\phi, \psi - U_T\psi) = 0$ or, equivalently, $(\phi,\psi) = (\phi, U_T\psi) = \big(U_T^*\phi,\psi\big)$. This means that $\big(U_T^*\phi - \phi,\psi\big) = 0$. Hence, $U_T^*\phi - \phi = 0$ or, equivalently, $\phi \in L^2_{inv}(\mu)$. Therefore,

$$G_U^\perp \subseteq L^2_{inv}(\mu).$$

Along with (2.31), this finishes the proof. ■

In subsequent chapters, we will need the following characterization of coboundaries.

Lemma 2.3.7 *If $T : X \longrightarrow X$ is a measurable map preserving a probability measure μ and if $g \in L^2(\mu)$, then the following two statements are equivalent.*

(a) *The function g is a coboundary, i.e., $g = u - u \circ T$ for some $u \in L^2(\mu)$.*
(b) *The sequence $(S_n g)_1^\infty$ is bounded in the Hilbert space $L^2(\mu)$.*

Proof (a)⇒(b). For every $n \geq 1$, we have that $S_n g = u - u \circ T^n$. Therefore,

$$\|S_n g\|_2 = \|u - u \circ T^n\|_2 \leq \|u\|_2 + \|u \circ T^n\|_2 = 2\|u\|_2$$

and, thus, the implication (a)⇒(b) is established.
(b)⇒(a). By our hypothesis, there exists $M \geq 1$ such that $\|S_n g\|_2 \leq M$ for all $n \geq 1$. Consequently,

$$\left\| \frac{1}{n} \sum_{j=1}^{n} S_j g \right\|_2 \leq M$$

for all $n \geq 1$. Since $L^2(\mu)$, as a Hilbert space, is reflexive, its closed ball $\overline{B}(0, M)$ is weakly compact in $L^2(\mu)$. There, thus, exist $u \in \overline{B}(0, M)$ and $(n_k)_1^\infty$, an increasing sequence of positive integers, such that the sequence $\left(n_k^{-1} \sum_{j=1}^{n_k} S_j g\right)_1^\infty$ converges weakly to u. But then

$$\lim_{k\to\infty} \frac{1}{n_k} \sum_{j=1}^{n_k} S_j(g \circ T) = u \circ T \quad \text{weakly in} \quad L^2(\mu) \tag{2.32}$$

and

$$\begin{aligned}\frac{1}{n_k} \sum_{j=1}^{n_k} S_j(g \circ T) &= \frac{1}{n_k} \left(\sum_{j=1}^{n_k} S_{j+1} g - n_k g \right) \\ &= \frac{1}{n_k} \sum_{j=1}^{n_k} S_j g + \frac{1}{n_k} \left(S_{n_k+1} g - g \right) - g.\end{aligned}$$

Taking the weak limits of both sides of this equation and invoking (2.32), we get $u \circ T = u - g$ or, equivalently, $g = u - u \circ T$. The proof of the implication (b)⇒(a) and the whole lemma is complete. ■

Corollary 2.3.8 *If $T : X \longrightarrow X$ is a measurable map preserving a probability measure μ and if $g \in L^2(\mu)$, then the following three statements are equivalent.*

(a) *The function g is a coboundary, i.e., $g = u - u \circ T$ for some $u \in L^2(\mu)$.*
(b) *There exists $l \geq 1$ such that $S_l g$ is a coboundary with respect to the dynamical system (T^l, μ), i.e., $S_l g = u - u \circ T^l$ for some $u \in L^2(\mu)$.*
(c) *For every $l \geq 1$, S_l is a coboundary with respect to the dynamical system (T^l, μ), i.e., $S_l g = u - u \circ T^l$ for some $u \in L^2(\mu)$.*

Proof (a)⇒(c). We have that

$$S_l g = \sum_{j=0}^{l-1} g \circ T^j = \sum_{j=0}^{l-1} \left(u \circ T^j - u \circ T^{j+1} \right) = u - u \circ T^l$$

and the implication (a)⇒(c) is established.

The implication (c)⇒(b) is obvious.

We shall prove (b)⇒(a).

Indeed, given an integer $n \geq 0$, write uniquely $n = lk_n + r_n$, $k = k_n \geq 0$, $0 \leq r_n \leq l - 1$.Then

$$S_n g = S_{k_n}^{(l)}(S_l g) + \sum_{j-n-r_n}^{n-1} g \circ f^j = S_k^{(l)}(S_l g) + S_r(g \circ f^{n-r_n}), \tag{2.33}$$

where $S_n^{(l)}(h)$ is the nth Birkhoff sum of the function $h\colon X \to \mathbb{R}$ with respect to the dynamical system $T^l\colon X \to X$. It follows from Lemma 2.3.7 that the norms $|S_{k_n}^{(l)}(S_l g)\|_2$ are uniformly bounded from above, say by M. It then follows from (2.33) that

$$\|S_n g\|_2 \leq M + \|S_r(g \circ f^{n-r_n})\|_2 \leq M + r\|g\|_2 \leq M + (l-1)\|g_2\|.$$

So, we conclude from Lemma 2.3.7 again that g is a coboundary in $L^2(\mu)$. The proof is complete. ■

Coming back to Theorem 2.3.2, as we have already said, it implies that the *time average* exists for μ-a.e. $x \in X$. If, additionally, T is ergodic, Theorem 2.3.2 says that the *time average is equal to the space average.*

As a consequence of Theorem 2.3.2, its ergodic part, we get the following statement about visiting a measurable set of positive measure.

Theorem 2.3.9 *Suppose that $(X, \mathfrak{F}, \mu)$ is a probability space and $T\colon X \longrightarrow X$ is a measurable ergodic map-preserving measure μ. Fix $F \in \mathfrak{F}$ with $\mu(F) > 0$. Then, for μ-a.e. $x \in X$,*

$$\lim_{n\to\infty} \frac{1}{n} \#\{0 \leq j \leq n-1\colon T^j(x) \in F\} = \mu(F).$$

In particular, the set

$$\{n \geq 0\colon T^n(x) \in F\}$$

is infinite.

Proof The first assertion is an immediate consequence of Theorem 2.3.2 applied to the function $\varphi := 1\!\!1_F$. The second assertion is an immediate consequence of the first one. ■

We shall now prove two further small, technical, slightly surprising, but useful consequences of Theorem 2.3.2

Proposition 2.3.10 *Suppose that $(X, \mathfrak{F}, \mu)$ is a probability space and $T: X \longrightarrow X$ is a measurable ergodic map-preserving measure μ. Fix $F \in \mathfrak{F}$ with $\mu(F) > 0$. For every $x \in X$, let $(k_n)_{n=1}^{\infty}$ be the sequence of consecutive visits of x to F under the action T. Then, for μ-a.e. $x \in X$,*

$$\lim_{n\to\infty} \frac{k_{n+1}}{k_n} = 1.$$

Proof Note that $S_{k_n} 1\!\!1_F(x) = n$. It, therefore, follows from Theorem 2.3.2 and ergodicity of T that

$$\begin{aligned}\lim_{n\to\infty} \frac{k_{n+1}}{k_n} &= \lim_{n\to\infty} \left(\frac{n}{k_n}\frac{k_{n+1}}{n+1}\right) = \lim_{n\to\infty} \frac{1}{k_n} S_{k_n} 1\!\!1_F(x) \frac{1}{\frac{1}{k_{n+1}} S_{k_{n+1}} 1\!\!1_F(x)} \\ &= \frac{\lim_{n\to\infty}\left(\frac{1}{k_n} S_{k_n} 1\!\!1_F(x)\right)}{\lim_{n\to\infty}\left(\frac{1}{k_{n+1}} S_{k_{n+1}} 1\!\!1_F(x)\right)} \\ &= \frac{\mu(F)}{\mu(F)} = 1.\end{aligned}$$

We are done. ■

Proposition 2.3.11 *Suppose that $(X, \mathfrak{F}, \mu)$ is a probability space and $T: X \longrightarrow X$ is a measurable map-preserving measure μ. If $f \in L^1(\mu)$, then, for μ-a.e. $x \in X$,*

$$\lim_{n\to\infty} \frac{1}{n} f(T^n(x)) = 0.$$

Proof It follows from Theorem 2.3.2 that, for μ-a.e. $x \in X$,

$$\begin{aligned}\lim_{n\to\infty} \frac{f(T^n(x))}{n} &= \lim_{n\to\infty} \frac{f(T^n(x))}{n+1} = \lim_{n\to\infty} \frac{S_{n+1}f(x) - S_nf(x)}{n+1} \\ &= \lim_{n\to\infty} \frac{S_{n+1}f(x)}{n+1} - \lim_{n\to\infty} \frac{S_nf(x)}{n+1} \\ &= \lim_{n\to\infty} \frac{S_{n+1}f(x)}{n+1} - \lim_{n\to\infty} \frac{S_nf(x)}{n} \\ &= 0.\end{aligned}$$

We are done. ■

As an application of Theorem 2.3.2 to the ergodic theory of measurable maps preserving an infinite measure, we shall prove the following, remarkable and useful in applications, theorem.

Theorem 2.3.12 (Hopf Ergodic Theorem). *Suppose that $(X, \mathfrak{F}, \mu)$ is a σ-finite measure space and $T: X \longrightarrow X$ is a measurable ergodic and conservative map-preserving measure μ.*

Consider two measurable functions $f, g \in L^1(\mu)$, where $g \geq 0$ μ–a.e. is such that $\mu(g^{-1}((0, +\infty))) > 0$, so that, in particular, $\int g d\mu > 0$. Then

$$\lim_{n\to\infty} \frac{S_n f(x)}{S_n g(x)} = \frac{\int f d\mu}{\int g d\mu}.$$

Proof Since the measure μ is σ-finite, there are countably many mutually disjoint sets $\{X_j\}_{j=1}^{\infty}$ such that, for all $j \geq 1$, $0 < \mu(X_j) < +\infty$ and $\mu\big(X \setminus \bigcup_{j=1}^{\infty} X_j\big) = 0$. Let

$$T_j = T_{X_j},$$

where, we recall, $T_{X_j}: X_j \to X_j$ is the first return map of T from X_j to X_j. For all $\phi: X \to \mathbb{R}$, let $\phi_j := \phi_{X_j}: X_j \to \mathbb{R}$. Given $x \in X$, let

$$S_n^{(j)} \phi_j(x) := \sum_{i=0}^{n-1} \phi_j\big(T_j^i(x)\big).$$

If $x \in X_j$ and $n \geq 1$, let $j_n \geq 1$ be the largest integer $k \geq 0$ such that

$$\sum_{i=0}^{k-1} \tau_{X_j}\big(T_j^i(x)\big) \leq n.$$

Then, for all $x \in X_j$, we have that

$$S_n\phi(x) = S_{T_i^{j_n}}^{(j)} \phi_j(x) + S_{\Delta n}\phi_j(x),$$

where

$$\Delta_n := n - \sum_{i=0}^{k-1} \tau_{X_j}\big(T_j^i(x)\big) \geq 0.$$

Then

$$\frac{S_n\phi(x)}{j_n} = \frac{S_{T_i^{j_n}}^{(j)} \phi(x)}{j_n} + \frac{S_{\Delta n}\phi(x)}{j_n}. \tag{2.34}$$

Now

$$\left| \frac{S_{\Delta n}\phi(x)}{j_n} \right| \leq \frac{1}{j_n} S_{\Delta n}|\phi|(x) \leq \frac{1}{j_n} |\phi|\big(T_j^{j_n}(x)\big).$$

It, therefore, follows from Proposition 2.3.11 and Proposition 2.2.9 that if $\phi \in L^1(\mu)$, then, for μ-a.e. $x \in X_j$,

$$\lim_{n\to\infty} \left| \frac{S_{\Delta n}\phi(x)}{j_n} \right| = 0.$$

Therefore, applying Theorem 2.3.2, Proposition 2.2.9, Proposition 2.2.11, and (2.34), we get, for μ-a.e. $x \in X_j$ and all $n \geq 1$, that

$$\frac{S_n f(x)}{S_n g(x)} = \frac{\frac{S_n f(x)}{j_n}}{\frac{S_n g(x)}{j_n}} = \frac{\frac{S_{j_n}^{(j)} f_j(x)}{j_n} + \frac{S_{\Delta n} f(x)}{j_n}}{\frac{S_{j_n}^{(j)} g_j(x)}{j_n} + \frac{S_{\Delta n} g(x)}{j_n}} \longrightarrow \frac{\int_{X_j} f_j d\mu_{X_j}}{\int_{X_j} g_j d\mu_{X_j}} = \frac{\int_X f d\mu}{\int_X g d\mu}.$$

We are done. ■

As we shall seee in Section 5.1, this theorem, being powerful and interesting in itself, somewhat surprisingly, rules out any hope for a more direct version of the Birkhoff Ergodic Theorem in the case of infinite measures. The world of infinite invariant measures is indeed very different from the one of finite measures.

Now we want to derive one additional consequence of Theorem 2.3.2. We shall prove the following.

Theorem 2.3.13 *Suppose that $\mathfrak{F}$ is a σ-algebra on a set X, $T: X \longrightarrow X$ is a measurable map, and μ_1 and μ_2 are two σ-finite measures invariant under T.*

If $T: X \longrightarrow X$ is ergodic and conservative with respect to both measures μ_1 and μ_2, then either μ_1 and μ_2 are mutually singular or else they coincide up to a positive multiplicative constant (if both are probabilistic, then they are equal).

Proof Suppose first that μ_1 and μ_2 are both probability measures. If $\mu_1 \neq \mu_2$, then there exists a set $F \in \mathfrak{F}$ such that $\mu_1(F) \neq \mu_2(F)$. Let, for $i = 1, 2$,

$$X_i = \left\{ x \in X : \frac{1}{n} S_n \mathbb{1}_F(x) \longrightarrow \int_X \mathbb{1}_F d\mu_i = \mu_i(F) \right\}.$$

Since $\mu_1(F) \neq \mu_2(F)$, we have that

$$X_1 \cap X_2 = \emptyset.$$

But, by Theorem 2.3.2, $\mu_1(X_1) = 1$ and $\mu_2(X_2) = 1$. So, $\mu_1(X_2) = 0$ and $\mu_2(X \backslash X_2) = 0$. Thus, μ_1 and μ_2 are mutually singular and we are done in this case. Now consider the general case. Since measures μ_1 and μ_2 are both σ-finite, there are countably many measurable disjoint sets $(Y_n)_{n=0}^{\infty}$ such that

$$X = \bigcup_{n=1}^{\infty} Y_n$$

and

$$\mu_1(Y_n), \quad \mu_2(Y_n) < +\infty$$

for all $n \geq 1$. Assume that without loss of generality neither measure μ_1 nor measure μ_2 vanishes. Suppose first that, for some $n \geq 1$ $\mu_1(Y_n) > 0$, and μ_1, μ_2 coincide on Y_n up to a positive multiplicative constant. We may assume without loss of generality that $n = 1$ and that

$$\mu_{1|Y_1} = \mu_{2|Y_1}.$$

It then immediately follows from Proposition 2.2.8 that $\mu_1 = \mu_2$ and we are done in this case.

Now assume that μ_1 and μ_2 do not coincide on X up to a positive multiplicative constant. If $\mu_1(Y_n) = 0$, set $Z_n = Y_n$; if $\mu_2(Y_n) = 0$, set $Z_n = \emptyset$.

Now consider the case when

$$\mu_1(Y_n), \mu_2(Y_n) \neq 0.$$

From what we have already proved, it follows that $\mu_{1|Y_1}$ and $\mu_{2|Y_1}$ do not coincide up to any positive multiplicative factor. Hence,

$$\mu_{1|Y_1} \neq \mu_{2|Y_1}.$$

Combining Proposition 2.2.11 and the already proved case of probability measures, we, thus, conclude that the measures $\mu_{1|Y_1}$ and $\mu_{2|Y_1}$ are mutually singular. This means that there exists a set $Z_n \subseteq Y_n$ such that $\mu_1(Z_n) = 0$ and $\mu_2(Y_n \backslash Z_n) = 0$. Setting

$$Z := \bigcup_{n=1}^{\infty} Z_n,$$

we, thus, have that

$$\mu_1(Z) = \sum_{n=1}^{\infty} \mu_1(Z_n) = 0$$

and

$$\mu_2(X\backslash Z) = \sum_{n=1}^{\infty} \mu_2(Y_n\backslash Z) = \sum_{n=1}^{\infty} \mu_2(Y_n\backslash Z_n) = 0.$$

So, the measures μ_1 and μ_2 are mutually singular and we are done. ■

Theorem 2.3.14 *Suppose that $\mathfrak{F}$ is a σ-algebra on a set X and $T: X \longrightarrow X$ is a measurable map. Let μ be σ-finite measure invariant under T.*

If $T: X \longrightarrow X$ is conservative with respect to μ, then μ is ergodic if and only if there is no σ-finite measure ν on $(X, \mathfrak{F})$ invariant under T such that

$$\nu \prec\!\!\prec \mu \quad \text{and} \quad \nu \neq t\mu$$

for every real $t > 0$.

Proof First, suppose that μ is ergodic. Let ν be a σ-finite measure on $(X, \mathfrak{F})$ invariant under T and such that

$$\nu \prec\!\!\prec \mu.$$

We claim that ν is ergodic too. Indeed, suppose, by way of contradiction, that there exists $A \in \mathfrak{F}$ such that

$$T^{-1}(A) = A$$

with

$$\nu(A) > 0 \quad \text{and} \quad \nu(X\backslash A) > 0.$$

Since $\nu \prec\!\!\prec \mu$, it follows that $\mu(A) > 0$ and $\mu(X\backslash A) > 0$. This contradicts the ergodicity of μ. So ν is ergodic.

Since μ is conservative and ν is absolutely continuous with respect to μ, the T-invariant measure ν is conservative too. So, if $\nu \neq t\mu$ for every real $t > 0$, then Theorem 2.3.13 asserts that $\nu \perp \mu$. This contradicts the hypothesis that $\nu \prec\!\!\prec \mu$. Hence,

$$\nu = t\mu$$

for some real $t > 0$.

For the converse implication, suppose that μ is not ergodic (but still T-invariant and conservative by hypothesis). Then there exists $A \in \mathfrak{F}$ such that

$$T^{-1}(A) = A$$

with

$$\mu(A) > 0 \quad \text{and} \quad \mu(X\backslash A) > 0.$$

Let μ_A^* be the σ-finite measure on $(X, \mathfrak{F})$ defined by the formula

$$\mu_A^*(F) = \mu(F \cap A)$$

for all $F \in \mathfrak{F}$. One immediately verifies that μ_A^* is a T-invariant probability measure such that

$$\mu_A^* \prec\prec \mu \quad \text{and} \quad \mu_A^* \neq t\mu$$

for any real $t > 0$. ■

2.4 Absolutely Continuous σ-Finite Invariant Measures: Marco Martens's Approach

In this section, we establish a very useful, relatively easy to verify, sufficient condition for a quasi-invariant measure to admit an absolutely continuous σ-finite invariant measure. This condition goes back to the work [**Mar**] of Marco Martens. It has been used many times, notably in [**KU2**], and obtained its final form in [**SU**]. In contrast to Martens's paper [**Mar**], where σ-compact metric spaces form the setting, the sufficient condition in [**SU**] is stated for abstract measure spaces, and the proof utilizes the concept of Banach limits rather than weak convergence. We start with the following uniqueness result.

Theorem 2.4.1 *Suppose that $T\colon (X, \mathfrak{F}, m) \longrightarrow (X, \mathfrak{F}, m)$ is a measurable map of a σ-finite measure space X. Suppose also that the measure m is quasi-invariant. Then, up to a positive multiplicative constant, there exists at most one nonzero σ-finite T-invariant measure μ absolutely continuous with respect to the measure m.*

Proof Suppose that μ_1 and μ_2 are σ-finite nonzero T-invariant measures absolutely continuous with respect to the measure m. Since m is ergodic and conservative, so are the measures μ_1 and μ_2. It now follows from Theorem 2.3.13 that if μ_1 and μ_2 do not coincide up to a positive multiplicative constant, then these two measures are mutually singular. But this means that there exists a measurable set $Y \subseteq X$ such that $\mu_1(Y) = 0$ and $\mu_2(X \backslash Y) = 0$. So,

$$\mu_1\left(\bigcup_{n=0}^{\infty} T^{-n}(Y)\right) \leq \sum_{n=0}^{\infty} \mu_1(T^{-n}(Y)) \leq \sum_{n=0}^{\infty} 0 = 0. \tag{2.35}$$

On the other hand, $\mu_2(Y) > 0$, so $m(Y) > 0$ and $m\big(X \backslash \bigcup_{n=0}^{\infty} T^{-n}(Y)\big) = 0$ by virtue of Theorem 2.1.10. Since μ_1 is absolutely continuous with respect to

m, this implies that $\mu_1\big(X\backslash\bigcup_{n=0}^{\infty}T^{-n}(Y)\big)=0$. Along with (2.35), this gives that $\mu_1(X)=0$. This contradiction finishes the proof. ■

We now introduce the concept of a Marco Martens map.

Definition 2.4.2 Let $T\colon (X,\mathfrak{F}) \longrightarrow (X,\mathfrak{F})$ be a measurable transformation. Also let m be a quasi-invariant probability measure on $(X,\mathfrak{F})$ with respect to T. The transformation T is called a Marco Martens map if it admits a countable family $\{X_n\}_{n=0}^{\infty}$ of subsets of X with the following properties.

(a) $X_n \in \mathfrak{F}$ for all $n \geq 0$.
(b) $m\big(X\backslash\bigcup_{n=0}^{\infty}X_n\big)=0$.
(c) For all $m,n \geq 0$, there exists $j \geq 0$ such that $m(X_m \cap T^{-j}(X_n)) > 0$.
(d) For all $j \geq 0$, there exists $K_j \geq 1$ such that, for all $A, B \in \mathfrak{F}$ with $A \cup B \subseteq X_j$ and for all $n \geq 0$,

$$m(T^{-n}(A))\,m(B) \leq K_j\,m(A)\,m(T^{-n}(B)).$$

(e) $\displaystyle\sum_{n=0}^{\infty} m(T^{-n}(X_0)) = +\infty.$
(f) $\displaystyle T\left(\bigcup_{j=l}^{\infty}Y_j\right) \in \mathfrak{F}$ for all $j \geq 0$, where $\displaystyle Y_j := X_j\Big\backslash\bigcup_{i<j}X_i.$
(g) $\displaystyle\lim_{l\to\infty} m\left(T\left(\bigcup_{j=l}^{\infty}Y_j\right)\right)=0.$

The family $\{X_n\}_{n=0}^{\infty}$ is called a Marco Martens cover.

Remark 2.4.3 Let us record the following observations.

(1) Of course, condition (b) follows from the stronger hypothesis that $\bigcup_{n=0}^{\infty}X_n = X$.
(2) Condition (c) implies that $m(X_n) > 0$ for all $n \geq 0$.
(3) In light of Corollary 2.1.9, if T is conservative with respect to μ, then condition (e) is fulfilled.
(4) In conditions (f–g), note that

$$\bigcup_{j=l}^{\infty}Y_j = \bigcup_{j=0}^{\infty}X_j\backslash\bigcup_{i<l}X_i \subseteq X\backslash\bigcup_{i<l}X_i.$$

(5) If the map $T: X \longrightarrow X$ is finite-to-one, then condition (g) is satisfied. For then,

$$\bigcap_{l=1}^{\infty} T\left(\bigcup_{j=l}^{\infty} Y_j\right) = \emptyset.$$

Let l^∞ denote the Banach space of all bounded real-valued sequences $x = (x_n)_{n=1}^\infty$ with norm

$$\|x\|_\infty := \sup\{|x_n|: n \geq 1\}.$$

Recall that a Banach limit is a shift-invariant positive bounded/continuous linear functional

$$l_B: l^\infty \longrightarrow \mathbb{R} \tag{2.36}$$

that extends the usual limits. More precisely, for all sequences

$$x = (x_n)_{n=1}^\infty,\ y = (y_n)_{n=1}^\infty \in l^\infty$$

and $\alpha, \beta \in \mathbb{R}$, the following properties hold.

(a) $l_B(\alpha x + \beta y) = \alpha\, l_B(x) + \beta\, l_B(y)$ (linearity).
(b) $\|l_B\| := \sup\{|l_B(x)|: \|x\|_\infty \leq 1\} < \infty$ (continuity/boundedness).
(c) If $x \geq 0$, i.e., $x_n \geq 0$ for all $n \in \mathbb{N}$, then $l_B(x) \geq 0$ (positivity).
(d) $l_B(\sigma(x)) = l_B(x)$, where $\sigma: l^\infty \to l^\infty$ is the (left) shift map defined by $(\sigma(x))_n = x_{n+1}$ for all $n \in \mathbb{N}$ (shift invariance).
(e) If x is a convergent sequence, then $l_B(x) = \lim_{n\to\infty} x_n$.

It follows from properties (a), (c), and (e) that a Banach limit also satisfies:

(f) $\liminf_{n\to\infty} x_n \leq l_B(x) \leq \limsup_{n\to\infty} x_n$.
(g) If $x \leq y$, i.e., $x_n \leq y_n$ for all $n \in \mathbb{N}$, then $l_B(x) \leq l_B(y)$.

Theorem 2.4.4 *Let $(X, \mathfrak{F}, m)$ be a probability space and $T: X \longrightarrow X$ a Marco Martens map with Marco Martens cover $\{X_j\}_{j=0}^\infty$. Then*

- *There exists a σ-finite T-invariant measure μ on X which is equivalent to m.*
- *In addition, $0 < \mu(X_j) < \infty$ for each $j \geq 0$.*

A measure μ with the above properties can be constructed as follows.

- *Let $l_B: l^\infty \longrightarrow \mathbb{R}$ be a Banach limit and let $Y_j := X_j \setminus \bigcup_{i<j} X_i$ for every $j \geq 0$.*

- *For each* $A \in \mathfrak{F}$, *set*

$$m_n(A) := \frac{\sum_{k=0}^{n} m(T^{-k}(A))}{\sum_{k=0}^{n} m(T^{-k}(X_0))}. \tag{2.37}$$

- *If* $A \in \mathcal{A}$ *and* $A \subseteq Y_j$ *for some* $j \geq 0$, *then* $(m_n(A))_{n=1}^{\infty} \in l^{\infty}$ *and set*

$$\mu(A) := l_B\big((m_n(A))_{n=1}^{\infty}\big). \tag{2.38}$$

- *For a general* $A \in \mathfrak{F}$, *set*

$$\mu(A) := \sum_{j=0}^{\infty} \mu(A \cap Y_j).$$

- *In addition, if* $(m_n(A))_{n=1}^{\infty} \in l^{\infty}$ *for some* $A \in \mathfrak{F}$, *then*

$$\mu(A) = l_B\big((m_n(A))_{n=1}^{\infty}\big) - \lim_{l\to\infty} l_B\left(\left(m_n\left(A \cap \bigcup_{j=l}^{\infty} Y_j\right)\right)_{n=0}^{\infty}\right). \tag{2.39}$$

- *In particular, if* $A \in \mathfrak{F}$ *is contained in a finite union of sets* X_j, $j \geq 0$, *then*

$$\mu(A) = l_B\big((m_n(A))_{n=1}^{\infty}\big). \tag{2.40}$$

Finally, if T *is ergodic and conservative with respect to* m, *then* μ *is also ergodic and conservative, and unique up to a positive multiplicative constant.*

In order to prove Theorem 2.4.4, we need several lemmas.

Lemma 2.4.5 *Let* $(Z, \mathcal{F})$ *be a measurable space such that:*

(a) $Z = \bigcup_{j=0}^{\infty} Z_j$ *for some mutually disjoint sets* $Z_j \in \mathcal{F}$; *and*

(b) ν_j *is a finite measure on* Z_j *for each* $j \geq 0$.

Then the set function $\nu \colon \mathcal{F} \longrightarrow [0, \infty]$, *defined by*

$$\nu(F) := \sum_{j=0}^{\infty} \nu_j(F \cap Z_j),$$

is a σ*-finite measure on* Z.

Proof Clearly, $\nu(\emptyset) = 0$. Let $F \in \mathcal{F}$ and $\{F_n\}_{n=1}^{\infty}$ be a partition of F into sets in $\mathcal{F}$. Then

$$\begin{aligned}\nu(F) &= \sum_{j=0}^{\infty} \nu_j(F \cap Z_j) = \sum_{j=0}^{\infty} \nu_j\left(\bigcup_{n=1}^{\infty}(F_n \cap Z_j)\right)\\ &= \sum_{j=0}^{\infty}\sum_{n=1}^{\infty} \nu_j(F_n \cap Z_j) = \sum_{n=1}^{\infty}\sum_{j=0}^{\infty} \nu_j(F_n \cap Z_j)\\ &= \sum_{n=1}^{\infty} \nu(F_n),\end{aligned}$$

where the order of summation could be changed since all terms involved are nonnegative. Thus, ν is a measure. Moreover, by definition,

$$Z = \bigcup_{j=0}^{\infty} Z_j$$

and $\nu(Z_j) = \nu_j(Z_j) < \infty$ for all $j \geq 0$. Therefore, ν is σ-finite. ■

From this point on, all lemmas rely on the same main hypotheses as Theorem 2.4.4.

Lemma 2.4.6 *For all $n, j \geq 0$ and all $A, B \in \mathfrak{F}$ with $A \cup B \subseteq X_j$, we have that*

$$m_n(A)\, m(B) \leq K_j\, m(A)\, m_n(B).$$

Proof This follows directly from the definition of m_n and condition (d) of Definition 2.4.2. ■

Lemma 2.4.7 *For every $j \geq 0$, we have that $(m_n(X_j))_{n=1}^{\infty} \in l^{\infty}$ and $\mu(Y_j) \leq \mu(X_j) < \infty$.*

Proof Fix $j \geq 0$. By virtue of condition (c) of Definition 2.4.2, there exists $q \geq 0$ such that $m\big(X_j \cap T^{-q}(X_0)\big) > 0$. By Lemma 2.4.6 and the definition of m_n, for all $n \geq 0$, we have that

$$\begin{aligned}m_n(Y_j) \leq m_n(X_j) &\leq K_j \frac{m(X_j)}{m\big(X_j \cap T^{-q}(X_0)\big)} m_n\big(X_j \cap T^{-q}(X_0)\big)\\ &\leq K_j \frac{m(X_j)}{m\big(X_j \cap T^{-q}(X_0)\big)} m_n(T^{-q}(X_0))\\ &= K_j \frac{m(X_j)}{m\big(X_j \cap T^{-q}(X_0)\big)} \frac{\sum_{k=0}^{n+q} m(T^{-k}(X_0))}{\sum_{k=0}^{n} m(T^{-k}(X_0))}\end{aligned}$$

$$= K_j \frac{m(X_j)}{m\big(X_j \cap T^{-q}(X_0)\big)} \left[1 + \frac{\sum_{k=n+1}^{n+q} m(T^{-k}(X_0))}{\sum_{k=0}^{n} m(T^{-k}(X_0))}\right]$$

$$\leq K_j \frac{m(X_j)}{m\big(X_j \cap T^{-q}(X_0)\big)} \left[1 + \frac{q}{m(X_0)}\right]. \qquad (2.41)$$

Consequently, $(m_n(X_j))_{n=1}^{\infty} \in l^{\infty}$ and properties (g) and (e) of a Banach limit yield that

$$\mu(Y_j) \leq K_j \frac{m(X_j)}{m\big(X_j \cap T^{-q}(X_0)\big)} \left[1 + \frac{q}{m(X_0)}\right] < +\infty.$$

Since $X_j = \bigcup_{i=0}^{j} Y_i$ and the Ys are mutually disjoint, we deduce that

$$\mu(Y_j) \leq \sum_{i=0}^{j} \mu(X_j \cap Y_i) = \sum_{i=0}^{\infty} \mu(X_j \cap Y_i) =: \mu(X_j) \leq \sum_{i=0}^{j} \mu(Y_i) < +\infty.$$

■

Now, for every $j \geq 0$, set $\mu_j := \mu|_{Y_j}$.

Lemma 2.4.8 *For every $j \geq 0$ such that $\mu(Y_j) > 0$ and for every measurable set $A \subseteq Y_j$, we have that*

$$K_j^{-1} \frac{\mu(Y_j)}{m(Y_j)} m(A) \leq \mu_j(A) \leq K_j \frac{\mu(Y_j)}{m(Y_j)} m(A).$$

Proof This follows from the definition of the measure μ and by setting $B = Y_j$ in Lemma 2.4.6 and using properties (a) and (g) of a Banach limit. ■

Lemma 2.4.9 *For each $j \geq 0$, μ_j is a finite measure on Y_j.*

Proof Let $j \geq 0$. Assume without loss of generality that $\mu_j(Y_j) > 0$. Let $A \subseteq Y_j$ be a measurable set and $(A_k)_{k=1}^{\infty}$ a countable measurable partition of A. Using termwise operations on sequences, for every $l \in \mathbb{N}$, we have that

$$\left(\sum_{k=1}^{\infty} m_n(A_k)\right)_{n=1}^{\infty} - \sum_{k=1}^{l} (m_n(A_k))_{n=1}^{\infty}$$

$$= \left(\sum_{k=1}^{\infty} m_n(A_k)\right)_{n=1}^{\infty} - \left(\sum_{k=1}^{l} m_n(A_k)\right)_{n=1}^{\infty}$$

$$= \left(\sum_{k=l+1}^{\infty} m_n(A_k)\right)_{n=1}^{\infty}.$$

It, therefore, follows from Lemma 2.4.6 (with $A = A_k$ and $B = Y_j$) that

$$\begin{aligned}
&\left\| \left(\sum_{k=1}^{\infty} m_n(A_k) \right)_{n=1}^{\infty} - \sum_{k=1}^{l} (m_n(A_k))_{n=1}^{\infty} \right\|_{\infty} \\
&= \left\| \left(\sum_{k=l+1}^{\infty} m_n(A_k) \right)_{n=1}^{\infty} \right\|_{\infty} \\
&\leq \left\| \frac{K_j}{m(Y_j)} \left(m_n(Y_j) \sum_{k=l+1}^{\infty} m(A_k) \right)_{n=1}^{\infty} \right\|_{\infty} \\
&= \frac{K_j}{m(Y_j)} \left\| \left(m_n(Y_j) \sum_{k=l+1}^{\infty} m(A_k) \right)_{n=1}^{\infty} \right\|_{\infty}.
\end{aligned}$$

Since $(m_n(Y_j))_{n=1}^{\infty} \in l^{\infty}$ by Lemma 2.4.7 and since $\lim_{l\to\infty} \sum_{k=l+1}^{\infty} m(A_k) = 0$, we conclude that

$$\lim_{l\to\infty} \left\| \left(\sum_{k=1}^{\infty} m_n(A_k) \right)_{n=1}^{\infty} - \sum_{k=1}^{l} (m_n(A_k))_{n=1}^{\infty} \right\|_{\infty} = 0.$$

This means that

$$\left(\sum_{k=1}^{\infty} m_n(A_k) \right)_{n=1}^{\infty} = \sum_{k=1}^{\infty} (m_n(A_k))_{n=1}^{\infty}$$

in l^{∞}. Hence, using the continuity of the Banach limit $l_B : l^{\infty} \longrightarrow \mathbb{R}$, we get

$$\begin{aligned}
\mu(A) = l_B\big((m_n(A))_{n=1}^{\infty}\big) &= l_B\left(\left(m_n\left(\bigcup_{k=1}^{\infty} A_k \right) \right)_{n=1}^{\infty} \right) \\
&= l_B\left(\left(\sum_{k=1}^{\infty} m_n(A_k) \right)_{n=1}^{\infty} \right) \\
&= \sum_{k=1}^{\infty} l_B\Big((m_n(A_k))_{n=1}^{\infty} \Big) \\
&= \sum_{k=1}^{\infty} \mu(A_k).
\end{aligned}$$

So, μ_j is countably additive. Also, $\mu_j(\emptyset) = 0$. Thus, μ_j is a measure. By Lemma 2.4.7, μ_j is finite. ■

Combining Lemmas 2.4.5, 2.4.7, 2.4.8, and 2.4.9, and condition (b) of Definition 2.4.2, we get the following.

Lemma 2.4.10 *We have that μ is a σ-finite measure on X equivalent to m. In addition,*

$$\mu(Y_j) \le \mu(X_j) < \infty \quad \text{and} \quad \mu(X_j) > 0$$

for all integers $j \ge 0$.

Lemma 2.4.11 *Formula* (2.39) *holds.*

Proof Fix $A \in \mathfrak{F}$ such that $(m_n(A))_{n=1}^{\infty} \in l^{\infty}$. Then, for every $l \in \mathbb{N}$, we have that

$$\begin{aligned} &l_B\big((m_n(A))_{n=1}^{\infty}\big)\\ &\quad = l_B\left(\sum_{j=0}^{l}(m_n(A\cap Y_j))_{n=1}^{\infty}\right) + l_B\left(\left(m_n\left(\bigcup_{j=l+1}^{\infty} A\cap Y_j\right)\right)_{n=1}^{\infty}\right)\\ &\quad = \sum_{j=0}^{l} l_B\big((m_n(A\cap Y_j))_{n=1}^{\infty}\big) + l_B\left(\left(m_n\left(A\cap \bigcup_{j=l+1}^{\infty} Y_j\right)\right)_{n=1}^{\infty}\right). \end{aligned}$$

Letting $l \to \infty$, we, thus, obtain that

$$\begin{aligned} &l_B\big((m_n(A))_{n=1}^{\infty}\big)\\ &\quad = \sum_{j=0}^{\infty} l_B\big((m_n(A\cap Y_j))_{n=1}^{\infty}\big) + \lim_{l\to\infty} l_B\left(\left(m_n\left(A\cap \bigcup_{j=l+1}^{\infty} Y_j\right)\right)_{n=1}^{\infty}\right)\\ &\quad = \sum_{j=0}^{\infty} \mu(A\cap Y_j) + \lim_{l\to\infty} l_B\left(\left(m_n\left(A\cap \bigcup_{j=l}^{\infty} Y_j\right)\right)_{n=1}^{\infty}\right)\\ &\quad = \mu(A) + \lim_{l\to\infty} l_B\left(\left(m_n\left(A\cap \bigcup_{j=l}^{\infty} Y_j\right)\right)_{n=1}^{\infty}\right). \end{aligned}$$

This establishes formula (2.39). In particular, if $A \subseteq \bigcup_{j=0}^{k} X_j$ for some $k \in \mathbb{N}$, then

$$A\cap \bigcup_{j=l}^{\infty} Y_j \subseteq \left(\bigcup_{j=0}^{k} X_j\right) \cap \left(X\backslash \bigcup_{i<l} X_i\right) = \emptyset$$

for all $l > k$. In that case, the equation above reduces to

$$l_B\big((m_n(A))_{n=1}^{\infty}\big) = \mu(A).$$

■

Lemma 2.4.12 *The σ-finite measure μ is T-invariant.*

Proof Let $i \geq 0$ be such that $m(Y_i) > 0$. Fix a measurable set $A \subset Y_i$. By definition,

$$\mu(A) = l_B\big((m_n(A))_{n=1}^{\infty}\big).$$

Furthermore, for all $n \geq 0$, notice that

$$\big|m_n(T^{-1}(A)) - m_n(A)\big| = \frac{\big|m\big(T^{-(n+1)}(A)\big) - m(A)\big|}{\sum_{k=0}^{n} m(T^{-k}(X_0))} \leq \frac{1}{\sum_{k=0}^{n} m(T^{-k}(X_0))}.$$

Thus, $(m_n(T^{-1}(A)))_{n=1}^{\infty} \in l^{\infty}$ because $(m_n(A))_{n=1}^{\infty} \in l^{\infty}$. Moreover, by condition (e) of Definition 2.4.2, it follows from the above and properties (a), (e), and (g) of a Banach limit that

$$l_B\big((m_n(T^{-1}(A)))_{n=1}^{\infty}\big) = l_B\big((m_n(A))_{n=1}^{\infty}\big) = \mu(A).$$

Keep A a measurable subset of Y_i. Fix $l \in \mathbb{N}$. We then have that

$$\begin{aligned}
m_n&\left(T^{-1}(A) \cap \bigcup_{j=l}^{\infty} Y_j\right) \\
&= \frac{\sum_{k=0}^{n} m\left(T^{-k}\big(T^{-1}(A) \cap \bigcup_{j=l}^{\infty} Y_j\big)\right)}{\sum_{k=0}^{n} m(T^{-k}(X_0))} \\
&\leq \frac{\sum_{k=0}^{n} m\left(T^{-(k+1)}\big(A \cap T\big(\bigcup_{j=l}^{\infty} Y_j\big)\big)\right)}{\sum_{k=0}^{n} m(T^{-k}(X_0))} \\
&\leq m_{n+1}\left(A \cap T\left(\bigcup_{j=l}^{\infty} Y_j\right)\right) \cdot \frac{\sum_{k=0}^{n+1} m(T^{-k}(X_0))}{\sum_{k=0}^{n} m(T^{-k}(X_0))} \\
&\leq K_i \frac{m_{n+1}(Y_i)}{m(Y_i)} \cdot m\left(A \cap T\left(\bigcup_{j=l}^{\infty} Y_j\right)\right) \cdot \frac{\sum_{k=0}^{n+1} m(T^{-k}(X_0))}{\sum_{k=0}^{n} m(T^{-k}(X_0))},
\end{aligned}$$

where the last inequality sign holds by Lemma 2.4.6 since $A \subseteq Y_i$. When $n \to \infty$, the last quotient on the right-hand side approaches 1. Therefore,

$$0 \leq l_B\left(\left(m_n\left(T^{-1}(A) \cap \bigcup_{j=l}^{\infty} Y_j\right)\right)_{n=1}^{\infty}\right) \leq K_i \frac{\mu(Y_i)}{m(Y_i)} m\left(T\left(\bigcup_{j=l}^{\infty} Y_j\right)\right).$$

Hence, by virtue of condition (g) of Definition 2.4.2,

$$
\begin{aligned}
0 &\le \lim_{l\to\infty} l_B\left(\left(m_n\left(T^{-1}(A)\cap\bigcup_{j=l}^{\infty}Y_j\right)\right)_{n=1}^{\infty}\right)\\
&\le K_i\frac{\mu(Y_i)}{m(Y_i)}\lim_{l\to\infty} m\left(T\left(\bigcup_{j=l}^{\infty}Y_j\right)\right)=0.
\end{aligned}
$$

So,

$$
\lim_{l\to\infty} l_B\left(\left(m_n\left(T^{-1}(A)\cap\bigcup_{j=l}^{\infty}Y_j\right)\right)_{n=1}^{\infty}\right)=0.
$$

It, thus, follows from Lemma 2.4.11 that

$$
\mu(T^{-1}(A)) = l_B\big((m_n(T^{-1}(A)))_{n=1}^{\infty}\big) = l_B\big((m_n(A))_{n=1}^{\infty}\big) = \mu(A).
$$

For an arbitrary $A \in \mathfrak{F}$, write $A = \bigcup_{j=0}^{\infty} A \cap Y_j$ and observe that

$$
\begin{aligned}
\mu(T^{-1}(A)) &= \mu\left(\bigcup_{j=0}^{\infty} T^{-1}(A\cap Y_j)\right) = \sum_{j=0}^{\infty}\mu\big(T^{-1}(A\cap Y_j)\big)\\
&= \sum_{j=0}^{\infty}\mu(A\cap Y_j) = \mu(A).
\end{aligned}
$$

We are done. ■

Proof of Theorem 2.4.4. Combining Lemmas 2.4.7, 2.4.10, 2.4.11, 2.4.12, and 2.4.1, we obtain the statement of Theorem 2.4.4. ■

Remark 2.4.13 In the course of the proof of Theorem 2.4.4, we have shown that

$$
0 < \inf\{m_n(A)\colon n \ge 1\} \le \sup\{m_n(A)\colon n \ge 1\} < +\infty
$$

for all $j \ge 0$ and all measurable sets $A \subseteq X_j$ such that $m(A) > 0$.

3

Probability (Finite) Invariant Measures: Basic Properties and Existence

3.1 Basic Properties of Probability Invariant Measures

Invoking Theorem 2.2.5, as an immediate consequence of Theorem 2.3.13, we get the following.

Theorem 3.1.1 *Suppose that $\mathfrak{F}$ is a σ-algebra on a set X, $T: X \longrightarrow X$ is a measurable map, and μ_1 and μ_2 are two T-invariant probability measures on $\mathfrak{F}$.*

If $T: X \longrightarrow X$ is ergodic with respect to both measures μ_1 and μ_2, then either μ_1 and μ_2 are mutually singular or they are equal.

Invoking Theorem 2.2.5 again, as an immediate consequence of Theorem 2.3.14, we get the following.

Theorem 3.1.2 *Suppose that $\mathfrak{F}$ is a σ-algebra on a set X and $T: X \longrightarrow X$ is a measurable map. If μ is a T-invariant probability measure on $\mathfrak{F}$, then μ is ergodic if and only if there is no T-invariant probability measure ν on $\mathfrak{F}$ such that*

$$\nu \prec\prec \mu \quad \text{and} \quad \nu \neq \mu.$$

Recall that, in a vector space, the extreme points of a convex set are those points which cannot be represented as a nontrivial convex combination of two distinct points of the set. More precisely, let V be a vector space and C be a convex subset of V. A vector $v \in C$ is called an extreme point of C if and only if the only combination of distinct vectors $v_1, v_2 \in C$ such that

$$v = \alpha v_1 + (1 - \alpha) v_2$$

for some $\alpha \in [0, 1]$ is a combination with $\alpha = 0$ or $\alpha = 1$.

If $T: (X, \mathfrak{F}) \longrightarrow (X, \mathfrak{F})$ is a measurable transformation, then $M(X, \mathfrak{F})$ is a convex subset of the vector space $SM(X, \mathfrak{F})$ of all signed measures on $(X, \mathfrak{F})$, and $M(T, \mathfrak{F})$ is a convex subset of $M(X, \mathfrak{F})$, thus a convex subset of the vector space $SM(X, \mathfrak{F})$.

Theorem 3.1.3 *Let $T: (X, \mathfrak{F}) \longrightarrow (X, \mathfrak{F})$ be a measurable transformation. The set $E(T, \mathfrak{F})$ of ergodic T-invariant measures of T coincides with the set of extreme points of the set of T-invariant probability measures $M(T, \mathfrak{F})$.*

Proof Suppose for a contradiction that some measure $\mu \in E(T, \mathfrak{F})$ is not an extreme point of $M(T, \mathfrak{F})$. Then there exist measures $\mu_1 \neq \mu_2$ in $M(T, \mathfrak{F})$ and $0 < \alpha < 1$ such that

$$\mu = \alpha\mu_1 + (1 - \alpha)\mu_2.$$

It follows immediately that

$$\mu_1 \ll \mu \quad \text{and} \quad \mu_2 \ll \mu.$$

By Theorem 3.1.2, we deduce from the ergodicity of μ that

$$\mu_1 = \mu = \mu_2.$$

This contradicts the fact that $\mu_1 \neq \mu_2$. Thus, μ is an extreme point of $M(T, \mathfrak{F})$.

In order to prove the converse implication, let $\mu \in M(T, \mathfrak{F}) \backslash E(T, \mathfrak{F})$. We want to show that μ is not an extreme point of $M(T, \mathfrak{F})$. Since μ is not ergodic, there exists a set $A \in \mathfrak{F}$ such that

$$T^{-1}(A) = A$$

with

$$\mu(A) > 0 \quad \text{and} \quad \mu(X \backslash A) > 0.$$

Observe that now μ can be written as the following nontrivial convex combination of the T-invariant conditional measures μ_A and $\mu_{X \backslash A}$. Indeed, for every $B \in \mathfrak{F}$, we have that

$$\begin{aligned}
\mu(B) &= \mu(A \cap B) + \mu((X \backslash A) \cap B) = \mu(A)\, \mu_A(B) + \mu(X \backslash A)\, \mu_{X \backslash A}(B) \\
&= \mu(A)\, \mu_A(B) + (1 - \mu(A))\, \mu_{X \backslash A}(B).
\end{aligned}$$

Thus,

$$\mu = \mu(A)\, \mu_A + (1 - \mu(A))\, \mu_{X \backslash A}.$$

So, μ is a nontrivial convex combination of two distinct T-invariant probability measures and, thus, is not an extreme point of $M(T, \mathfrak{F})$. ■

3.2 Existence of Borel Probability Invariant Measures: Bogolyubov–Krylov Theorem

In general, a given measurable transformation $T\colon (X,\mathfrak{F}) \longrightarrow (X,\mathfrak{F})$ may not admit any invariant measure. However, in one situation, quite common, although in general failing in this book, especially in Volume 2, namely, for X a compact metrizable space and the Borel σ-algebra $\mathcal{B} = \mathcal{B}_X$, on X, every continuous transformation does have a probability invariant measure. This is the celebrated Bogolyubov–Krylov Theorem. We call all continuous self-maps of compact metrizable spaces topological dynamical systems. Before proving the Bogolyubov–Krylov Theorem, we briefly examine the map

$$M(X) \ni \mu \longmapsto \mu \circ T^{-1} \ni M(X).$$

By the Riesz Representation Theorem, the set $M(X)$ can be identified with a subset of the normed space $C^*(X)$, the space of all bounded linear functionals on X. The set $M(X)$ is obviously convex. If $C^*(X)$ is endowed with the weak* topology, then by the Banach–Alaoglu Theorem, the set $M(X)$ becomes compact and metrizable. The following lemma will be helpful in proving the existence of invariant measures.

Lemma 3.2.1 *Let $T\colon X \longrightarrow X$ be a continuous transformation of a compact metrizable space X. Then the map $S\colon M(X) \longrightarrow M(X)$, given by the formula*

$$S(\mu) := \mu \circ T^{-1},$$

is a continuous affine map.

Proof The proof of affinity is left to the reader; here, we concentrate on the continuity of S. Let $(\mu_n)_{n=1}^{\infty}$ be a sequence in $M(X)$ which weak* converges to μ. Then, for any $f \in C(X)$, we have that

$$\begin{aligned}\lim_{n\to\infty}\int_X f\,d(S(\mu_n)) &= \lim_{n\to\infty}\int_X f\,d(\mu_n\circ T^{-1}) = \lim_{n\to\infty}\int_X f\circ T\,d\mu_n\\ &= \int_X f\circ T\,d\mu = \int_X f\,d(\mu\circ T^{-1})\\ &= \int_X f\,d(S(\mu)).\end{aligned}$$

Since f was chosen arbitrarily in $C(X)$, the sequence $(S(\mu_n))_{n=1}^{\infty}$ weak* converges to $S(\mu)$. Thus, S is continuous. ■

We come now to the main result of this section; namely, showing that a continuous map on a compact metric space admits at least one invariant measure. This theorem is not very difficult to prove, but it is obviously important.

For this reason, we provide two different proofs. The first one depends on some basic functional analysis, whereas the second one is rather more constructive.

Theorem 3.2.2 (Bogolyubov–Krylov Theorem) *Any topological dynamical system, i.e., any continuous transformation $T: X \longrightarrow X$ of a compact metrizable space X admits a T-invariant Borel probability measure.*

Proof We treat $M(X)$ as a subspace of the vector topological space $C^*(X)$ endowed with the weak* topology. By Lemma 3.2.1, we know that the map given by the formula

$$S(\mu) = \mu \circ T^{-1}$$

is a continuous affine map of the convex, compact set $M(X)$. Thus, by Schauder Fixed Point Theorem (see Theorem V.10.5 in [**DS**]) the map S has a fixed point. In other words, there exists $\mu \in M(X)$ such that $\mu \circ T^{-1} = \mu$. ■

An Alternative Proof. Suppose that μ_0 is an arbitrary Borel probability measure on X (e.g., a Dirac point mass supported at a point of X, defined in Example 3.3.4 below). Construct the sequence of Borel probability measures $(\mu_n)_{n=1}^{\infty}$, where

$$\mu_n = \frac{1}{n}\sum_{j=0}^{n-1} \mu_0 \circ T^{-j}.$$

The set of Borel probability measures on a compact metric space X being weak* compact, the sequence $(\mu_n)_{n=1}^{\infty}$ has at least one weak* limit point. Let μ_∞ be such a limit point. We claim that the measure μ_∞ is T-invariant. To show this, let $(\mu_{n_k})_{k=1}^{\infty}$ be a subsequence of the sequence $(\mu_n)_{n=1}^{\infty}$, which weak* converges to μ_∞. The weak* convergence of the subsequence means that, for any function $f \in C(X)$, we have that

$$\int_X f\, d\mu_\infty = \lim_{k\to\infty} \int_X f\, d\mu_{n_k}.$$

On the other hand,

$$\begin{aligned}
&\left| \int_X f \circ T\, d\mu_{n_k} - \int_X f\, d\mu_{n_k} \right| \\
&= \left| \frac{1}{n_k}\sum_{j=0}^{n_k-1} \int_X f \circ T\, d\big(\mu_0 \circ T^{-j}\big) - \frac{1}{n_k}\sum_{j=0}^{n_k-1} \int_X f\, d\big(\mu_0 \circ T^{-j}\big) \right|
\end{aligned}$$

$$
\begin{aligned}
&= \frac{1}{n_k}\left|\sum_{j=0}^{n_k-1}\left(\int_X f\circ T^{j+1}\,d\mu_0 - \int_X f\circ T^j\,d\mu_0\right)\right| \\
&= \frac{1}{n_k}\left|\int_X f\circ T^{n_k}\,d\mu_0 - \int_X f\,d\mu_0\right| \\
&\le \frac{2}{n_k}\|f\|_\infty.
\end{aligned}
$$

Recall that, since X is a compact metrizable space, the function $f \in C(X)$ is necessarily bounded. Therefore, passing to the limit as k tends to infinity, we obtain that

$$
\begin{aligned}
\left|\int_X f\circ T\,d\mu_\infty - \int_X f\,d\mu_\infty\right| &= \left|\lim_{k\to\infty}\int_X f\circ T\,d\mu_{n_k} - \lim_{k\to\infty}\int_X f\,d\mu_{n_k}\right| \\
&= \lim_{k\to\infty}\left|\int_X f\circ T\,d\mu_{n_k} - \int_X f\,d\mu_{n_k}\right| \\
&\le 2\|f\|_\infty \lim_{k\to\infty}\frac{1}{n_k} \\
&= 0.
\end{aligned}
$$

Thus,

$$
\int_X f\,d(\mu_\infty\circ T^{-1}) = \int_X f\circ T\,d\mu_\infty = \int_X f\,d\mu_\infty
$$

for all $f \in C(X)$; so, we obtain that

$$
\mu_\infty \circ T^{-1} = \mu_\infty. \qquad \blacksquare
$$

We now invoke the Krein–Milman Theorem, a classical theorem of functional analysis, to deduce from Bogolyubov–Krylov Theorem somewhat more; namely, that every topological dynamical system admits a Borel probability invariant ergodic measure.

Recall that the convex hull of a subset F, denoted by $co(F)$, of a vector space V is the set convex subset of V containing F. It is equal to the intersection of all convex subsets of V containing F and coincides with the set of all convex combinations of vectors from F.

The proof of the following theorem can be found as Theorem V.8.4 in Dunford and Schwartz's book [**DS**].

Theorem 3.2.3 (Krein–Milman Theorem) *If K is a compact subset of a locally convex topological vector space V and $E(K)$ is the set of its all extreme points, then*

$$
\overline{co}(E(K)) \supseteq K.
$$

Consequently,

$$\overline{co}(E(K)) = \overline{co}(K).$$

In particular, if K is convex then

$$\overline{co}(E(K)) = K.$$

We can now prove the following.

Theorem 3.2.4 *Any topological dynamical system, i.e., any continuous transformation $T\colon X \longrightarrow X$ of a compact metrizable space X admits at least one T-invariant Borel probability ergodic measure on X.*

Proof It follows from Theorem 3.2.2 and Theorem 3.2.3 that the set $E(T, \mathcal{B})$ of all ergodic T-invariant Borel probability measures on X is nonempty. ■

3.3 Examples of Invariant and Ergodic Measures

In order to prove that a given measure μ is invariant for some measurable dynamical system T, it is frequently useful to be able to reduce checking the equation $mu(T^{1}(A)) = \mu(A)$ for some smaller and convenient class of sets. One, especially important, class of such sets is given by the concept of π-systems. We define it now and prove its fundamental, and extremely useful, property below.

Definition 3.3.1 Let X be a set. A nonempty family $\mathcal{P} \subseteq \mathcal{P}(X)$ is a π-system on X if and only if

$$P_1 \cap P_2 \in \mathcal{P}$$

for all $P_1, P_2 \in \mathcal{P}$. In other words, a π-system is a collection that is closed under finite intersections.

The following proposition is classical; see, for example, Theorem 3.3 in [**Bil2**] for its proof.

Proposition 3.3.2 *Let X be a set and $\mathcal{P}$ be a π-system on X. Let μ and ν be probability measures on $(X, \sigma(\mathcal{P}))$. Then*

$$\mu = \nu \iff \mu(P) = \nu(P)\ \forall\, P \in \mathcal{P}.$$

As its immediate consequence, we get the following.

Proposition 3.3.3 *Let $T\colon (X, \mathfrak{F}, \mu) \longrightarrow (X, \mathfrak{F}, \mu)$ be a measurable endomorphism of a probability space $(X, \mathfrak{F}, \mu)$. If a π-system $\mathcal{P}$ generates the σ-algebra $\mathfrak{F}$, then*

$$\mu \circ T^{-1} = \mu \iff \mu \circ T^{-1}(P) = \mu(P) \, \forall \, P \in \mathcal{P}.$$

We now may deal with actual examples.

Example 3.3.4 Let X be a set and let $T: X \longrightarrow X$ be a map. Fix an arbitrary point $\xi \in X$. This is the Dirac point mass supported at ξ, also called the δ-Dirac measure supported at ξ. It is formally defined by the following formula:

$$\delta_\xi(A) := \begin{cases} 1 & \text{if } \xi \in A \\ 0 & \text{if } \xi \notin A \end{cases}$$

for every $A \subseteq X$.

Now assume that ξ is a fixed point of T, i.e., $T(\xi) = \xi$. Then the measure δ_ξ is T-invariant, i.e.,

$$\delta_\xi(T^{-1}(A)) = \delta_\xi(A)$$

for each $A \in \mathcal{A}$, since $\xi \in T^{-1}(A)$ if and only if $\xi \in A$. This example easily generalizes to invariant measures supported on periodic orbits. Namely, if $T^k(\xi) = \xi$ with some $k \geq 1$, then the measure

$$\frac{1}{k}\sum_{j=0}^{k-1} \delta_{T^j(\xi)}$$

is T-invariant. It is also obvious that this measure is ergodic.

Example 3.3.5 Let $S^1 := [0,1]/(0 = 1)$. Let $\alpha \in \mathbb{R}$ and define the map $T_\alpha : S^1 \longrightarrow S^1$ by

$$T_\alpha(x) = x + \alpha \ (\text{mod} 1).$$

Thus, T_α is the rotation of the unit circle by the angle 2α. The topological dynamics of T_α is radically different depending on whether the number α is rational or irrational. In particular, the ergodicity of T_α with respect to the Lebesgue measure λ will. However, it is fairly easy to see that T_α preserves Lebesgue measure λ, irrespective of the nature of α. Indeed,

$$T_\alpha^{-1}(x) = x - \alpha \ (\text{mod} 1),$$

and so $|\det DT_\alpha^{-1}(x)| = 1$ for all $x \in S^1$. Therefore,

$$\lambda(T_\alpha^{-1}(B)) = \int_B |\det DT_\alpha^{-1}(x)| \, d\lambda(x) = \int_B \mathbb{1} d\lambda = \lambda(B)$$

for all $B \in \mathcal{B}(S^1)$, i.e., T_α preserves λ. We shall prove the following.

Theorem 3.3.6 *If $T_\alpha : S^1 \longrightarrow S^1$ is the map of the circle defined as above, i.e.,*

$$T_\alpha(x := x + \alpha(\text{mod } 1),$$

then T_α is ergodic with respect to the Lebesgue measure λ if and only if $\alpha \in \mathbb{R}\backslash\mathbb{Q}$.

Proof First, assume that $\alpha \notin \mathbb{Q}$. We want to show that λ is ergodic with respect to T_α. For this, it is sufficient to show that if

$$f \circ T_\alpha = f$$

and $f \in L^2(\lambda)$, then f is λ-a.e. constant. Consider the Fourier series representation of f, which is given by

$$f(x) = \sum_{k\in\mathbb{Z}} a_k e^{2\pi ikx}.$$

Then

$$f \circ T_\alpha(x) = \sum_{k\in\mathbb{Z}} a_k e^{2\pi ik(x+\alpha)} = \sum_{k\in\mathbb{Z}} a_k e^{2\pi ik\alpha} e^{2\pi ikx}.$$

Since we assumed that $f \circ T_\alpha = f$, we deduce from the uniqueness of the Fourier series representation that

$$a_k e^{2\pi ik\alpha} = a_k$$

for all $k \in \mathbb{Z}$. Hence, for each k, we have that either

$$a_k = 0 \quad \text{or} \quad e^{2\pi ik\alpha} = 1.$$

The latter equality holds if and only if $k\alpha \in \mathbb{Z}$. Since $\alpha \notin \mathbb{Q}$, this occurs only when $k = 0$. Thus, $f(x) = a_0$ for λ-a.e. $x \in S^1$, i.e., f is λ-a.e. constant. This implies that λ is ergodic.

The converse implication is obvious. ■

Example 3.3.7 Fix $q \in \mathbb{N}$ and consider the map $T_q : S^1 \longrightarrow S^1$ defined by the formula

$$T_q(x) := qx \text{ (mod1)}.$$

We claim that λ is T-invariant. Let I be a proper subinterval of $\S^1$. Then $T_q^{-1}(I)$ consists of q mutually disjoint intervals (arcs) of length $\frac{1}{q}\lambda(I)$. Consequently,

$$\lambda(T_q^{-1}(I)) = q \cdot \frac{1}{q}\lambda(I) = \lambda(I).$$

Since the family of all proper subintervals of S^1 forms a π-system which generates the Borel σ-algebra $\mathcal{B}(S^1)$ and since T_q preserves the Lebesgue measure of all proper subintervals, Lemma 3.3.3 asserts that T_q preserves λ.

We shall prove the following.

Theorem 3.3.8 *For every integer $q \geq 2$, the map $T_q : S^1 \longrightarrow S^1$ is ergodic with respect to the Lebesgue measure λ.*

Proof It is possible to demonstrate the ergodicity of $T_q : S^1 \longrightarrow S^1$ with respect to λ in a way that is similar to what we did for T_α in Theorem 3.3.6. However, in this example, we will provide a different proof which is more geometric and whose idea is applicable to many more classes of examples, particularly to full Markov maps, which are dealt with in the next example.

Let $A \in \mathcal{B}(S^1)$ be a Borel set such that

$$T_q^{-1}(A) = A \quad \text{and} \quad \lambda(A) > 0.$$

The surjectivity of T_q then implies that $T_n(A) = A$. In order to establish ergodicity, we need to show that $\lambda(A) = 1$. By the Lebesgue Density Theorem, Theorem 1.3.7, we have for λ-almost every point of $x \in A$ that

$$\lim_{r \to 0} \frac{\lambda\big(A \cap B(x,r)\big)}{\lambda(B(x,r))} = 1.$$

Since $\lambda(A) > 0$, there is at least one which we denote by x. Set

$$r_k := 1/(2q^k).$$

Then T_q^k is injective on each open arc of length $2r_k$. So, the map $T_q^k|_{B(x,r_k)}$ is injective. Also,

$$T_q^k(B(x,r_k)) = S^1 \backslash \big\{T_q^k(x + r_k)\big\}.$$

Thus,

$$\lambda\big(T_n^k(B(x,r_k))\big) = 1.$$

Therefore,

$$\begin{aligned}\lambda(A) = \lambda\big(T_q^k(A)\big) \geq \frac{\lambda\big(T_q^k(A \cap B(x,r_k))\big)}{\lambda\big(T_q^k(B(x,r_k))\big)} &= \frac{q^k \lambda\big(A \cap B(x,r_k)\big)}{q^k \lambda\big(B(x,r_k)\big)} \\ &= \frac{\lambda\big(A \cap B(x,r_k)\big)}{q\lambda\big(B(x,r_k)\big)} \xrightarrow[k \to \infty]{} 1.\end{aligned}$$

Consequently, $\lambda(A) = 1$. This proves the ergodicity of λ. ∎

Example 3.3.9 The tent map $T: [0,1] \to [0,1]$ is defined by the formula

$$T(x) := \begin{cases} 2x & \text{if } x \in [0,1/2] \\ 2-2x & \text{if } x \in [1/2,1]. \end{cases}$$

The family of all intervals $\{[a,b),(a,b): 0 < a < b < 1\}$ forms a π-system that generates the Borel σ-algebra $\mathcal{B}([0,1])$. Since the preimage of any such interval consists of two disjoint subintervals (one on each side of the tent) of half the length of the original interval, one readily sees, by applying Proposition 3.3.3, that the Lebesgue measure on the interval $[0,1]$ is invariant under the tent map.

Example 3.3.10 In fact, the last two examples naturally generalize to a much larger family of maps. Let $T: [0,1] \longrightarrow [0,1]$ be a piecewise linear map of the unit interval that admits a "partition" $\mathcal{P} = \{p_j\}_{j=0}^q$, where $1 \le q < +\infty$ and $0 = p_0 < p_1 < \cdots < p_{q-1} < p_q = 1$, with the following properties.

(1)

$$[0,1] = I_1 \cup \cdots \cup I_q,$$

where $I_j = [p_{j-1}, p_j]$s are the successive intervals of monotonicity of T.

(2)

$$T(I_j) = [0,1]$$

for all $1 \le j \le q$.

(3) The map T is linear on I_j for all $1 \le j \le q$.

Such a map T will be called a *full Markov map*. We claim that such a T preserves the Lebesgue measure λ, i.e.,

$$\lambda \circ T^{-1} = \lambda.$$

Indeed, it is easy to see that the absolute value of the slope of the restriction of T to the interval I_j is $1/(p_j - p_{j-1})$. Therefore, the absolute value of the slope of the corresponding inverse branch of T is $p_j - p_{j-1}$. Let $J \subseteq (0,1)$ be any interval. Then

$$T^{-1}(J) = \bigcup_{j=1}^{q} T|_{I_j}^{-1}(J),$$

where $T|_{I_j}^{-1}(J)$ is a subinterval of $\mathrm{Int}(I_j)$ of length $(p_j - p_{j-1}) \cdot \lambda(I)$. Since

$$\mathrm{Int}(I_j) \cap \mathrm{Int}(I_k) = \emptyset$$

for all $1 \le j < k \le q$, it follows that

$$\begin{aligned}\lambda(T^{-1}(J)) &= \sum_{j=1}^{q} \lambda\big(T|_{I_j}^{-1}(J)\big) \\ &= \sum_{j=1}^{q} (p_j - p_{j-1}) \cdot \lambda(I) = (p_q - p_0)\lambda(J) = \lambda(J).\end{aligned}$$

In addition,

$$0 \le \lambda(T^{-1}(\{0\})) \le \lambda(\{p_j : 0 \le j \le k\}) = 0.$$

So,

$$\lambda(T^{-1}(\{0\})) = 0 = \lambda(\{0\}).$$

Similarly,

$$\lambda(T^{-1}(\{1\})) = 0 = \lambda(\{1\}).$$

It follows that $\lambda(T^{-1}(J)) = \lambda(J)$ for every interval $J \subseteq [0,1]$. Since the family of all intervals forms a π-system that generates $\mathcal{B}([0,1])$, Lemma 3.3.3 asserts that the Lebesgue measure is invariant under any full Markov map.

We shall prove the following.

Theorem 3.3.11 *Every full Markov map* $T\colon [0,1] \longrightarrow [0,1]$ *is ergodic with respect to Lebesgue measure* λ *on* $[0,1]$.

Proof As we did in the proof of Theorem 3.3.8, we would like to use the Lebesgue Density Theorem. However, now, for all $r > 0$ and all $k \in \mathbb{N}$, the restriction of T^k to the ball $B(p_j, r)$ is not one-to-one when $p_j \ne 0, 1$ is a point of continuity for a full Markov map T, e.g., the point $1/2$ for the tent map. Nevertheless, despite the potential lack of injectivity, the proof presented below is motivated by the one above using the Lebesgue Density Theorem.

For each $n \in \mathbb{N}$, let

$$\mathcal{P}_n := \big\{ I_j^{(n)} \mid 1 \le j \le q^n \big\}$$

be the partition of the interval $[0,1]$ into the successive intervals of monotonicity of T^n. In particular, $I_j^{(1)} = I_j$ for all $1 \le j \le q$. For each $1 \le j < q^n$, let $p_j^{(n)}$ be the unique point in $I_j^{(n)} \cap I_{j+1}^{(n)}$. For all $x \in [0,1] \backslash \{p_j^{(n)} : 1 \le j < q^n\}$, let $I^{(n)}(x)$ be the unique element of $\mathcal{P}_n$ containing x. We will need two claims. First:

Claim 1°. For every $n \in \mathbb{N}$, the map $T^n\colon [0,1] \longrightarrow [0,1]$ is a full Markov map under the partition $\mathcal{P}_n$. In addition, $\mathcal{P}_{n+1}$ is finer than $\mathcal{P}_n$.

Proof We will proceed by induction. Suppose that $T^n\colon [0,1] \to [0,1]$ is a full Markov map under the partition $\mathcal{P}_n$. It is obvious that T^{n+1} is piecewise linear. Fix $I_j^{(n)} \in \mathcal{P}_n$. For all $1 \le i \le q$, consider

$$I_{j,i}^{(n+1)} := T|_{I_i}^{-1}\left(I_j^{(n)}\right).$$

Define

$$\widetilde{\mathcal{P}}_{n+1} := \left\{I_{j,i}^{(n+1)} : 1 \le j \le q^n \text{ and } 1 \le i \le q\right\}.$$

Then

$$\begin{aligned}
\bigcup_{j=1}^{q^n}\bigcup_{i=1}^{q} I_{j,i}^{(n+1)} &= \bigcup_{j=1}^{q^n}\bigcup_{i=1}^{q} T|_{I_i}^{-1}\left(I_j^{(n)}\right) = \bigcup_{i=1}^{q}\bigcup_{j=1}^{q^n} T|_{I_i}^{-1}\left(I_j^{(n)}\right) \\
&= \bigcup_{i=1}^{q} T|_{I_i}^{-1}\left(\bigcup_{j=1}^{q^n} I_j^{(n)}\right) = \bigcup_{i=1}^{q} T|_{I_i}^{-1}([0,1]) \\
&= \bigcup_{i=1}^{q} I_i = [0,1].
\end{aligned}$$

Thus, $\widetilde{\mathcal{P}}_{n+1}$ is a cover of $[0,1]$. Obviously, the interiors of the intervals in $\widetilde{\mathcal{P}}_{n+1}$ are mutually disjoint and $\widetilde{\mathcal{P}}_{n+1}$ is finer than $\mathcal{P}_n$. For all $1 \le j \le q^n$ and all $1 \le i \le q$, we further have that

$$\begin{aligned}
T^{n+1}\left(I_{j,i}^{(n+1)}\right) &= T^n\left(T\left(I_{j,i}^{(n+1)}\right)\right) = T^n\left(T\left(T|_{I_i}^{-1}\left(I_j^{(n)}\right)\right)\right) \\
&= T^n\left(I_j^{(n)}\right) = [0,1].
\end{aligned}$$

So $\mathcal{P}_{n+1} = \widetilde{\mathcal{P}}_{n+1}$ and T^{n+1} is a full Markov map under the partition $\mathcal{P}_{n+1}$, which is finer than $\mathcal{P}_n$. This completes the inductive step. Since the claim clearly holds when $n = 1$, Claim 1° has been established for all $n \in \mathbb{N}$. ■

Claim 2°. If $A \in \mathcal{B}([0,1])$, then

$$\lim_{n\to\infty} \frac{\lambda\left(A \cap I^{(n)}(x)\right)}{\lambda\left(I^{(n)}(x)\right)} = 1 \text{ for } \lambda\text{-a.e. } x \in A.$$

Proof Let

$$m := \min\left\{|\text{slope}(T|_{I_i})| : 1 \le i \le q\right\} > 1.$$

Then

$$\operatorname{diam}(\mathcal{P}_n) := \sup\left\{\operatorname{diam}\left(I_j^{(n)}\right) : 1 \le j \le q^n\right\} \le m^{-n} \xrightarrow[n\to\infty]{} 0.$$

Therefore, the σ-algebra $\sigma\left(\bigcup_{n=1}^{\infty}\mathcal{P}_n\right)$ contains all open, relative to $[0,1]$, subintervals of $[0,1]$. Hence,

$$\sigma\left(\bigcup_{n=1}^{\infty}\mathcal{P}_n\right) = \mathcal{B}([0,1]).$$

Claim 2° now directly follows from this formula, the last assertion of Claim 1°, and Theorem 1.4.11. ■

Concluding the proof of Theorem 3.3.11, let A be a Borel subset of $[0,1]$ such that $T^{-1}(A) = A$ and $\lambda(A) > 0$. By the surjectivity of T, we know that $T(A) = A$. Fix any $x \in A$ satisfying Claim 2°. For each $n \in \mathbb{N}$, let m_n be the slope of $T^n|_{I^{(n)}(x)}$. Using both claims, we obtain that

$$\begin{aligned}
\lambda(A) = \lambda(T^n(A)) &\geq \frac{\lambda\big(T^n(A \cap I^{(n)}(x))\big)}{\lambda\big(T^n(I^{(n)}(x))\big)} \\
&= \frac{m_n\,\lambda\big(A \cap I^{(n)}(x)\big)}{m_n\,\lambda\big(I^{(n)}(x)\big)} = \frac{\lambda\big(A \cap I^{(n)}(x)\big)}{\lambda\big(I^{(n)}(x)\big)} \xrightarrow[n\to\infty]{} 1.
\end{aligned}$$

Consequently, $\lambda(A) = 1$. This proves the ergodicity of λ. ■

Example 3.3.12 Let $T\colon (X,\mathcal{A}) \to (X,\mathcal{A})$ and $S\colon (Y,\mathcal{B}) \to (Y,\mathcal{B})$ be measurable transformations for which there exists a measurable transformation $h\colon (X,\mathcal{A}) \to (Y,\mathcal{B})$ such that $h \circ T = S \circ h$. We will show that every T-invariant measure generates an S-invariant push down under h. Let μ be a T-invariant measure on $(X,\mathcal{A})$. Recall that the push down of μ under h is the measure $\mu \circ h^{-1}$ on $(Y,\mathcal{B})$. It follows from the T-invariance of μ that

$$\begin{aligned}
(\mu \circ h^{-1}) \circ S^{-1} &= \mu \circ (S \circ h)^{-1} = \mu \circ (h \circ T)^{-1} \\
&= (\mu \circ T^{-1}) \circ h^{-1} = \mu \circ h^{-1}.
\end{aligned}$$

That is, $\mu \circ h^{-1}$ is S-invariant.

Example 3.3.13 Let $(X,\mathcal{A},\mu)$ and $(Y,\mathcal{B},\nu)$ be two probability spaces, and let $T\colon X \longrightarrow X$ and $S\colon Y \longrightarrow Y$ be two measure-preserving dynamical systems. The Cartesian product of T and S is the map $T \times S\colon X \times Y \longrightarrow X \times Y$ defined by the formula:

$$(T \times S)(x,y) = (T(x), S(y)).$$

The direct product σ-algebra $\sigma(\mathcal{A} \times \mathcal{B})$ on $X \times Y$ is the σ-algebra generated by the π-system of measurable rectangles

$$\mathcal{A} \times \mathcal{B} := \{A \times B\colon A \in \mathcal{A},\ B \in \mathcal{B}\}.$$

The direct product measure, commonly defined in measure theory, $\mu \otimes \nu$ on $(X \times Y, \sigma(\mathcal{A} \times \mathcal{B}))$ is the measure uniquely determined by its values on its generating π-system:

$$(\mu \otimes \nu)(A \times B) := \mu(A)\nu(B).$$

We claim that the product map

$$T \times S\colon (X \times Y, \sigma(\mathcal{A} \times \mathcal{B}), \mu \otimes \nu) \to (X \times Y, \sigma(\mathcal{A} \times \mathcal{B}), \mu \times \nu)$$

is measure preserving, i.e.,

$$(\mu \otimes \nu) \circ (T \times S)^{-1}(F) = (\mu \otimes \nu)(F)$$

for every set $F \in \sigma(\mathcal{A} \times \mathcal{B})$.

Owing to Proposition 3.3.3, it suffices to show that

$$(\mu \otimes \nu) \circ (T \times S)^{-1}(A \times B) = (\mu \otimes \nu)(A \times B)$$

for all $A \times B \in \mathcal{A} \times \mathcal{B}$. And, indeed,

$$\begin{aligned}(\mu \otimes \nu) \circ (T \times S)^{-1}(A \times B) &= (\mu \otimes \nu)\big(T^{-1}(A) \times S^{-1}(B)\big)\\ &= \mu(T^{-1}(A))\, \nu(S^{-1}(B))\\ &= \mu(A)\nu(B) = (\mu \otimes \nu)(A \times B).\end{aligned}$$

Our final example pertains to the shift map introduced by formula (3.2) below. In the special case when the set E is countable, it will be treated at length in Section 11.1. In fact, we now will look at this map in a more general context.

Example 3.3.14 Let $(E, \mathfrak{F}, P)$ be a probability space. Consider the one-sided product set

$$E^{\mathbb{N}} := \prod_{k=1}^{\infty} E.$$

The product σ-algebra $\mathfrak{F}^{\infty}$ on E^{∞} is the σ-algebra generated by the π-system $\mathcal{S}$ of all (finite) cylinders (also called rectangles), i.e., the sets of the form

$$\prod_{k=1}^{n} E_k \times \prod_{l=n+1}^{\infty} E = \Big\{\omega = (\omega_j)_{j=1}^{\infty} \in E^N : \omega_k \in E_k \ \forall\, 1 \le k \le n\Big\},$$

where $n \in \mathbb{N}$ and $E_k \in \mathfrak{F}$ for all $1 \le k \le n$. Commonly in measure theory, the product measure μ_P on $\mathcal{F}_{\infty}$, frequently referred to as the Bernoulli measure generated by P, is the unique probability measure that gives a cylinder the value

$$\mu_P\left(\prod_{k=1}^{n} E_k \times \prod_{l=n+1}^{\infty} E\right) = \prod_{k=1}^{n} P(E_k). \tag{3.1}$$

Let $\sigma : E^{\mathbb{N}} \longrightarrow E^{\mathbb{N}}$ be the left shift map, which is defined by the formula

$$\sigma\left((\omega_n)_{n=1}^{\infty}\right) := (\omega_{n+1})_{n=1}^{\infty}. \tag{3.2}$$

As we have said above, in the special case when the set E is countable, it will be treated at length in Section 11.1. We shall prove that the product measure μ_P is σ-invariant.

Indeed, since the cylinder sets form a π-system which generates the product σ-algebra $\mathfrak{F}^{\infty}$, in light of Lemma 3.3.3, it is sufficient to show that

$$\mu_P(\sigma^{-1}(S)) = \mu_P(S)$$

for all cylinder sets $S \in \mathcal{S}$. And we have that

$$\begin{aligned}
\mu_P \circ \sigma^{-1}\left(\prod_{k=1}^{n} E_k \times \prod_{l=n+1}^{\infty} E\right) &= \mu_P\left(E \times \prod_{k=1}^{n} E_k \times \prod_{l=n+2}^{\infty} E\right) \\
&= P(E)\prod_{k=1}^{n} P(E_k) \\
&= \prod_{k=1}^{n} P(E_k) = \mu_P\left(\prod_{k=1}^{n} E_k \times \prod_{l=n+1}^{\infty} E\right).
\end{aligned}$$

This completes the proof that the product measure is shift invariant.

We shall prove the following.

Theorem 3.3.15 *If $(E, \mathfrak{F}, P)$ is a probability space, then the shift map $\sigma : E^{\mathbb{N}} \longrightarrow E^{\mathbb{N}}$ is ergodic with respect to the Bernoulli probability measure shift invariant measure μ_P generated by P.*

Proof Given an integer $k \geq 1$, let $\mathfrak{F}_k'$ be the σ-algebra on E^k generated by all cylinders of length k, i.e., by all sets of the form

$$\prod_{j=1}^{k} E_j,$$

where $E_j \in \mathfrak{F}$ for all $1 \le j \le k$. Furthermore, let $\mathfrak{F}_k$ be the sub-σ-algebra of $\mathfrak{F}^\infty$ generated by all sets of the form

$$F \times \prod_{l=k+1}^{\infty} E,$$

where $F \in \mathfrak{F}'$. Let $A \in \mathfrak{F}^\infty$ be a set such that

$$\sigma^{-1}(A) = A \quad \text{and} \quad \mu_P(A) > 0.$$

Our task is to show that $\mu_P(A) = 1$. By the Martingale Convergence Theorem for Conditional Expectations, i.e., by Theorem 1.4.10, we have that

$$E_{\mu_P}(1\!\!1_A|\mathfrak{F}_k)(x) \xrightarrow[k\to\infty]{} 1\!\!1_A(x)$$

for μ_P-a.e. $x \in E^{\mathbb{N}}$. For every $k \ge 1$, let

$$A_k := \left\{x \in E^{\mathbb{N}} : \left|E_{\mu_P}(1\!\!1_A|\mathfrak{F}_k)(x) - 1\right| \le \frac{1}{4}\right\} \in \mathfrak{F}_k.$$

So,

$$\forall\, \varepsilon \in (0, 1/8) \ \ \exists N_\varepsilon \quad \forall\, k \ge N_\varepsilon$$

$$\mu_P\left(\left\{x \in E^{\mathbb{N}} : \left|E_{\mu_P}(1\!\!1_A|\mathfrak{F}_k)(x) - 1\!\!1_A(x)\right| \ge \frac{1}{8}\right\}\right) \le \frac{\varepsilon}{2}.$$

Denote the latter set by B_k. If $x \in A\backslash A_k$, then

$$\left|E_{\mu_P}(1\!\!1_A|\mathfrak{F}_k)(x) - 1\!\!1_A(x)\right| > \frac{1}{4}.$$

Hence, $x \in B_k$. This means that $A\backslash A_k \subseteq B_k$. Therefore,

$$\mu_P(A\backslash A_k) \le \mu_P(B_k) \le \frac{\varepsilon}{2}. \tag{3.3}$$

If, on the other hand, $x \in A_k\backslash A$, then

$$\left|E_{\mu_P}(1\!\!1_A|\mathfrak{F}_k)(x) - 1\!\!1_A(x)\right| = \left|E_{\mu_P}(1\!\!1_A|\mathfrak{F}_k(x))\right| \ge \frac{3}{4} > \frac{1}{8}.$$

Hence, $x \in B_k$. This means that $A_k\backslash A \subseteq B_k$. Therefore,

$$\mu_P(A_k\backslash A) \le \mu_P(B_k) \le \frac{\varepsilon}{2}.$$

Together with (3.3), this gives that

$$\mu_P(A_k \triangle A) \le \varepsilon. \tag{3.4}$$

Now, if $n \geq k+1$, then, slightly abusing notation, we have that $\sigma^{-n}(A_k) = E^n \times A_k \in \mathfrak{F}_n$. So,

$$\mu_P(A_k \cap \sigma^{-n}(A_k)) = \mu_P(A_k) \cdot \mu_P(\sigma^{-n}(A_k)) = \mu_P^2(A_k). \tag{3.5}$$

Hence,

$$\begin{aligned}
|\mu_P^2(A) - \mu_P(A)| &\leq |\mu_P^2(A) - \mu_P^2(A_k)| + |\mu_P^2(A_k) - \mu_P(A)| \\
&\leq (\mu_P(A) + \mu_P(A_k))|\mu_P(A) - \mu_P(A_k)| \\
&\quad + \left|\mu_P\left(A_k \cap \sigma^{-n}(A)\right) - \mu(A)\right| \\
&\leq 2\mu_P(A_k \,\triangle\, A) + \mu_P(A \,\triangle\, (A_k \cap \sigma^{-n}(A_k))) \\
&\leq 2\mu_P(A_k \,\triangle\, A) + \mu_P(A \backslash A_k) \\
&\quad + \mu_P(A \backslash \sigma^{-n}(A)) + \mu_P\left((A_k \cap \sigma^{-n}(A_k)) \backslash A\right) \\
&= 2\mu_P(A_k \,\triangle\, A) + \mu_P(A \backslash A_k) \\
&\quad + \mu_P(\sigma^{-n}(A) \backslash \sigma^{-n}(A_k)) + \mu_P\left((A_k \cap \sigma^{-n}(A_k)) \backslash A\right) \\
&= 2\mu_P(A_k \,\triangle\, A) + \mu_P(A \backslash A_k) \\
&\quad + \mu_P\left(\sigma^{-n}(A) \backslash A_k\right) + \mu_P\left((A_k \cap \sigma^{-n}(A_k)) \backslash A\right) \\
&= 2\mu_P(A_k \,\triangle\, A) + \mu_P(A \backslash A_k) \\
&\quad + \mu_P(A \backslash A_k) + \mu_P\left((A_k \cap \sigma^{-n}(A_k))\right) \backslash A) \\
&\leq 4\mu_P(A_k \,\triangle\, A) + \mu_P(A_k \backslash A) \\
&\leq 5\mu_P(A_k \,\triangle\, A) = 5\varepsilon.
\end{aligned}$$

So, letting $\varepsilon \searrow 0$, we get that $\mu_P^2(A) = \mu_P(A)$. As $\mu_{\underline{p}}(A) > 0$, this yields $\mu_P(A) = 1$. The proof is complete. ■

Out of Example 3.3.7 and/or Example 3.3.14, we easily get the classical result of Borel that almost all Lebesgue numbers in the interval $[0, 1]$ are normal. This is done in the following example.

Example 3.3.16 Fix an integer $q \geq 2$. Set

$$E_q := \{0, 1, 2, \ldots, q-1\}$$

and consider the map $\pi : E_q^{\mathbb{N}} \longrightarrow [0, 1]$, called the coding map, defined by the formula

$$\pi(\omega) := \sum_{k=1}^{\infty} \frac{\omega_k}{q^k}.$$

Since, for each integer $k \geq 1$, the function $p_k \colon E_q^{\mathbb{N}} \longrightarrow [0,1]$, defined by the formula

$$p_k(\omega) := \frac{\omega_k}{q^k},$$

is continuous, since

$$\pi = \sum_{k=1}^{\infty} p_k,$$

and since this series is uniformly convergent, we conclude that the function $\pi \colon E_q^{\mathbb{N}} \longrightarrow [0,1]$ is continuous, thus Borel measurable. Let Z be the (countable) set of the endpoints of all elements of all partitions $\mathcal{P}_n$, $n \geq 1$, coming from Example 3.3.7. Then $E^{\mathbb{N}} \backslash \pi^{-1}(Z)$ consists of all sequences in $E^{\mathbb{N}}$ that have no tail consisting either of 0s only or of $(q-1)$s only, and for each $x \in [0,1] \backslash Z$ the set $\pi^{-1}(x)$ is a singleton. In particular, the map

$$\pi|_{E^{\mathbb{N}} \backslash \pi^{-1}(Z)} \colon E^{\mathbb{N}} \backslash \pi^{-1}(Z) \longrightarrow [0,1]$$

is one-to-one. Let now $F \in \bigcup_{n=1}^{\infty} \mathcal{P}_n$ be arbitrary. Then there exists an integer $n \geq 1$ such that $F \in \mathcal{P}_n$, meaning that

$$F = \left[\frac{i}{q^n}, \frac{i+1}{q^n}\right]$$

for some integer $i \in \{0, 1, 2, \ldots, q^n - 1\}$. Then, up to a countable set,

$$\pi^{-1}(F) = [\omega],$$

where $\omega \in E^n$ is such that $\pi(\omega 0^\infty) = i/q^n$. Hence,

$$\lambda(F) = \frac{1}{q^n} = \mu_{P_q}([\omega]) = \mu_{P_q} \circ \pi^{-1}(F),$$

where P_q is the probability measure on E_q given by the formula

$$P_q(B) := \frac{\#B}{\#E_q} = \frac{\#B}{q}.$$

Since $\bigcup_{n=1}^{\infty} \mathcal{P}_n$ is a π-system generating $\mathcal{B}([0,1])$, we conclude from Proposition 3.3.2 that

$$\mu_{P_q} \circ \pi^{-1} = \lambda. \tag{3.6}$$

Now fix $i \in \{0, 1, 2, \ldots, q-1\}$. Then Theorem 2.3.9 affirms that

$$\begin{aligned}\lim_{n\to\infty} \frac{1}{n} \#\{1 \leq j \leq n \colon \omega_j = i\} &= \lim_{n\to\infty} \frac{1}{n} \#\{0 \leq j \leq n-1 \colon \sigma^j(\omega) \in [i]\} \\ &= \mu_{P_q}([i]) = \frac{1}{q}\end{aligned}$$

for μ_{P_q}-a.e. $\omega \in E_q^{\mathbb{N}}$. Since the coding map is measure preserving, we deduce that λ-almost every number between 0 and 1 has a q-adic expansion whose digits are equal to i with an asymptotic frequency of $1/q$. Such numbers are normal to base q. Since the intersection of any countable family of sets of measure 1 on a probability space is a set of measure 1, we, thus, have the following main result of this example.

Theorem 3.3.17 *Lebesgue almost every number in* $[0,1]$ *is normal with respect to every base* $q \geq 2$.

Of course, this theorem has obvious refinements such as the appropriate value of asymptotic frequencies of fixed digits placed at fixed $(\mathrm{mod}\, q)$ positions. The ultimate generality is the Birkhoff Ergodic Theorem, i.e., Theorem 2.3.2.

Example 3.3.18 The map $G\colon [0,1] \to [0,1]$ defined by

$$G(x) := \begin{cases} 0 & \text{if} \quad x = 0 \\ \left\{\dfrac{1}{x}\right\} & \text{if} \quad x > 0, \end{cases}$$

where $\{x\}$ denotes the fractional part of a nonnegative number x, is called the *Gauss map*. This map does not preserve the Lebesgue measure λ on $[0,1]$. However, let μ_G be the Borel probability measure on $[0,1]$ defined by the formula

$$\mu_G(B) := \frac{1}{\log 2} \int_B \frac{1}{1+x}\, dx$$

for every Borel set $B \subseteq [0,1]$. This means that $\mu_G(B)$ is uniquely determined by the property that

$$\frac{\mathrm{d}\mu_G}{\mathrm{d}\lambda}(x) = \frac{1}{\log 2} \frac{1}{1+x}$$

for all $x \in [0,1]$. The measure μ_G is known as the *Gauss measure*.

Theorem 3.3.19 *The Borel probability measure* μ_G *(equivalent to the Lebesgue measure* λ*) on* $[0,1]$ *is* G*-invariant.*

Proof Let $[a,b] \subseteq (0,1)$ be arbitrary. Then

$$G^{-1}([a,b]) = \bigcup_{n=1}^{\infty} \left[\frac{1}{b+n}, \frac{1}{a+n}\right].$$

Hence,

$$
\begin{aligned}
\mu_G(G^{-1}([a,b])) &= \frac{1}{\log 2}\sum_{n=1}^{\infty}\int_{\frac{1}{b+n}}^{\frac{1}{a+n}} \frac{1}{1+x}\,dx \\
&= \frac{1}{\log 2}\sum_{n=1}^{\infty}\left[\log\left(1+\frac{1}{a+n}\right) - \log\left(1+\frac{1}{b+n}\right)\right] \\
&= \frac{1}{\log 2}\sum_{n=1}^{\infty}\big[\log(a+n+1) - \log(a+n) \\
&\qquad\qquad - \log(b+n+1) + \log(b+n)\big] \\
&= \frac{1}{\log 2}\big[\log(b+1) - \log(a+1)\big] \\
&= \frac{1}{\log 2}\int_a^b \frac{1}{1+x}\,dx \\
&= \mu_G([a,b]).
\end{aligned}
$$

So, by applying Proposition 3.3.3, we conclude that the Gauss measure μ_G is G-invariant under the Gauss map. ■

The Gauss measure μ_G is G also ergodic with respect to G. This can be proved directly by improving the reasoning from Theorem 3.3.11. The new difficulty is, however, now two-fold: although the map is "full" and "Markov," it is not linear and there are infinitely many pieces of monotonicity. Ergodicity also follows from Theorem 11.6.6. In fact, the Gauss map G and, more precisely, its inverse branches

$$
[0,1] \ni x \longmapsto g_n(x) := \frac{1}{x+n}, \quad n \in \mathbb{N},
$$

form a conformal iterated function system in the sense of Chapter 11, and all proved therein applies to the system $\{g_n\}_{n\in\mathbb{N}}$.

4

Probability (Finite) Invariant Measures: Finer Properties

In this chapter, we deal with stochastic laws for measurable endomorphisms preserving a finite measure that are finer than the mere Birkhoff Ergodic Theorem. Of course, in order to obtain them, we need more hypotheses. First, we must make it clear that, having such a measure, we can always normalize it, i.e., multiply it by a constant factor, to produce a probability measure. So, from now on throughout the entire chapter, any invariant measure is understood to be a probability one. In the first section of this chapter, we deal with the Law of the Iterated Logarithm, while the second section is devoted to describing the method of Young towers, which Young developed in [**LSY2**] and [**LSY3**]; see also [**Go1**] for further progress. With appropriate assumptions imposed on the first return time, her construction yields the exponential decay of correlations, the Central Limit Theorem, and the Law of the Iterated Logarithm follows too.

4.1 The Law of the Iterated Logarithm

We shall show in this section that, under mild hypotheses, if a first return map satisfies the Law of the Iterated Logarithm, then so does the original map. More precisely, let

$$T: X \longrightarrow X$$

be a measurable dynamical system preserving a probability measure μ on X. We say that a μ-integrable function

$$g: X \longrightarrow \mathbb{R}$$

satisfies the Law of the Iterated Logarithm if there exists a positive number A_g such that

$$\limsup_{n\to\infty} \frac{S_n g(x) - n \int g d\mu}{\sqrt{n \log \log n}} = A_g$$

for μ-a.e. $x \in X$. From now on, we assume without loss of generality that

$$\mu(g) = \int g d\mu = 0.$$

Keep a measurable set $A \subseteq X$ with $\mu(A) > 0$. Given a point $x \in A$, the sequence $(\tau_n(x))_{n=1}^{\infty}$ is then defined as follows:

$$\tau_1(x) := \tau_A(x) \quad \text{and} \quad \tau_n(x) = \tau_{n-1}(x) + \tau_A\big(T^{\tau_{n-1}(x)}(x)\big),\ n \geq 2,$$

where, we recall, $\tau_A(x) \geq 1$ is the first return time of x to A. We are working toward the following theorem. Its proof is taken from [**SkU1**] and [**SUZ**].

Theorem 4.1.1 *Let $T: X \longrightarrow X$ be a measurable dynamical system preserving a probability measure μ on X. Assume that the dynamical system (T, μ) is ergodic. Fix A, a measurable subset of X having a positive measure μ. Let $g: X \longrightarrow \mathbb{R}$ be a measurable function such that the function $g_A: A \longrightarrow \mathbb{R}$ (see* (2.13) *for its definition) satisfies the Law of the Iterated Logarithm with respect to the dynamical system (T_A, μ_A). If, in addition,*

$$\int_A |g_A|^{2+\gamma} d\mu < +\infty \tag{4.1}$$

with some $\gamma > 0$, then the function $g: X \longrightarrow \mathbb{R}$ satisfies the Law of the Iterated Logarithm with respect to the original dynamical system (T, μ) and $A_g = A_{g_A}$.

Proof Since the Law of the Iterated Logarithm holds for a point $x \in X$ if and only if it holds for $T(x)$, by virtue of ergodicity of T it suffices to prove our theorem for almost all points in A. By our assumptions,

$$\limsup_{j\to\infty} \frac{S_{\tau_j(x)} g(x)}{\sqrt{j \log \log j}} = A_{g_A}. \tag{4.2}$$

Now, for every $n \in \mathbb{N}$ and (almost) every $x \in A$, let $k = k(x, n)$ be the positive integer uniquely determined by the condition that

$$\tau_k(x) \leq n < \tau_{k+1}(x)$$

for μ_A-a.e. $x \in A$. Since, by Kac's Lemma,

$$\lim_{j\to\infty} \frac{\tau_j(x)}{j} = \int_X \tau_A d\mu = \int_A \tau_A d\mu = 1$$

for μ_A-a.e. $x \in A$, we, thus, have that

$$\lim_{n\to\infty} \frac{\tau_{k+1}(x)}{\tau_k(x)} = 1;$$

consequently,

$$\lim_{n\to\infty} \frac{n}{k} = 1.$$

Therefore, by also using (4.2), we get that

$$\limsup_{n\to\infty} \frac{S_{\tau_k(x)}g(x)}{\sqrt{n\log\log n}} = \limsup_{n\to\infty} \frac{S_{\tau_k(x)}g(x)}{\sqrt{k\log\log k}} = A_{g_A}. \tag{4.3}$$

Since

$$S_n g(x) = S_{\tau_k(x)}g(x) + S_{n-\tau_k(x)}g\big(T^{\tau_k(x)}(x)\big),$$

we have that

$$\frac{S_n g(x)}{\sqrt{n\log\log n}} = \frac{S_{\tau_k(x)}g(x)}{\sqrt{n\log\log n}} + \frac{S_{n-\tau_k(x)}g(x)}{\sqrt{n\log\log n}}.$$

Because of this and because of (4.3), we are only left to show that

$$\lim_{n\to\infty} \frac{S_{n-\tau_k(x)}g(x)}{\sqrt{n\log\log n}} = 0 \tag{4.4}$$

μ_A-a.e. on A. To do this, note first that

$$\frac{S_{\tau_{k+1}(x)-\tau_k(x)}|g|\big(T^{\tau_k}(x)\big)}{\sqrt{k\log\log k}} = \frac{|g_A|\big(T_A^k(x)\big)}{\sqrt{k\log\log k}}. \tag{4.5}$$

Take an arbitrary $\varepsilon > 0$. Since, by Tchebyschev's Inequality,

$$\begin{aligned} \mu\Big(\Big\{x \in A\colon |g_A|\big(T_A^k(x)\big) \geq \varepsilon\sqrt{k\log\log k}\Big\}\Big) &= \\ &= \mu\Big(\Big\{x \in A\colon |g_A|(x) \geq \varepsilon\sqrt{k\log\log k}\Big\}\Big) \\ &= \mu\Big(\Big\{x \in A\colon |g_A|^{2+\gamma}(x) \geq \varepsilon^{2+\gamma}(k\log\log k)^{1+\gamma/2}\Big\}\Big) \\ &\leq \frac{\int |g_A|^{2+\gamma}\,d\mu}{\varepsilon^{2+\gamma}(k\log\log k)^{1+\gamma/2}}, \end{aligned} \tag{4.6}$$

and using (4.1), we conclude that

$$\sum_{k=1}^{\infty} \mu\Big(\Big\{x \in A\colon |g_A|\big(T_A^k(x)\big) \geq \varepsilon\sqrt{k\log\log k}\Big\}\Big) < +\infty.$$

So, applying (4.5) and the Borel–Cantelli Lemma, (4.4) follows. We are done. ∎

4.2 Decay of Correlations and the Central Limit Theorems: Lai-Sang Young Towers

Let $T: X \longrightarrow X$ be a measurable dynamical system preserving a probability measure μ on X. We say that a μ-integrable function $g: X \longrightarrow \mathbb{R}$ satisfies the Central Limit Theorem if there exists $\sigma > 0$ such that

$$\frac{\sum_{j=0}^{n-1} g \circ T^j - n \int g d\mu}{\sqrt{n}} \underset{n\to\infty}{\longrightarrow} \mathcal{N}(0,\sigma)$$

in distribution. $\mathcal{N}(0,\sigma)$ is here the normal (Gaussian) distribution with 0 mean and variance σ. This precisely means that, for every $t \in \mathbb{R}$,

$$\mu\left(\left\{x \in X : \frac{\sum_{j=0}^{n-1} g \circ T^j(x) - n \int g d\mu}{\sqrt{n}} \leq t\right\}\right) \underset{n\to\infty}{\longrightarrow} \frac{1}{\sigma\sqrt{2\pi}} \int_{-\infty}^{t} \exp\left(-u^2/2\sigma^2\right) du.$$

Equivalently, for every Borel set $B \subseteq \mathbb{R}$ with $S(\partial B) = 0$ (we recall that S denotes here the Lebesgue measure on $\mathbb{R}$),

$$\mu\left(\left\{x \in X : \frac{\sum_{j=0}^{n-1} g \circ T^j(x) - n \int g d\mu}{\sqrt{n}} \in B\right\}\right) \underset{n\to\infty}{\longrightarrow} \frac{1}{\sigma\sqrt{2\pi}} \int_{B} \exp\left(-u^2/2\sigma^2\right) du.$$

Another important stochastic feature of a dynamical system is the rate of decay of the correlations it yields. Let ψ_1 and ψ_2 be real square μ-integrable functions on X. For every positive integer n, the *nth correlation* of the pair ψ_1, ψ_2 is the number

$$C_n(\psi_1, \psi_2) := \int_X \psi_1 \cdot (\psi_2 \circ f^n) d\mu - \int \psi_1 \, d\mu_\phi \int \psi_2 \, d\mu, \tag{4.7}$$

provided that the above integrals exist. Notice that, because of the T-invariance of μ, we can also write

$$C_n(\psi_1, \psi_2) = \int_X (\psi_1 - E\psi_1)\left((\psi_2 - E\psi_2) \circ T^n\right) d\mu,$$

where we put

$$E\psi := \int_X \psi \, d\mu.$$

We shall now describe a powerful tool, commonly referred to as the Young tower technique, which provides a way to prove the Central Limit Theorem and to estimate from above the decay of correlations in many sufficiently "regular" dynamical systems exhibiting some sufficient expanding or hyperbolic features. Let

- $(\Delta_0, \mathcal{B}_0, m)$ be a measure space with a finite measure m.
- $\mathcal{P}_0$ be a countable measurable partition of Δ_0.
- $T_0 \colon \Delta_0 \longrightarrow \Delta_0$ be a measurable map such that, for every $\Gamma \in \mathcal{P}_0$, the map

$$T_0 \colon \Gamma \longrightarrow \Delta_0$$

  is injective and $T_0(\Gamma)$ can be represented as a union of elements of $\mathcal{P}_0$.
- (Big Images Property)

$$\inf\{m(T_0(\Gamma)) \colon \Gamma \in \mathcal{P}_0\} > 0.$$

Furthermore,

- We assume that the partition $\mathcal{P}_0$ is generating, i.e., for every $x, y \in \Delta_0$ there exists $s \geq 0$ such that the points $T_0^s(x)$ and $T_0^s(y)$ are in different elements of the partition $\mathcal{P}_0$.
- We denote by $s = s(x, y)$ the smallest integer with this property and we call it a separation time for the pair x, y.
- We assume also that, for each $\Gamma \in \mathcal{P}_0$, the map $(T_0|_\Gamma)^{-1}$ is measurable and that $Jac_m(T_0)$, the Jacobian of T_0 with respect to the measure m, is well defined and positive a.e. in Γ.
- (Bounded Distortion Property) With some constants $0 < \beta < 1$ and $C > 0$,

$$\left| \frac{Jac_m T_0(x)}{Jac_m T_0(y)} - 1 \right| \leq C\beta^{s(x,y)} \tag{4.8}$$

  for all $\Gamma \in \mathcal{P}_0$ and all $x, y \in \Gamma$.
- We also have a function $R \colon \Delta_0 \longrightarrow \{1, 2, \dots\}$ (the first return time to Δ_0) which is constant on each element of the partition $\mathcal{P}_0$.
- Finally, let

$$\Delta := \{(z, n) \in \Delta_0 \times \mathbb{N} \cup \{0\} \colon 0 \leq n < R(z)\},$$

  where each point $z \in \Delta_0$ is identified with $(z, 0)$.

  The map $T : \Delta \longrightarrow \Delta$ is defined as follows:

$$T(z, n) := \begin{cases} (z, n+1) & \text{if } n+1 < R(z) \\ (T_0(z), 0) & \text{if } n+1 = R(z). \end{cases}$$

  We assume that the map $T \colon \Delta \longrightarrow \Delta$ is topologically mixing, meaning that, for all $A \in \mathcal{P}_0$, all $k \in \{0, 1, \dots, R(A) - 1\}$, all $B \in \mathcal{P}_0$, and all $l \in \{0, 1, \dots, R(B) - 1\}$, there exists an integer $N \geq 0$ such that

$$T^n(A \times \{k\}) \supseteq B \times \{l\}$$

  for all $n \geq N$.

The measure m is spread over the whole space Δ by putting

$$\tilde{m}_{|\Delta_0} := m \quad \text{and} \quad \tilde{m}_{|\Gamma\times\{j\}} := m_{|\Gamma} \circ \pi_j^{-1} \tag{4.9}$$

for all $\Gamma \in \mathcal{P}_0$, where $\pi_j(z,0) = (z,j)$. Thus, we have the following.

Observation 4.2.1 The measure $\tilde{m}$ is finite if and only if

$$\int_{\Delta_0} R dm < +\infty.$$

The separation time

$$s((x,k),(y,l))$$

between any two points (x,k) and (y,l) in Δ is defined to be $s(x,y)$ if $k = l$ and x, y are in the same set of the partition $\mathcal{P}$. Otherwise, we set $s(x,y) = 0$. We define the space

$$C_\beta(\Delta) := \big\{\varphi\colon \Delta \to \mathbb{R} : \exists\, C_\varphi > 0 \text{ such that} \\ |\varphi(x) - \varphi(y)| \le C_\varphi \beta^{s(x,y)} \ \forall\, x, y \in \Delta\big\}.$$

Note that, in particular,

$$|\varphi(x) - \varphi(y)| \le C_\varphi$$

for all $x, y \in \Delta$. In particular, all functions in $C_\beta(\Delta)$ are uniformly bounded.

We refer to the pentuple $\mathcal{Y} = (\Delta_0, m, T_0, \mathcal{P}_0, R)$ as a Young tower. The following basic result has been essentially proved in [**LSY3**]; see [**Go1**] for a refinement which we incorporated in our hypotheses.

Theorem 4.2.2 *If $\mathcal{Y} = (\Delta_0, m, T_0, \mathcal{P}_0, R)$ is a Young tower and $\int_{\Delta_0} R dm < +\infty$, then*

- *There exists a unique probability T-invariant measure ν, absolutely continuous with respect to $\tilde{m}$.*
- *The Radon–Nikodym derivative $d\nu/d\tilde{m}$ is bounded from below by a positive constant.*
- *The dynamical system (T,ν) is metrically exact, thus ergodic.*

In order to formulate the further results, we need one notion more. This is the concept of the first entrance time. In our context, this is the function $E\colon \Delta \longrightarrow \{0,1,2,\dots,\infty\}$, given by

$$E_T(x) := \min\big\{n \in \{0,1,2,\dots,\infty\} : \ T^n(z) \in \Delta_0\big\}. \tag{4.10}$$

Its distribution is determined by the distribution of $R\colon \Delta_0 \longrightarrow \{0,1,2,\dots\}$; for every integer $n \geq 1$,

$$\tilde{m}\big(\{x \in \Delta\colon E_T(x) > n\}\big) = \sum_{k=n+1}^{\infty} m\big(\{x \in \Delta_0\colon R(x) > k\}\big). \tag{4.11}$$

The following finer stochastic properties of the dynamical system (T,ν) have also been essentially proved in [**LSY3**]; see [**Go1**] for a refinement which we incorporated in our hypotheses.

Theorem 4.2.3 *Let $\mathcal{Y} = (\Delta_0, m, T_0, \mathcal{P}_0, R)$ be a Young tower. Then the following hold.*

1. *If $\tilde{m}\big(\{x \in \Delta\colon E_T(x) > n\}\big) \preceq \mathcal{O}(n^{-\alpha})$ with some $\alpha > 0$, then*

$$|C_n(\psi,g)| = \left|\int (\psi \circ T^n) g\,d\nu - \int \psi\,d\nu \int g\,d\nu\right| = \mathcal{O}(n^{-\alpha}) \tag{4.12}$$

   for all functions $\psi \in L^\infty$ and $g \in C_\beta(\Delta)$.

2. *If $\tilde{m}\big(\{x \in \Delta\colon E_T(x) > n\}\big) \preceq \mathcal{O}(\theta^n)$ (implied by $m\big(\{x \in \Delta_0\colon R(x) > n\}\big) \preceq \theta^n$) for some $0 < \theta < 1$, then there exists $0 < \tilde{\theta} < 1$ such that, for all functions $\psi \in L^\infty$ and $g \in C_\beta$, we have that*

$$|C_n(\psi,g)| = \left|\int (\psi \circ T^n) g\,d\nu - \int \psi\,d\nu \int g\,d\nu\right| = \mathcal{O}(\tilde{\theta}^n). \tag{4.13}$$

3. *If $\tilde{m}\big(\{x \in \Delta\colon E_T(x) > n\}\big) \preceq n^{-\alpha}$ with some $\alpha > 1$ (in particular, if $m(R > n) \preceq \theta^n$), then the Central Limit Theorem is satisfied for all functions $g \in C_\beta(\Delta)$ that are not cohomologous to a constant in $L^2(\nu)$. This means, we recall, that there exists $\sigma > 0$ such that*

$$\frac{\sum_{j=0}^{n-1} g \circ T^j - n\int g\,d\nu}{\sqrt{n}} \xrightarrow[n\to\infty]{} \mathcal{N}(0,\sigma)$$

   in distribution, where, as noted above, $\mathcal{N}(0,\sigma)$ is here the normal (Gaussian) distribution with 0 mean and variance σ.

We also have the following.

Theorem 4.2.4 *Let $\mathcal{Y} = (\Delta_0, m, T_0, \mathcal{P}_0, R)$ be a Young tower. If*

$$m\big(\{x \in \Delta_0\colon R(x) > n\}\big) \preceq n^{-\alpha}$$

with some $\alpha > 2$ (in particular, if $m(R > n) \preceq \theta^n$) with some $\theta \in (0,1)$, then the Law of the Iterated Logarithm holds for all functions $g \in C_\beta(\Delta)$ that *are not cohomologous to a constant in* $L^2(\nu)$. *This means, we recall, that there exists a constant* $A_g \in (0, +\infty)$ *such that*

$$\limsup_{n\to\infty} \frac{S_n g(x) - n\int g d\nu}{\sqrt{n \log\log n}} = A_g$$

for ν*-a.e.* $x \in \Delta$.

Proof Take $\gamma > 0$ so small that

$$2 + \gamma < \alpha.$$

Recalling that each function in $g \in C_\beta(\Delta)$ is uniformly bounded, we then get that

$$\begin{aligned}\int_{\Delta_0} |g_{\Delta_0}|^{2+\gamma}\, d\nu &\preceq \int_{\Delta_0} R^{2+\gamma}\, d\nu \preceq \int_{\Delta_0} R^{2+\gamma}\, dm = \int_0^{+\infty} m\big(R^{2+\gamma} > t\big)\, dt \\ &= \int_0^{+\infty} m\left(R > t^{\frac{1}{2+\gamma}}\right) dt \preceq \int_0^{+\infty} t^{-\frac{\alpha}{2+\gamma}}\, dt \\ &< +\infty.\end{aligned}$$

A direct application of Theorem 4.1.1 and stochastic law results in [**MU2**] (for countable alphabet subshifts of finite type) then completes the proof. ■

The status of this theorem is, however, somewhat different than that of Theorem 4.2.3. It was not proved in [**LSY2**] or [**LSY3**] but was concluded in [**SUZ**] from Theorem 4.1.1 and the stochastic law results in [**MU2**] exactly as we did above. It also follows from [**MN**] and [**Go2**], where the Invariance Principle Almost Surely was established for Young towers, from which the Law of the Iterated Logarithm follows in a well-known standard way.

4.3 Rokhlin Natural Extension

The main theorem of this section, Theorem 4.3.6, permits one to sometimes replace an endomorphism $T: X \to X$, preserving some probability measure that is not necessarily invertible with the (invertible) automorphism $\widetilde{\varphi}: \widetilde{X} \to \widetilde{X}$. This may turn out to be of great advantage in some proofs, since dealing with invertible maps is in many cases easier than dealing with noninvertible ones. Natural extensions share many properties with the original maps. These can be defined in the context of Lebesgue spaces. Lebesgue spaces form the core of ergodic theory and are central in the descriptive set theory. The following theorem charactrizes them.

Theorem 4.3.1 *If (X_1,ρ_1) and (X_2,ρ_2) are two Polish (separable, completely metrizable) topological spaces and μ_1 and μ_2 are two atomless Borel probability measures, respectively, on X_1 and X_2, then the measure spaces $(X_1,\mathcal{B}_{X_1},\mu_1)$ and $(X_2,\mathcal{B}_{X_2},\mu_2)$ are isomorphic. Also, their completions $(X_1,\tilde{\mathcal{B}}_{X_1},\tilde{\mu}_1)$ and $(X_2,\tilde{\mathcal{B}}_{X_2},\tilde{\mu}_2)$ are isomorphic.*

This theorem leads us to the following.

Definition 4.3.2 Any complete probability space which is isomorphic to a convex combination of a (complete) measure space appearing in Theorem 4.3.1 and a collection of (countably many) atoms is called a Lebesgue space. In other words, a Lebesgue space is any probability space which is isomorphic to a convex combination of Lebesgue measures on $[0,1]$ and a collection of (countably many) Dirac δ measures supported at atoms in $[0,1]$.

Let $(X,\mathcal{B},\mu)$ be a Lebesgue space. In addition, let $T\colon X \longrightarrow X$ be a surjective measurable map which preserves the measure μ. We will shortly define a new map related to T, but first we form the space

$$\widetilde{X} := \left\{(x_n)_{n=0}^{\infty} \in X^{\mathbb{N}_0} : T(x_{n+1}) = x_n \text{ for all } n \geq 0\right\} \subset X^{\mathbb{N}_0}.$$

Then, for every $k \geq 0$, let $\pi_k\colon \widetilde{X} \longrightarrow X$ denote the projection onto the kth coordinate of $\widetilde{X}$, i.e.,

$$\pi_k\left((x_n)_{n=0}^{\infty}\right) := x_k.$$

Finally, endow the space $\widetilde{X}$ with the smallest σ-algebra $\widetilde{\mathcal{B}}$ for which every projection $\pi_k\colon \widetilde{X} \to X$ is measurable.

Definition 4.3.3 Define the measurable map $\widetilde{T}\colon \widetilde{X} \longrightarrow \widetilde{X}$ by setting

$$\widetilde{T}\left((x_n)_{n=0}^{\infty}\right) := (T(x_0), x_0, x_1, x_2, \ldots).$$

We refer to the map $\widetilde{T}$ as the *Rokhlin natural extension of* T.

The Rokhlin natural extension has the following properties.

Theorem 4.3.4 *If $(X,\mathcal{B},\mu)$ is a Lebesgue space and $T\colon X \longrightarrow X$ is a surjective measurable map which preserves the measure μ, then the map $\widetilde{T}\colon \widetilde{X} \longrightarrow \widetilde{X}$ is invertible and its inverse $\widetilde{T}^{-1}\colon \widetilde{X} \longrightarrow \widetilde{X}$ is given by the formula*

$$\widetilde{T}^{-1}\left((x_n)_{n=0}^{\infty}\right) := (x_{n+1})_{n=0}^{\infty}.$$

Furthermore, for each $n \geq 0$, the following diagram commutes:

$$\begin{array}{ccc} \widetilde{X} & \xrightarrow{\widetilde{T}} & \widetilde{X} \\ {\scriptstyle \pi_n}\downarrow & & \downarrow{\scriptstyle \pi_n} \\ X & \xrightarrow{T} & X \end{array}$$

This theorem is obvious. After passing to completed σ-algebras, we deduce the following result.

Theorem 4.3.5 *If $(X, \mathcal{B}, \mu)$ is a Lebesgue space and $T: X \longrightarrow X$ is a surjective measurable map which preserves the measure μ, then there exists a unique probability measure $\tilde{\mu}$ on the space $(\widetilde{X}, \widetilde{\mathcal{B}})$ such that*

$$\tilde{\mu} \circ \pi_n^{-1} = \mu, \ \textit{for every } n \geq 0.$$

Proof This follows directly from the Daniel–Kolmogorov Consistency Theorem (see Theorem 3.6.4 in [**Par**]). ■

In light of Theorem 4.3.5, for every set $A \in \mathcal{B}(X)$ and every $n \geq 0$, we have that

$$\begin{aligned} \tilde{\mu} \circ \widetilde{T}^{-1}(\pi_n^{-1}(A)) &= \tilde{\mu} \circ (\pi_n \circ \widetilde{T})^{-1}(A) = \tilde{\mu} \circ (T \circ \pi_n)^{-1} \\ &= \tilde{\mu} \circ \pi_n^{-1} \circ T^{-1}(A) \\ &= \mu \circ T^{-1}(A) = \mu(A) \\ &= \tilde{\mu}(\pi_n^{-1}(A)). \end{aligned}$$

It, therefore, follows from the definition of $\widetilde{\mathcal{B}}$ that $\tilde{\mu} \circ \widetilde{\varphi}^{-1} = \tilde{\mu}$. Considering this along with Theorem 4.3.4 and Theorem 4.3.5, we obtain the following result.

Theorem 4.3.6 *Let $(X, \mathcal{B}, \mu)$ be a Lebesgue space and let $T: X \longrightarrow X$ be a surjective measurable map which preserves the measure μ. If $\widetilde{T}: \widetilde{X} \longrightarrow \widetilde{X}$ is the Rokhlin natural extension of $T: X \to X$, then we obtain the following.*

(a) *For every $n \geq 0$, the following diagram commutes:*

$$\begin{array}{ccc} \widetilde{X} & \xrightarrow{\widetilde{T}} & \widetilde{X} \\ {\scriptstyle \pi_n}\downarrow & & \downarrow{\scriptstyle \pi_n} \\ X & \xrightarrow{T} & X \end{array}$$

(b) *There exists a unique measure $\tilde{\mu}$ on the space $(\widetilde{X}, \widetilde{F})$ such that*

$$\tilde{\mu} \circ \pi_n^{-1} = \mu, \ \textit{for all } n \geq 0.$$

(c) *We have that $\tilde{\mu} \circ \widetilde{T}^{-1} = \tilde{\mu}$, i.e., the probability measure $\tilde{\mu}$ is $\widetilde{T}$-invariant.*

Note that the surjectivity assumption is, in fact, not essential, since the sets $T^n(X)$, for each $n \geq 0$, are measurable (as X is a Lebesgue space), $\mu\left(\bigcap_{n=0}^{\infty} T^n(X)\right) = 1$, and the map

$$T\colon \bigcap_{n=0}^{\infty} T^n(X) \longrightarrow \bigcap_{n=0}^{\infty} T^n(X)$$

is clearly surjective.

This theorem permits one to sometimes replace the endomorphism T, which is not necessarily invertible, with the automorphism $\widetilde{\varphi}\colon \widetilde{X} \to \widetilde{X}$. This may turn out to be of great advantage in some proofs, since dealing with invertible maps is frequently easier than dealing with noninvertible ones. Natural extensions share many properties with the original maps, e.g., ergodicity, as is shown in the next theorem.

Theorem 4.3.7 *If $(X,\mathcal{B},\mu)$ is a Lebesgue space and $T\colon X \longrightarrow X$ is a measurable map preserving the measure μ, then the natural extension measure $\tilde{\mu}$ on $\widetilde{X}$ is ergodic with respect to $\widetilde{T}$ if and only if the measure μ is ergodic with respect to the map $T\colon X \longrightarrow X$.*

Proof Suppose first that $T\colon X \to X$ is not ergodic. Then there exists a set $A \in \mathfrak{F}$ such that $T^{-1}(A) = A$ and $0 < \mu(A) < 1$. It then follows from Theorem 4.3.6(b) that $\tilde{\mu}(\pi_0^{-1}(A)) = \mu(A) \in (0,1)$. Further, from Theorem 4.3.6(a), applied with $n = 0$, it follows that

$$\widetilde{T}^{-1}(\pi_0^{-1}(A)) = (\pi_0 \circ \widetilde{T})^{-1}(A) = (T \circ \pi_0)^{-1}(A) = \pi_0^{-1}(T^{-1}(A)) = \pi_0^{-1}(A).$$

Therefore, $\widetilde{\varphi}$ is not ergodic either.

Now assume that $T\colon X \to X$ is ergodic. The Birkhoff Ergodic Theorem then yields that

$$\lim_{n\to\infty} \frac{1}{n} \sum_{j=0}^{n-1} g \circ T^j = \int_X g \, d\mu$$

for every function $g \in L^1(X,\mathcal{B},\mu)$ and the convergence is understood in the Banach space $L^1(X,\mathcal{B},\mu)$. Invoking Theorem 4.3.6, this implies that

$$\lim_{n\to\infty} \frac{1}{n} \sum_{j=0}^{n-1} G \circ \widetilde{T}^j = \int_X G \, d\tilde{\mu}, \tag{4.14}$$

where $G\big((x_n)_{n=0}^{\infty}\big) := g(x_0) = g \circ \pi_0\big((x_n)_{n=0}^{\infty}\big)$ and the convergence is understood in the Banach space $L^1(\widetilde{X},\widetilde{\mathcal{B}},\tilde{\mu})$. Now, for every $n \geq 0$, let $\widetilde{\mathcal{B}}_n := \pi_n^{-1}(\mathcal{B})$. Since $T \circ \pi_{n+1} = \pi_n$, we obtain that

$$\widetilde{\mathcal{B}}_{n+1} = \pi_{n+1}^{-1}(\mathcal{B}) \supseteq \pi_{n+1}^{-1}(T^{-1}(\mathcal{B})) = (T \circ \pi_{n+1})^{-1}(\mathcal{B}) = \widetilde{\mathcal{B}}_n.$$

Moreover, by definition, $\widetilde{\mathcal{B}}$ is the σ-algebra generated by the sequence $(\widetilde{\mathfrak{F}}_n)_{n=0}^{\infty}$. It, therefore, follows from the Martingale Convergence Theorem that, for any $F \in L^1(\widetilde{X}, \widetilde{\mathcal{B}}, \tilde{\mu})$,

$$F = \lim_{n\to\infty} E_{\tilde{\mu}}(F|\widetilde{\mathcal{B}}_n), \tag{4.15}$$

where the convergence is again understood in the Banach space $L^1(\widetilde{X}, \widetilde{\mathcal{B}}, \tilde{\mu})$. However, for each $n \geq 0$, the function $E_{\tilde{\mu}}(F|\widetilde{\mathcal{B}}_n)$ depends only upon the nth coordinate of a point in $\widetilde{X}$ and can, therefore, be represented as $f_n \circ \pi_n$, where $f_n \in L^1(X, \mathcal{B}, \mu)$. Now, for any $j \geq 0$, we deduce that

$$f_n \circ \pi_n \circ \widetilde{T}^{j+n} = f_n \circ (\pi_n \circ \widetilde{T}^n) \circ \widetilde{T}^j = f_n \circ \pi_0 \circ \widetilde{T}^j;$$

it, therefore, follows from (4.14) that

$$\lim_{k\to\infty} \frac{1}{k} \sum_{j=0}^{k-1} E_{\tilde{\mu}}(F|\widetilde{\mathcal{B}}_n) \circ \widetilde{T}^j = \int_X E_{\tilde{\mu}}(F|\widetilde{\mathcal{B}}_n) d\tilde{\mu} = \int_X F\, d\tilde{\mu},$$

where the convergence is understood in the Banach space $L^1(\widetilde{X}, \widetilde{\mathcal{B}}, \tilde{\mu})$. Combining this with (4.15), we conclude that

$$\lim_{k\to\infty} \frac{1}{k} \sum_{j=0}^{k-1} F \circ \widetilde{T}^j = \int_X F\, d\tilde{\mu},$$

where once more the convergence is understood in the Banach space $L^1(\widetilde{X}, \widetilde{\mathcal{B}}, \tilde{\mu})$. Therefore, the transformation $\widetilde{\varphi}\colon \widetilde{X} \longrightarrow \widetilde{X}$ is ergodic and we are done. ■

5

Infinite Invariant Measures: Finer Properties

In this chapter, we deal with invariant measures that are infinite. The concepts introduced and explored here are meaningful only for such measures. Even if they make sense for finite measures, they are trivial then.

5.1 Counterexamples to Ergodic Theorems

The Hopf Ergodic Theorem (Theorem 2.3.12), being powerful and interesting in itself, somewhat surprisingly rules out any hope for a more direct version of the Birkhoff Ergodic Theorem in the case of infinite measures. The world of infinite invariant measures is indeed very different than the one of finite measures. The following proposition provides the first indication of this.

Proposition 5.1.1 *Suppose that $(X, \mathfrak{F}, \mu)$ is a σ-finite measure space and that $T: X \longrightarrow X$ is a measurable map preserving the measure μ. If $\mu(X) = +\infty$, then, for every function $f \in L^1(\mu)$, we have that*

$$\lim_{n\to\infty} \frac{1}{n} S_n f = 0 \quad \mu\text{-}a.e.$$

Proof Since the measure μ is σ-finite and $\mu(X) = \infty$, there exists a sequence $(A_n)_{n=1}^{\infty}$ of measurable sets such that $0 < \mu(A_n) < +\infty$ for all $n \geq 1$ and such that

$$\lim_{n\to\infty} \mu(A_n) = \infty. \tag{5.1}$$

Since $|f| \in L^1(\mu)$, we conclude from Theorem 2.3.12 that, for every $k \geq 1$, we have that

$$\varlimsup_{n\to\infty} \frac{1}{n} S_n|f| \leq \varlimsup_{n\to\infty} \frac{S_n|f|}{S_n 1\!\!1_{A_k}} = \frac{\int |f| d\mu}{\int 1\!\!1_{A_k} d\mu} = \frac{\int |f| \, d\mu}{\mu(A_k)} \qquad \mu\text{-}a.e.$$

So, by 5.1, $\limsup_{n\to\infty} S_n|f| = 0$ a.e. Since $\left|\frac{1}{n}S_n f\right| \le \frac{1}{n}S_n|f|$, we are done. ■

As a matter of fact, the last two results of this section show that, no matter how much it is desired, the Hopf Ergodic Theorem, i.e., Theorem 2.3.12, rules out any hope of even much weaker forms of Theorem 2.3.2 in the case of infinite invariant measures. Indeed, for all $f \in L^1(\mu)$, Proposition 5.1.1 (a corollary of the Hopf Ergodic Theorem) implies that

$$\lim_{n\to\infty} \frac{1}{a_n} S_n f(x) = 0 \text{ for } \mu\text{-a.e. } x \in X$$

if there exists $C > 0$ such that $a_n \ge Cn$ for all n. We will now show that there are no constants $a_n > 0$ such that

$$\lim_{n\to\infty} \frac{1}{a_n} S_n f(x) = \int_X f\, d\mu \text{ for } \mu\text{-a.e. } x \in X \quad \forall f \in L^1(\mu).$$

We will accomplish this in two steps. Their statements and proofs are inspired by Aaronson's book [**Aa**]. The first step will require the following proposition. Since we will simultaneously deal with Birkhoff's sum with respect to the maps $T: X \longrightarrow X$ and the first return maps $\tau_F: F \longrightarrow F$, in order to discern between them we put

$$S_{F,n}(g) := \sum_{j=0}^{n-1} g \circ T_F^j, \tag{5.2}$$

where $g: F \longrightarrow \mathbb{R}$ is a measurable function.

Proposition 5.1.2 *Suppose that $(X, \mathfrak{F}, \mu)$ is a probability space, $T: X \longrightarrow X$ is an ergodic map preserving the measure μ, and $f: X \longrightarrow \mathbb{R}$ is measurable. Let $a: [0, +\infty) \to [0, +\infty)$ be a continuous, strictly increasing function satisfying $\frac{a(x)}{x} \searrow 0$ as $x \to +\infty$. If*

$$\int_X a(|f|)d\mu < +\infty,$$

then

$$\lim_{n\to\infty} \frac{a(|S_n f|)}{n} = 0 \quad \mu\text{-a.e. in } X.$$

Proof We first claim that

$$a(x+y) \le a(x) + a(y) \tag{5.3}$$

for all $x, y \geq 0$. Indeed, assume without loss of generality that $x \leq y$. Then

$$a(x+y) \leq \frac{a(y)}{y}(x+y) = a(y) + \frac{a(y)}{y}x \leq a(y) + \frac{a(x)}{x}x = a(x) + a(y)$$

and (5.3) is proved. Dealing directly with the proposition, we first establish it under the additional assumption that $a(0) = 0$. Given $\varepsilon > 0$, we will prove that

$$\limsup_{n\to\infty} \frac{a(|S_n f|)}{n} \leq \varepsilon \quad \text{a.e.}$$

In order to do this, we first claim that there are two measurable functions $g, h\colon X \to \mathbb{R}$ such that $|f| = g + h$, $\sup_X(g) < +\infty$, and $\int_X a(h)\, d\mu < \varepsilon$. This is because, for the sets $A_M := \{x \in X\colon \ |f(x)| \geq M\}$, we have that

$$\lim_{M\to+\infty} a(|f|\mathbb{1}_{A_M}) = 0 \quad \text{a.e.},$$

whence, by the Lebesgue Dominated Convergence Theorem, which is applicable since $a(|f|)$ is integrable, we have that

$$\lim_{M\to+\infty} \int_X a(|f|\mathbb{1}_{A_M}) d\mu = 0,$$

and so, for a sufficiently large $M > 0$, setting

$$g := |f|\mathbb{1}_{A_M^c} \quad \text{and} \quad h := |f|\mathbb{1}_{A_M}$$

gives the required representation of $|f|$. Using (5.3) and the Birkhoff Ergodic Theorem (Theorem 2.3.2), we have that

$$\begin{aligned}
\frac{a(|S_n f|)}{n} &\leq \frac{a(|S_n g|)}{n} + \frac{a(|S_n h|)}{n} \\
&\leq \frac{a(|Mn|)}{n} + \frac{S_n a(h)}{n} \xrightarrow[n\to\infty]{} \int_X a(h) d\mu < \varepsilon.
\end{aligned}$$

Hence, letting $\varepsilon \searrow 0$, our assertion follows (under the assumption that $a(0) = 0$). We now shall establish the proposition without assuming that $a(0) = 0$. To do this, we fix any $m > 0$ such that $a(m)/m = \alpha \in (0, 1)$. We then define

$$\tilde{a}(x) := \begin{cases} \alpha x & \text{if } 0 \leq x \leq m \\ a(x) & \text{if } x \geq m. \end{cases}$$

It is straightforward to verify that $\tilde{a}$ is continuous, increasing, $a \equiv \tilde{a}$ on $[m, \infty)$, and that $\frac{\tilde{a}}{x} \searrow 0$ as $x \to \infty$ and $\tilde{a}(0) = 0$. Therefore, by the previous step ($a(0) = 0$), we have that

$$\lim_{n\to\infty} \frac{\tilde{a}(|S_n f|)}{n} = 0$$

a.e. in X. Since $a(x) \le \max\{a(m), \tilde{a}(x)\} \le a(m) + \tilde{a}(x)$, we, thus, get that

$$0 \le \limsup_{n\to\infty} \frac{a(|S_n f|)}{n} \le \limsup_{n\to\infty} \frac{a(m) + \tilde{a}(|S_n f|)}{n} = \lim_{n\to\infty} \frac{\tilde{a}(|S_n f|)}{n} = 0.$$

The proof is complete. ■

Let $L^1_+(\mu)$ denote the set of functions $f \in L^1(\mu)$ such that $\int_X f d\mu > 0$. The next two theorems rule out any hope for any pointwise version of the Birkhoff Ergodic Theorem in the context of infinite invariant measures.

Theorem 5.1.3 *Suppose that $(X, \mathfrak{F}, \mu)$ is a measurable space with a σ-finite infinite measure μ and that $T : X \longrightarrow X$ is an ergodic conservative transformation preserving μ. If $a(n) \nearrow \infty$ and $\frac{a(n)}{n} \searrow 0$ as $n \to \infty$, then the following hold.*

(1) *If there exists a set $F \in \mathfrak{F}$ such that $0 < \mu(F) < \infty$ and $\int_F a(\tau_F)\, d\mu < +\infty$, then*

$$\lim_{n\to\infty} \frac{S_n f}{a(n)} = +\infty \quad \mu\text{-a.e. } \forall\, f \in L^1(\mu)_+.$$

(2) *Otherwise,*

$$\liminf_{n\to\infty} \frac{S_n f}{a(n)} = 0 \quad \mu\text{-a.e. } \forall\, f \in L^1(\mu)_+.$$

Proof We begin with proving (1). Suppose that $F \in \mathfrak{F}$, $0 < \mu(F) < +\infty$, and that $\int_F a(\tau_F) d\mu < +\infty$. Then, by Proposition 5.1.2,

$$\lim_{n\to\infty} \frac{a(S_{F,n}((\mathbb{1}_F)_F))}{n} = 0 \quad \text{a.e. on } F. \tag{5.4}$$

For every $x \in F_\infty$ and every $n \ge 1$, fix the only integer $k_n(x) \ge 1$ such that

$$\sum_{k=0}^{k_n(x)-1} \tau_F \circ T_F^k(x) \le n < \sum_{k=0}^{k_n(x)} \tau_F \circ T_F^k(x).$$

Since also $S_{F,k_n(x)}((\mathbb{1}_F)_F) \equiv k_n(x)$, we get that

$$\begin{aligned}
\frac{S_n \mathbb{1}_F}{a(n)(x)} &\ge \frac{S_{F,k_n(x)}((\mathbb{1}_F)_F)(x)}{a(S_{F,k_n(x)+1}((\mathbb{1}_F)_F))(x)} \\
&= \frac{k_n(x)}{a(S_{F,k_n(x)+1}((\mathbb{1}_F)_F))(x)} \xrightarrow[n\to\infty]{} +\infty,
\end{aligned}$$

where the divergence to $+\infty$ is due to (5.4). Clearly, the set of all points $x \in X$ on which this divergence occurs is T invariant, and, as it contains F_∞, it must be equal to X modulo μ. Thus,

$$\lim_{n\to\infty} \frac{S_n 1\!\!1_F}{a(n)} = +\infty$$

μ-a.e. on X, and (1) follows from the Hopf Ergodic Theorem (Theorem 2.3.12).

We now prove (2). To this end, we note, first, that, for every $f \in L^1_+(\mu)$,

$$\frac{S_n \circ T}{a(n)} \leq \left(1 + \frac{1}{n}\right) \frac{S_{n+1} f}{a(n+1)}.$$

Hence,

$$\liminf_{n\to\infty} \frac{S_n f}{a(n)} \circ T \leq \liminf_{n\to\infty} \frac{S_n f}{a(n)}.$$

So, this lower limit is μ-a.e constant on X by ergodicity of T with respect to the measure μ. Denote it by $2\varepsilon \geq 0$. Now, seeking a contradiction, suppose that the hypothesis of condition (2) is satisfied but its assretion fails for the function f. Then $\varepsilon > 0$. Fix arbitrarily a set $F \in \mathfrak{F}$ such that $0 < \mu(F) < +\infty$. It then follows from Egorov's Theorem that there exists a measurable set $G \subseteq F$ such that $0 < \mu(G) < \infty$, and

$$S_n 1\!\!1_F(x) \geq \varepsilon a(n)$$

for all $x \in G$ and all $n \geq 1$ large enough. Hence,

$$S_{\tau_G(x)} 1\!\!1_F(x) \geq \varepsilon a(\tau_G(x))$$

for all $x \in G$. Using the Hopf Ergodic Theorem (Theorem 2.3.12) this implies that, for μ-a.e. $x \in G$, we have that

$$\begin{aligned}
\frac{1}{n} S_{n,G}(a \circ \tau_G)(x) &= \frac{1}{n} \sum_{k=0}^{n-1} a\left(\tau_G\left(T^k(x)\right)\right) \\
&\leq \frac{1}{n\varepsilon} \sum_{k=0}^{n-1} S_{\tau_G(T^k(x))} 1\!\!1_F\left(T_G^k(x)\right) \\
&= \frac{1}{n\varepsilon} S_{S_n \tau_G(x)} 1\!\!1_F(x) \\
&= \frac{1}{\varepsilon} \frac{S_{S_n \tau_G(x)} 1\!\!1_F(x)}{S_{S_n \tau_G(x)} 1\!\!1_F(x)} \xrightarrow[n\to\infty]{} \frac{\mu(F)}{\varepsilon \mu(G)}.
\end{aligned}$$

It, thus, follows that $\int_G a(\tau_G) d\mu < \infty$, contradicting the hypothesis (2). ■

Theorem 5.1.4 *Suppose that $(X, \mathfrak{F}, \mu)$ is a σ-finite measure space and $T\colon X \longrightarrow X$ is an ergodic conservative transformation preserving measure μ. Let $(a_n)_{n=1}^{\infty}$ be a sequence such that $a_n > 0$ for all $n \in \mathbb{N}$. Then either*

(1)

$$\liminf_{n\to\infty} \frac{S_n f}{a_n} = 0 \quad \forall\, f \in L^1(\mu)_+$$

or

(2) *there exists a diverging to infinity sequence $(n_k)_{k=1}^{\infty}$ consisting of positive integers such that*

$$\liminf_{n\to\infty} \frac{S_{n_k} f}{a_{n_k}} = +\infty \;\; a.e. \;\; \forall\, f \in L^1(\mu)_+.$$

Proof If the sequence $(a_n)_{n=1}^{\infty}$ is bounded, then (2) holds. Indeed, let $f \in L_+^1(\mu)$. Since $\int_X f\, d\mu > 0$, there exists $\varepsilon > 0$ and $B \in \mathfrak{F}$ such that

$$\mu(B) > 0 \quad \text{and} \quad f_B \geq \varepsilon.$$

As T is conservative and ergodic, Theorem 2.1.10 affirms that $\mu(X \backslash B_\infty) = 0$. Therefore, for μ-a.e. $x \in X$, there exists a sequence $(n_k(x))_{k=1}^{\infty}$ such that $n_k(x) \nearrow \infty$ and

$$T^{n_k(x)}(x) \in B.$$

Hence, $S_{n_k(x)} f(x) \geq k\varepsilon$. In fact, $S_n f(x) \geq k\varepsilon$ for any $n \geq n_k(x)$ since $f \in L_+^1(\mu)$. Therefore,

$$\lim_{n\to\infty} S_n f(x) = +\infty.$$

As $(a_n)_{n=1}^{\infty}$ is bounded, it follows that

$$\lim_{n\to\infty} \frac{S_n f(x)}{a_n} = \infty$$

for μ-a.e. $x \in X$. So (2) holds for this function f with $(n_k)_{k=1}^{\infty} = (n)_{n=1}^{\infty}$. By the Hopf Ergodic Theorem, this actually holds for every $f \in L_+^1(\mu)$, with the same sequence.

We can, thereby, restrict our attention to the case

$$\limsup_{n\to\infty} a_n = +\infty.$$

Suppose that (1) does not hold. That is, there exists a set $A \in \mathcal{A}$ with $\mu(A) > 0$ and a function $f \in L_+^1(\mu)$ such that

$$F(x) := \liminf_{n\to\infty} \frac{S_n f(x)}{a_n} > 0 \;\text{ for } \mu\text{-a.e. } x \in A. \tag{5.5}$$

By the Hopf Ergodic Theorem, this actually holds for every $f \in L^1_+(\mu)$, with the same set A. Then, for μ-a.e. $x \in A$, we get that

$$\begin{aligned}
0 \le \limsup_{n\to\infty} \frac{a_n}{n} &= \limsup_{n\to\infty} \left[\frac{a_n}{S_n f(x)} \frac{S_n f(x)}{n} \right] \\
&\le \limsup_{n\to\infty} \frac{a_n}{S_n f(x)} \limsup_{n\to\infty} \frac{S_n f(x)}{n} \\
&= \left[\liminf_{n\to\infty} \frac{S_n f(x)}{a_n} \right]^{-1} \lim_{n\to\infty} \frac{S_n f(x)}{n} = 0
\end{aligned}$$

by (5.5) and Proposition 5.1.1. Thus, $a_n = o(n)$ as $n \to \infty$. For every $n \in \mathbb{N}$, set

$$\overline{a}_n = \max_{1 \le k \le n} a_k.$$

Clearly, $a_n \le \overline{a}_n$ for all $n \in \mathbb{N}$ and $\overline{a}_n \nearrow \infty$ as $n \nearrow \infty$. Moreover, for each $n \in \mathbb{N}$, there is $1 \le k(n) \le n$ such that $\overline{a}_n = a_{k(n)}$. Note that $k(n) \to \infty$ as $n \to \infty$. Then

$$\liminf_{n\to\infty} \frac{S_n f}{a_n} \ge \liminf_{n\to\infty} \frac{S_n f}{\overline{a}_n} = \liminf_{n\to\infty} \frac{S_n f}{a_{k(n)}} \ge \liminf_{n\to\infty} \frac{S_{k(n)} f}{a_{k(n)}} \ge \liminf_{n\to\infty} \frac{S_n f}{a_n},$$

where the last inequality follows from the fact that the lim inf of a subsequence of a sequence is greater than or equal to the lim inf of the full sequence. Hence,

$$\liminf_{n\to\infty} \frac{S_n f}{\overline{a}_n} = \liminf_{n\to\infty} \frac{S_n f}{a_n} > 0 \quad \mu\text{-a.e. on } A \quad \forall f \in L^1_+(\mu). \tag{5.6}$$

Next, set

$$f_n := \frac{\overline{a}_n}{n}.$$

Let $1 = n_0 < n_1 < \cdots$ be defined by

$$\{n_k\}_{k\in\mathbb{N}} = \left\{ j \ge 2 \colon\ f_i > f_j \ \forall\, 1 \le i \le j-1 \right\}.$$

For every $k \ge 0$,

$$f_{n_k} > f_{n_{k+1}}, \quad n_k f_{n_k} \le n_{k+1} f_{n_{k+1}},$$

whence

$$0 < \frac{n_k}{n_{k+1}} \le \frac{f_{n_{k+1}}}{f_{n_k}} < 1.$$

Thus, there exists $\alpha_k \in (0,1]$ such that

$$\left(\frac{n_k}{n_{k+1}}\right)^{\alpha_k} = \frac{f_{n_{k+1}}}{f_{n_k}}.$$

Define

$$f(x) = \frac{f_{n_k} n_k^{\alpha_k}}{x^{\alpha_k}}, \quad x \in [n_k, n_{k+1}],\ k \in \mathbb{N},$$

and

$$a(x) = x f(x).$$

Evidently,

$$a(n_k) = \overline{a}_{n_k} \quad \forall k \in \mathbb{N}.$$

By definition of n_k, we have that, for $k \in \mathbb{N}$, $n \in [n_k, n_{k+1})$,

$$f_n \geq f_{n_k}; \text{ hence, } f_n \geq f(n),$$

whereby

$$a(n) \leq \overline{a}_n \quad \forall n \in \mathbb{N}.$$

Hence, following (5.6),

$$\liminf_{n\to\infty} \frac{S_n f}{a(n)} > 0, \quad \mu\text{-a.e. on } A \quad \forall f \in L_+^1(\mu).$$

It is evident that

$$a(n) \nearrow \infty, \quad \frac{a(n)}{n} \searrow 0 \quad \text{as } n \nearrow \infty.$$

So, by Theorem 5.1.3,

$$\lim_{n\to\infty} \frac{S_n f}{a(n)} = +\infty \quad \mu\text{-a.e. on } X \quad \forall f \in L_+^1(\mu).$$

Then (2) follows since $a_{n_k} \leq \overline{a}_{n_k} = a(n_k)$. ■

5.2 Weak Ergodic Theorems

Although the previous section has ruled out any hope for a direct counterpart of the Birkhoff Ergodic Theorem in the realm of infinite invariant measures, there are nevertheless some weak versions of this theorem. By saying weak, we mean the convergence in the sense of distribution and along a subsequence.

We provide below some such theorems taken from various papers of Aaronson, primarily in Aaronson's book [**Aa**] and the papers cited therein.

If $(X, \mathfrak{F}, \nu)$ is a probability space and

$$\big(Y_n : X \longrightarrow [-\infty, +\infty]\big)_{n=1}^{\infty}$$

is a sequence of random variables (measurable functions) with respect to the σ-algebra $\mathfrak{F}$, then $(Y_n)_{n=1}^{\infty}$ is said to converge in distribution to some probability distribution P on $\mathbb{R}$ if and only if the sequence $\big(\nu \circ Y_n^{-1}\big)_1^{\infty}$ of distributions of Y_n, $n \geq 1$, converges weakly to P. We then write

$$Y_n \xrightarrow{\ w\ } P.$$

If P is the probability distribution of some random variable $Y : (\Omega, \mathcal{A}, Q) \longrightarrow [-\infty, +\infty]$ defined on some probability space $(\Omega, \mathcal{A}, Q)$, then we also say that the sequence $(Y_n)_{n=1}^{\infty}$ converges in distribution to Y and we write

$$Y_n \xrightarrow{\ w\ } Y.$$

Explicitly, this means that

$$\lim_{n\to\infty} \int_X g \circ Y_n \, dm = E(g \circ Y) := \int_\Omega g \circ Y \, dQ$$

for every continuous function $g : [-\infty, \infty] \longrightarrow \mathbb{R}$.

If $(X, \mathfrak{F}, m)$ is a measure space (m can be infinite), then a sequence $(Y_n)_{n=1}^{\infty}$ of real-valued measurable functions on X is said to converge strongly in distribution to some probability distribution P on $\mathbb{R}$ if and only if the sequence $(\nu \circ Y_n^{-1})_{n=1}^{\infty}$ converges weakly to P for every probability measure ν absolutely continuous with respect to m. We then write

$$Y_n \xrightarrow{\ \mathcal{L}\ } P.$$

As above, if P is the probability distribution of some random variable

$$Y : (\Omega, \mathcal{A}, Q) \longrightarrow \mathbb{R}$$

defined on some probability space $(\Omega, \mathcal{A}, Q)$, then we also say that the sequence $(Y_n)_{n=1}^{\infty}$ converges strongly in distribution to Y and we write

$$Y_n \xrightarrow{\ \mathcal{L}\ } Y.$$

Explicitly, this means that

$$\lim_{n\to\infty} \int_X h g \circ Y_n \, dm = E(g \circ Y) := \int_\Omega g \circ Y \, dQ$$

for every continuous function $g\colon [-\infty,\infty] \longrightarrow \mathbb{R}$ and every integrable function $h\colon X \longrightarrow [0,+\infty)$ such that $\int_X h\,dm = 1$. Furthermore, by the standard approximation argument, this means that

$$\lim_{n\to\infty}\int_X hg\circ Y_n\,dm = E(g\circ Y) := \int_\Omega g\circ Y\,dQ \tag{5.7}$$

for every continuous function $g\colon [-\infty,\infty] \longrightarrow \mathbb{R}$ and every m-essentially bounded measurable function $h\colon X \longrightarrow [0,+\infty)$ such that $\int_X h\,dm = 1$. This, in turn, is equivalent to having that

$$\lim_{n\to\infty}\int_X hg\circ Y_n\,dm = E(g\circ Y) := \int_\Omega g\circ Y\,dQ\int_X h\,dm \tag{5.8}$$

for every continuous function $g\colon [-\infty,\infty] \longrightarrow \mathbb{R}$ and every function $h\in L^\infty(m)$.

Theorem 5.2.1 *Let $T\colon (X,\mathfrak{F},m) \longrightarrow (X,\mathfrak{F},\mu)$ be an ergodic transformation of a σ-finite measure space $(X,\mathfrak{F},\mu)$ and $f\colon X\longrightarrow\mathbb{R}$ be a bounded measurable function.*

Suppose that $(n_k)_1^\infty$ is a sequence of positive integers diverging to $+\infty$ and that $(d_k)_1^\infty$ is a sequence of positive real numbers.

Then there exist a subsequence $\left(n_{k_l}\right)_{l=1}^\infty$ diverging to $+\infty$ and a random variable Y taking values in $[-\infty,\infty]$ such that

$$\frac{1}{d_{k_l}}\sum_{j=0}^{n_{k_l}-1} f\circ T^j \xrightarrow{\ \mathcal{L}\ } Y.$$

Proof As usual,

$$S_nf = \sum_{j=0}^{n-1} f\circ T^j.$$

Because of (5.8) and because, for every $h\in L^\infty(m)$, the function

$$L^\infty(\mu)\ni k\longmapsto \int_X hk\,d\mu\in\mathbb{R}$$

belongs to $L^\infty(\mu)^*$, in order to prove our theorem it is sufficient to obtain a random variable Y taking values in $[-\infty,+\infty]$ and $m_l := n_{k_l} \to \infty$ such that, for each continuous function $g\colon [-\infty,+\infty]\to\mathbb{R}$,

$$g\circ\left(\frac{S_{n_{k_l}}f}{d_{k_l}}\right) \xrightarrow[l\to\infty]{} E(g\circ Y) \quad \text{weak * in } L^\infty(m).$$

For any $g \in C([-\infty, +\infty])$, the Banach space of real-valued continuous functions defined on $[-\infty, +\infty]$ endowed with the supremum norm $\|\cdot\|_\infty$, we have, for all integers $k \geq 1$, that

$$\left\| g \circ \left(\frac{S_{n_k}}{d_k} \right) \right\|_{L^\infty(\mu)} \leq \|g\|_\infty.$$

Hence, because of the Banach–Aaloglou Theorem there exists an increasing subsequence $k' \to \infty$ and a function $\lambda(g) \in L^\infty(\mu)$ such that

$$g \circ \left(\frac{S_{n_{k'}}}{d_{k'}} \right) \xrightarrow[k' \to \infty]{} \lambda(g) \quad \text{weak * in } L^\infty(\mu).$$

We claim that the limit function $\lambda(g)$ is T-invariant. Indeed, putting, for every ε,

$$\omega_g(\varepsilon) := \sup \left\{ |g(y+h) - g(y)| : y, h \in \mathbb{R}, |h| < \varepsilon \right\},$$

we have, by the uniform continuity of g, that

$$\lim_{\varepsilon \to 0} \omega_g(\varepsilon) = 0.$$

Hence,

$$\begin{aligned} \left\| g \circ \left(\frac{S_{n_k}}{d_k} \right) \circ T - g \circ \left(\frac{S_{n_k}}{d_k} \right) \right\|_\infty &= \left\| g \circ \left(\frac{S_{n_k}}{d_k} + \frac{f \circ T^{n_k} - f}{d_k} \right) - g \circ \left(\frac{S_{n_k}}{d_k} \right) \right\|_\infty \\ &\leq \omega_g \left(\frac{2\|f\|_\infty}{d_k} \right) \xrightarrow[k \to \infty]{} 0. \end{aligned}$$

Thus, $\lambda(g) \circ T = \lambda(g)$. By ergodicity of T, this yields that the function $\lambda(g)$ is constant.

Since the Banach space $C([-\infty, +\infty])$ is separable, there exists $\mathcal{G} \subseteq C([-\infty, \infty])$, a countable, dense set in $C([-\infty, +\infty])$. A standard diagonalization argument yields an unbounded increasing susequence $(n_{k_l})_{l=1}6\infty$ and constants $\{\lambda(g) \in \mathbb{R} : g \in \mathcal{G}\}$ such that

$$g \circ \left(\frac{S_{n_{k_l}} f}{d_{k_l}} \right) \xrightarrow[l \to \infty]{} \lambda(g) \quad \text{weak * in } L^\infty(m) \tag{5.9}$$

for all $g \in \mathcal{G}$. We claim that this property can be extended to all elements of $C([-\infty, +\infty])$. More precisely, we claim that, for all $h \in C([-\infty, +\infty])$, there exists $\lambda(h) \in \mathbb{R}$ such that

$$h \circ \left(\frac{S_{n_{k_l}} f}{d_{k_l}} \right) \xrightarrow[l \to \infty]{} \lambda(h) \quad \text{weak * in } L^\infty(m)$$

for all $h \in C([-\infty, +\infty])$.

In order to prove this, let $h \in C([-\infty, +\infty])$. Since the set $\mathcal{G}$ is dense in $C([-\infty, +\infty])$, there is a sequence $(g_j)_{j=1}^\infty \in \mathcal{G}$ such that

$$\lim_{j \to +\infty} \|g_j - h|_\infty = 0.$$

Because of (5.9), for any probability measure ν on X absolutely continuous with respect to μ with the Radon–Nikodym derivative $d\nu/d\mu \in L^\infty(\mu)$ and all $i, j \geq 1$, we have that

$$|\lambda(g_i) - \lambda(g_j)| \xleftarrow{l \to \infty} \left| \int_X \left(g_i \left(\frac{S_{n_{k_l}}}{d_{k_l}} \right) - g_j \left(\frac{S_{n_{k_l}}}{d_{k_l}} \right) \right) d\nu \right| \leq \|g_i - g_j\|_\infty.$$

Since, as convergent, the sequence $(g_j)_{j=1}^\infty$ is fundamental (Cauchy), we, therefore, conclude that the sequence $\left(\lambda(g_j)\right)_{j=1}^\infty$ is fundamental in $\mathbb{R}$. Thus, there exists a limit

$$\lambda(h) =: \lim_{j \to \infty} \lambda(g_j) \in \mathbb{R}$$

and

$$|\lambda(g_j) - \lambda(h)| \leq \|g_j - h\|_\infty$$

for all $j \geq 1$. For any probability measure ν on X absolutely continuous with respect to μ with the Radon–Nikodym derivative $d\nu/d\mu \in L^\infty(\mu)$ and all integers $j \geq 1$, by applying the triangle inequality, we get that

$$\begin{aligned}
&\left| \int_X h \circ \left(\frac{S_{n_{k_l}}}{d_{k_l}} \right) d\nu - \lambda(h) \right| \\
&\quad \leq \left| \int_X h \circ \left(\frac{S_{n_{k_l}}}{d_{k_l}} \right) d\nu - \int_X g_j \circ \left(\frac{S_{n_{k_l}}}{d_{k_l}} \right) d\nu \right| \\
&\qquad + \left| \int_X g_j \left(\frac{S_{n_{k_l}}}{d_{k_l}} \right) d\nu - \lambda(g_j) \right| + |\lambda(g_j) - \lambda(h)| \\
&\quad \leq 2\|h - g_j\|_\infty + \left| \int_X g_j \circ \left(\frac{S_{n_{k_l}}}{d_{k_l}} \right) d\nu - \lambda(g_j) \right| \xrightarrow{l \to \infty} 2\|h - g_j\|_\infty.
\end{aligned}$$

By letting $j \to \infty$, we, thus, conclude that

$$h \circ \left(\frac{S_{n_{k_l}}}{d_{k_l}} \right) \xrightarrow[n \to \infty]{} \lambda(h) \quad \text{weak * in } L^\infty(\mu).$$

It also follows from this that $\lambda\colon C([-\infty, +\infty]) \longrightarrow \mathbb{R}$ is linear and positive, whence there exists a random variable Y taking values in $[-\infty, \infty]$ such that

$$\lambda(h) = E(h(Y)),$$

for all $h \in C([-\infty, +\infty])$. The proof is complete. ■

Corollary 5.2.2 *Let $T\colon (X, \mathfrak{F}, \mu) \longrightarrow (X, \mathfrak{F}, \mu)$ be a conservative ergodic measure preserving transformation of a σ-finite measure space $(X, \mathfrak{F}, m)$. Suppose that $(n_k)_1^\infty$ is a sequence of positive integers diverging to $+\infty$ and that $(d_k)_1^\infty$ is a sequence of positive real numbers.*

Then there exist a subsequence $\left(n_{k_l}\right)_{l=1}^\infty$ diverging to $+\infty$ and a random variable Y taking values in $[-\infty, \infty]$ such that

$$\frac{1}{d_{k_l}} \sum_{j=0}^{n_{k_l}-1} f \circ T^j \xrightarrow{\mathcal{L}} \left(\int_X f \, d\mu\right) Y.$$

Proof Fix a set $A \in \mathfrak{F}$ such that $\mu(A) = 1$. By Proposition 5.2.1, there exist a subsequence $\left(n_{k_l}\right)_{l=1}^\infty$ diverging to $+\infty$ and a random variable Y taking values in $[-\infty, \infty]$ such that

$$\frac{1}{d_{k_l}} \sum_{j=0}^{n_{k_l}-1} \mathbb{1}_A \circ T^j \xrightarrow{\mathcal{L}} Y.$$

Since also, by the Hopf Ergodic Theorem (Theorem 2.3.12),

$$\lim_{l \to +\infty} \frac{\frac{1}{d_{k_l}} \sum_{j=0}^{n_{k_l}-1} f \circ T^j}{\frac{1}{d_{k_l}} \sum_{j=0}^{n_{k_l}-1} \mathbb{1}_A \circ T^j} = \frac{\int_X f \, d\mu}{\int_X \mathbb{1}_A d\mu} = \int_X f \, d\mu$$

μ-a.e. on X, the statement of our corollary follows. ■

5.3 Darling–Kac Theorem: Abstract Version

In Section 5.1, we have provided evidence for any reasonable version of the Birkhoff Ergodic Theorem (i.e., pointwise) not to hold in the realm of infinite invariant measures. In Section 5.2, we provided some very weak counterparts of the Birkhoff Ergodic Theorem; namely, convergence in distribution along some subsequences of “times.” In this section, we will strengthen the results of the previous one by stating, see Theorems 5.3.2 and 5.3.4, due to [**TZ**], the distributional convergence of weighted averages along the whole sequence of “times” and also some pointwise lim sup results for such weighted averages.

There is, however, a cost: we need to assume the existence of some special sets; namely, the sets Y of Theorems 5.3.2 and 5.3.4 for distributional convergence and the so-called Darling–Kac sets for lim sup results. It is particularly difficult to prove that the latter exist, even in the simplest case of parabolic Markov maps of an interval. It is easier to prove that the former exist, and we will do this for parabolic elliptic functions in Section 22.8 in Volume 2. We would like to mention that, in the early papers by Denker and Aaronson (see [**Aa**] and references therein), Darling–Kac sets were also used to prove the so-called Pointwise Uniform Dual Ergodic Theorem. More recently, Melbourne and Theresiu (see [**MT**]) have provided fairly simple sufficient conditions that do not require the existence of Darling–Kac sets for this theorem to hold. Because our book is already quite large, we will not (unfortunately) deal with Dual Ergodic Theorems any further. We would like, however, to add that Darling–Kac Theorems have also been the objects of intensive research over the last several decades, they have attracted attention from the research community (see, for example, [**Aa**], [**TZ**]), and research on them continues.

A function $u\colon [a,+\infty) \longrightarrow (0,+\infty)$ is called a regularly varying index of $s \in \mathbb{R}$ if

$$\lim_{t\to\infty} \frac{u(ct)}{u(t)} = c^s$$

for all $c > 0$. We then write that $u \in \mathcal{R}_s$. Given a sequence $(a_n)_1^\infty$ of positive numbers, we form the function $\hat{a}\colon [1,+\infty) \longrightarrow (0,+\infty)$ as

$$\hat{a}(t) = a_{[t]}$$

and we say that the sequence $(a_n)_1^\infty$ is regularly varying with index $s \in \mathbb{R}$ if the function $\hat{a}$ enjoys this property. We then also write that

$$(a_n)_1^\infty \in \mathcal{R}_s.$$

The regularly varying functions and sequences of index 0 are also referred to as slowly varying functions.

If $(X,\mathfrak{F},\nu)$ is a probability space and $f\colon X \longrightarrow \mathbb{R}$ is an $\mathfrak{F}$-measurable function, following probabilistic custom, we call f a random variable and call the probability measure $\nu \circ f^{-1}$ on $\mathbb{R}$ the distribution of the random variable f. Frequently, any Borel probability measure on $\mathbb{R}$ is referred to as a distribution. Notice that every distribution ν on $\mathbb{R}$ is uniquely determined by its moments, i.e., the integrals

$$\int_{-\infty}^{+\infty} x^k d\nu(x), \quad k = 1,2,3,\ldots.$$

Given $\alpha \in [0, 1]$, the normalized Mittag–Leffler distribution $\mathcal{M}_\alpha$ of order α is characterized by its moments

$$\int_{-\infty}^{+\infty} x^k d\mathcal{M}_\alpha(x) = k!\,\frac{(\Gamma(1+\alpha))^k}{\Gamma(1+k\alpha)}, \quad k = 1, 2, 3, \ldots.$$

Given a Borel probability measure ν on $\mathbb{R}$, we denote by $a\nu + b$, $a > 0$ the probability distribution of the random variable $aX_\nu + b$, where X_ν is any random variable whose distribution is equal to ν, i.e.,

$$(a\nu + b)(F) = \nu\left(\frac{F-b}{a}\right)$$

for any real-valued ν-integrable function $F\colon X \to \mathbb{R}$.

Every distribution of the form $a\mathfrak{M}_\alpha + b$, $a > 0$, is called a Mittag–Leffler distribution. Every Mittag–Leffler distribution is stable, meaning that the convolution of two Mittag–Leffler distributions of the same order $\alpha \in [0, 1]$ is again a Mittag–Leffler distribution of order α. In other words, the distribution of the sum of two independent random variables with Mittag–Leffler distribution is again a Mittag–Leffler distribution.

Now let $(X, \mathfrak{F}, \mu)$ be a σ-finite measure space. Assume that

$$\mu(X) = +\infty.$$

Let $T\colon X \to X$ be an $\mathfrak{F}$-measurable map preserving measure μ. For every measurable set F, the wandering rates of F are defined to be

$$w_n(F) := \mu\left(\bigcup_{k=0}^{n} T^{-k}(F)\right), \quad n \geq 0. \tag{5.10}$$

We shall now prove the following lemma, which establishes some formula for wandering rates that is useful in many calculations.

Lemma 5.3.1 *If $(X, \mathfrak{F}, \mu)$ is a σ-finite measure space and $T\colon X \longrightarrow X$ is an ergodic conservative measurable map preserving measure μ, then*

$$w_n(F) = \mu(F)\sum_{k=0}^{n} \mu_F(\tau_F > k)$$

for every $F \in \mathfrak{F}$ with $0 < \mu(F) < +\infty$ and all $n \geq 0$.

Proof Since

$$\bigcup_{k=0}^{n} T^{-k}(F) = \bigcup_{k=0}^{n} T^{-k}(F) \cap \left\{x \in X\colon \tau_F\left(T^k(x)\right) > n-k\right\}$$

and the constituents of the latter union are mutually disjoint, we get that

$$\begin{aligned}
w_n(F) &= \mu\left(\bigcup_{k=0}^{n} T^{-k}(F)\right) = \mu\left(\bigcup_{k=0}^{n} T^{-k}(F) \cap \{x \in X : \tau_F(T^k(x)) > n-k\}\right) \\
&= \sum_{k=0}^{n} \mu(T^{-k}(F) \cap \{x \in X : \tau_F\big(T^k(x)\big) > n-k\}) \\
&= \sum_{k=0}^{n} \mu(T^{-k}(F) \cap T^{-k}(\{x \in X : \tau_F(x) > n-k\})) \\
&= \sum_{k=0}^{n} \mu(T^{-k}(F \cap \{x \in X : \tau_F(x) > n-k\})) \\
&= \sum_{k=0}^{n} \mu(T^{-k}(\{x \in F : \tau_F(x) > n-k\})) \\
&= \sum_{k=0}^{n} \mu(\{x \in F : \tau_F(x) > n-k\}) \\
&= \sum_{j=0}^{n} \mu(\{x \in F : \tau_F(x) > j\}) \\
&= \sum_{j=0}^{n} \mu(\tau_F > j) \\
&= \mu(F) \sum_{j=0}^{n} \mu_F(\tau_F > j).
\end{aligned}$$

We are done. ■

We say that two number sequnces $(a_n)_1^\infty$ and $(b_n)_1^\infty$ are asymtotically equivalent if

$$\lim_{n\to\infty} \frac{a_n}{b_n} = 1.$$

We then write that

$$a_n \sim b_n.$$

A measurable function $f : X \longrightarrow [0, +\infty)$ is said to be supported on F if

$$f^{-1}(0) \supset X \backslash F.$$

Furthermore, it is called uniformly sweeping on F if there exists an integer $N \geq 0$ such that

$$\inf_F \left\{ \sum_{n=0}^{N} \mathcal{L}_\mu^n(f) \right\} > 0,$$

where, we recall,

$$\mathcal{L}_\mu : L_\mu^1 \longrightarrow L_\mu^1$$

is the transfer operator of the map T with respect to the measure μ defined in Section 2.1. We shall now formulate the theorem that is commonly referred to as the Darling–Kac Theorem. As we have already said, this theorem is in a sense a weak version of Theorem 2.3.2 for (special) systems with infinite invariant measure, and it has a long history. It took on a particularly simple, clear, and elegant form in [**TZ**]. Thaler and Zweimüler weakened the assumptions so much that these became in a sense local, in particular getting rid of dual ergodicity, an annoying hypothesis present in Aaronson's book [**Aa**]. We now quote Theorem 3.1 from [**TZ**].

Theorem 5.3.2 (Darling–Kac Theorem I) Let $(X, \mathfrak{F}, \mu)$ be a σ-finite measure space and $T: X \longrightarrow X$ be a measurable map preserving measure μ. Assume that there exists some set $Y \in \mathfrak{F}$ with the following properties.

(a) $0 < \mu(Y) < +\infty$.
(b) $(w_n(Y))_{n=1}^\infty \in \mathcal{R}_{1-\alpha}$ with some $\alpha \in [0, 1]$.
(c) There exists a uniformly sweeping function $H: X \longrightarrow [0, +\infty)$ on Y such that

$$\frac{1}{w_n(Y)} \sum_{k=0}^{n-1} \mathcal{L}_\mu^k\left(1\!\!1_{Y_k^c}\right) \xrightarrow[n\to\infty]{} H,$$

uniformly on Y, where

$$Y_k^c = Y_k^c(T) := T^{-k}(Y) \cap \bigcap_{j=0}^{k-1} T^{-j}(X \backslash Y)$$

for $k \geq 1$ and $Y_0^c := Y$.

Then, for every function $f \in L^1(\mu)$ with $\int f d\mu \neq 0$, the sequence $\left(\frac{1}{a_n} S_n f\right)_1^\infty$ converges strongly, with respect to the measure μ, to the Mittag–Leffler distribution $(\int f d\mu)\mathcal{M}_\alpha$; in symbols,

$$\frac{1}{a_n} S_n f \xrightarrow{\mathcal{L}} \left(\int f d\mu\right) \mathcal{M}_\alpha,$$

where

$$a_n := \frac{1}{\mu(Y)} \int_Y S_n(\mathbb{1}_Y)\, d\mu_Y \sim \frac{1}{\Gamma(1+\alpha)\Gamma(2-\alpha)} \cdot \frac{n}{w_n(Y)} \quad \text{as} \quad n \longrightarrow \infty.$$

A proof of this theorem can be found in [**TZ**]. This paper also contains some other stochastic asymptotic laws (like arcsin law) that are relevant in the context of infinite measures.

For every $k \geq 1$, let

$$Y_k = Y_k(T) := \tau_Y^{-1}(k) = \{x \in Y : \tau_Y(x) = k\}. \tag{5.11}$$

We shall prove the following, quite often relatively easily verifiable, sufficient condition for requirement Theorem 5.3.2(c) to hold.

Lemma 5.3.3 *Let $(X, \mathfrak{F}, \mu)$ be a σ-finite measure space and let $T : X \longrightarrow X$ be a measurable map preserving measure μ. If there exists a (measurable) function $\hat{H} : X \to [0, +\infty)$, supported on some set $Y \in \mathfrak{F}$, such that the sequence*

$$\left(\frac{1}{\mu_Y(Y_k)} \mathcal{L}_\mu^k(\mathbb{1}_{Y_k}) \right)_{k=1}^{\infty}$$

converges uniformly on Y to $\hat{H}|_Y$, then condition Theorem 5.3.2(c) holds with the function

$$H := (\mu(Y))^{-1} \hat{H} : X \longrightarrow [0, +\infty),$$

although we do not claim that $H : X \to [0, +\infty)$ is uniformly sweeping. Obviously, however, if $\hat{H}$ is uniformly sweeping, then so is H.

Proof By virtue of (2.7), we have, for every set $F \in \mathfrak{F}$ and any $g \in L^1(\mu)$, that

$$\begin{aligned} \int g \mathbb{1}_F d\mu &= \int g \circ T \cdot \mathbb{1}_F \circ T d\mu \\ &= \int g \circ T \cdot \mathbb{1}_{T^{-1}(F)} d\mu = \int g \mathcal{L}_\mu\big(\mathbb{1}_{T^{-1}(F)}\big) d\mu. \end{aligned}$$

This implies that

$$\mathbb{1}_F = \mathcal{L}_\mu\big(\mathbb{1}_{T^{-1}(F)}\big) \quad \mu\text{-a.e.} \tag{5.12}$$

Hence, for every $n \geq 0$, we have μ-a.e. that

$$\begin{aligned} \mathbb{1}_{Y_n^c} &= \mathcal{L}_\mu\left(\mathbb{1}_{T^{-1}(Y_n^c)}\right) = \mathcal{L}_\mu\left(\mathbb{1}_{Y_{n+1}^c \cup Y_{n+1}}\right) = \mathcal{L}_\mu\left(\mathbb{1}_{Y_{n+1}} + \mathbb{1}_{Y_{n+1}^c}\right) \\ &= \mathcal{L}_\mu\left(\mathbb{1}_{Y_{n+1}}\right) + \mathcal{L}_\mu\left(\mathbb{1}_{Y_{n+1}^c}\right). \end{aligned} \tag{5.13}$$

By Remark 2.2.7, we have that

$$\lim_{n\to\infty} \mu(Y_n^c) = 0. \tag{5.14}$$

Since $Y_{n+1}^c \subseteq T^{-1}(Y_n^c)$ and since $\mathcal{L}_\mu$ is a positive operator, we get that

$$\mathcal{L}_\mu\left(1\!\!1_{Y_{n+1}^c}\right) \le \mathcal{L}_\mu\left(1\!\!1_{T^{-1}(Y_n^c)}\right). \tag{5.15}$$

It follows, by immediate induction, from (5.13) that

$$1\!\!1_{Y_n^c} = \sum_{j=1}^{k} \mathcal{L}_\mu^j\left(1\!\!1_{Y_{n+j}}\right) + \mathcal{L}_\mu^k\left(1\!\!1_{Y_{n+k}^c}\right) \tag{5.16}$$

for all integers $k \ge 1$. By (5.15) and (5.12), we have that

$$\begin{aligned}\mathcal{L}_\mu^{k+1}\left(1\!\!1_{Y_{n+k+1}^c}\right) &\le \mathcal{L}_\mu^{k+1}\left(1\!\!1_{T^{-1}\left(Y_{n+k}^c\right)}\right)\\ &= \mathcal{L}_\mu^k\left(\mathcal{L}_\mu\left(1\!\!1_{T^{-1}\left(Y_{n+k}^c\right)}\right)\right) = \mathcal{L}_\mu^k\left(1\!\!1_{T^{-1}\left(Y_{n+k+1}^c\right)}\right),\end{aligned}$$

meaning that the sequence $(\mathcal{L}_\mu^k(1\!\!1_{Y_{n+k}^c}))_{k=1}^\infty$ is decreasing. Let

$$g := \lim_{k\to\infty} \mathcal{L}_\mu^k\left(1\!\!1_{Y_{n+k}^c}\right) \ge 0.$$

By the Lebesgue Monotone Convergence Theorem and (5.14), we, thus, have that

$$\int g\,d\mu = \lim_{k\to\infty}\int \mathcal{L}_\mu^k\left(1\!\!1_{Y_{n+k}^c}\right)\,d\mu = \lim_{k\to\infty}\int 1\!\!1_{Y_{n+k}^c}\,d\mu = \lim_{k\to\infty}\mu\left(1\!\!1_{Y_{n+k}^c}\right) = 0.$$

Hence, $g = 0$ μ-a.e. and (5.16) yields

$$1\!\!1_{Y_n^c} = \sum_{j=1}^{\infty} \mathcal{L}_\mu^j\left(1\!\!1_{Y_{n+j}^c}\right)$$

μ-a.e. on X. Therefore,

$$\frac{1}{\mu_Y(\tau_Y^{-1}([n+1,\infty)))}\mathcal{L}_\mu^n\left(1\!\!1_{Y_n^c}\right) = \frac{1}{\mu_Y(\tau_Y > k)}\sum_{j=1}^{\infty}\mathcal{L}_\mu^{n+j}\left(1\!\!1_{Y_{n+j}}\right). \tag{5.17}$$

Put

$$\hat{H} := \lim_{n\to\infty}\frac{1}{\mu(Y_n)}\mathcal{L}_\mu^n\left(1\!\!1_{Y_n}\right)\mu - \text{a.e.}X \longrightarrow [0,+\infty).$$

Then, for every $\varepsilon > 0$, there exists $n_\varepsilon \ge 1$ such that, for all $n \ge n_\varepsilon$,

$$\hat{H}(x) - \varepsilon \le \frac{1}{\mu_Y(Y_{n+j})}\mathcal{L}_\mu^{n+j}\left(1\!\!1_{Y_{n+j}}\right)(x) \le \hat{H}(x) + \varepsilon$$

for all $j \geq 1$ and all $x \in Y$. Equivalently,

$$(\hat{H}(x) - \varepsilon)\mu_Y(Y_{n+j}) \leq \mathcal{L}_\mu^{n+j}(1\!\!1_{Y_{n+j}})(x) \leq (\hat{H}(x) + \varepsilon)\mu_Y(Y_{n+j}).$$

Summing up over all $j = 1, 2, \ldots,$ we, thus, obtain that

$$(\hat{H}(x) - \varepsilon)\mu_Y(\tau_Y > n) \leq \sum_{j=1}^{\infty} \mathcal{L}_\mu^{n+j}(1\!\!1_{Y_{n+j}})(x) \leq (\hat{H}(x) + \varepsilon)\mu_Y(\tau_Y > n).$$

By virtue of (5.17), this means that

$$(H(x) - \varepsilon)\mu_Y(\tau_Y > n) \leq \mathcal{L}_\mu^n(1\!\!1_{Y_n^c})(x) \leq (H(x) + \varepsilon)\mu_Y(\tau_Y > n)$$

for all $n \geq n_\varepsilon$ and all $x \in Y$. Hence, fixing $l \geq n_\varepsilon + 1$ and summing up over all $n = n_\varepsilon, n_\varepsilon + 1, \ldots, l - 1$, we get for all $x \in Y$ that

$$\begin{aligned}(\hat{H}(x) - \varepsilon) \sum_{n=n_\varepsilon}^{l-1} \mu_Y(\tau_Y > n) &\leq \sum_{n=n_\varepsilon}^{l-1} \mathcal{L}_\mu^n(1\!\!1_{Y_n^c})(x) \\ &\leq (\hat{H}(x) + \varepsilon) \sum_{n=n_\varepsilon}^{l-1} \mu_Y(\tau_Y > n).\end{aligned}$$

Dividing by $w_l(Y)$ and using Lemma 5.3.1, we then get that

$$\begin{aligned}&\frac{(\hat{H}(x) - \varepsilon)}{\mu(Y)} \cdot \frac{\sum_{n=n_\varepsilon}^{l-1} \mu_Y(\tau_Y > n)}{\sum_{n=0}^{l-1} \mu_Y(\tau_Y > n)} \\ &\leq \frac{1}{w_l(Y)} \sum_{n=n_\varepsilon}^{l-1} \mathcal{L}_\mu^n(1\!\!1_{Y_n^c})(x) \cdot \frac{\sum_{n=n_\varepsilon}^{l-1} \mathcal{L}_\mu^n(1\!\!1_{Y_n^c})(x)}{\sum_{n=0}^{l-1} \mathcal{L}_\mu(1\!\!1_{Y_n^c})(x)} \\ &\leq \frac{(\hat{H}(x) + \varepsilon)}{\mu(Y)} \cdot \frac{\sum_{n=n_\varepsilon}^{l-1} \mu_Y(\tau_Y > n)}{\sum_{n=0}^{l-1} \mu_Y(\tau_Y > n)}.\end{aligned}$$

Since $\lim_{l\to\infty} w_l(Y) = +\infty$, taking $l \geq l_\varepsilon \geq n_\varepsilon$ large enough, we will have that

$$1 - \varepsilon \leq \frac{\sum_{n=n_\varepsilon}^{l-1} \mu_Y(\tau_Y > n)}{\sum_{n=0}^{l-1} \mu_Y(\tau_Y > n)} \leq 1$$

and

$$1 - \varepsilon \leq \frac{\sum_{n=n_\varepsilon}^{l-1} \mathcal{L}_\mu^n(1\!\!1_{Y_n^c})(x)}{\sum_{n=0}^{l-1} \mathcal{L}_\mu^n(1\!\!1_{Y_n^c})(x)} \leq 1$$

for all $x \in Y$. Thus,

$$(1-\varepsilon)\frac{(\hat{H}(x)-\varepsilon)}{\mu(Y)} \leq \frac{1}{w_l(Y)}\sum_{n=0}^{l-1}\mathcal{L}_\mu^n\left(\mathbb{1}_{Y_n^c}\right)(x) \leq \frac{1}{(1-\varepsilon)}\frac{(\hat{H}(x)+\varepsilon)}{\mu(Y)}$$

for all $x \in X$. Hence, the sequence

$$\left(\frac{1}{w_l(Y)}\sum_{n=0}^{l-1}\mathcal{L}_\mu^n\left(\mathbb{1}_{Y_n^c}\right)(x)\right)_{l=1}^{\infty}$$

converges to $(\mu(Y))^{-1}\hat{H}|_Y$ uniformly on Y. We are done. ■

The Darling–Kac Theorem also holds if the hypotheses of Theorem 5.3.2 are verified for some iterate of T only. Indeed, we have the following.

Theorem 5.3.4 (Darling-Kac Theorem II) Let $(X, \mathfrak{F}, \mu)$ be a σ-finite measure space and let $T: X \to X$ be a measurable map preserving the measure μ. Assume that there exists some integer $q \geq 1$ and some set $Y \in \mathfrak{F}$ with the following properties.

(a) $0 < \mu(Y) < +\infty$.
(b) $(w_n(T^q, Y))_{n=1}^{\infty} \in \mathcal{R}_{1-\alpha}$ with some $\alpha \in [0, 1]$.
(c) There exists a function $H: X \longrightarrow [0, +\infty)$, uniformly sweeping on Y with respect to T^q, such that

$$\frac{1}{w_n(T^q, Y)}\sum_{k=0}^{n-1}\mathcal{L}_\mu^{qk}\left(\mathbb{1}_{Y_k^c(T^q)}\right) \xrightarrow[n\to\infty]{} H|_Y$$

uniformly on Y, where

$$Y_k^c(T^q) := T^{-kq}(Y) \cap \bigcap_{j=0}^{k-1} T^{-qj}(X\backslash Y)$$

for $k \geq 1$ and $Y_0^c(T^q) := Y$.

Then, for every function $f \in L^1(\mu)$ with $\int f\,d\mu \neq 0$, the sequence $\left(\frac{1}{b_n}S_n f\right)_1^{\infty}$ converges strongly, with respect to the measure μ, to the Mittag–Leffler distribution $(\int f\,d\mu)\mathcal{M}_\alpha$; in symbols,

$$\frac{1}{a_n}S_n f \xrightarrow{\mathcal{L}} \left(\int f\,d\mu\right)\mathcal{M}_\alpha,$$

where

$$b_n := \frac{q}{\mu(Y)} \int_Y S_{E(n/q)}(\mathbb{1}_Y) d\mu_Y \sim \frac{1}{\Gamma(1+\alpha)\Gamma(2-\alpha)} \cdot \frac{n}{w_{E(n/q)}(T^q, Y)}. \tag{5.18}$$

Proof Let

$$a_n = \frac{1}{\mu(Y)} \int_Y S_n^{T^q}(\mathbb{1}_Y) \sim \frac{1}{\Gamma(1+\alpha)\Gamma(2-\alpha)} \cdot \frac{n}{w_n(T^q, A)}.$$

Theorem 5.3.2 tells us that the sequence $\frac{1}{a_n} S_n^{T^q}(S_q f)$ converges strongly with respect to the measure μ to the Mittag–Leffler distribution $\big(\int S_q(f)\, d\mu\big)\mathcal{M}_\alpha$. This means that, for every probability measure ν absolutely continuous with respect to μ and every bounded uniformly continuous function $\phi\colon \mathbb{R} \to \mathbb{R}$, we have that

$$\lim_{n\to\infty} \int_X \phi \circ \left(\frac{1}{a_n} S_n^{T^q}(S_q(f))\right) d\nu = \int_X \phi\, d\big(\mu(S_q(f))\mathcal{M}_\alpha\big).$$

Equivalently,

$$\lim_{n\to\infty} \int_X \phi \circ \left(\frac{1}{a_n} S_{qn}(f)\right) d\nu = \int_X \phi\, d\big(\mu(S_q(f))\mathcal{M}_\alpha\big).$$

Writing this equality for the function $\phi \circ \frac{1}{q}$, which is also bounded and uniformly continuous, rather than ϕ, we get that

$$\begin{aligned}
\lim_{n\to\infty} \int_X \phi\left(\frac{1}{qa_n} S_{qn}(f)\right) d\nu &= \lim_{n\to\infty} \int_X \left(\phi \circ \frac{1}{q}\right) \circ \left(\frac{1}{qa_n} S_{qn}(f)\right) d\nu \\
&= \int_X \left(\phi \circ \frac{1}{q}\right) d\big(\mu(S_q(f))\mathcal{M}_\alpha\big) \\
&= \int_X \phi(x)\, d\big(q\mu(f)\mathcal{M}_\alpha(qx)\big) \\
&= \int_X \phi\, d\left(q\mu(f)\frac{1}{q}\mathcal{M}_\alpha\right) \\
&= \int_X \phi\, d\big(\mu(f)\mathcal{M}_\alpha\big).
\end{aligned} \tag{5.19}$$

Now, for every integer $n \geq q$, write uniquely $n = k_n q + r_n$, with integers $k \geq 1$ and $0 \leq r_n \leq q-1$. Note that $k_n = E(n/q)$ and

$$\begin{aligned}
\left\| \frac{1}{qa_{E(n/q)}} S_n(f) - \frac{1}{qa_{k_n}} S_{qk_n}(f) \right\|_\infty &= \left\| \frac{1}{qa_{k_n}} S_{r_n}\big(f \circ T^{qk_n}\big) \right\|_\infty \\
&\leq \frac{r_n \|f\|_\infty}{qa_{k_n}} \leq \frac{\|f\|_\infty}{a_{k_n}} = \frac{\|f\|_\infty}{a_{E(n/q)}}.
\end{aligned}$$

As $\lim_{j\to\infty} a_j = \infty$, it follows from this formula, applied in (5.19), that

$$\lim_{n\to\infty} \int_X \phi \circ \left(\frac{1}{q a_{E(n/q)}} S_n(f) \right) d\nu = \int_X \phi \, d(\mu(f)) \mathcal{M}_\alpha).$$

This precisely means that the sequence $\frac{1}{b_n} S_n(f)$, $b_n = q a_{E(n/q)}$, converges strongly to the Mittag–Leffler distribution $\mu(f)\mathcal{M}_\alpha$. We are done. ■

This is precisely the form of Darling–Kac Theorem that we will prove in Section 22.8 in Volume 2 for some classes of parabolic elliptic functions. Lemma 5.3.3 now takes on the following form.

Lemma 5.3.5 *Let $(X, \mathfrak{F}, \mu)$ be a σ-finite measure space and let $T: X \longrightarrow X$ be a measurable map preserving measure μ. If there exists a (measurable) function $\hat{H}: X \longrightarrow [0, +\infty)$, supported on some set $Y \in \mathfrak{F}$, such that the sequence*

$$\left(\frac{1}{\mu_Y(Y_k(f^q))} \mathcal{L}_\mu^{qk} \left(1\!\!1_{Y_k(f^q)} \right) \right)_{k=1}^{\infty}$$

converges uniformly on Y to $\hat{H}|_Y$, then condition Theorem 5.3.4(c) holds with the function $H := (\mu(Y))^{-1} \hat{H}: X \longrightarrow [0, +\infty)$, although we do not claim that $H: X \longrightarrow [0, +\infty)$ is uniformly sweeping. Obviously, however, if $\hat{H}$ is uniformly sweeping, then so is H.

We end this section by stating without proof a weak pointwise version of ergodic theorem in the realm of infinite invariant measures. The existence of Darling–Kac sets is needed now, and also some kind of mixing. We say that $A \in \mathfrak{F}$ is called a Darling–Kac set if and only if $0 < \mu(A) < +\infty$ and there exists a sequence $(a_n)_{n=1}$ of positive reals such that

$$\lim_{n\to\infty} \frac{1}{a_n} \sum_{j=0}^{n-1} \mathcal{L}_\mu^j (1\!\!1_A) = \mu(A)$$

uniformly on A.

Theorem 5.3.6 *Let $(X, \mathfrak{F}, \mu)$ be a σ-finite measure space and let $T: X \longrightarrow X$ be a measurable map preserving measure μ. Assume that a Darling–Kac set exists. Suppose that $(\varphi(n))_{n=1}^{\infty}$ is an increasing sequence of positive real numbers such that the sequence $\left(\frac{\varphi(n)}{n}\right)_{n=1}^{\infty}$ is decreasing. Then*

(a) *If* $\sum_{n=1}^{\infty} \frac{1}{n} \exp(-\beta\varphi(n)) < +\infty$ *for all* $\beta > 1$, *then*

$$\limsup_{n\to\infty} \frac{1}{n^\alpha h\left(\frac{n}{\varphi(n)}\right)\varphi(n)^{1-\alpha}} \sum_{k=1}^{n} f \circ T^k \le K_\alpha \int_X f\,d\mu$$

μ-*a.e. for every* $f \in L^1_+$.

(b) *If* $\sum_{n=1}^{\infty} \frac{1}{n} \exp(-\beta\varphi(n)) = +\infty$ *for all* $\beta < 1$, *then*

$$\limsup_{n\to\infty} \frac{1}{n^\alpha h\left(\frac{n}{\varphi(n)}\right)\varphi(n)^{1-\alpha}} \sum_{k=1}^{n} f \circ T^k \ge K_\alpha \int_X f\,d\mu$$

μ-*a.e. for every* $f \in L^1_+$.

(c)

$$\limsup_{n\to\infty} \frac{1}{n^\alpha h\left(\frac{n}{L_2(n)}\right)L_2(n)^{1-\alpha}} \sum_{k=1}^{n} f \circ T^k = K_\alpha \int_X f\,d\mu$$

μ-*a.e. for every* $f \in L^1_+$.

5.4 Points of Infinite Condensation: Abstract Setting

In this very short section, we provide a framework for dealing with infinite invariant measures in a topological setting. We pay special attention to the concept of points of infinite condensation introduced in [**U4**]. So, assume that X is a separable locally compact metrizable space and that

$$T: X \longrightarrow \hat{X}$$

is a continuous map, where

$$\hat{X} := X \cup \{\infty\}$$

is the Alexander (one-point) compactification of X. We want to emphasize that we by no means assume that $T: X \to \hat{X}$ is extendable to a continuous map from $\hat{X}$ to $\hat{X}$. We, however, set $T(\infty) = \infty$.

Given a Borel σ-finite measure μ on X, let $X_\mu(\infty)$ be the set of all points $x \in X$ such that $\mu(U) = +\infty$ for every open set U containing x. Following [**U4**], we call $X_\mu(\infty)$ the set of points of infinite condensation of μ. Notice that $X_\mu(\infty)$ is a closed subset of $\hat{X}$ and $X_\mu(\infty) = \emptyset$ if and only if the measure μ is finite.

Denote by $\mathcal{M}_T^\infty$ the family of all Borel ergodic conservative T-invariant measures μ for which

$$\mu(X \backslash X_\mu(\infty)) > 0. \tag{5.20}$$

Note that if $\mu \in \mathcal{M}_T^\infty$, then $X_\mu(\infty)$ is forward invariant, i.e.,

$$T(X_\mu(\infty)) \subseteq X_\mu(\infty). \tag{5.21}$$

The following simple proposition shows that $X_\mu(\infty)$ is measurably negligible and, in the topologically transitive case, it is also topologically negligible.

Proposition 5.4.1 *If X is a locally compact separable metric space, if $T: X \longrightarrow \hat{X}$ is a continuous map, and if $\mu \in \mathcal{M}_T^\infty$, then $\mu(X_\mu(\infty)) = 0$. If, in addition, $T: X \longrightarrow \hat{X}$ is topologically transitive, then $X_\mu(\infty)$ is a nowhere dense subset of X.*

Proof Proving the first assertion of this proposition, suppose on the contrary that

$$\mu(X_\mu(\infty)) \neq 0.$$

Then, by ergodicity and conservativity of μ, we have that

$$\mu\left(X \backslash \bigcup_{n=0}^{\infty} T^n(X_\mu(\infty))\right) = 0.$$

Along with (5.21), this implies that

$$\mu\big(X \backslash X_\mu(\infty)\big) = 0.$$

This contradicts (5.20) and finishes the proof of the first assertion of the proposition.

Suppose now that the map $T: X \to X$ is topologically transitive. Seeking contradiction, assume that the set $X_\mu(\infty)$ is not nowhere dense. This means that $\mathrm{Int}\big(X_\mu(\infty)\big) \neq \emptyset$. Hence, $\mathrm{Int}\big(X_\mu(\infty)\big)$ contains a transitive point and we conclude that

$$\bigcup_{n=0}^{\infty} T^n(X_\mu(\infty))$$

is a dense subset of X. Since, however,

$$\overline{\bigcup_{n=0}^{\infty} T^n(X_\mu(\infty))} \subseteq X_\mu(\infty)$$

and since $X \backslash X_\mu(\infty) \neq \emptyset$ (by (5.20)), we get a contradiction that finishes the proof. ∎

Proposition 5.4.2 *If $T : X \longrightarrow \hat{X}$ is a continuous map, X is a locally compact separable metric space, and $\mu \in \mathcal{M}_T^\infty$, then μ is σ-finite.*

Proof By the definition of $X_\mu(\infty)$, for every point $x \in X \backslash X_\mu(\infty)$, there exists an open ball $B_x \subseteq X \backslash X_\mu(\infty)$ such that $x \in B_x$ and

$$\mu(B_x) < +\infty.$$

Since $X \backslash X_\mu(\infty)$ is a separable metric space, Lindelöf's Theorem implies that there exists a countable set $Y \subseteq X$ such that

$$\bigcup_{y \in Y} B_y = X \backslash X_\mu(\infty).$$

Thus,

$$\{X_\mu(\infty)\} \cup \{B_y\}_{y \in Y}$$

forms a countable cover of X by Borel sets of finite measure as $\mu(X_\mu(\infty)) = 0$ by Proposition 5.4.1. ■

6

Measure-Theoretic Entropy

In this chapter, we shall deal with measure-theoretic entropy of a measurable transformation preserving a probability space. Measure-theoretic entropy is also sometimes known as metric entropy or Kolmogorov–Sinai metric entropy. It was introduced by Kolmogorov and Sinai in the late 1950s; see [**Sin1**]. Since then its account has been presented in virtually every textbook on ergodic theory. Its introduction to dynamical systems was motivated by the concept of Boltzmann entropy of statistical mechanics and Shannon's work on information theory; see [**Sh1**] and [**Sh2**].

We will encounter three stages in the definition of metric entropy. It is defined by partitioning the underlying measurable space with measurable sets. Indeed, whereas one cannot generally partition a topological space into open sets, it is no problem to partition a measurable space into measurable sets.

6.1 Partitions

Let $(X, \mathfrak{F})$ be a measurable space and $\mathcal{A}$ be a countable measurable partition of X, i.e., $\mathcal{A} = \{A_k\}_{k\geq 1}$ with each $A_k \in \mathfrak{F}$ such that

- $A_i \cap A_j = \emptyset$ for all $i \neq j$
  and
- $\bigcup_{k\geq 1} A_k = X$.

For each $x \in X$, denote by $\mathcal{A}(x)$ the unique element (atom) of the partition $\mathcal{A}$ that contains the point x. In the remainder of this chapter, we shall use the calligraphic letters $\mathcal{A}, \mathcal{B}, \mathcal{C}, \ldots$ to denote partitions, with the notable exception of $\mathfrak{F}$, which will denote a σ-algebra on the space X.

Also, if X is a metrizable topological space, then, as usual, $\mathcal{B}(X)$ denotes the Borel σ-algebra on X.

We shall denote the set of all countable measurable partitions on the space $(X, \mathfrak{F})$ by $\text{Part}(X, \mathfrak{F})$. Moreover, it will always be implicitly understood that partitions are countable (finite or infinite) and measurable.

Definition 6.1.1 Let $(X, \mathfrak{F})$ be a measurable space and $\mathcal{A}, \mathcal{B} \in \text{Part}(X, \mathfrak{F})$. We say that partition $\mathcal{B}$ is *finer* than partition $\mathcal{A}$, or that $\mathcal{A}$ is *coarser* than $\mathcal{B}$, which will be denoted by $\mathcal{A} \leq \mathcal{B}$, if, for every atom $B \in \mathcal{B}$, there exists some atom $A \in \mathcal{A}$ such that $B \subseteq A$.

Equivalently, $\mathcal{B}$ is a refinement of $\mathcal{A}$ if $\mathcal{B}(x) \subseteq \mathcal{A}(x)$ for all $x \in X$. We now introduce, for partitions, the analogue of the join of two covers.

Definition 6.1.2 Given $\mathcal{A}, \mathcal{B} \in \text{Part}(X, \mathfrak{F})$, the partition

$$\mathcal{A} \vee \mathcal{B} := \{A \cap B : A \in \mathcal{A}, B \in \mathcal{B}\}$$

is called the *join* of $\mathcal{A}$ and $\mathcal{B}$.

The basic properties of the join are given in the following lemma. Their proofs are left to the reader as an exercise.

Lemma 6.1.3 *Let* $\mathcal{A}, \mathcal{B}, \mathcal{C}, \mathcal{D} \in \text{Part}(X, \mathfrak{F})$. *Then*

(a) $\mathcal{A} \vee \{X\} = \mathcal{A}$.
(b) $\mathcal{A} \vee \mathcal{B} = \mathcal{B} \vee \mathcal{A}$.
(c) $\mathcal{A} \leq \mathcal{A} \vee \mathcal{B}$ *and* $\mathcal{B} \leq \mathcal{A} \vee \mathcal{B}$.
(d) $\mathcal{A} \leq \mathcal{B}$ *if and only if* $\mathcal{A} \vee \mathcal{B} = \mathcal{B}$.
(e) *If* $\mathcal{A} \leq \mathcal{C}$ *and* $\mathcal{B} \leq \mathcal{D}$, *then* $\mathcal{A} \vee \mathcal{B} \leq \mathcal{C} \vee \mathcal{D}$.

6.2 Information and Conditional Information Functions

Let $(X, \mathfrak{F}, \mu)$ be a probability space. As such, the set X may be construed as the set of all possible states (or outcomes) of an experiment, while the σ-algebra $\mathfrak{F}$ consists of the set of all possible events and $\mu(E)$ is the probability that event $E \in \mathfrak{F}$ takes place. Imagine that this experiment is conducted using an instrument that, because of some limitation, can only provide measurements that are accurate up to the atoms of a partition $\mathcal{A} = \{A_k\}_{k \geq 1} \in \text{Part}(X, \mathfrak{F})$. In other words, this instrument can only tell us which atom of $\mathcal{A}$ the outcome of the experiment falls into. Any observation made through this instrument will therefore be of the form A_k for a unique k. If the experiment were conducted today, the probability that its outcome belongs to A_k, i.e., the probability that the experimental result in observation A_k being made with our instrument, would be given by $\mu(A_k)$.

We would like to introduce a function that describes the information that our instrument would give us about the outcome of the experiment. So, let $x \in X$. Intuitively, the smaller the atom of the partition to which x belongs, the more information our instrument provides us about x. In particular, if x lies in an atom of full measure, then our instrument gives us essentially no information about x. Moreover, because our instrument does not distinguish points that belong to a common atom of the partition, the sought information function must be constant on every atom.

Definition 6.2.1 Let $(X, \mathfrak{F}, \mu)$ be a probability space and $\mathcal{A} \in \text{Part}(X, \mathfrak{F})$. The function $I_\mu(\mathcal{A})\colon X \longrightarrow [0, \infty]$ defined by

$$I_\mu(\mathcal{A})(x) := -\log \mu(\mathcal{A}(x))$$

is called the *information function* of the partition $\mathcal{A}$.

As the function $t \mapsto -\log t$ is a decreasing function, for any $x \in X$, the smaller $\mu(\mathcal{A}(x))$ is, the larger $I_\mu(\mathcal{A})(x)$ is, i.e., the smaller the measure of the atom $\mathcal{A}(x)$ is, the more information the partition $\mathcal{A}$ gives us about x. In particular, the finer the partition, the more information it gives us about every point in the space.

We now enumerate some of the basic properties of the information function. Their proofs are straightforward and are again left to the reader.

Lemma 6.2.2 *Let $(X, \mathfrak{F}, \mu)$ be a probability space and $Meas(X, \mathfrak{F})$ be the set of all measurable functions on $(X, \mathfrak{F})$.*

(a) *The map*

$$\begin{array}{rccc} I_\mu\colon & \text{Part}(X,\mathfrak{F}) & \longrightarrow & Meas(X,\mathfrak{F}) \\ & \mathcal{A} & \longmapsto & I_\mu(\mathcal{A}) \end{array}$$

is an increasing function. In other words, if $\mathcal{A} \le \mathcal{B}$, then $I_\mu(\mathcal{A}) \le I_\mu(\mathcal{B})$.

(b) $I_\mu(\mathcal{A})(x) = 0$ *if and only if* $\mu(\mathcal{A}(x)) = 1$.

(c) $I_\mu(\mathcal{A})(x) = \infty$ *if and only if* $\mu(\mathcal{A}(x)) = 0$.

(d) $I_\mu(\mathcal{A})(x) = I_\mu(\mathcal{A})(y)$ *if* $\mathcal{A}(x) = \mathcal{A}(y)$, *i.e.,* $I_\mu(\mathcal{A})$ *is constant over each atom of* $\mathcal{A}$.

More advanced properties of the information function will be presented below. Meanwhile, we introduce a function which describes the information given by a partition $\mathcal{A}$ given that a partition $\mathcal{B}$ has already been applied.

Definition 6.2.3 The *conditional information function* of a partition $\mathcal{A}$ given a partition $\mathcal{B}$ is defined by

$$I_\mu(\mathcal{A}|\mathcal{B})(x) := -\log \mu_{\mathcal{B}(x)}(\mathcal{A}(x)).$$

Note that

$$\begin{aligned} I_\mu(\mathcal{A}|\mathcal{B})(x) &= -\log \frac{\mu\big(\mathcal{A}(x) \cap \mathcal{B}(x)\big)}{\mu(\mathcal{B}(x))} = -\log \frac{\mu\big((\mathcal{A} \vee \mathcal{B})(x)\big)}{\mu(\mathcal{B}(x))} \\ &= I_\mu(\mathcal{A} \vee \mathcal{B})(x) - I_\mu(\mathcal{B})(x). \end{aligned}$$

It is implicitly understood that $\frac{0}{0} = 0$ and $\infty - \infty = \infty$.

For any partition $\mathcal{A}$, observe that

$$I_\mu(\mathcal{A}|\{X\}) = I_\mu(\mathcal{A}),$$

i.e., the information function coincides with the conditional information function with respect to the trivial partition $\{X\}$. Note further that $I_\mu(\mathcal{A}|\mathcal{B})$ is constant over each atom of $\mathcal{A} \vee \mathcal{B}$.

We shall now provide some advanced properties of the conditional information function. Note that some of these properties hold pointwise, while others hold atomwise only, i.e., after integrating over atoms. In particular, the reader should compare statements (8) and (9).

Theorem 6.2.4 *Let $(X, \mathfrak{F}, \mu)$ be a probability space and $\mathcal{A}, \mathcal{B}, \mathcal{C} \in \mathrm{Part}(X, \mathfrak{F})$. The following statements hold.*

(1) $I_\mu(\mathcal{A} \vee \mathcal{B}|\mathcal{C}) = I_\mu(\mathcal{A}|\mathcal{C}) + I_\mu(\mathcal{B}|\mathcal{A} \vee \mathcal{C})$.
(2) $I_\mu(\mathcal{A} \vee \mathcal{B}) = I_\mu(\mathcal{A}) + I_\mu(\mathcal{B}|\mathcal{A})$.
(3) *If* $\mathcal{A} \leq \mathcal{B}$, *then* $I_\mu(\mathcal{A}|\mathcal{C}) \leq I_\mu(\mathcal{B}|\mathcal{C})$.
(4) *If* $\mathcal{A} \leq \mathcal{B}$, *then* $I_\mu(\mathcal{A}) \leq I_\mu(\mathcal{B})$.
(5) *If* $\mathcal{B} \leq \mathcal{C}$, *then* $\displaystyle\int_{A \cap B} I_\mu(\mathcal{A}|\mathcal{B})\, d\mu \geq \int_{A \cap B} I_\mu(\mathcal{A}|\mathcal{C})\, d\mu, \ \forall\, A \in \mathcal{A}, \forall\, B \in \mathcal{B}$.
Note: In general, $\mathcal{B} \leq \mathcal{C} \not\Rightarrow I_\mu(\mathcal{A}|\mathcal{B}) \geq I_\mu(\mathcal{A}|\mathcal{C})$.
(6) $\displaystyle\int_C I_\mu(\mathcal{A} \vee \mathcal{B}|\mathcal{C})\, d\mu \leq \int_C I_\mu(\mathcal{A}|\mathcal{C})\, d\mu + \int_C I_\mu(\mathcal{B}|\mathcal{C})\, d\mu, \ \forall\, C \in \mathcal{C}$.
Note: In general, $I_\mu(\mathcal{A} \vee \mathcal{B}|\mathcal{C}) \not\leq I_\mu(\mathcal{A}|\mathcal{C}) + I_\mu(\mathcal{B}|\mathcal{C})$.
(7) $\displaystyle\int_X I_\mu(\mathcal{A} \vee \mathcal{B})\, d\mu \leq \int_X I_\mu(\mathcal{A})\, d\mu + \int_X I_\mu(\mathcal{B})\, d\mu$.
Note: In general, $I_\mu(\mathcal{A} \vee \mathcal{B}) \not\leq I_\mu(\mathcal{A}) + I_\mu(\mathcal{B})$.
(8) $\displaystyle\int_{A \cap B} I_\mu(\mathcal{A}|\mathcal{C})\, d\mu \leq \int_{A \cap B} I_\mu(\mathcal{A}|\mathcal{B})\, d\mu + \int_{A \cap B} I_\mu(\mathcal{B}|\mathcal{C})\, d\mu, \forall\, A \in \mathcal{A}, \forall\, B \in \mathcal{B}$.
Note: In general, $I_\mu(\mathcal{A}|\mathcal{C}) \not\leq I_\mu(\mathcal{A}|\mathcal{B}) + I_\mu(\mathcal{B}|\mathcal{C})$.

(9) $I_\mu(\mathcal{A}) \leq I_\mu(\mathcal{A}|\mathcal{B}) + I_\mu(\mathcal{B})$.

Proof First, notice that (2) follows directly from (1) by setting $\mathcal{C} = \{X\}$. Similarly, (4) follows directly from (3), and (7) from (6). It is also easy to see that (6) follows upon combining (1) and (5), since $\mathcal{C} \leq \mathcal{A} \vee \mathcal{C}$. It, therefore, only remains to prove parts (1), (3), (5), (8), and (9).

(1) Let $x \in X$. Then

$$\begin{aligned}
I_\mu(\mathcal{A} \vee \mathcal{B}|\mathcal{C})(x) &= -\log \frac{\mu\big((\mathcal{A} \vee \mathcal{B} \vee \mathcal{C})(x)\big)}{\mu(\mathcal{C}(x))} \\
&= -\log \frac{\mu\big(\mathcal{B}(x) \cap (\mathcal{A} \vee \mathcal{C})(x)\big)}{\mu(\mathcal{C}(x))} \\
&= -\log \left(\frac{\mu\big(\mathcal{B}(x) \cap (\mathcal{A} \vee \mathcal{C})(x)\big)}{\mu\big((\mathcal{A} \vee \mathcal{C})(x)\big)} \cdot \frac{\mu\big((\mathcal{A} \vee \mathcal{C})(x)\big)}{\mu(\mathcal{C}(x))} \right) \\
&= -\log \frac{\mu\big(\mathcal{B}(x) \cap (\mathcal{A} \vee \mathcal{C})(x)\big)}{\mu\big((\mathcal{A} \vee \mathcal{C})(x)\big)} - \log \frac{\mu\big((\mathcal{A} \vee \mathcal{C})(x)\big)}{\mu(\mathcal{C}(x))} \\
&= I_\mu(\mathcal{B}|\mathcal{A} \vee \mathcal{C})(x) + I_\mu(\mathcal{A}|\mathcal{C})(x).
\end{aligned}$$

(3) If $\mathcal{A} \leq \mathcal{B}$, then $\mathcal{A} \vee \mathcal{B} = \mathcal{B}$. It follows from (1) that

$$I_\mu(\mathcal{B}|\mathcal{C}) = I_\mu(\mathcal{A} \vee \mathcal{B}|\mathcal{C}) = I_\mu(\mathcal{A}|\mathcal{C}) + I_\mu(\mathcal{B}|\mathcal{A} \vee \mathcal{C}) \geq I_\mu(\mathcal{A}|\mathcal{C}).$$

(5) Suppose that $\mathcal{B} \leq \mathcal{C}$. Let $A \in \mathcal{A}$ and $B \in \mathcal{B}$. The function $k\colon [0,1] \to [0,\infty)$ defined by $k(t) = -t \log t$ when $t \in (0,1]$ and $k(0) = 0$ is concave, i.e.,

$$k\big(tx + (1-t)y\big) \geq tk(x) + (1-t)k(y)$$

for all $t \in [0,1]$ and all $x, y \in [0,1]$. Consequently,

$$k\left(\sum_{n=1}^{\infty} a_n b_n\right) \geq \sum_{n=1}^{\infty} a_n k(b_n)$$

whenever $\sum_{n=1}^{\infty} a_n = 1$ and $a_n \geq 0$ for all $n \geq 1$. Then

$$k\left(\sum_{C \in \mathcal{C}} \mu_B(C) \frac{\mu(A \cap C)}{\mu(C)}\right) \geq \sum_{C \in \mathcal{C}} \mu_B(C) k\left(\frac{\mu(A \cap C)}{\mu(C)}\right).$$

Since $\mathcal{B} \le \mathcal{C}$, either $C \cap B = C$ or $C \cap B = \emptyset$. Thus, either $\mu_B(C) = \frac{\mu(C)}{\mu(B)}$ or $\mu_B(C) = 0$. So the left-hand side of the previous inequality simplifies to

$$k\left(\sum_{C\in\mathcal{C}} \mu_B(C)\frac{\mu(A\cap C)}{\mu(C)}\right) = k\left(\sum_{C\subseteq B} \frac{\mu(A\cap C)}{\mu(B)}\right) = k\left(\frac{\mu(A\cap B)}{\mu(B)}\right)$$
$$= -\frac{\mu(A\cap B)}{\mu(B)}\log\frac{\mu(A\cap B)}{\mu(B)},$$

whereas the right-hand side reduces to

$$\sum_{C\in\mathcal{C}} \mu_B(C)k\left(\frac{\mu(A\cap C)}{\mu(C)}\right) = \sum_{C\subseteq B}\frac{\mu(C)}{\mu(B)}k\left(\frac{\mu(A\cap C)}{\mu(C)}\right)$$
$$= \sum_{C\subseteq B} -\frac{\mu(A\cap C)}{\mu(B)}\log\frac{\mu(A\cap C)}{\mu(C)}.$$

Hence, the inequality becomes

$$-\frac{\mu(A\cap B)}{\mu(B)}\log\frac{\mu(A\cap B)}{\mu(B)} \ge \sum_{C\subseteq B} -\frac{\mu(A\cap C)}{\mu(B)}\log\frac{\mu(A\cap C)}{\mu(C)}.$$

Multiplying both sides by $\mu(B)$ yields

$$-\mu(A\cap B)\log\frac{\mu(A\cap B)}{\mu(B)} \ge \sum_{C\subseteq B} -\mu(A\cap C)\log\frac{\mu(A\cap C)}{\mu(C)}.$$

Therefore,

$$\begin{aligned}\int_{A\cap B} I_\mu(\mathcal{A}|\mathcal{B})\,d\mu &= -\mu(A\cap B)\log\frac{\mu(A\cap B)}{\mu(B)}\\ &\ge \sum_{C\subseteq B} -\mu(A\cap C)\log\frac{\mu(A\cap C)}{\mu(C)}\\ &= \sum_{C\subseteq B}\int_{A\cap C} I_\mu(\mathcal{A}|\mathcal{C})\,d\mu\\ &= \int_{A\cap B} I_\mu(\mathcal{A}|\mathcal{C})\,d\mu.\end{aligned}$$

(8) Using parts (3) and (1) in turn, we have that

$$I_\mu(\mathcal{A}|\mathcal{C}) \le I_\mu(\mathcal{A}\vee\mathcal{B}|\mathcal{C}) = I_\mu(\mathcal{B}|\mathcal{C}) + I_\mu(\mathcal{A}|\mathcal{B}\vee\mathcal{C}).$$

Since $\mathcal{B} \le \mathcal{B}\vee\mathcal{C}$, part (5) ensures that

$$\int_{A\cap B} I_\mu(\mathcal{A}|\mathcal{B})\,d\mu \ge \int_{A\cap B} I_\mu(\mathcal{A}|\mathcal{B}\vee\mathcal{C})\,d\mu, \forall\, A\in\mathcal{A}, \forall\, B\in\mathcal{B}.$$

Therefore,

$$\int_{A\cap B} I_\mu(\mathcal{A}|\mathcal{C})\,d\mu \le \int_{A\cap B} I_\mu(\mathcal{A}|\mathcal{B})\,d\mu + \int_{A\cap B} I_\mu(\mathcal{B}|\mathcal{C})\,d\mu, \forall\, A \in \mathcal{A}, \forall\, B \in \mathcal{B}.$$

(9) Using parts (4) and (2) in succession, we get that

$$I_\mu(\mathcal{A}) \le I_\mu(\mathcal{A} \vee \mathcal{B}) = I_\mu(\mathcal{A}|\mathcal{B}) + I_\mu(\mathcal{B}).$$

■

6.3 Entropy and Conditional Entropy for Partitions

The information function associated with a partition gives us the amount of information that can be gathered from the partition about each and every outcome of the experiment. It it obviously useful to encompass the information given by a partition within a single number rather than a function. A natural way to achieve this is to calculate the average information given by the partition. This means integrating the information function over the entire space. The resulting integral is called the entropy of the partition. This is the first stage in the definition of the entropy of an endomorphism.

Definition 6.3.1 Let $(X, \mathfrak{F}, \mu)$ be a probability space and $\mathcal{A} \in \text{Part}(X, \mathfrak{F})$. The *entropy of* $\mathcal{A}$ with respect to the measure μ is defined to be

$$\text{H}_\mu(\mathcal{A}) := \int_X I_\mu(\mathcal{A})\,d\mu = \sum_{A \in \mathcal{A}} -\mu(A) \log \mu(A),$$

where it is implicitly understood that $0 \cdot \infty = 0$, since null sets do not contribute to the integral.

The entropy of a partition is equal to zero if and only if the partition has an atom of full measure (which implies that all other atoms are null). In particular, $\text{H}_\mu(\{X\}) = 0$. Moreover, the entropy of a partition is small if the partition contains one atom with nearly full measure (so all other atoms have small measure). Using calculus, it is also possible to show that if the partition $\mathcal{A}$ is finite, then

$$0 \le \text{H}_\mu(\mathcal{A}) \le \log \#\mathcal{A} \tag{6.1}$$

and that

$$\text{H}_\mu(\mathcal{A}) = \log \#\mathcal{A}$$

if and only if

$$\mu(A) = 1/\#\mathcal{A}$$

for all $A \in \mathcal{A}$. In other words, on average, we gain the most information from carrying out an experiment when the potential events have an equal probability of occurring. Similarly, the conditional entropy of a partition $\mathcal{A}$ given a partition $\mathcal{B}$ is the average conditional information of $\mathcal{A}$ given $\mathcal{B}$.

Definition 6.3.2 Let $(X, \mathfrak{F}, \mu)$ be a probability space and $\mathcal{A}, \mathcal{B} \in \mathrm{Part}(X, \mathfrak{F})$. The *conditional entropy of $\mathcal{A}$ given $\mathcal{B}$* is defined to be

$$\mathrm{H}_\mu(\mathcal{A}|\mathcal{B}) := \int_X I_\mu(\mathcal{A}|\mathcal{B})d\mu = \sum_{A\in\mathcal{A}}\sum_{B\in\mathcal{B}} -\mu(A\cap B)\log\frac{\mu(A\cap B)}{\mu(B)}.$$

Note that $\mathrm{H}_\mu(\mathcal{A}) = \mathrm{H}_\mu(\mathcal{A}|\{X\})$. Also, if $B \in \mathfrak{F}$, we define a new partition $\mathcal{A}|_B$ of B by setting

$$\mathcal{A}|_B := \{A \cap B : A \in \mathcal{A}\}.$$

Then

$$\mathrm{H}_\mu(\mathcal{A}|\mathcal{B}) = \sum_{B\in\mathcal{B}} \mathrm{H}_{\mu_B}(\mathcal{A}|_B)\mu(B).$$

Indeed,

$$\begin{aligned}
\mathrm{H}_\mu(\mathcal{A}|\mathcal{B}) &= \sum_{A\in\mathcal{A}}\sum_{B\in\mathcal{B}} -\mu(A\cap B)\log\frac{\mu(A\cap B)}{\mu(B)} \\
&= \sum_{B\in\mathcal{B}}\sum_{A\in\mathcal{A}} -\frac{\mu(A\cap B)}{\mu(B)}\log\frac{\mu(A\cap B)}{\mu(B)}\cdot\mu(B) \\
&= \sum_{B\in\mathcal{B}}\sum_{A\in\mathcal{A}} -\mu_B(A)\log\mu_B(A)\cdot\mu(B) \\
&= \sum_{B\in\mathcal{B}} \mathrm{H}_{\mu_B}(\mathcal{A}|_B)\mu(B).
\end{aligned}$$

Hence, the conditional entropy of $\mathcal{A}$ given $\mathcal{B}$ is equal to the weighted average of the entropies of the partitions of each atom $B \in \mathcal{B}$ into the sets $A \cap B$, $A \in \mathcal{A}$.

Of course, the properties of entropy (resp. conditional entropy) are inherited from the properties of the information function (resp. the conditional information function) via integration.

Theorem 6.3.3 *Let $(X, \mathfrak{F}, \mu)$ be a probability space and $\mathcal{A}, \mathcal{B}, \mathcal{C} \in \mathrm{Part}(X, \mathfrak{F})$. The following statements hold.*

(1) $\mathrm{H}_\mu(\mathcal{A} \vee \mathcal{B}|\mathcal{C}) = \mathrm{H}_\mu(\mathcal{A}|\mathcal{C}) + \mathrm{H}_\mu(\mathcal{B}|\mathcal{A} \vee \mathcal{C})$.
(2) $\mathrm{H}_\mu(\mathcal{A} \vee \mathcal{B}) = \mathrm{H}_\mu(\mathcal{A}) + \mathrm{H}_\mu(\mathcal{B}|\mathcal{A})$.

(3) *If* $\mathcal{A} \le \mathcal{B}$, *then* $\mathrm{H}_\mu(\mathcal{A}|\mathcal{C}) \le \mathrm{H}_\mu(\mathcal{B}|\mathcal{C})$.
(4) *If* $\mathcal{A} \le \mathcal{B}$, *then* $\mathrm{H}_\mu(\mathcal{A}) \le \mathrm{H}_\mu(\mathcal{B})$.
(5) *If* $\mathcal{B} \le \mathcal{C}$, *then* $\mathrm{H}_\mu(\mathcal{A}|\mathcal{B}) \ge \mathrm{H}_\mu(\mathcal{A}|\mathcal{C})$.
(6) $\mathrm{H}_\mu(\mathcal{A} \vee \mathcal{B}|\mathcal{C}) \le \mathrm{H}_\mu(\mathcal{A}|\mathcal{C}) + \mathrm{H}_\mu(\mathcal{B}|\mathcal{C})$.
(7) $\mathrm{H}_\mu(\mathcal{A} \vee \mathcal{B}) \le \mathrm{H}_\mu(\mathcal{A}) + \mathrm{H}_\mu(\mathcal{B})$.
(8) $\mathrm{H}_\mu(\mathcal{A}|\mathcal{C}) \le \mathrm{H}_\mu(\mathcal{A}|\mathcal{B}) + \mathrm{H}_\mu(\mathcal{B}|\mathcal{C})$.
(9) $\mathrm{H}_\mu(\mathcal{A}) \le \mathrm{H}_\mu(\mathcal{A}|\mathcal{B}) + \mathrm{H}_\mu(\mathcal{B})$.

Proof All the statements follow from their counterparts in Theorem 6.2.4 after integration or summation over atoms. For instance, let us prove (5). If $\mathcal{B} \le \mathcal{C}$, then it follows from Theorem 6.2.4(5) that

$$\begin{aligned}
\mathrm{H}_\mu(\mathcal{A}|\mathcal{B}) = \int_X I_\mu(\mathcal{A}|\mathcal{B})\,d\mu &= \sum_{A\in\mathcal{A}}\sum_{B\in\mathcal{B}}\int_{A\cap B} I_\mu(\mathcal{A}|\mathcal{B})\,d\mu \\
&\ge \sum_{A\in\mathcal{A}}\sum_{B\in\mathcal{B}}\int_{A\cap B} I_\mu(\mathcal{A}|\mathcal{C})\,d\mu \\
&= \int_X I_\mu(\mathcal{A}|\mathcal{C})\,d\mu \\
&= \mathrm{H}_\mu(\mathcal{A}|\mathcal{C}).
\end{aligned}$$

■

6.4 Entropy of a (Probability) Measure-Preserving Endomorphism

So far in this chapter, we have studied partitions of a space, but, of course, this is not our ultimate objective. Our ultimate aim is, however, to study measure-theoretic dynamical systems. So let

$$T : X \longrightarrow X$$

be a measure-preserving endomorphism of a probability space $(X, \mathfrak{F}, \mu)$ and $\mathcal{A} = \{A_k\}_{k\ge1}$ be a countable measurable partition of X. Observe that

$$T^{-1}\mathcal{A} := \{T^{-1}(A) : A \in \mathcal{A}\}$$

is also a countable measurable partition of X.

Recall that the set X can be thought of as representing the set of all possible outcomes (or states) of an experiment, while the σ-algebra $\mathfrak{F}$ consists of the set of all possible events and $\mu(E)$ is the probability that event E takes place. Recall also that a partition $\mathcal{A} = \{A_k\}_{k\ge1}$ can be thought of as the set of all observations that can be made with a given instrument. The action of T on

$(X, \mathfrak{F}, \mu)$ may be conceived as the passage of one unit of time (e.g., a day). Today would naturally be taken as the reference point for time 0. Suppose that we conduct the experiment with our instrument tomorrow. The resulting observation would be one of the A_ks, say A_{k_1}, on day 1. Because of the passage of time (in other words, one iteration of T), in order to make observation A_{k_1} at time 1, our measure-theoretic system would have to be in one of the states of $T^{-1}(A_{k_1})$ today. The probability of making observation A_{k_1} on day 1 is, thus, $\mu(T^{-1}(A_{k_1}))$. Assume now that we conduct the same experiment for n consecutive days, starting today. What is the probability that we make the sequence of observations $A_{k_0}, A_{k_1}, \ldots, A_{k_{n-1}}$ on those successive days? We would make those observations precisely if our system is in one of the states of

$$\bigcap_{m=0}^{n-1} T^{-m}(A_{k_m})$$

today. Therefore, the probability that our observations are, respectively $A_{k_0}, A_{k_1}, \ldots, A_{k_{n-1}}$ on n successive days starting today is

$$\mu\left(\bigcap_{m=0}^{n-1} T^{-m}(A_{k_m})\right).$$

Given this discussion, we claim, it is, thus, natural to introduce, for all $0 \le m \le n-1$, the partitions

$$\mathcal{A}_m^n := \bigvee_{i=m}^{n-1} T^{-i}\mathcal{A} = T^{-m}\mathcal{A} \vee \cdots \vee T^{-(n-1)}\mathcal{A}.$$

If $m \ge n$, we define $\mathcal{A}_m^n$ to be the trivial partition $\{X\}$. To shorten the notation, we shall write $\mathcal{A}^n$ in lieu of $\mathcal{A}_0^n$. Let us give the basic properties of the operator T^{-1} on partitions.

Lemma 6.4.1 *Let $T: X \longrightarrow X$ be a measurable transformation of a measurable space $(X, \mathfrak{F})$ and $\mathcal{A}, \mathcal{B} \in \mathrm{Part}(X, \mathfrak{F})$. Then the following statements hold.*

(1) *The operator T^{-1} commutes with the operator $\vee$, i.e., $T^{-1}(\mathcal{A} \vee \mathcal{B}) = T^{-1}\mathcal{A} \vee T^{-1}\mathcal{B}$.*
(2) *$T^{-1}(\mathcal{A}_m^n) = (T^{-1}\mathcal{A})_m^n$ for all $m, n \ge 0$.*
(3) *$(\mathcal{A} \vee \mathcal{B})_m^n = \mathcal{A}_m^n \vee \mathcal{B}_m^n$ for all $m, n \ge 0$.*
(4) *$(\mathcal{A}_k^l)_m^n = \mathcal{A}_{k+m}^{l+n-1}$.*
(5) *T^{-1} preserves the order $\le$, i.e., if $\mathcal{A} \le \mathcal{B}$, then $T^{-1}\mathcal{A} \le T^{-1}\mathcal{B}$.*
(6) *More generally, if $\mathcal{A} \le \mathcal{B}$, then $\mathcal{A}_m^n \le \mathcal{B}_m^n$ for all $m, n \ge 0$.*

(7) $(T^{-1}\mathcal{A})(x) = T^{-1}\big(\mathcal{A}(T(x))\big)$ *for all* $x \in X$.

Proof The proof of assertions (1) and (5) is left to the reader as an exercise. In order to prove (2) by using (1) repeatedly, we get that

$$\begin{aligned} T^{-1}(\mathcal{A}_m^n) = T^{-1}\left(\bigvee_{i=m}^{n-1} T^{-i}\mathcal{A}\right) &= \bigvee_{i=m}^{n-1} T^{-1}(T^{-i}\mathcal{A}) \\ &= \bigvee_{i=m}^{n-1} T^{-i}(T^{-1}\mathcal{A}) = (T^{-1}\mathcal{A})_m^n. \end{aligned}$$

(3) Again, by using (1) repeatedly, we obtain that

$$\begin{aligned} (\mathcal{A} \vee \mathcal{B})_m^n &= \bigvee_{i=m}^{n-1} T^{-i}(\mathcal{A} \vee \mathcal{B}) = \bigvee_{i=m}^{n-1} (T^{-i}\mathcal{A} \vee T^{-i}\mathcal{B}) \\ &= \left(\bigvee_{i=m}^{n-1} T^{-i}\mathcal{A}\right) \vee \left(\bigvee_{i=m}^{n-1} T^{-i}\mathcal{B}\right) \\ &= \mathcal{A}_m^n \vee \mathcal{B}_m^n. \end{aligned}$$

Dealing with (4), by using (1), it follows that

$$\begin{aligned} (\mathcal{A}_k^l)_m^n = \bigvee_{j=m}^{n-1} T^{-j}(\mathcal{A}_k^l) &= \bigvee_{j=m}^{n-1} T^{-j}\left(\bigvee_{i=k}^{l-1} T^{-i}\mathcal{A}\right) \\ &= \bigvee_{j=m}^{n-1}\bigvee_{i=k}^{l-1} T^{-(i+j)}\mathcal{A} = \bigvee_{s=k+m}^{l+n-2} T^{-s}\mathcal{A} = \mathcal{A}_{k+m}^{l+n-1}. \end{aligned}$$

(6) Suppose that $\mathcal{A} \le \mathcal{B}$. Using (5) repeatedly and Lemma 6.1.3(5), we obtain that

$$\mathcal{A}_m^n = \bigvee_{i=m}^{n-1} T^{-i}\mathcal{A} \le \bigvee_{i=m}^{n-1} T^{-i}\mathcal{B} = \mathcal{B}_m^n.$$

(7) Let $x \in X$. Let $T^{-1}(A) = (T^{-1}\mathcal{A})(x)$, i.e., $A \in \mathcal{A}$ is such that $x \in T^{-1}(A)$. Then $T(x) \in A$, i.e., $A = \mathcal{A}(T(x))$. Hence,

$$(T^{-1}\mathcal{A})(x) = T^{-1}(A) = T^{-1}\big(\mathcal{A}(T(x))\big).$$ ■

We now describe the behavior of the operator T^{-1} with respect to the information function.

Lemma 6.4.2 *Let* $T\colon X \longrightarrow X$ *be a measure-preserving endomorphism of a probability space* $(X, \mathfrak{F}, \mu)$ *and* $\mathcal{A}, \mathcal{B} \in \mathrm{Part}(X, \mathfrak{F})$. *Then the following statements hold.*

(1) $I_\mu(T^{-1}\mathcal{A}|T^{-1}\mathcal{B}) = I_\mu(\mathcal{A}|\mathcal{B}) \circ T$.
(2) $I_\mu(T^{-1}\mathcal{A}) = I_\mu(\mathcal{A}) \circ T$.

Proof It is clear that (2) follows from (1) by setting $\mathcal{B} = \{X\}$. To prove (1), let $x \in X$. By Lemma 6.4.1(1) and (7) and the assumption that μ is T-invariant, we have that

$$\begin{aligned} I_\mu(T^{-1}\mathcal{A}|T^{-1}\mathcal{B})(x) &= -\log \frac{\mu\big((T^{-1}\mathcal{A} \vee T^{-1}\mathcal{B})(x)\big)}{\mu\big((T^{-1}\mathcal{B})(x)\big)} \\ &= -\log \frac{\mu\big((T^{-1}(\mathcal{A} \vee \mathcal{B}))(x)\big)}{\mu\big((T^{-1}\mathcal{B})(x)\big)} \\ &= -\log \frac{\mu\big(T^{-1}\big((\mathcal{A} \vee \mathcal{B})(T(x))\big)\big)}{\mu\big(T^{-1}\big(\mathcal{B}(T(x))\big)\big)} \\ &= -\log \frac{\mu\big((\mathcal{A} \vee \mathcal{B})(T(x))\big)}{\mu\big(\mathcal{B}(T(x))\big)} \\ &= I_\mu(\mathcal{A}|\mathcal{B})(T(x)) \\ &= I_\mu(\mathcal{A}|\mathcal{B}) \circ T(x). \end{aligned}$$

■

A more intricate property of the information function is given in the following lemma.

Lemma 6.4.3 *Let $T : X \longrightarrow X$ be a measure-preserving endomorphism of a probability space $(X, \mathfrak{F}, \mu)$. If $\mathcal{A} \in \mathrm{Part}(X, \mathfrak{F})$, then*

$$I_\mu(\mathcal{A}^n) = \sum_{j=1}^{n} I_\mu\big(\mathcal{A}|\mathcal{A}_1^j\big) \circ T^{n-j}$$

for all integers $n \geq 1$.

Proof We will prove this lemma by induction. For $n = 1$, since $\mathcal{A}_1^1$ is by definition equal to the trivial partition $\{X\}$, we have that

$$I_\mu(\mathcal{A}^1) = I_\mu(\mathcal{A}) = I_\mu(\mathcal{A}|\{X\}) = I_\mu\big(\mathcal{A}|\mathcal{A}_1^1\big) = I_\mu\big(\mathcal{A}|\mathcal{A}_1^1\big) \circ T^{1-1}.$$

Now suppose that the lemma holds for some $n \geq 1$. Then, in light of Theorem 6.2.4(2) and Lemma 6.4.2(2), we obtain that

$$\begin{aligned} I_\mu\big(\mathcal{A}^{n+1}\big) &= I_\mu\big(\mathcal{A}_1^{n+1} \vee \mathcal{A}\big) = I_\mu\big(\mathcal{A}_1^{n+1}\big) + I_\mu\big(\mathcal{A}|\mathcal{A}_1^{n+1}\big) \\ &= I_\mu\big(T^{-1}(\mathcal{A}^n)\big) + I_\mu\big(\mathcal{A}|\mathcal{A}_1^{n+1}\big) \\ &= I_\mu(\mathcal{A}^n) \circ T + I_\mu\big(\mathcal{A}|\mathcal{A}_1^{n+1}\big) \end{aligned}$$

$$= \sum_{j=1}^{n} I_\mu(\mathcal{A}|\mathcal{A}_1^j) \circ T^{n-j} \circ T + I_\mu(\mathcal{A}|\mathcal{A}_1^{n+1})$$

$$= \sum_{j=1}^{n} I_\mu(\mathcal{A}|\mathcal{A}_1^j) \circ T^{n+1-j} + I_\mu(\mathcal{A}|\mathcal{A}_1^{n+1}) \circ T^{n+1-(n+1)}$$

$$= \sum_{j=1}^{n+1} I_\mu(\mathcal{A}|\mathcal{A}_1^j) \circ T^{n+1-j}.$$

■

We now turn our attention to the effect that our measure-theoretic dynamical system T has on entropy. In particular, observe that, because the system is measure preserving, conducting the experiment today or tomorrow (or at any time in the future) gives us the same amount of average information about the outcome. This is the meaning of the second of the following properties of entropy.

Lemma 6.4.4 *If $T: X \longrightarrow X$ is a measure-preserving endomorphism of a probability space $(X, \mathfrak{F}, \mu)$ and $\mathcal{A}, \mathcal{B} \in \text{Part}(X, \mathfrak{F})$, then*

(1) $\text{H}_\mu(T^{-1}\mathcal{A}|T^{-1}\mathcal{B}) = \text{H}_\mu(\mathcal{A}|\mathcal{B})$.
(2) $\text{H}_\mu(T^{-1}\mathcal{A}) = \text{H}_\mu(\mathcal{A})$.
(3) $\text{H}_\mu(\mathcal{A}^n|\mathcal{B}^n) \le n\text{H}_\mu(\mathcal{A}|\mathcal{B})$.

Proof Part (2) follows from (1) by taking $\mathcal{B} = \{X\}$. (1) Using Lemma 6.4.2(1) and the T-invariance of μ, we obtain that

$$\begin{aligned}\text{H}_\mu(T^{-1}\mathcal{A}|T^{-1}\mathcal{B}) &= \int_X I_\mu(T^{-1}\mathcal{A}|T^{-1}\mathcal{B})\,d\mu = \int_X I_\mu(\mathcal{A}|\mathcal{B}) \circ T\,d\mu \\ &= \int_X I_\mu(\mathcal{A}|\mathcal{B})\,d\mu \\ &= \text{H}_\mu(\mathcal{A}|\mathcal{B}).\end{aligned}$$

(3) We first prove that

$$\text{H}_\mu(\mathcal{A}^n|\mathcal{B}^n) \le \sum_{j=0}^{n-1} \text{H}_\mu(T^{-j}\mathcal{A}|T^{-j}\mathcal{B}).$$

This statement clearly holds when $n = 1$. Suppose that it holds for some $n \ge 1$. Then, using Theorem 6.3.3(1) and (5), we have that

$$\begin{aligned}\text{H}_\mu(\mathcal{A}^{n+1}|\mathcal{B}^{n+1}) &= \text{H}_\mu\big(\mathcal{A}^n \vee T^{-n}\mathcal{A}|\mathcal{B}^n \vee T^{-n}\mathcal{B}\big) \\ &= \text{H}_\mu\big(\mathcal{A}^n|\mathcal{B}^n \vee T^{-n}\mathcal{B}\big) + \text{H}_\mu\big(T^{-n}\mathcal{A}|\mathcal{A}^n \vee \mathcal{B}^n \vee T^{-n}\mathcal{B}\big) \\ &\le \text{H}_\mu(\mathcal{A}^n|\mathcal{B}^n) + \text{H}_\mu(T^{-n}\mathcal{A}|T^{-n}\mathcal{B})\end{aligned}$$

$$\leq \sum_{j=0}^{n-1} \mathrm{H}_\mu(T^{-j}\mathcal{A}|T^{-j}\mathcal{B}) + \mathrm{H}_\mu(T^{-n}\mathcal{A}|T^{-n}\mathcal{B})$$
$$= \sum_{j=0}^{n} \mathrm{H}_\mu(T^{-j}\mathcal{A}|T^{-j}\mathcal{B}).$$

By induction, the above statement holds for all $n \geq 1$. By (1), we get that

$$\mathrm{H}_\mu(\mathcal{A}^n|\mathcal{B}^n) \leq \sum_{j=0}^{n-1} \mathrm{H}_\mu(T^{-j}\mathcal{A}|T^{-j}\mathcal{B}) = \sum_{j=0}^{n-1} \mathrm{H}_\mu(\mathcal{A}|\mathcal{B}) = n\mathrm{H}_\mu(\mathcal{A}|\mathcal{B}).$$

■

The average information gained by conducting an experiment on n consecutive days, using the partition $\mathcal{A}$, is given by the entropy $\mathrm{H}_\mu(\mathcal{A}^n)$ since $\mathcal{A}^n$ has, for atoms, the sets

$$\bigcap_{m=0}^{n-1} T^{-m}(A_{k_m}),$$

where $A_{k_m} \in \mathcal{A}$ for all m. Not surprisingly, the average information gained by conducting the experiment on n consecutive days using the partition $\mathcal{A}$ is equal to the sum of the average conditional information gained by performing $\mathcal{A}$ on day $j+1$ given that the outcome of performing $\mathcal{A}$ over the previous j days is known, summing from the first day to the last day. This is summarized in the next lemma.

Lemma 6.4.5 *If $T: X \longrightarrow X$ is a measure-preserving endomorphism of a probability space $(X, \mathfrak{F}, \mu)$ and $\mathcal{A} \in \mathrm{Part}(X, \mathfrak{F})$, then*

$$\mathrm{H}_\mu(\mathcal{A}^n) = \sum_{j=1}^{n} \mathrm{H}_\mu(\mathcal{A}|\mathcal{A}_1^j)$$

for all integers $n \geq 1$.

Proof We deduce from Lemma 6.4.3 and the T-invariance of μ that

$$\mathrm{H}_\mu(\mathcal{A}^n) = \int_X I_\mu(\mathcal{A}^n)\,d\mu = \sum_{j=1}^{n} \int_X I_\mu(\mathcal{A}|\mathcal{A}_1^j) \circ T^{n-j}\,d\mu$$
$$= \sum_{j=1}^{n} \int_X I_\mu(\mathcal{A}|\mathcal{A}_1^j)\,d\mu = \sum_{j=1}^{n} \mathrm{H}_\mu(\mathcal{A}|\mathcal{A}_1^j).$$

■

In view of Theorem 6.3.3(5), observe that

$$\mathrm{H}_\mu\big(\mathcal{A}|\mathcal{A}_1^{j+1}\big) \le \mathrm{H}_\mu\big(\mathcal{A}|\mathcal{A}_1^{j}\big),$$

since $\mathcal{A}_1^{j+1} \ge \mathcal{A}_1^{j}$. So the sequence

$$\big(\mathrm{H}_\mu\big(\mathcal{A}|\mathcal{A}_1^{j}\big)\big)_{j\ge 1}$$

decreases to some limit, which we shall denote by

$$\mathrm{H}_\mu(T,\mathcal{A}).$$

Consequently, the corresponding sequence of Cesàro averages

$$\left(\frac{1}{n}\sum_{j=1}^{n}\mathrm{H}_\mu\big(\mathcal{A}|\mathcal{A}_1^{j}\big)\right)_{n\ge 1} = \left(\frac{1}{n}\mathrm{H}_\mu(\mathcal{A}^n)\right)_{n\ge 1}$$

decreases to the same limit. Thus, the following definition makes sense. This is the second stage in the definition of the entropy of an endomorphism.

Definition 6.4.6 If $T\colon X \longrightarrow X$ is a measure-preserving endomorphism of a probability space $(X,\mathfrak{F},\mu)$ and $\mathcal{A} \in \mathrm{Part}(X,\mathfrak{F})$, then the quantity $\mathrm{H}_\mu(T,\mathcal{A})$ defined by

$$\begin{aligned}\mathrm{h}_\mu(T,\mathcal{A}) := \lim_{n\to\infty}\mathrm{H}_\mu\big(\mathcal{A}|\mathcal{A}_1^{n}\big) &= \lim_{n\to\infty}\frac{1}{n}\mathrm{H}_\mu(\mathcal{A}^n)\\ &= \inf_{n\ge 1}\big\{\mathrm{H}_\mu\big(\mathcal{A}|\mathcal{A}_1^{n}\big)\big\} = \inf_{n\ge 1}\left\{\frac{1}{n}\mathrm{H}_\mu(\mathcal{A}^n)\right\},\end{aligned}$$

is called the *entropy of T with respect to $\mathcal{A}$*.

The following theorem lists some of the basic properties of $\mathrm{H}_\mu(T,\cdot)$.

Theorem 6.4.7 *Let $T: X \longrightarrow X$ be a measure-preserving endomorphism of a probability space $(X,\mathfrak{F},\mu)$. If $\mathcal{A},\mathcal{B} \in \mathrm{Part}(X,\mathfrak{F})$, then the following statements hold.*

(1) $\mathrm{h}_\mu(T,\mathcal{A}) \le \mathrm{H}_\mu(\mathcal{A})$.
(2) $\mathrm{h}_\mu(T,\mathcal{A}\vee\mathcal{B}) \le \mathrm{h}_\mu(T,\mathcal{A}) + \mathrm{h}_\mu(T,\mathcal{B})$.
(3) *If $\mathcal{A} \le \mathcal{B}$, then* $\mathrm{H}_\mu(T,\mathcal{A}) \le \mathrm{h}_\mu(T,\mathcal{B})$.
(4) $\mathrm{h}_\mu(T,\mathcal{A}) \le \mathrm{h}_\mu(T,\mathcal{B}) + \mathrm{H}_\mu(\mathcal{A}|\mathcal{B})$.
(5) $\mathrm{h}_\mu(T,T^{-1}\mathcal{A}) = \mathrm{h}_\mu(T,\mathcal{A})$.
(6) *For all $k \ge 1$,* $\mathrm{h}_\mu(T,\mathcal{A}^k) = \mathrm{h}_\mu(T,\mathcal{A})$.
(7) *For all $k \ge 1$,* $\mathrm{H}_\mu(T^k,\mathcal{A}^k) = k\mathrm{H}_\mu(T,\mathcal{A})$.
(8) *If T is invertible and $k \ge 1$, then* $\mathrm{h}_\mu(T,\mathcal{A}) = \mathrm{h}_\mu\big(T,\bigvee_{i=-k}^{k}T^i\mathcal{A}\big)$.

(9) *If* $\mathcal{B}_1, \mathcal{B}_2, \ldots \in \text{Part}(X, \mathfrak{F})$ *are such that* $\lim_{n\to\infty} \text{H}_\mu(\mathcal{A}|\mathcal{B}_n) = 0$, *then*

$$\text{h}_\mu(T, \mathcal{A}) \le \liminf_{n\to\infty} \text{h}_\mu(T, \mathcal{B}_n).$$

(10) *If* $\lim_{n\to\infty} \text{H}_\mu(\mathcal{A}|\mathcal{B}^n) = 0$, *then* $\text{h}_\mu(T, \mathcal{A}) \le \text{h}_\mu(T, \mathcal{B})$.

Proof (1) This follows from the fact that $\text{h}_\mu(T, \mathcal{A}) = \lim_{n\to\infty} \frac{1}{n}\text{H}_\mu(\mathcal{A}^n)$ and that $\left(\frac{1}{n}\text{H}_\mu(\mathcal{A}^n)\right)_{n\ge1}$ is a decreasing sequence with the first term given by $\text{H}_\mu(\mathcal{A})$.

(2) Using Theorem 6.3.3(7), we get that

$$\begin{aligned}
\text{h}_\mu(T, \mathcal{A} \vee \mathcal{B}) &= \lim_{n\to\infty} \frac{1}{n}\text{H}_\mu((\mathcal{A} \vee \mathcal{B})^n) = \lim_{n\to\infty} \frac{1}{n}\text{H}_\mu(\mathcal{A}^n \vee \mathcal{B}^n) \\
&\le \lim_{n\to\infty} \frac{1}{n}\Big[\text{H}_\mu(\mathcal{A}^n) + \text{H}_\mu(\mathcal{B}^n)\Big] \\
&= \lim_{n\to\infty} \frac{1}{n}\text{H}_\mu(\mathcal{A}^n) + \lim_{n\to\infty} \frac{1}{n}\text{H}_\mu(\mathcal{B}^n) \\
&= \text{h}_\mu(T, \mathcal{A}) + \text{h}_\mu(T, \mathcal{B}).
\end{aligned}$$

(3) If $\mathcal{A} \le \mathcal{B}$, then $\mathcal{A}^n \le \mathcal{B}^n$ for all $n \ge 1$. Consequently, $\text{H}_\mu(\mathcal{A}^n) \le \text{H}_\mu(\mathcal{B}^n)$ for all $n \ge 1$. Therefore,

$$\text{h}_\mu(T, \mathcal{A}) = \lim_{n\to\infty} \frac{1}{n}\text{H}_\mu(\mathcal{A}^n) \le \lim_{n\to\infty} \frac{1}{n}\text{H}_\mu(\mathcal{B}^n) = \text{h}_\mu(T, \mathcal{B}).$$

(4) Calling upon Theorem 6.3.3(9) and Lemma 6.4.4(3), we obtain that

$$\begin{aligned}
\text{h}_\mu(T, \mathcal{A}) &= \lim_{n\to\infty} \frac{1}{n}\text{H}_\mu(\mathcal{A}^n) \le \liminf_{n\to\infty} \frac{1}{n}\Big(\text{H}_\mu(\mathcal{A}^n|\mathcal{B}^n) + \text{H}_\mu(\mathcal{B}^n)\Big) \\
&= \liminf_{n\to\infty} \frac{1}{n}\text{H}_\mu(\mathcal{A}^n|\mathcal{B}^n) + \lim_{n\to\infty} \frac{1}{n}\text{H}_\mu(\mathcal{B}^n) \\
&\le \text{H}_\mu(\mathcal{A}|\mathcal{B}) + \text{h}_\mu(T, \mathcal{B}).
\end{aligned}$$

(5) By Lemma 6.4.1(2), we know that $(T^{-1}\mathcal{A})^n = T^{-1}(\mathcal{A}^n)$ for all $n \ge 1$. Then, using Lemma 6.4.4(2), we deduce that

$$\begin{aligned}
\text{h}_\mu(T, T^{-1}\mathcal{A}) &= \lim_{n\to\infty} \frac{1}{n}\text{H}_\mu\big((T^{-1}\mathcal{A})^n\big) \\
&= \lim_{n\to\infty} \frac{1}{n}\text{H}_\mu\big(T^{-1}(\mathcal{A}^n)\big) = \lim_{n\to\infty} \frac{1}{n}\text{H}_\mu(\mathcal{A}^n) = \text{h}_\mu(T, \mathcal{A}).
\end{aligned}$$

(6) By Lemma 6.4.1(4), we know that $(\mathcal{A}^k)^n = \mathcal{A}^{n+k-1}$ and, hence,

$$\begin{aligned}
\mathrm{h}_\mu(T,\mathcal{A}^k) &= \lim_{n\to\infty}\frac{1}{n}\mathrm{H}_\mu\big((\mathcal{A}^k)^n\big) = \lim_{n\to\infty}\frac{1}{n}\mathrm{H}_\mu\big(\mathcal{A}^{n+k-1}\big)\\
&= \lim_{n\to\infty}\frac{n+k-1}{n}\cdot\frac{1}{n+k-1}\mathrm{H}_\mu\big(\mathcal{A}^{n+k-1}\big)\\
&= \lim_{n\to\infty}\frac{n+k-1}{n}\cdot\lim_{n\to\infty}\frac{1}{n+k-1}\mathrm{H}_\mu\big(\mathcal{A}^{n+k-1}\big)\\
&= \lim_{m\to\infty}\frac{1}{m}\mathrm{H}_\mu(\mathcal{A}^m) = \mathrm{h}_\mu(T,\mathcal{A}).
\end{aligned}$$

(7) Let $k \geq 1$. Using part (6), we have that

$$\begin{aligned}
\mathrm{h}_\mu(T^k,\mathcal{A}^k) &= \lim_{n\to\infty}\frac{1}{n}\mathrm{H}_\mu\left(\bigvee_{j=0}^{n-1}T^{-kj}(\mathcal{A}^k)\right)\\
&= \lim_{n\to\infty}\frac{1}{n}\mathrm{H}_\mu\left(\bigvee_{j=0}^{n-1}T^{-kj}\left(\bigvee_{i=0}^{k-1}T^{-i}\mathcal{A}\right)\right)\\
&= \lim_{n\to\infty}\frac{1}{n}\mathrm{H}_\mu\left(\bigvee_{l=0}^{kn-1}T^{-l}\mathcal{A}\right)\\
&= k\lim_{n\to\infty}\frac{1}{kn}\mathrm{H}_\mu(\mathcal{A}^{kn}) = k\,\mathrm{h}_\mu(T,\mathcal{A}).
\end{aligned}$$

(8) The proof is similar to that of part (6) and is, thus, left to the reader as an exercise.

(9) By part (4), for each $n \geq 1$, we have that

$$\mathrm{h}_\mu(T,\mathcal{A}) \leq \mathrm{h}_\mu(T,\mathcal{B}_n) + \mathrm{H}_\mu(\mathcal{A}|\mathcal{B}_n).$$

So, if $(\mathcal{B}_n)_{n\geq 1}$ are partitions such that $\lim_{n\to\infty}\mathrm{H}_\mu(\mathcal{A}|\mathcal{B}_n) = 0$, then

$$\begin{aligned}
\mathrm{h}_\mu(T,\mathcal{A}) &\leq \liminf_{n\to\infty}\big[\mathrm{h}_\mu(T,\mathcal{B}_n) + \mathrm{H}_\mu(\mathcal{A}|\mathcal{B}_n)\big]\\
&= \liminf_{n\to\infty}\mathrm{h}_\mu(T,\mathcal{B}_n) + \lim_{n\to\infty}\mathrm{H}_\mu(\mathcal{A}|\mathcal{B}_n) = \liminf_{n\to\infty}\mathrm{h}_\mu(T,\mathcal{B}_n).
\end{aligned}$$

(10) Suppose that $\lim_{n\to\infty}\mathrm{H}_\mu(\mathcal{A}|\mathcal{B}^n) = 0$. By parts (9) and (6), we have that

$$\mathrm{h}_\mu(T,\mathcal{A}) \leq \liminf_{n\to\infty}\mathrm{h}_\mu(T,\mathcal{B}^n) = \lim_{n\to\infty}\mathrm{h}_\mu(T,\mathcal{B}) = \mathrm{h}_\mu(T,\mathcal{B}).$$ ■

The entropy of T is defined in a similar way to topological entropy. The third and last stage in the definition of the entropy of an endomorphism consists of passing to a supremum.

Definition 6.4.8 If $T: X \longrightarrow X$ is a measure-preserving endomorphism of a probability space $(X, \mathfrak{F}, \mu)$, then the *measure-theoretic entropy* of T, denoted $\mathrm{h}_\mu(T)$, is defined by

$$\mathrm{h}_\mu(T) := \sup \left\{ \mathrm{h}_\mu(T, \mathcal{A}) : \mathcal{A} \text{ is a finite partition of } X \right\}.$$

The following theorem is a useful tool for calculating the entropy of an endomorphism.

Theorem 6.4.9 *If $T: X \to X$ is a measure-preserving endomorphism of a probability space $(X, \mathfrak{F}, \mu)$, then*

(1) *For all $k \geq 1$, $\mathrm{h}_\mu(T^k) = k\mathrm{h}_\mu(T)$.*
(2) *If T is invertible, then $\mathrm{H}_\mu(T^{-1}) = \mathrm{H}_\mu(T)$.*

Proof (1) Let $k \geq 1$. Then, by Theorem 6.4.7(7),

$$\begin{aligned} k\mathrm{h}_\mu(T) &= \sup\{k\mathrm{h}_\mu(T, \mathcal{A}) : \mathcal{A} \text{ a finite partition}\} \\ &= \sup\{\mathrm{h}_\mu(T^k, \mathcal{A}^k) : \mathcal{A} \text{ a finite partition}\} \\ &\leq \sup\{\mathrm{h}_\mu(T^k, \mathcal{B}) : \mathcal{B} \text{ a finite partition}\} = \mathrm{h}_\mu(T^k). \end{aligned}$$

On the other hand, by Theorem 6.4.7(3) and (7),

$$\mathrm{h}_\mu(T^k, \mathcal{A}) \leq \mathrm{h}_\mu(T^k, \mathcal{A}^k) = k\mathrm{h}_\mu(T, \mathcal{A}).$$

Passing to the supremum over all finite partitions $\mathcal{A}$ of X on both sides, we obtain the desired inequality, namely $\mathrm{h}_\mu(T^k) \leq k\mathrm{h}_\mu(T)$.

(2) To distinguish the action of T from the action of T^{-1} on a partition, we shall use the respective notations $\mathcal{A}_T^n$ and $\mathcal{A}_{T^{-1}}^n$. Using Lemmas 6.4.4(2) and 6.4.1(2) in turn, we deduce that

$$\begin{aligned} \mathrm{h}_\mu(\mathcal{A}_{T^{-1}}^n) &= \mathrm{H}_\mu\left(\bigvee_{i=0}^{n-1} (T^{-1})^{-i}\mathcal{A}\right) = \mathrm{H}_\mu\left(\bigvee_{i=0}^{n-1} T^i\mathcal{A}\right) \\ &= \mathrm{H}_\mu\left(T^{-(n-1)}\left(\bigvee_{i=0}^{n-1} T^i\mathcal{A}\right)\right) \\ &= \mathrm{H}_\mu\left(\bigvee_{i=0}^{n-1} T^{-(n-1-i)}\mathcal{A}\right) \\ &= \mathrm{H}_\mu\left(\bigvee_{j=0}^{n-1} T^{-j}\mathcal{A}\right) = \mathrm{H}_\mu(\mathcal{A}_T^n). \end{aligned}$$

It follows that $\mathrm{h}_\mu(T^{-1}, \mathcal{A}) = \mathrm{h}_\mu(T, \mathcal{A})$ for every partition $\mathcal{A}$ and, thus, passing to the supremum on both sides, we conclude that $\mathrm{h}_\mu(T^{-1}) = \mathrm{h}_\mu(T)$. ■

Our goal now is to provide tools for calculating the entropy of an endomorphism. Its very definition requires us to take the supremum over a huge set of all finite partitions. Our task is to reduce this to some sequences of partitions or even to a single partition. The following result, toward this end, is purely measure-theoretic. It says that, given a finite partition $\mathcal{A}$ of a compact metric space X and given any partition $\mathcal{C}$ of X of sufficiently small diameter, we can group the atoms of $\mathcal{C}$ together in such a way that we nearly construct partition $\mathcal{A}$. It is worth noticing that $\mathcal{C}$ may be countably infinite.

Lemma 6.4.10 *Suppose that μ is a Borel probability measure on a compact metric space X. Suppose further that $\mathcal{A} = \{A_1, A_2, \ldots, A_n\}$ is a finite partition of X into Borel sets. Then, for all $\varepsilon > 0$, there exists $\delta > 0$ so that, for every Borel partition $\mathcal{C}$ with* $\mathrm{diam}(\mathcal{C}) < \delta$, *there is a Borel partition $\mathcal{B} = \{B_1, B_2, \ldots, B_n\} \leq \mathcal{C}$ such that*

$$\mu(B_i \Delta A_i) < \varepsilon$$

for all $1 \leq i \leq n$.

Proof Fix $\varepsilon > 0$. Since μ is regular, there exists for each $1 \leq i \leq n$ a compact set $K_i \subseteq A_i$ such that $\mu(A_i \backslash K_i) < \varepsilon/n$. As usual, let d denote the metric on X and let

$$\theta := \min\{d(K_i, K_j) : i \neq j\}.$$

Then $\theta > 0$, as the sets K_i are compact and disjoint. Let $\delta = \theta/2$ and $\mathcal{C}$ be a partition with $\mathrm{diam}(\mathcal{C}) < \delta$. For each $1 \leq i \leq n$, define

$$B_i := \bigcup_{\substack{C \in \mathcal{C} \\ C \cap K_i \neq \emptyset}} C.$$

Clearly, the B_is are Borel sets and $B_i \supseteq K_i$ for each i. Moreover, because of the choice of δ, $B_i \cap B_j = \emptyset$ for all $i \neq j$. However, the family of pairwise disjoint Borel sets $\{B_i\}_{i=1}^n$ may not cover X completely. Indeed, there may be some sets $C \in \mathcal{C}$ such that $C \cap \bigcup_{i=1}^n K_i = \emptyset$. Simply take all those sets and put them into one of the B_is, say B_1. Then the resulting family $\mathcal{B} := \{B_i\}_{i=1}^n$ is a Borel partition of X. Clearly, $\mathcal{B} \leq \mathcal{C}$. It remains to show that

$$\mu(B_i \Delta A_i) < \varepsilon$$

for all $1 \leq i \leq n$. But

$$\begin{aligned}
\mu(B_i \Delta A_i) = \mu(B_i \backslash A_i) + \mu(A_i \backslash B_i) &= \mu\left(\left(X \backslash \bigcup_{j \neq i} B_j\right) \backslash A_i\right) + \mu(A_i \backslash K_i) \\
&\leq \mu\left(\left(X \backslash \bigcup_{j \neq i} K_j\right) \backslash A_i\right) + \mu(A_i \backslash K_i) \\
&= \mu\left(\left(\bigcup_{k=1}^{n} A_k \backslash \bigcup_{j \neq i} K_j\right) \backslash A_i\right) + \mu(A_i \backslash K_i) \\
&= \mu\left(\bigcup_{k \neq i} A_k \backslash \cup_{j \neq i} K_j\right) + \mu(A_i \backslash K_i) \\
&\leq \mu\left(\bigcup_{j \neq i} A_j \backslash K_j\right) + \mu(A_i \backslash K_i) \\
&= \sum_{j=1}^{n} \mu(A_j \backslash K_j) < n \cdot \frac{\varepsilon}{n} = \varepsilon.
\end{aligned}$$

■

From the above result, we will show that the conditional entropy of a partition $\mathcal{A}$ given a partition $\mathcal{C}$ is as small as desired provided that $\mathcal{C}$ has a small enough diameter. Indeed, from Theorem 6.3.3(5), given partitions $\mathcal{A}, \mathcal{B}$, and $\mathcal{C}$ as in the above lemma, we have that $\mathrm{H}_\mu(\mathcal{A}|\mathcal{C}) \leq \mathrm{H}_\mu(\mathcal{A}|\mathcal{B})$, where the partition $\mathcal{B}$ is designed to resemble the partition $\mathcal{A}$. In order to estimate the conditional entropy $\mathrm{H}_\mu(\mathcal{A}|\mathcal{B})$, we must estimate the contribution of all atoms of the partition $\mathcal{A} \vee \mathcal{B}$. There are essentially two kinds of atoms to be taken into account; namely, atoms of the form $A_i \cap B_i$ and atoms of the form $A_i \cap B_j$, for $i \neq j$. Intuitively, because A_i looks like B_i (after all, $\mu(A_i \Delta B_i)$ is small), the information provided by A_i given that measurement $\mathcal{B}$ resulted in B_i is small. On the other hand, since A_i is nearly disjoint from B_j when $i \neq j$ (after all, A_i is close to B_i and $B_i \cap B_j = \emptyset$), the information obtained from getting A_i given that observation B_j occurred is small. This is what we now prove rigorously. First, let us make one definition which will prove useful here and throughout both volumes of the book. Recall that a function $\psi : (a,b) \to \mathbb{R}$, where $-\infty \leq a < b \leq \infty$, is concave if and only if

$$\psi\big(tx + (1-t)y\big) \geq t\psi(x) + (1-t)\psi(y)$$

for all $t \in [0,1]$ and all $x, y \in (a,b)$.

Definition 6.4.11 Let the function $k\colon [0,1] \to [0,1]$ be defined by

$$k(t) := \begin{cases} 0 & \text{if } x = 0 \\ -t \log t & \text{if } t \in (0,1]. \end{cases}$$

Note that the function k is continuous, concave, increasing on the interval $[0, e^{-1}]$, and decreasing on the interval $[e^{-1}, 1]$.

Lemma 6.4.12 *Let μ be a Borel probability measure on a compact metric space X and $\mathcal{A}$ be a finite Borel partition of X. Then, for every $\varepsilon > 0$, there exists $\delta > 0$ such that*

$$\mathrm{H}_\mu(\mathcal{A}|\mathcal{C}) < \varepsilon$$

for every Borel partition $\mathcal{C}$ with $\operatorname{diam}(\mathcal{C}) < \delta$.

Proof Let $\mathcal{A} = \{A_1, A_2, \ldots, A_n\}$ be a finite partition of X. Fix $\varepsilon > 0$ and let $0 < \overline{\varepsilon} < \min\{e^{-1}, 1 - e^{-1}\}$ be so small that

$$\max\left\{k(\overline{\varepsilon}), k(1 - \overline{\varepsilon})\right\} < \varepsilon/n.$$

Then there exists $\tilde{\varepsilon} > 0$ such that

$$0 < \frac{\tilde{\varepsilon}}{\mu(A_i) - \tilde{\varepsilon}} < \overline{\varepsilon} \quad \text{and} \quad \frac{\mu(A_i) - \tilde{\varepsilon}}{\mu(A_i) + \tilde{\varepsilon}} > 1 - \overline{\varepsilon}$$

for all $1 \le i \le n$ such that $\mu(A_i) > 0$. Let $\delta > 0$ be the number ascribed to $\tilde{\varepsilon}$ in Lemma 6.4.10. Let $\mathcal{C}$ be a partition with $\operatorname{diam}(\mathcal{C}) < \delta$ and

$$\mathcal{B} = \{B_1, B_2, \ldots, B_n\} \le \mathcal{C}$$

be such that

$$\mu(A_i \Delta B_i) \le \tilde{\varepsilon}$$

for all $1 \le i \le n$, also as prescribed in Lemma 6.4.10. Then, for all $1 \le i \le n$, we have that

$$|\mu(A_i) - \mu(B_i)| \le \mu(A_i \Delta B_i) \le \tilde{\varepsilon}.$$

Therefore,

$$\begin{aligned} 0 < \mu(A_i) - \tilde{\varepsilon} \le \mu(A_i) - \mu(A_i \Delta B_i) &\le \mu(B_i) \\ &\le \mu(A_i) + \mu(A_i \Delta B_i) \le \mu(A_i) + \tilde{\varepsilon} \end{aligned}$$

for all i. Moreover,

$$\mu(A_i \cap B_i) = \mu(A_i) - \mu(A_i \backslash B_i) \ge \mu(A_i) - \mu(A_i \Delta B_i) \ge \mu(A_i) - \tilde{\varepsilon}$$

for all i. Hence,

$$\frac{\mu(A_i \cap B_i)}{\mu(B_i)} \geq \frac{\mu(A_i) - \tilde{\varepsilon}}{\mu(A_i) + \tilde{\varepsilon}} > 1 - \overline{\varepsilon}$$

for all i such that $\mu(A_i) > 0$. By our choice of $\overline{\varepsilon}$, the function k is decreasing on the interval $[1 - \overline{\varepsilon}, 1]$ and, thus,

$$k\left(\frac{\mu(A_i \cap B_i)}{\mu(B_i)}\right) \leq k(1 - \overline{\varepsilon}) < \frac{\varepsilon}{n} \tag{6.2}$$

for all i such that $\mu(A_i) > 0$. Suppose now that $i \neq j$. Since $\mathcal{A} = \{A_k\}_{k=1}^n$ is a partition of X, we know that $A_i \cap B_j \subseteq B_j \backslash A_j \subseteq A_j \Delta B_j$. Hence,

$$\frac{\mu(A_i \cap B_j)}{\mu(B_j)} \leq \frac{\mu(A_j \Delta B_j)}{\mu(A_j) - \mu(A_j \Delta B_j)} \leq \frac{\tilde{\varepsilon}}{\mu(A_j) - \tilde{\varepsilon}} < \overline{\varepsilon}$$

for all i such that $\mu(A_i) > 0$. By our choice of $\overline{\varepsilon}$, the function k is increasing on the interval $[0, \overline{\varepsilon}]$ and, hence,

$$k\left(\frac{\mu(A_i \cap B_j)}{\mu(B_j)}\right) \leq k(\overline{\varepsilon}) < \frac{\varepsilon}{n} \tag{6.3}$$

for all i such that $\mu(A_i) > 0$. Furthermore, note that $k\left(\frac{\mu(A_i \cap B_j)}{\mu(B_j)}\right) = 0$ if $\mu(A_i) = 0$. Then, by Theorem 6.3.3(5) and (6.2) and (6.3), we have that

$$\begin{aligned}
\mathrm{H}_\mu(\mathcal{A}|\mathcal{C}) \leq \mathrm{H}_\mu(\mathcal{A}|\mathcal{B}) &= \sum_{A \in \mathcal{A}} \sum_{B \in \mathcal{B}} -\mu(A \cap B) \log \frac{\mu(A \cap B)}{\mu(B)} \\
&= \sum_{i,j=1}^{n} \mu(B_j) k\left(\frac{\mu(A_i \cap B_j)}{\mu(B_j)}\right) \\
&= \sum_{i=1}^{n} \mu(B_i) k\left(\frac{\mu(A_i \cap B_i)}{\mu(B_i)}\right) + \sum_{\substack{i,j=1 \\ i \neq j}}^{n} \mu(B_j) k\left(\frac{\mu(A_i \cap B_j)}{\mu(B_j)}\right) \\
&< \sum_{i=1}^{n} \mu(B_i) \frac{\varepsilon}{n} + \sum_{i=1}^{n} \sum_{j=1}^{n} \mu(B_j) \frac{\varepsilon}{n} \\
&= \frac{\varepsilon}{n} + n \cdot \frac{\varepsilon}{n} \\
&< 2\varepsilon.
\end{aligned}$$

■

From the above lemma, we can infer that any sequence of partitions whose diameters tend to 0 provides asymptotically as much information as any given partition can.

Corollary 6.4.13 *Let μ be a Borel probability measure on a compact metric space X. If $(\mathcal{A}_n)_{n\geq 1}$ is a sequence of Borel partitions of X such that*

$$\lim_{n\to\infty} \operatorname{diam}(\mathcal{A}_n) = 0,$$

then

$$\lim_{n\to\infty} \mathrm{H}_\mu(\mathcal{A}|\mathcal{A}_n) = 0$$

for every finite Borel partition $\mathcal{A}$ of X.

Proof Let $\mathcal{A}$ be a finite Borel partition of X. Then, by Lemma 6.4.12, for every $\varepsilon > 0$, there exists a $\delta > 0$ such that if $\operatorname{diam}(\mathcal{C}) < \delta$, then $\mathrm{H}_\mu(\mathcal{A}|\mathcal{C}) < \varepsilon$. Since $\operatorname{diam}(\mathcal{A}_n) \to 0$, it follows that $\mathrm{H}_\mu(\mathcal{A}|\mathcal{A}_n) \to 0$. ■

This result about conditional entropy of partitions allows us to deduce the following fact about entropy of endomorphisms.

Theorem 6.4.14 *Let μ be a Borel probability measure on a compact metric space X. If $T\colon X \longrightarrow X$ is a measure-preserving transformation of $(X, \mathcal{B}(X), \mu)$ and $(\mathcal{A}_n)_{n\geq 1}$ is a sequence of finite Borel partitions of X such that*

$$\lim_{n\to\infty} \operatorname{diam}(\mathcal{A}_n) = 0,$$

then

$$\mathrm{h}_\mu(T) = \lim_{n\to\infty} \mathrm{h}_\mu(T, \mathcal{A}_n).$$

Proof Let $\mathcal{A}$ be a finite partition consisting of Borel sets. By Corollary 6.4.13, we know that $\lim_{n\to\infty} \mathrm{H}_\mu(\mathcal{A}|\mathcal{A}_n) = 0$. So, by Theorem 6.4.7(9), it follows that

$$\mathrm{h}_\mu(T, \mathcal{A}) \leq \liminf_{n\to\infty} \mathrm{h}_\mu(T, \mathcal{A}_n) \leq \limsup_{n\to\infty} \mathrm{h}_\mu(T, \mathcal{A}_n) \leq \mathrm{H}_\mu(T).$$

Since this is true for any finite Borel partition $\mathcal{A}$ of X, we deduce from a passage to the supremum that

$$\mathrm{h}_\mu(T) \leq \liminf_{n\to\infty} \mathrm{h}_\mu(T, \mathcal{A}_n) \leq \limsup_{n\to\infty} \mathrm{h}_\mu(T, \mathcal{A}_n) \leq \mathrm{h}_\mu(T).$$

■

Corollary 6.4.15 *Let μ be a Borel probability measure on a compact metric space X. If $T\colon X \longrightarrow X$ is a measure-preserving transformation of $(X, \mathcal{B}(X), \mu)$ and $\mathcal{A}$ is a finite Borel partition of X such that*

$$\lim_{n\to\infty} \operatorname{diam}(\mathcal{A}^n) = 0,$$

then

$$\mathrm{h}_\mu(T) = \mathrm{h}_\mu(T, \mathcal{A}).$$

Proof By Theorems 6.4.14 and 6.4.7(6), we have that

$$\mathrm{h}_\mu(T) = \lim_{n\to\infty} \mathrm{h}_\mu(T, \mathcal{A}^n) = \lim_{n\to\infty} \mathrm{h}_\mu(T, \mathcal{A}) = \mathrm{h}_\mu(T, \mathcal{A}). \qquad \blacksquare$$

We now will introduce a notion, very classical and interesting in itself, that guarantees the hypotheses of Corollary 6.4.15 to be satisfied and that will play an important role later on, notably when dealing with the variational principle and equilibrium states.

Definition 6.4.16 Let (X,d) be a compact metric space. A continuous dynamical system $T\colon (X,d) \longrightarrow (X,d)$ is said to be positively expansive provided that there exists $\delta > 0$ such that, for every $x, y \in X$, $x \neq y$, there exists an integer $n = n(x, y) \geq 0$ with

$$d\big(T^n(x), T^n(y)\big) > \delta.$$

The constant δ is called an expansive constant for T and T is then also said to be δ-expansive. Equivalently, T is δ-expansive if

$$\sup\left\{d\big(T^n(x), T^n(y)\big) : n \geq 0\right\} \leq \delta \implies x = y.$$

In other words, δ-expansiveness means that two forward T-orbits that remain forever within a distance δ from each other originate from the same point (and are, therefore, only one orbit).

Remark 6.4.17 Let us record the following.

(1) If T is δ-expansive, then T is δ'-expansive for any $0 < \delta' < \delta$.
(2) The expansiveness of T is independent of topologically equivalent metrics, although particular expansive constants generally depend on the metric chosen. That is, if two metrics d and d' generate the same topology on X, then T is expansive when X is equipped with the metric d if and only if T is expansive when X is equipped with the metric d'.

We record now the following fact, which will follow from a somewhat stronger fact; namely, Corollary 7.1.40, proven in the next chapter, asserts the same, but for covers and not merely partitions.

Proposition 6.4.18 *If (X,d) is a compact metric space and $T\colon X \longrightarrow X$ is a positively expansive continuous map with an expansive constant $\delta > 0$, then*

$$\lim_{n\to\infty} \mathrm{diam}(\mathcal{A}^n) = 0$$

for every partition $\mathcal{A}$ *with* $\operatorname{diam}(\mathcal{A}) \le \delta$.

As an immediate consequence of this proposition and Corollary 6.4.15, we get the following.

Corollary 6.4.19 *Let* $T\colon X \longrightarrow X$ *be an expansive dynamical system preserving a Borel probability measure* μ. *If* $\mathcal{A}$ *is a finite partition with* $\operatorname{diam}(\mathcal{A}) \le \delta$, *where* $\delta > 0$ *is an expansive constant for* T, *then*

$$\mathrm{h}_\mu(T) = \mathrm{h}_\mu(T, \mathcal{A}).$$

6.5 Shannon–McMillan–Breiman Theorem

The sole goal of this section is to prove Shannon–McMillan–Breiman Theorem, i.e., Theorem 6.5.4. It sheds a lot of light on what entropy really is and provides a very useful tool both for further theoretic investigations of entropy and for its actual calculations.

We begin this section with the following purely measure-theoretic result.

Lemma 6.5.1 *Let* $T\colon X \longrightarrow X$ *be a measure-preserving endomorphism of a probability space* $(X, \mathfrak{F}, \mu)$. *Let* $\mathcal{A} \in \operatorname{Part}(X, \mathfrak{F})$. *Let*

$$f_n := I_\mu(\mathcal{A} | \mathcal{A}_1^n)$$

for each $n \ge 1$ *and*

$$f^* := \sup_{n \ge 1} f_n.$$

Then, for all $\lambda \in \mathbb{R}$ *and all* $A \in \mathcal{A}$, *we have that*

$$\mu\big(\{x \in A\colon f^*(x) > \lambda\}\big) \le \min\{\mu(A), e^{-\lambda}\}.$$

Proof Let $A \in \mathcal{A}$ and fix $n \ge 1$. Let also

$$f_n^A := -\log E(\chi_A | \sigma(\mathcal{A}_1^n)),$$

where $\sigma(\mathcal{A}_1^n)$ is the sub-σ-algebra generated by the countable partition $\mathcal{A}_1^n$. Let $x \in A$. Then

$$\begin{aligned} f_n^A(x) &= -\log E\big(\chi_A | \sigma(\mathcal{A}_1^n)\big)(x) = -\log \frac{\int_{\mathcal{A}_1^n(x)} \chi_A \, d\mu}{\mu(\mathcal{A}_1^n(x))} \\ &= -\log \frac{\mu(A \cap \mathcal{A}_1^n(x))}{\mu(\mathcal{A}_1^n(x))} \end{aligned}$$

$$\begin{aligned}
&= -\log \frac{\mu\big(\mathcal{A}(x) \cap \mathcal{A}_1^n(x)\big)}{\mu\big(\mathcal{A}_1^n(x)\big)} \\
&= I_\mu(\mathcal{A}|\mathcal{A}_1^n)(x) \\
&= f_n(x).
\end{aligned}$$

Hence, $f_n = \sum_{A\in\mathcal{A}} \chi_A f_n^A$. Fix $A \in \mathcal{A}$ and, for $n \geq 1$ and $\lambda \in \mathbb{R}$, consider the set

$$B_n^{A,\lambda} := \Big\{x \in X\colon \max_{1\leq i<n} f_i^A(x) \leq \lambda, f_n^A(x) > \lambda\Big\}.$$

The family $\{B_n^{A,\lambda}\}_{n\geq 1}$ consists of mutually disjoint sets. Also, recall that $\mathcal{A}_1^n \leq \mathcal{A}_1^{n+1}$ and, thus, $\sigma\big(\mathcal{A}_1^n\big) \subseteq \sigma\big(\mathcal{A}_1^{n+1}\big)$ for each $n \geq 1$. By definition, each function f_n^A is measurable with respect to $\sigma\big(\mathcal{A}_1^n\big)$. Consequently, $B_n^{A,\lambda} \in \sigma\big(\mathcal{A}_1^n\big)$. Then

$$\begin{aligned}
\mu\big(B_n^{A,\lambda} \cap A\big) &= \int_{B_n^{A,\lambda}} \chi_A \, d\mu = \int_{B_n^{A,\lambda}} E\big(\chi_A|\sigma\big(\mathcal{A}_1^n\big)\big) \, d\mu \\
&= \int_{B_n^{A,\lambda}} \exp\big(-f_n^A\big) \, d\mu \leq \int_{B_n^{A,\lambda}} e^{-\lambda} \, d\mu \\
&= e^{-\lambda}\mu\big(B_n^{A,\lambda}\big).
\end{aligned}$$

Since

$$\begin{aligned}
\mu(\{x \in A\colon f^*(x) > \lambda\}) &= \mu(\{x \in A\colon \exists\, n \geq 1 \text{ such that } f_n(x) > \lambda\}) \\
&= \mu(\{x \in A\colon \exists\, n \geq 1 \text{ such that } f_n^A(x) > \lambda\}),
\end{aligned}$$

we have that

$$\begin{aligned}
\mu(\{x \in A\colon f^*(x) > \lambda\}) &= \mu\left(\bigcup_{n=1}^{\infty} B_n^{A,\lambda} \cap A\right) = \sum_{n=1}^{\infty} \mu\left(B_n^{A,\lambda} \cap A\right) \\
&\leq \sum_{n=1}^{\infty} e^{-\lambda}\mu\left(B_n^{A,\lambda}\right) = e^{-\lambda} \sum_{n=1}^{\infty} \mu\left(B_n^{A,\lambda}\right) \\
&= e^{-\lambda}\mu\left(\bigcup_{n=1}^{\infty} B_n^{A,\lambda}\right) \\
&\leq e^{-\lambda}.
\end{aligned}$$

■

Corollary 6.5.2 *Let $T\colon X \to X$ be a measure-preserving endomorphism of a probability space $(X, \mathfrak{F}, \mu)$. Let $\mathcal{A}$ be a partition of X with finite entropy. Let*

$$f_n := I_\mu\big(\mathcal{A}|\mathcal{A}_1^n\big)$$

for all $n \geq 1$ and

$$f^* := \sup_{n\geq 1} f_n.$$

Then the function f^* belongs to $L^1(X, \mathfrak{F}, \mu)$ and

$$\int_X f^* \, d\mu \leq \mathrm{H}_\mu(\mathcal{A}) + 1.$$

Proof Since $f^* \geq 0$, we have that $\int_X |f^*| \, d\mu = \int_X f^* \, d\mu$. Thus,

$$\begin{aligned}
\int_X f^* \, d\mu &= \sum_{A\in\mathcal{A}} \int_A f^* \, d\mu = \sum_{A\in\mathcal{A}} \int_0^\infty \mu(\{x \in A \colon f^*(x) > \lambda\}) \, d\lambda \\
&\leq \sum_{A\in\mathcal{A}} \int_0^\infty \min\{\mu(A), e^{-\lambda}\} \, d\lambda \\
&= \sum_{A\in\mathcal{A}} \left[\int_0^{-\log\mu(A)} \mu(A) \, d\lambda + \int_{-\log\mu(A)}^\infty e^{-\lambda} \, d\lambda \right] \\
&= \sum_{A\in\mathcal{A}} \left[-\mu(A)\log\mu(A) + \left[-e^{-\lambda}\right]_{-\log\mu(A)}^\infty \right] \\
&= \sum_{A\in\mathcal{A}} -\mu(A)\log\mu(A) + \sum_{A\in\mathcal{A}} \mu(A) \\
&= \mathrm{H}_\mu(\mathcal{A}) + 1 < +\infty.
\end{aligned}$$

■

Corollary 6.5.3 *The sequence $(f_n)_{n\geq 1}$, defined in the previous corollary, converges μ-a.e. and also in $L^1(X, \mathfrak{F}, \mu)$.*

Proof Recall that $\mathcal{A}_1^n \leq \mathcal{A}_1^{n+1}$ and, thus, $\sigma(\mathcal{A}_1^n) \subseteq \sigma(\mathcal{A}_1^{n+1})$ for each $n \geq 1$. For any $x \in A \in \mathcal{A}$, we have that

$$f_n(x) = f_n^A(x) = -\log E_\mu(\chi_A | \sigma(\mathcal{A}_1^n))(x)$$

and the Martingale Convergence Theorem for Conditional Expectations, i.e., Theorem 1.4.10, guarantees that the limit

$$\lim_{n\to\infty} E_\mu(\chi_A | \sigma(\mathcal{A}_1^n))$$

exists μ-a.e. Hence, the sequence of nonnegative functions $(f_n)_{n\geq 1}$ converges to some limit function $\hat{f} \geq 0$ μ-a.e.

Since $|f_n| = f_n \leq f^*$ for all n, we have that $|\hat{f}| = \hat{f} \leq f^*$ and, thus,

$$|f_n - \hat{f}| \leq 2f^*$$

μ-a.e. So, by applying the Lebesgue Dominated Convergence Theorem to the sequence $(|f_n - \hat{f}|)_{n\geq 1}$, we obtain that

$$\lim_{n\to\infty} \|f_n - \hat{f}\|_1 = \lim_{n\to\infty} \int_X |f_n - \hat{f}|\, d\mu = \int_X \lim_{n\to\infty} |f_n - \hat{f}|\, d\mu = 0,$$

i.e., $f_n \to \hat{f}$ in $L^1(X, \mathfrak{F}, \mu)$. ■

We can now prove the main result of this section, i.e., the famous Shannon–McMillan–Breiman Theorem.

Theorem 6.5.4 (Shannon–McMillan–Breiman Theorem) *Let $T: X \longrightarrow X$ be a measure-preserving endomorphism of a probability space $(X, \mathfrak{F}, \mu)$. Let $\mathcal{A}$ be a partition of X with finite entropy. Then the following limits exist:*

$$f := \lim_{n\to\infty} I_\mu\big(\mathcal{A}|\mathcal{A}_1^n\big) \text{ and } \lim_{n\to\infty} \frac{1}{n}\sum_{j=0}^{n-1} f \circ T^j = E_\mu(f|\mathcal{I}_\mu) \ \mu\text{-a.e.}$$

Moreover, the following hold:

(1) $\lim_{n\to\infty} \frac{1}{n} I_\mu(\mathcal{A}^n) = E_\mu(f|\mathcal{I}_\mu)$ *in $L^1(\mu)$ and μ-a.e.*

(2) $\mathrm{h}_\mu(T, \mathcal{A}) = \lim_{n\to\infty} \frac{1}{n}\mathrm{H}_\mu(\mathcal{A}^n) = \int_X E_\mu(f|\mathcal{I}_\mu)\, d\mu = \int_X f\, d\mu.$

Proof The first sequence of functions

$$(f_n)_{n\geq 1} = \big(I_\mu\big(\mathcal{A}|\mathcal{A}_1^n\big)\big)_{n\geq 1}$$

converges to an integrable function f by Corollary 6.5.3. The second limit exists by virtue of the Birkhoff Ergodic Theorem.

In order to prove the remaining two statements, let us first assume that (1) holds and derive (2) from it. Then we will prove (1). In fact, a.e. convergence in (1) is not necessary to deduce (2). So, suppose that

$$\lim_{n\to\infty} \frac{1}{n} I_\mu(\mathcal{A}^n) = E_\mu(f|\mathcal{I}_\mu)$$

in $L^1(\mu)$. The convergence in L^1 entails the convergence of the corresponding integrals, i.e.,

$$\lim_{n\to\infty} \int_X \frac{1}{n} I_\mu(\mathcal{A}^n)\, d\mu = \int_X E_\mu(f|\mathcal{I}_\mu)\, d\mu.$$

Then

$$\begin{aligned}
\mathrm{h}_\mu(T,\mathcal{A}) &= \lim_{n\to\infty}\frac{1}{n}\mathrm{H}_\mu(\mathcal{A}^n) = \lim_{n\to\infty}\int_X \frac{1}{n}I_\mu(\mathcal{A}^n)\,d\mu \\
&= \int_X E_\mu(f|\mathcal{I}_\mu)\,d\mu = \int_X f\,d\mu.
\end{aligned}$$

This establishes (2).

In order to prove (1), first notice that, by Lemma 6.4.3, we obtain that

$$I_\mu(\mathcal{A}^n) = \sum_{k=1}^{n} I_\mu\big(\mathcal{A}|\mathcal{A}_1^k\big)\circ T^{n-k} = \sum_{j=0}^{n-1} I_\mu\big(\mathcal{A}|\mathcal{A}_1^{n-j}\big)\circ T^j = \sum_{j=0}^{n-1} f_{n-j}\circ T^j.$$

Then, by the triangle inequality,

$$\begin{aligned}
\left|\frac{1}{n}I_\mu(\mathcal{A}^n) - E_\mu(f|\mathcal{I}_\mu)\right| &= \left|\frac{1}{n}\sum_{j=0}^{n-1}\big(f_{n-j}\circ T^j - f\circ T^j\big)\right. \\
&\qquad \left. + \frac{1}{n}\sum_{j=0}^{n-1} f\circ T^j - E(f|\mathcal{I}_\mu)\right| \\
&\le \left|\frac{1}{n}\sum_{j=0}^{n-1}(f_{n-j} - f)\circ T^j\right| + \left|\frac{1}{n}S_n f - E(f|\mathcal{I}_\mu)\right| \\
&\le \frac{1}{n}\sum_{j=0}^{n-1}|f_{n-j} - f|\circ T^j + \left|\frac{1}{n}S_n f - E_\mu(f|\mathcal{I}_\mu)\right|.
\end{aligned}$$

The second term on the right-hand side tends to 0 μ-a.e. by the Birkhoff Ergodic Theorem. Furthermore, observe that

$$\begin{aligned}
\int_X \frac{1}{n}S_n f\,d\mu &= \frac{1}{n}\sum_{j=0}^{n-1}\int_X f\circ T^j\,d\mu = \frac{1}{n}\sum_{j=0}^{n-1}\int_X f\,d\mu \\
&= \int_X f\,d\mu \qquad (6.4)\\
&= \int_X E_\mu(f|\mathcal{I}_\mu)d\mu.
\end{aligned}$$

Thus, the second term converges to 0 in $L^1(\mu)$. Let us now investigate the first term on the right-hand side. Set $g_n := |f_n - f|$. Then

$$\begin{aligned}\lim_{n\to\infty}\left\|\frac{1}{n}\sum_{j=0}^{n-1} g_{n-j}\circ T^j - 0\right\|_1 &= \lim_{n\to\infty}\int_X \left|\frac{1}{n}\sum_{j=0}^{n-1} g_{n-j}\circ T^j - 0\right| d\mu \\ &= \lim_{n\to\infty}\frac{1}{n}\sum_{j=0}^{n-1}\int_X g_{n-j}\circ T^j\, d\mu \\ &= \lim_{n\to\infty}\frac{1}{n}\sum_{j=0}^{n-1}\int_X g_{n-j}\, d\mu \\ &= \lim_{n\to\infty}\frac{1}{n}\sum_{i=1}^{n-1}\int_X g_i\, d\mu \\ &= 0,\end{aligned}$$

where the last equality sign was written since $f_i \to f$ in $L^1(X, \mathfrak{F}, \mu)$ according to Corollary 6.5.3, whence $g_i \to 0$ in $L^1(X, \mathfrak{F}, \mu)$ and so do the Cesàro averages of the g_is. This ensures the convergence of the functions

$$\frac{1}{n}\sum_{j=0}^{n-1} g_{n-j}\circ T^j$$

to zero in L^1 and, thus, the convergence of the first term on the right-hand side to 0 in L^1. It only remains to show convergence μ-a.e. of the same term. To this end, for each $N \geq 1$, let

$$G_N := \sup_{n\geq N} g_n.$$

The sequence of functions $(G_N)_{N\geq 1}$ is decreasing and bounded below by 0, so it converges to some function. As $f_n \to f$ μ-a.e., we know that $g_n = |f_n - f| \to 0$ μ-a.e. It follows that

$$G_N \xrightarrow[N\to\infty]{} 0$$

μ-a.e. Also, the functions G_N are bounded above by an integrable function since

$$0 \leq G_N \leq G_1 = \sup_{n\geq 1} g_n \leq \sup_{n\geq 1}\big(|f_n| + |f|\big) \leq 2f^* \in L^1(\mu).$$

Fix momentarily $N \geq 1$. Then, for any $n > N$, we have that

$$\frac{1}{n}\sum_{j=0}^{n-1} g_{n-j} \circ T^j = \frac{1}{n}\sum_{j=0}^{n-N} g_{n-j} \circ T^j + \frac{1}{n}\sum_{j=n-N+1}^{n-1} g_{n-j} \circ T^j$$

$$\leq \frac{n-N}{n} \cdot \frac{1}{n-N}\sum_{j=0}^{n-N} G_N \circ T^j + \frac{1}{n}\sum_{j=n-N+1}^{n-1} G_1 \circ T^j.$$

Let $F_N = \sum_{j=0}^{N-2} G_1 \circ T^j$. Using the Birkhoff Ergodic Theorem, we deduce that

$$\limsup_{n\to\infty} \frac{1}{n}\sum_{j=0}^{n-1} g_{n-j} \circ T^j \leq \lim_{n\to\infty} \frac{1}{n-N}\sum_{j=0}^{n-N} G_N \circ T^j + \limsup_{n\to\infty} \frac{1}{n} F_N \circ T^{n-N+1}$$

$$= E(G_N|\mathcal{I}_\mu) + \limsup_{n\to\infty} \frac{1}{n} F_N \circ T^{n-N+1} \quad \mu\text{-a.e.}$$

$$= E(G_N|\mathcal{I}_\mu) \quad \mu\text{-a.e.}$$

But, since each G_N is a nonnegative function uniformly bounded by $G_0 \in L^1(\mu)$ and since the sequence $(G_N)_{N\geq 1}$ converges to zero μ-a.e., the Lebesgue Dominated Convergence Theorem implies that

$$\lim_{N\to\infty} \int_X E(G_N|\mathcal{I}_\mu)\, d\mu = \lim_{N\to\infty} \int_X G_N\, d\mu = \int_X \lim_{N\to\infty} G_N\, d\mu = 0.$$

However, the sequence of expected values $(E(G_N|\mathcal{I}_\mu))_{N\geq 1}$ is decreasing since the sequence $(G_N)_{N\geq 1}$ is decreasing. Therefore,

$$\int_X E_\mu(G_N|\mathcal{I}_\mu)\, d\mu \searrow 0,$$

as $n \to \infty$, and, hence,

$$E_\mu(G_N|\mathcal{I}_\mu) \xrightarrow[N\to\infty]{} 0$$

μ-a.e. Thus,

$$\limsup_{n\to\infty} \frac{1}{n}\sum_{j=0}^{n-1} g_{n-j} \circ T^j = 0 \quad \mu\text{-a.e.},$$

thereby establishing the a.e. convergence of the first term on the right-hand side. ■

As an immediate consequence of this theorem and the Birkhoff Ergodic Theorem (Theorem 2.3.2), we get the following.

Corollary 6.5.5 (Ergodic case of Shannon–McMillan–Breiman Theorem) *Let $T\colon X \to X$ be an ergodic endomorphism of a probability space $(X, \mathfrak{F}, \mu)$. Let $\mathcal{A}$ be a partition of X with finite entropy. Then*

$$\mathrm{h}_\mu(T, \mathcal{A}) = \lim_{n\to\infty} \frac{1}{n} I_\mu(\mathcal{A}^n)(x) \quad \textit{for } \mu\textit{-a.e. } x \in X.$$

The right-hand side in the above equality can be viewed as a local entropy at a point x. The theorem then states that at μ-a.e. $x \in X$ the local entropy exists and is equal to the (global) entropy of the endomorphism . Moreover, the theorem affirms that if $\mathrm{H}_\mu(T, \mathcal{A}) > 0$, then $\mu(\mathcal{A}^n(x)) \to 0$ with exponential rate $e^{-\mathrm{H}_\mu(T,\mathcal{A})}$ for μ-a.e. $x \in X$.

6.6 Abramov's Formula and Krengel's Entropy (Infinite Measures Allowed)

In previous chapters, we have devoted a good amount of time to studying induced maps. There is the celebrated Abramov's Formula that relates the entropy of an induced system and the original one. It was originally proved by Abramov in [**Ab**]. We quote it here without a proof.

Theorem 6.6.1 (Abramov's Formula) *If $T\colon X \to X$ is an ergodic measure-preserving transformation of a probability space $(X, \mathfrak{F}, \mu)$, then, for every set $F \in \mathfrak{F}$ with $0 < \mu(F) < +\infty$, we have that*

$$\mathrm{h}_{\mu_F}(T_F) = \frac{1}{\mu(F)} \mathrm{h}_\mu(T).$$

As an immediate consequence (take $X := F \cup G$) of this theorem, we get the following.

Corollary 6.6.2 (Krengel's Entropy) *If $T\colon X \to X$ is a conservative ergodic measure-preserving transformation of a measure space $(X, \mathfrak{F}, \mu)$, then, for all sets F and G in $\mathfrak{F}$ with $0 < \mu(F), \mu(G) < +\infty$, we have that*

$$\mu(F)\mathrm{h}_{\mu_F}(T_F) = \mu(G)\mathrm{h}_{\mu_G}(T_G).$$

This common value is called Krengel's entropy of the map $T\colon X \longrightarrow X$ and is denoted simply by $\mathrm{h}_\mu(T)$. In the case when the measure μ is probabilistic, it coincides with the standard entropy of T with respect to μ.

7
Thermodynamic Formalism

In this chapter, we introduce the fundamental concepts of thermodynamic formalism, such as topological pressure and topological entropy, and we establish their basic properties. We then, in the last section of this chapter, relate them to Kolmogorov–Sinai metric entropies by proving the Variational Principle, which is the cornerstone of thermodynamic formalism. This principle naturally leads to the concept of equilibrium states and measures of maximal entropy. We deal with them at length, particularly with the problem of existence of equilibrium states. We do not touch on the issue of its uniqueness as this requires a more involved and lengthy apparatus and holds only for some special systems, such as open transitive distance expanding maps in the sense of the book [**PU2**]. However, some considerations of the chapter touch on the issue of uniqueness, though in a somewhat different setting.

Thermodynamic formalism originated in the late 1960s with the works of Ruelle. Its foundations, classical concepts, and theorem were obtained throughout the 1970s in the works of Ruelle [**Ru1**], Bowen [**Bow1**], Walters [**Wa1**], and Sinai [**Sin2**]. The more recent and modern expositions can be found, for example, in [**Wa2**], [**PU2**], or [**Ru1**]. We should also mention the paper [**Mis**] by Misiurewicz, who provided an elegant, short, and simple proof of the Variational Principle. This is the proof we reproduce in the last section of this chapter.

7.1 Topological Pressure

7.1.1 Covers of a Set

Let X be a nonempty set. Recall that a family $\mathcal{U}$ of subsets of X is said to form a cover of X if and only if

$$X \subseteq \bigcup_{U \in \mathcal{U}} U.$$

Recall further that $\mathcal{V}$ is said to be *subcover* of $\mathcal{U}$ if $\mathcal{V}$ is itself a cover and

$$\mathcal{V} \subseteq \mathcal{U}.$$

We will always denote covers by calligraphic letters, $\mathcal{U}, \mathcal{V}, \mathcal{W}$ and so on.

Let us begin by introducing a useful way of obtaining a new cover from two existing covers.

Definition 7.1.1 If $\mathcal{U}$ and $\mathcal{V}$ are covers of X, then their *join*, denoted $\mathcal{U} \vee \mathcal{V}$, is the cover

$$\mathcal{U} \vee \mathcal{V} := \{U \cap V : U \in \mathcal{U}, V \in \mathcal{V}\}.$$

Remark 7.1.2 The join operation is commutative (i.e., $\mathcal{U} \vee \mathcal{V} = \mathcal{V} \vee \mathcal{U}$) and associative; in other words,

$$(\mathcal{U} \vee \mathcal{V}) \vee \mathcal{W} = \mathcal{U} \vee (\mathcal{V} \vee \mathcal{W}).$$

Owing to the associativity of the join, this operation extends naturally to any finite collection $\{\mathcal{U}_j\}_{j=0}^{n-1}$ of covers of X. That is, we have that

$$\bigvee_{j=0}^{n-1} \mathcal{U}_j := \mathcal{U}_0 \vee \cdots \vee \mathcal{U}_{n-1} = \left\{ \bigcap_{j=0}^{n-1} U_j : U_j \in \mathcal{U}_j, 0 \le j < 1 \right\}.$$

It is also useful to be able to compare covers. For this purpose, we introduce the following relation on the collection of all covers of a set X.

Definition 7.1.3 Let $\mathcal{U}$ and $\mathcal{V}$ be covers of X. We say that $\mathcal{V}$ is *finer* than, or a *refinement* of, the cover $\mathcal{U}$, and denote this by

$$\mathcal{U} \prec \mathcal{V}$$

if and only if every element of $\mathcal{V}$ is a subset of an element of $\mathcal{U}$. That is, for every set $V \in \mathcal{V}$, there exists a set $U \in \mathcal{U}$ such that $V \subseteq U$. It is also sometimes said that $\mathcal{V}$ is *inscribed* in $\mathcal{U}$.

Lemma 7.1.4 *Let $\mathcal{U}$, $\mathcal{V}$, $\mathcal{W}$, and $\mathcal{X}$ be covers of X. Then:*

(a) *The refinement relation $\prec$ is reflexive (i.e., $\mathcal{U} \prec \mathcal{U}$) and transitive (i.e., if $\mathcal{U} \prec \mathcal{V}$ and $\mathcal{V} \prec \mathcal{W}$, then $\mathcal{U} \prec \mathcal{W}$).*

(b) $\mathcal{U} \prec \mathcal{U} \vee \mathcal{V}$.

(c) *If $\mathcal{V}$ is a subcover of $\mathcal{U}$, then $\mathcal{U} \prec \mathcal{V}$.*

(d) $\mathcal{U}$ *is a subcover of* $\mathcal{U} \vee \mathcal{U}$. *Hence, from* (*b*) *and* (*c*), *we deduce that*

$$\mathcal{U} \prec \mathcal{U} \vee \mathcal{U} \prec \mathcal{U}.$$

Nevertheless, $\mathcal{U}$ *is not equal to* $\mathcal{U} \vee \mathcal{U}$ *in general.*

(e) *If* $\mathcal{U} \prec \mathcal{V}$ *or* $\mathcal{U} \prec \mathcal{W}$, *then* $\mathcal{U} \prec \mathcal{V} \vee \mathcal{W}$.

(f) *If* $\mathcal{U} \prec \mathcal{W}$ *and* $\mathcal{V} \prec \mathcal{W}$, *then* $\mathcal{U} \vee \mathcal{V} \prec \mathcal{W}$.

(g) *If* $\mathcal{U} \prec \mathcal{W}$ *and* $\mathcal{V} \prec \mathcal{X}$, *then* $\mathcal{U} \vee \mathcal{V} \prec \mathcal{W} \vee \mathcal{X}$.

Proof All of these properties can be proved directly and are left to the reader as an exercise. As a hint, observe that property (e) is a consequence of (b) and (a) (transitivity), while property (g) follows from (e) and (f). ■

Remark 7.1.5 The relation $\prec$ does not constitute a partial order relation on the collection of all covers of X. This is because, although it is reflexive and transitive, it is not antisymmetric, i.e., $\mathcal{U} \prec \mathcal{V} \prec \mathcal{U}$ does not necessarily imply that $\mathcal{U} = \mathcal{V}$; see Lemma 7.1.4(*d*).

If X is a metric space, then the maximum size of the elements of a cover is encompassed by the notion of diameter of the cover.

Definition 7.1.6 If (X,d) is a metric space, then the *diameter* of a cover $\mathcal{U}$ of X is defined by

$$\operatorname{diam}(\mathcal{U}) := \sup\{\operatorname{diam}(U) : U \in \mathcal{U}\},$$

where

$$\operatorname{diam}(U) := \sup\{d(x,y) : x, y \in U\}.$$

It is also often of interest to know that all balls of some specified radius are each contained in at least one element of a given cover. Such a radius is known as a Lebesgue number for the cover.

Definition 7.1.7 A number $\delta > 0$ is said to be a Lebesgue number for a cover $\mathcal{U}$ of a metric space (X,d) if every subset of X of diameter not exceeding 2δ is contained in an element of $\mathcal{U}$.

It is clear that if δ_0 is a Lebesgue number for a cover $\mathcal{U}$, then so is any $0 < \delta < \delta_0$. One can easily prove by contradiction that every open cover of a compact metric space admits such a number. Recall that an open cover is simply a cover whose elements are all open subsets of the space.

7.1.2 Dynamical Covers

In this section, we now add a dynamical aspect to the above discussion. Let X be a nonempty set and let $T: X \longrightarrow X$ be a map. We will define covers that are induced by the dynamics of the map T. First, let us define the preimage of a cover under a map.

Definition 7.1.8 Let X and Y be nonempty sets. Let $h: X \longrightarrow Y$ be a map and $\mathcal{V}$ be a cover of Y. The *preimage* of $\mathcal{V}$ under the map h is the cover consisting of all the preimages of the elements of $\mathcal{V}$ under h, i.e.,

$$h^{-1}\mathcal{V} := \{h^{-1}(V): V \in \mathcal{V}\}.$$

We will now show that, as far as set operations go, the operator h^{-1} behaves well with respect to cover operations.

Lemma 7.1.9 *Let $h: X \longrightarrow Y$ be a map and $\mathcal{U}$ and $\mathcal{V}$ be covers of Y. The following assertions hold.*

(a) *The operation h^{-1} preserves the refinement relation, i.e.,*

$$\mathcal{U} \prec \mathcal{V} \Longrightarrow h^{-1}\mathcal{U} \prec h^{-1}\mathcal{V}.$$

Moreover, if $\mathcal{V}$ is a subcover of $\mathcal{U}$, then $h^{-1}\mathcal{V}$ is a subcover of $h^{-1}\mathcal{U}$.

(b) *The map h^{-1} respects the join operation, i.e.,*

$$h^{-1}(\mathcal{U} \vee \mathcal{V}) = h^{-1}\mathcal{U} \vee h^{-1}\mathcal{V}.$$

By induction, operation h^{-n} for each $n \in \mathbb{N}$ also enjoys these properties.

Proof These assertions are straightforward to prove and are left to the reader as an exercise. ■

We now introduce covers that follow the orbits of a given map by indicating to which elements of a given cover the successive iterates of the map belong.

Definition 7.1.10 Let $T: X \longrightarrow X$ be a map and $\mathcal{U}$ be a cover of X. For every $n \in \mathbb{N}$, define the *dynamical cover*

$$\mathcal{U}^n := \bigvee_{j=0}^{n-1} T^{-j}\mathcal{U} = \mathcal{U} \vee T^{-1}\mathcal{U} \vee \cdots \vee T^{-(n-1)}\mathcal{U}.$$

A typical element of $\mathcal{U}^n$ is of the form

$$U_0 \cap T^{-1}(U_1) \cap T^{-2}(U_2) \cap \cdots \cap T^{-(n-1)}(U_{n-1})$$

for some $U_0, U_1, U_2, \ldots, U_{n-1} \in \mathcal{U}$. This element is the set of all points whose iterates fall successively into the elements $U_0, U_1, U_2, \ldots$, and U_{n-1}.

Lemma 7.1.11 *Let $T: X \longrightarrow X$ be a map and let $\mathcal{U}$ and $\mathcal{V}$ be some covers of X. Fix $n \in \mathbb{N}$. Then*

(a) *If $\mathcal{U} \prec \mathcal{V}$, then $\mathcal{U}^n \prec \mathcal{V}^n$.*
(b) $(\mathcal{U} \vee \mathcal{V})^n = \mathcal{U}^n \vee \mathcal{V}^n$.

Proof The first property follows directly from Lemmas 7.1.9(a) and 7.1.4(g). The second is a consequence of Lemma 7.1.9(b). ■

7.1.3 Definition of Topological Pressure via Open Covers

We are now closer to the definition of topological pressure. It will involve a potential.

Recall that a topological dynamical system $T: X \longrightarrow X$ is a self-transformation T of a compact metrizable space X. Let $\varphi: X \longrightarrow \mathbb{R}$ be a real-valued continuous function. In the context of topological pressure (for historical and physical reasons), such a function is usually referred to as a potential.

The topological pressure of a potential φ with respect to the transformation T is defined in two stages. The first stage is to define topological pressure relative to an open cover, and then to take the appropriate supremum over such covers.

7.1.4 First Stage: Pressure of a Potential Relative to an Open Cover

Recall that the nth Birkhoff sum of a potential φ at a point $x \in X$ is given by

$$S_n\varphi(x) = \sum_{j=0}^{n-1} \varphi(T^j(x)).$$

This is the sum of the values of the potential φ at the first n iterates of x under T.

Definition 7.1.12 For every set $Y \subseteq X$ and every $n \in \mathbb{N}$, define

$$\overline{S}_n\varphi(Y) := \sup\{S_n\varphi(y) : y \in Y\} \quad \text{and} \quad \underline{S}_n\varphi(Y) := \inf\{S_n\varphi(y) : y \in Y\}.$$

Now let $\mathcal{U}$ be an open cover of X. We start thermodynamic formalism with the following definition.

Definition 7.1.13 Let $T: X \longrightarrow X$ be a topological dynamical system and let $\varphi: X \longrightarrow \mathbb{R}$ be a potential. Let $\mathcal{U}$ be an open cover of X. For each $n \in \mathbb{N}$, define the nth level functions, frequently called partition functions, of $\mathcal{U}$ with respect to the potential φ by

$$Z_n(\varphi,\mathcal{U}) := \inf\left\{\sum_{V\in\mathcal{V}} e^{\overline{S}_n\varphi(V)} : \mathcal{V} \text{ is a subcover of } \mathcal{U}^n\right\}$$

and

$$z_n(\varphi,\mathcal{U}) := \inf\left\{\sum_{V\in\mathcal{V}} e^{\underline{S}_n\varphi(V)} : \mathcal{V} \text{ is a subcover of } \mathcal{U}^n\right\}.$$

Note that if $\varphi \equiv 0$, then both numbers $Z_n(0,\mathcal{U})$ and $z_n(0,\mathcal{U})$ are equal to the minimum number of elements of $\mathcal{U}^n$ required to cover X. We then frequently write simply $Z_n(\mathcal{U})$ for $Z_n(0,\mathcal{U})$.

Remark 7.1.14

(a) It is sufficient to take the infimum over all finite subcovers since the exponential function takes only positive values and every subcover has itself a finite subcover. However, this infimum may not be achieved if $\mathcal{U}$ is infinite.
(b) In general, $Z_n(\varphi,\mathcal{U}) \neq Z_1(\varphi,\mathcal{U}^n)$ and $z_n(\varphi,\mathcal{U}) \neq z_1(\varphi,\mathcal{U}^n)$.
(c) If $\varphi \equiv 0$, then $Z_n(0,\mathcal{U}) = z_n(0,\mathcal{U}) = Z_n(\mathcal{U})$ for all $n \in \mathbb{N}$ and any open cover $\mathcal{U}$ of X.
(d) If $\varphi \equiv c$ for some $c \in \mathbb{R}$, then $Z_n(c,\mathcal{U}) = z_n(c,\mathcal{U}) = e^{nc} Z_n(\mathcal{U})$ for all $n \in \mathbb{N}$ and every open cover $\mathcal{U}$ of X.
(e) For all open covers $\mathcal{U}$ of X and all $n \in \mathbb{N}$, we have that

$$e^{n\inf(\varphi)} Z_n(\mathcal{U}) \leq Z_n(\varphi,\mathcal{U}) \leq e^{n\sup(\varphi)} Z_n(\mathcal{U})$$

and

$$e^{n\inf(\varphi)} Z_n(\mathcal{U}) \leq z_n(\varphi,\mathcal{U}) \leq e^{n\sup(\varphi)} Z_n(\mathcal{U}).$$

We need the following.

Definition 7.1.15 The oscillation of φ with respect to an open cover $\mathcal{U}$ is defined to be

$$\operatorname{osc}(\varphi,\mathcal{U}) := \sup\left\{|\varphi(y) - \varphi(x)| : U \in \mathcal{U},\ x, y \in U\right\}.$$

Note that $\operatorname{osc}(\varphi,\cdot) \leq 2\|\varphi\|_\infty$. Also, $\operatorname{osc}(c,\cdot) = 0$ for all $c \in \mathbb{R}$.

Lemma 7.1.16 *Let $T: X \longrightarrow X$ be a topological dynamical system and let $\varphi: X \longrightarrow \mathbb{R}$ be a potential. Then, for every $n \in \mathbb{N}$ and every open cover $\mathcal{U}$ of X, we have that*

$$\operatorname{osc}\big(S_n\varphi, \mathcal{U}^n\big) \leq n \operatorname{osc}(\varphi, \mathcal{U}).$$

Proof Let

$$V := U_0 \cap \cdots \cap T^{-(n-1)}(U_{n-1}) \in \mathcal{U}^n$$

and $x, y \in V$. For each $0 \leq j \leq n-1$, we have that $T^j(x), T^j(y) \in U_j \in \mathcal{U}$. Hence, for all $0 \leq j \leq n-1$,

$$\big|\varphi(T^j(x)) - \varphi(T^j(y))\big| \leq \operatorname{osc}(\varphi, \mathcal{U}).$$

Therefore,

$$\big|S_n\varphi(x) - S_n\varphi(y)\big| \leq \sum_{j=0}^{n-1} \big|\varphi(T^j(x)) - \varphi(T^j(y))\big| \leq n \operatorname{osc}(\varphi, \mathcal{U}).$$

Since this is true for all $x, y \in V$ and all $V \in \mathcal{U}^n$, the result follows. ■

We now look at the relationship between the Z_ns and the z_ns.

Lemma 7.1.17 *Let $T: X \longrightarrow X$ be a topological dynamical system and let $\varphi: X \longrightarrow \mathbb{R}$ be a potential. Then, for all $n \in \mathbb{N}$ and all open covers $\mathcal{U}$ of X, the following inequalities hold:*

$$z_n(\varphi, \mathcal{U}) \leq Z_n(\varphi, \mathcal{U}) \leq e^{n \operatorname{osc}(\varphi, \mathcal{U})} z_n(\varphi, \mathcal{U}).$$

Proof The left inequality is obvious. To ascertain the right one, let $\mathcal{W}$ be a subcover of $\mathcal{U}^n$. Then

$$\begin{aligned}
\sum_{W \in \mathcal{W}} e^{\overline{S}_n\varphi(W)} &\leq \exp\Big(\sup\big\{\overline{S}_n\varphi(W) - \underline{S}_n\varphi(W) : W \in \mathcal{W}\big\}\Big) \sum_{W \in \mathcal{W}} e^{\underline{S}_n\varphi(W)} \\
&\leq e^{\operatorname{osc}(S_n\varphi, \mathcal{U}^n)} \sum_{W \in \mathcal{W}} e^{\underline{S}_n\varphi(W)} \\
&\leq e^{n \operatorname{osc}(\varphi, \mathcal{U})} \sum_{W \in \mathcal{W}} e^{\underline{S}_n\varphi(W)}.
\end{aligned}$$

Taking on both sides, the infimum over all subcovers of $\mathcal{U}^n$ results in the right inequality. ■

In the next few results, we will see that the Z_ns and the z_ns have distinct properties.

Lemma 7.1.18 *Let $T\colon X \longrightarrow X$ be a topological dynamical system and let $\varphi\colon X \longrightarrow \mathbb{R}$ be a potential. If $\mathcal{U} \prec \mathcal{V}$, then, for all $n \in \mathbb{N}$, we have that*

$$Z_n(\varphi,\mathcal{U})e^{-n\,\mathrm{osc}(\varphi,\mathcal{U})} \le Z_n(\varphi,\mathcal{V}) \qquad \textit{while} \qquad z_n(\varphi,\mathcal{U}) \le z_n(\varphi,\mathcal{V}).$$

Proof Fix $n \in \mathbb{N}$. Let $i : \mathcal{V} \to \mathcal{U}$ be a map such that $V \subseteq i(V)$ for all $V \in \mathcal{V}$. The map i induces a map $i_n : \mathcal{V}^n \to \mathcal{U}^n$ in the following way. For every

$$W := V_0 \cap \cdots \cap T^{-(n-1)}(V_{n-1}) \in \mathcal{V}^n,$$

define

$$i_n(W) := i(V_0) \cap \cdots \cap T^{-(n-1)}(i(V_{n-1})).$$

Observe that

$$W \subseteq i_n(W) \in \mathcal{U}^n$$

for all $W \in \mathcal{V}^n$. Moreover, if $x \in W$ and $y \in i_n(W)$, then, for each $0 \le j \le n-1$, we have that $T^j(x) \in V_j \subseteq i(V_j) \ni T^j(y)$. So $T^j(x), T^j(y) \in i(V_j)$ for all $0 \le j < n$. Hence, $x, y \in i_n(W) \in \mathcal{U}^n$, and, thus,

$$S_n\varphi(x) \ge S_n\varphi(y) - \mathrm{osc}\big(S_n\varphi,\mathcal{U}^n\big).$$

Taking the supremum over all $x \in W$ on the left-hand side and over all $y \in i_n(W)$ on the right-hand side yields

$$\overline{S}_n\varphi(W) \ge \overline{S}_n\varphi(i_n(W)) - \mathrm{osc}\big(S_n\varphi,\mathcal{U}^n\big).$$

Now, let $\mathcal{W}$ be a subcover of $\mathcal{V}^n$. Then $i_n(\mathcal{W}) := \{i_n(W) : W \in \mathcal{W}\}$ is a subcover of $\mathcal{U}^n$. Therefore,

$$\begin{aligned}\sum_{W\in\mathcal{W}} e^{\overline{S}_n\varphi(W)} &\ge e^{-\mathrm{osc}(S_n\varphi,\mathcal{U}^n)} \sum_{W\in\mathcal{W}} e^{\overline{S}_n\varphi(i_n(W))}\\ &= e^{-\mathrm{osc}(S_n\varphi,\mathcal{U}^n)} \sum_{Y\in i_n(\mathcal{W})} e^{\overline{S}_n\varphi(Y)}\\ &\ge e^{-\mathrm{osc}(S_n\varphi,\mathcal{U}^n)} Z_n(\varphi,\mathcal{U}).\end{aligned}$$

Taking the infimum over all subcovers $\mathcal{W}$ of $\mathcal{V}^n$ on the left-hand side and using Lemma 7.1.16, we conclude that

$$Z_n(\varphi,\mathcal{V}) \ge e^{-\mathrm{osc}(S_n\varphi,\mathcal{U}^n)} Z_n(\varphi,\mathcal{U}) \ge e^{-n\,\mathrm{osc}(\varphi,\mathcal{U})} Z_n(\varphi,\mathcal{U}).$$

The proof of the inequality for the z_n is left to the reader as an exercise. ■

The proof of the following lemma is left to the reader as an exercise.

Lemma 7.1.19 *Let $T\colon X \longrightarrow X$ be a topological dynamical system and $\varphi\colon X \longrightarrow \mathbb{R}$ be a potential. Let $\mathcal{U}$ and $\mathcal{V}$ be open covers of X and let $n \in \mathbb{N}$. Then*

$$Z_n(\varphi,\mathcal{U} \vee \mathcal{V}) \le \min\{Z_n(\varphi,\mathcal{U}) \cdot Z_n(\mathcal{V}), Z_n(\mathcal{U}) \cdot Z_n(\varphi,\mathcal{V})\}$$

and

$$z_n(\varphi,\mathcal{U} \vee \mathcal{V}) \le \min\{e^{n\operatorname{osc}(\varphi,\mathcal{U})} z_n(\varphi,\mathcal{U}) \cdot Z_n(\mathcal{V}), Z_n(\mathcal{U}) \cdot e^{n\operatorname{osc}(\varphi,\mathcal{V})} z_n(\varphi,\mathcal{V})\}.$$

In order to define topological pressure, we need the following well-known concept.

Definition 7.1.20 A sequence $\{a_n\}_{n=1}^{\infty}$ consisting of real numbers is said to be subadditive if and only if

$$a_{n+m} \le a_n + a_n$$

for all integers $m, n \ge 1$. Likewise, a sequence $\{b_n\}_{n=1}^{\infty}$ consisting of positive real numbers is said to be submultiplicative if and only if

$$b_{n+m} \le b_n b_n$$

for all integers $m, n \ge 1$.

We immediately get the following.

Observation 7.1.21 If $\{a_n\}_{n=1}^{\infty}$ is a submultiplicative sequence of positive real numbers, then $\{\log(a_n)\}_{n=1}^{\infty}$ is a subadditive sequence of real numbers.

Subadditive sequences of real numbers possess the following incredibly helpful property. The reader should take note that this benign-looking lemma is one of the foundation stones of the theory of topological pressure.

Lemma 7.1.22 *If $(a_n)_{n=1}^{\infty}$ is a subadditive sequence of real numbers, then the sequence $\left(\frac{1}{n}a_n\right)_{n=1}^{\infty}$ converges and*

$$\lim_{n\to\infty} \frac{1}{n}a_n = \inf_{n\in\mathbb{N}} \left\{\frac{1}{n}a_n\right\}.$$

If, moreover, $(a_n)_{n=1}^{\infty}$ is bounded from below, then $\inf_{n\in\mathbb{N}} \frac{1}{n}a_n \ge 0$.

Proof Fix $m \in \mathbb{N}$. By the division algorithm, every $n \in \mathbb{N}$ can be uniquely written in the form $n = km + r$, where $0 \le r < m$. The subadditivity of the sequence implies that

$$\frac{a_n}{n} = \frac{a_{km+r}}{km+r} \le \frac{a_{km} + a_r}{km+r} \le \frac{ka_m + a_r}{km} = \frac{a_m}{m} + \frac{a_r}{km}.$$

Notice that

$$-\infty < \min\left\{a_s : 0 \le s < m\right\} \le a_r \le \max\left\{a_s : 0 \le s < m\right\} < +\infty$$

for all $n \in \mathbb{N}$. Therefore, as n tends to infinity, k also tends to infinity and, therefore, a_r/k approaches 0 by the Sandwich Theorem. Hence,

$$\limsup_{n\to\infty} \frac{a_n}{n} \le \frac{a_m}{m}.$$

Since $m \in \mathbb{N}$ was chosen arbitrarily, taking the infimum over m yields that

$$\limsup_{n\to\infty} \frac{a_n}{n} \le \inf_{m\in\mathbb{N}} \frac{a_m}{m}.$$

Thus,

$$\limsup_{n\to\infty} \frac{a_n}{n} \le \inf_{m\in\mathbb{N}} \left\{\frac{a_m}{m}\right\} \le \liminf_{n\to\infty} \frac{a_n}{n} \le \limsup_{n\to\infty} \frac{a_n}{n}.$$

Consequently,

$$\lim_{n\to\infty} \frac{a_n}{n} = \inf_{m\in\mathbb{N}} \left\{\frac{a_m}{m}\right\}.$$

■

The application of this lemma in the current section is due to the following.

Lemma 7.1.23 *Let $T\colon X \longrightarrow X$ be a topological dynamical system and $\varphi\colon X \longrightarrow \mathbb{R}$ be a potential. If $\mathcal{U}$ is an open cover of X, the sequence $(Z_n(\varphi,\mathcal{U}))_{n=1}^{\infty}$ is submultiplicative.*

Proof Fix $m,n \in \mathbb{N}$. Let $\mathcal{V}$ be a subcover of $\mathcal{U}^m$ and $\mathcal{W}$ a subcover of $\mathcal{U}^n$. Note that $\mathcal{V} \vee T^{-m}(\mathcal{W})$ is a subcover of $\mathcal{U}^{m+n}$ since it is a cover and

$$\mathcal{V} \vee T^{-m}(\mathcal{W}) \subseteq \mathcal{U}^m \vee T^{-m}(\mathcal{U}^n) = \mathcal{U}^{m+n}.$$

Take arbitrary $V \in \mathcal{V}$ and $W \in \mathcal{W}$. Then, for every $x \in V \cap T^{-m}(W)$, we have, $x \in V$ and $T^m(x) \in W$; hence,

$$S_{m+n}\varphi(x) = S_m\varphi(x) + S_n\varphi(T^m(x)) \le \overline{S}_m\varphi(V) + \overline{S}_n\varphi(W).$$

Taking the supremum over all $x \in V \cap T^{-m}(W)$, we deduce that

$$\overline{S}_{m+n}\varphi(V \cap T^{-m}(W)) \le \overline{S}_m\varphi(V) + \overline{S}_n\varphi(W).$$

Therefore,

$$\begin{aligned} Z_{m+n}(\varphi,\mathcal{U}) &\le \sum_{E\in\mathcal{V}\vee T^{-m}(\mathcal{W})} e^{\overline{S}_{m+n}\varphi(E)} \le \sum_{V\in\mathcal{V}}\sum_{W\in\mathcal{W}} e^{\overline{S}_{m+n}\varphi(V\cap T^{-m}(W))} \\ &\le \sum_{V\in\mathcal{V}}\sum_{W\in\mathcal{W}} e^{\overline{S}_m\varphi(V)}e^{\overline{S}_n\varphi(W)} \\ &= \sum_{V\in\mathcal{V}} e^{\overline{S}_m\varphi(V)} \sum_{W\in\mathcal{W}} e^{\overline{S}_n\varphi(W)}. \end{aligned}$$

Taking the infimum of the right-hand side over all subcovers $\mathcal{V}$ of $\mathcal{U}^m$ and over all subcovers $\mathcal{W}$ of $\mathcal{U}^n$ gives

$$Z_{m+n}(\varphi,\mathcal{U}) \le Z_m(\varphi,\mathcal{U})Z_n(\varphi,\mathcal{U}). \qquad \blacksquare$$

We immediately get from this lemma and Observation 7.1.21 the following fact.

Corollary 7.1.24 *If $T: X \longrightarrow X$ is a topological dynamical system and $\varphi: X \longrightarrow \mathbb{R}$ is a potential, then the sequence*

$$\big(\log Z_n(\varphi,\mathcal{U})\big)_{n=1}^{\infty}$$

is subadditive for every open cover $\mathcal{U}$ of X.

Because of this fact, we can define the topological pressure of a potential with respect to an open cover. This constitutes the first step in the definition of the topological pressure of a potential.

Definition 7.1.25 Let $T: X \longrightarrow X$ be a topological dynamical system and $\varphi: X \longrightarrow \mathbb{R}$ be a potential. The topological pressure of the potential φ with respect to an open cover $\mathcal{U}$ of X, denoted by $\mathrm{P}(T,\varphi,\mathcal{U})$, is defined to be

$$\mathrm{P}(T,\varphi,\mathcal{U}) := \lim_{n\to\infty}\frac{1}{n}\log Z_n(\varphi,\mathcal{U}) = \inf_{n\in\mathbb{N}}\left\{\frac{1}{n}\log Z_n(\varphi,\mathcal{U})\right\},$$

where the existence of the limit and its equality with the infimum follow immediately from Lemma 7.1.22 and Corollary 7.1.24.

If $\varphi \equiv 0$, we simply write

$$\mathrm{h_{top}}(T,\mathcal{U})$$

for $P(T,0,\mathcal{U})$ and we call this quantity the topological entropy of T with respect to the cover $\mathcal{U}$.

It is also possible to define similar quantities using the $z_n(\varphi,\mathcal{U})$s rather than the $Z_n(\varphi,\mathcal{U})$s.

Definition 7.1.26 Let $T: X \longrightarrow X$ be a topological dynamical system. Given a potential $\varphi: X \longrightarrow \mathbb{R}$ and an open cover $\mathcal{U}$ of X, let

$$\underline{p}(T,\varphi,\mathcal{U}) := \liminf_{n\to\infty} \frac{1}{n} \log z_n(\varphi,\mathcal{U}) \text{ and } \overline{p}(T,\varphi,\mathcal{U}) := \limsup_{n\to\infty} \frac{1}{n} \log z_n(\varphi,\mathcal{U}).$$

Remark 7.1.27 Let $\mathcal{U}$ be an open cover of X.

(a) $\mathrm{P}(T,0,\mathcal{U}) = \underline{p}(T,0,\mathcal{U}) = \overline{p}(T,0,\mathcal{U}) = \mathrm{h_{top}}(T,\mathcal{U})$ by Remark 7.1.14*(c)*.
(b) By Remark 7.1.14*(e)*,

$$-\infty < \mathrm{h_{top}}(T,\mathcal{U}) + \inf\varphi \le \mathrm{P}(T,\varphi,\mathcal{U}) \le \mathrm{h_{top}}(T,\mathcal{U}) + \sup\varphi < \infty.$$

These inequalities also hold with $\mathrm{P}(T,\varphi,\mathcal{U})$ replaced by $\underline{p}(T,\varphi,\mathcal{U})$ and $\overline{p}(T,\varphi,\mathcal{U})$, respectively.
(c) Using Lemma 7.1.17,

$$\underline{p}(T,\varphi,\mathcal{U}) \le \overline{p}(T,\varphi,\mathcal{U}) \le \mathrm{P}(T,\varphi,\mathcal{U}) \le \underline{p}(T,\varphi,\mathcal{U}) + \mathrm{osc}(\varphi,\mathcal{U}).$$

The topological pressure respects the refinement relation and is subadditive with respect to the join operation. It has the following properties.

Proposition 7.1.28 *Let $T: X \longrightarrow X$ be a topological dynamical system and $\varphi: X \longrightarrow \mathbb{R}$ be a potential. Let $\mathcal{U}$ and $\mathcal{V}$ be open covers of X. If $\mathcal{U} \prec \mathcal{V}$, then*

(a)

$$\mathrm{P}(T,\varphi,\mathcal{U}) - \mathrm{osc}(\varphi,\mathcal{U}) \le \mathrm{P}(T,\varphi,\mathcal{V}),$$

while

$$\underline{p}(T,\varphi,\mathcal{U}) \le \underline{p}(T,\varphi,\mathcal{V}) \quad \text{and} \quad \overline{p}(T,\varphi,\mathcal{U}) \le \overline{p}(T,\varphi,\mathcal{V}).$$

(b)

$$\mathrm{P}(T,\varphi,\mathcal{U}\vee\mathcal{V}) \le \min\big\{\mathrm{P}(T,\varphi,\mathcal{U}) + \mathrm{h_{top}}(T,\mathcal{V}), \mathrm{P}(T,\varphi,\mathcal{V}) + \mathrm{h_{top}}(T,\mathcal{U})\big\},$$

whereas

$$\overline{p}(T,\varphi,\mathcal{U}\vee\mathcal{V}) \le \min\Big\{\overline{p}(T,\varphi,\mathcal{U}) + \mathrm{osc}(\varphi,\mathcal{U}) + \mathrm{h_{top}}(T,\mathcal{V}),$$
$$\overline{p}(T,\varphi,\mathcal{V}) + \mathrm{osc}(\varphi,\mathcal{V}) + \mathrm{h_{top}}(T,\mathcal{U})\Big\}.$$

Proof Part (a) is an immediate consequence of Lemma 7.1.18 while (b) follows directly from Lemma 7.1.19. ■

We shall prove the following.

Lemma 7.1.29 *Let $T: X \longrightarrow X$ be a topological dynamical system and $\varphi: X \longrightarrow \mathbb{R}$ be a potential. If $\mathcal{U}$ is an open cover of X, then*

$$\underline{p}(T,\varphi,\mathcal{U}^n) = \underline{p}(T,\varphi,\mathcal{U}) \quad \text{and} \quad \overline{p}(T,\varphi,\mathcal{U}^n) = \overline{p}(T,\varphi,\mathcal{U}),$$

whereas

$$\mathrm{P}(T,\varphi,\mathcal{U}^n) \le \mathrm{P}(T,\varphi,\mathcal{U})$$

for all $n \in \mathbb{N}$.

In addition, if $\mathcal{U}$ is an open partition of X, then

$$\mathrm{P}(T,\varphi,\mathcal{U}^n) = \mathrm{P}(T,\varphi,\mathcal{U})$$

for all $n \in \mathbb{N}$.

Proof Fix $n \in \mathbb{N}$. For all $k \in \mathbb{N}$ and all $x \in X$, we already know that

$$S_{k+n-1}\varphi(x) = S_k\varphi(x) + S_{n-1}\varphi(T^k(x)).$$

Therefore,

$$S_k\varphi(x) - \|S_{n-1}\varphi\|_\infty \le S_{k+n-1}\varphi(x) \le S_k\varphi(x) + \|S_{n-1}\varphi\|_\infty.$$

Hence, for any subset Y of X,

$$\overline{S}_k\varphi(Y) - \|S_{n-1}\varphi\|_\infty \le \overline{S}_{k+n-1}\varphi(Y) \le \overline{S}_k\varphi(Y) + \|S_{n-1}\varphi\|_\infty \tag{7.1}$$

and

$$\underline{S}_k\varphi(Y) - \|S_{n-1}\varphi\|_\infty \le \underline{S}_{k+n-1}\varphi(Y) \le \underline{S}_k\varphi(Y) + \|S_{n-1}\varphi\|_\infty. \tag{7.2}$$

We claim that

$$e^{-\|S_{n-1}\varphi\|_\infty} Z_k(\varphi,\mathcal{U}^n) \le Z_{k+n-1}(\varphi,\mathcal{U}) \tag{7.3}$$

and

$$e^{-\|S_{n-1}\varphi\|_\infty} z_k(\varphi,\mathcal{U}^n) \le z_{k+n-1}(\varphi,\mathcal{U}) \le e^{\|S_{n-1}\varphi\|_\infty} z_k(\varphi,\mathcal{U}^n). \tag{7.4}$$

Let us first prove (7.3). Recall that, according to Lemma 7.1.11(d), we have that

$$(\mathcal{U}^n)^k \prec \mathcal{U}^{k+n-1} \prec (\mathcal{U}^n)^k$$

for all $k \in \mathbb{N}$. However, this is insufficient to declare that a subcover $\mathcal{U}^{k+n-1}$ is also a subcover of $(\mathcal{U}^n)^k$, or vice versa. We need to remember that $\mathcal{U} \vee \mathcal{U} \supseteq \mathcal{U}$ and, thus, $(\mathcal{U}^n)^k \supseteq \mathcal{U}^{k+n-1}$, i.e., $\mathcal{U}^{k+n-1}$ is a subcover of $(\mathcal{U}^n)^k$. Let $\mathcal{V}$ be a subcover of $\mathcal{U}^{k+n-1}$. Then $\mathcal{V}$ is a subcover of $(\mathcal{U}^n)^k$. Using the left inequality in (7.1) with Y replaced by each $V \in \mathcal{V}$ successively, we obtain

$$e^{-\|S_{n-1}\varphi\|_\infty} Z_k(\varphi,\mathcal{U}^n) \le e^{-\|S_{n-1}\varphi\|_\infty} \sum_{V\in\mathcal{V}} e^{\overline{S}_k\varphi(V)} \le \sum_{V\in\mathcal{V}} e^{\overline{S}_{k+n-1}\varphi(V)}.$$

Taking the infimum over all subcovers $\mathcal{V}$ of $\mathcal{U}^{k+n-1}$ yields (7.3). Similarly, using the left inequality in (7.2), we get that

$$e^{-\|S_{n-1}\varphi\|_\infty} z_k(\varphi,\mathcal{U}^n) \le e^{-\|S_{n-1}\varphi\|_\infty} \sum_{V\in\mathcal{V}} e^{\underline{S}_k\varphi(V)} \le \sum_{V\in\mathcal{V}} e^{\underline{S}_{k+n-1}\varphi(V)}.$$

Taking the infimum over all subcovers $\mathcal{V}$ of $\mathcal{U}^{k+n-1}$ yields the left inequality in (7.4). Regarding the right inequality, since $\mathcal{U}^{k+n-1} \prec (\mathcal{U}^n)^k$, there exists a map $i\colon (\mathcal{U}^n)^k \to \mathcal{U}^{k+n-1}$ such that $W \subseteq i(W)$ for all $W \in (\mathcal{U}^n)^k$. Let $\mathcal{W}$ be a subcover of $(\mathcal{U}^n)^k$. Then $i(\mathcal{W})$ is a subcover of $\mathcal{U}^{k+n-1}$ and, using the right inequality in (7.2), we deduce that

$$\begin{aligned}\sum_{W\in\mathcal{W}} e^{\underline{S}_k\varphi(W)} &\ge \sum_{W\in\mathcal{W}} e^{\underline{S}_k\varphi(i(W))} = \sum_{Z\in i(\mathcal{W})} e^{\underline{S}_k\varphi(Z)} \\ &\ge \sum_{Z\in i(\mathcal{W})} e^{\underline{S}_{k+n-1}\varphi(Z)-\|S_{n-1}\varphi\|_\infty} \\ &\ge e^{-\|S_{n-1}\varphi\|_\infty} z_{k+n-1}(\varphi,\mathcal{U}).\end{aligned}$$

Taking the infimum over all subcovers of $(\mathcal{U}^n)^k$ on the left-hand side gives the right inequality in (7.4). So (7.3) and (7.4) always hold.

Moreover, if $\mathcal{U}$ is a partition then $\mathcal{U}\vee\mathcal{U} = \mathcal{U}$ and, thus, $(\mathcal{U}^n)^k = \mathcal{U}^{k+n-1}$ for all $k \in \mathbb{N}$. Let $\mathcal{W}$ be a subcover of $(\mathcal{U}^n)^k$. Using the right inequality in (7.1), we conclude that

$$\sum_{W\in\mathcal{W}} e^{\overline{S}_k\varphi(W)} \ge \sum_{W\in\mathcal{W}} e^{\overline{S}_{k+n-1}\varphi(W)-\|S_{n-1}\varphi\|_\infty} \ge e^{-\|S_{n-1}\varphi\|_\infty} Z_{k+n-1}(\varphi,\mathcal{U}).$$

Taking the infimum over all subcovers of $(\mathcal{U}^n)^k$ on the left-hand side gives

$$Z_k(\varphi,\mathcal{U}^n) \ge e^{-\|S_{n-1}\varphi\|_\infty} Z_{k+n-1}(\varphi,\mathcal{U}). \tag{7.5}$$

Finally, for the passage from the z_ns to $\overline{p}$, it follows from (7.4) that

$$\frac{k}{k+n-1}\cdot\frac{1}{k}\log z_k(\varphi,\mathcal{U}^n) - \frac{\|S_{n-1}\varphi\|_\infty}{k+n-1} \le \frac{1}{k+n-1}\log z_{k+n-1}(\varphi,\mathcal{U})$$

and

$$\frac{1}{k+n-1}\log z_{k+n-1}(\varphi,\mathcal{U}) \le \frac{k}{k+n-1}\cdot\frac{1}{k}\log z_k(\varphi,\mathcal{U}^n) + \frac{\|S_{n-1}\varphi\|_\infty}{k+n-1}.$$

Taking the lim sup as $k \to \infty$ in these two relations yields

$$\overline{p}(T,\varphi,\mathcal{U}^n) \le \overline{p}(T,\varphi,\mathcal{U}) \le \overline{p}(T,\varphi,\mathcal{U}^n).$$

Similarly, one deduces from (7.3) that $\mathrm{P}(T,\varphi,\mathcal{U}^n) \le \mathrm{P}(T,\varphi,\mathcal{U})$ and, when $\mathcal{U}$ is a partition, it follows from (7.5) that $\mathrm{P}(T,\varphi,\mathcal{U}^n) \ge \mathrm{P}(T,\varphi,\mathcal{U})$. ■

7.1.5 Second Stage: The Pressure of a Potential

We now give the definition of topological pressure of the potential.

Definition 7.1.30 Let $T\colon X \to X$ be a topological dynamical system and $\varphi\colon X \to \mathbb{R}$ be a potential. The topological pressure of the potential φ, denoted $\mathrm{P}(T,\varphi)$, is defined by

$$\mathrm{P}(T,\varphi) := \sup\big\{\mathrm{P}(T,\varphi,\mathcal{U}) - \mathrm{osc}(\varphi,\mathcal{U})\colon \mathcal{U} \text{ is an open cover of } X\big\}.$$

This definition may look a little bit awkward because of the term $\mathrm{osc}(\varphi,\mathcal{U})$ that appears in it. This is because of Proposition 7.1.28(a), due to which taking the supremum of the pressure relative to all covers does not always lead to a quantity that has natural desired properties. Our definition is nevertheless purely topological and if $\varphi \equiv 0$, then this term vanishes, thus disappears. We then write

$$\mathrm{h_{top}}(T) := \mathrm{P}(T,0) = \sup\big\{\mathrm{P}(T,0,\mathcal{U})\big\} = \sup\big\{\mathrm{h_{top}}(T,\mathcal{U})\big\},$$

where, as above, the supremum is taken over all open covers of X. The quantity $\mathrm{h_{top}}(T)$ is called the topological entropy of T.

In light of Proposition 7.1.28(a), we may define the counterparts $\underline{p}(T,\varphi)$ and $\overline{p}(T,\varphi)$ of $\mathrm{P}(T,\varphi)$ by simply taking the supremum over all covers.

Definition 7.1.31 Let $T\colon X \longrightarrow X$ be a topological dynamical system and $\varphi\colon X \longrightarrow \mathbb{R}$ be a potential. Define

$$\underline{p}(T,\varphi) := \sup\big\{\underline{p}(T,\varphi,\mathcal{U})\colon \mathcal{U} \text{ is an open cover of } X\big\}$$

and

$$\overline{p}(T,\varphi) := \sup\big\{\overline{p}(T,\varphi,\mathcal{U})\colon \mathcal{U} \text{ is an open cover of } X\big\}.$$

Clearly, $\underline{p}(T,\varphi) \le \overline{p}(T,\varphi)$. In fact, $\underline{p}(T,\varphi)$ and $\overline{p}(T,\varphi)$ are just other expressions of the topological pressure.

Theorem 7.1.32 *For any topological dynamical system $T\colon X \longrightarrow X$ and potential $\varphi\colon X \longrightarrow \mathbb{R}$, it turns out that*

$$\underline{p}(T,\varphi) = \overline{p}(T,\varphi) = \mathrm{P}(T,\varphi).$$

Proof From a rearrangement of the right inequality in Remark 7.1.27(c), it follows that

$$\mathrm{P}(T,\varphi) \le \underline{p}(T,\varphi) \le \overline{p}(T,\varphi).$$

In order to prove that

$$\overline{p}(T,\varphi) \le \mathrm{P}(T,\varphi),$$

let $(\mathcal{U}_n)_{n=1}^{\infty}$ be a sequence of open covers such that

$$\lim_{n\to\infty} \overline{p}(T,\varphi,\mathcal{U}_n) = \overline{p}(T,\varphi).$$

Each open cover $\mathcal{U}_n$ has a Lebesgue number $\delta_n > 0$. The compactness of X guarantees that there are finitely many open balls of radius $\min\{\delta_n, 1/(2n)\}$ that cover X. These balls thereby constitute a refinement of $\mathcal{U}_n$ of diameter at most $1/n$. Owing to Proposition 7.1.28(a), this means that we may assume without loss of generality that the sequence $(\mathcal{U}_n)_{n=1}^{\infty}$ is such that

$$\lim_{n\to\infty} \mathrm{diam}(\mathcal{U}_n) = 0.$$

Since φ is uniformly continuous, it implies that

$$\lim_{n\to\infty} \mathrm{osc}(\varphi,\mathcal{U}_n) = 0.$$

Consequently, using the left inequality in Remark 7.1.27(c), we conclude that

$$\begin{aligned}
\mathrm{P}(T,\varphi) &\ge \sup_{n\in\mathbb{N}}\big\{\mathrm{P}(T,\varphi,\mathcal{U}_n) - \mathrm{osc}(\varphi,\mathcal{U}_n)\big\} \ge \sup_{n\in\mathbb{N}}\big\{\overline{p}(T,\varphi,\mathcal{U}_n) - \mathrm{osc}(\varphi,\mathcal{U}_n)\big\} \\
&\ge \lim_{n\in\mathbb{N}}\big\{\overline{p}(T,\varphi,\mathcal{U}_n) - \mathrm{osc}(\varphi,\mathcal{U}_n)\big\} \\
&= \overline{p}(T,\varphi).
\end{aligned}$$

■

Remark 7.1.33 We want to record the following straightforward properties of topological pressure

(a) $\mathrm{P}(T,0) = \mathrm{h_{top}}(T)$.

(b) By Remark 7.1.27*(b)*,

$$\mathrm{h_{top}}(T) + \inf(\varphi) - \mathrm{osc}(\varphi, X) \le \mathrm{P}(T,\varphi) \le \mathrm{h_{top}}(T) + \sup(\varphi).$$

(c) $\mathrm{P}(T,\varphi) = +\infty$ if and only if $\mathrm{h_{top}}(T) = +\infty$, according to part *(b)*.

We now shall show that topological pressure does not increase under "factorization."

Proposition 7.1.34 *Suppose that* $S\colon Y \longrightarrow Y$ *is a topological factor of* $T\colon X \longrightarrow X$ *via the factor continuous surjection* $h\colon X \longrightarrow Y$*. Then, for every potential* $\varphi\colon Y \longrightarrow \mathbb{R}$*, we have that*

$$\mathrm{P}(S,\varphi) \le \mathrm{P}(T,\varphi \circ h).$$

Proof Let $\mathcal{V}$ be an open cover of Y. Observe that

$$h^{-1}(\mathcal{V}_S^n) = (h^{-1}(\mathcal{V}))_T^n$$

for all $n \in \mathbb{N}$. Moreover, letting C be the collection of all subcovers of $\mathcal{V}_S^n$, the map

$$C \ni \mathcal{C} \longmapsto h^{-1}(\mathcal{C})$$

defines a bijection between subcovers of $\mathcal{V}_S^n$ and subcovers of $h^{-1}(\mathcal{V}_S^n) = (h^{-1}(\mathcal{V}))_T^n$ that preserves cardinalities, i.e.,

$$\# h^{-1}(\mathcal{C}) = \#\mathcal{C}$$

for all $\mathcal{C} \in C$, since h is a surjection. So

$$Z_n(S,\varphi,\mathcal{V}) = Z_n(T,\varphi \circ h, h^{-1}(\mathcal{V}));$$

therefore,

$$\mathrm{P}(S,\varphi,\mathcal{V}) = \mathrm{P}(T,\varphi \circ h, h^{-1}(\mathcal{V})).$$

Also, observe that $\mathrm{osc}(\varphi \circ h, h^{-1}(\mathcal{V})) = \mathrm{osc}(\varphi,\mathcal{V})$. Then

$$\begin{aligned}\mathrm{P}(T,\varphi \circ h) &\ge \mathrm{P}(T,\varphi \circ h, h^{-1}(\mathcal{V})) - \mathrm{osc}(\varphi \circ h, h^{-1}(\mathcal{V}))\\ &= \mathrm{P}(S,\varphi,\mathcal{V}) - \mathrm{osc}(\varphi,\mathcal{V}).\end{aligned}$$

Passing to the supremum over all open covers $\mathcal{V}$ of Y yields that $\mathrm{P}(T,\varphi \circ h) \ge \mathrm{P}(S,\varphi)$. ■

An immediate but important consequence of this lemma is the following result.

Corollary 7.1.35 *If* $T\colon X \longrightarrow X$ *and* $S\colon Y \longrightarrow Y$ *are topologically conjugate dynamical systems via a conjugacy* $h\colon X \longrightarrow Y$*, then*

$$\mathrm{P}(S,\varphi) = \mathrm{P}(T,\varphi \circ h)$$

for all potentials $\varphi\colon Y \longrightarrow \mathbb{R}$.

We will now prove a result showing that topological pressure is determined by any sequence of covers whose diameters tend to zero.

Proposition 7.1.36 *If $T: X \longrightarrow X$ is a topological dynamical system and $\varphi: X \longrightarrow \mathbb{R}$ is a continuous potential, then all the following quantities are all equal.*

(a) $\mathrm{P}(T,\varphi)$.

(b) $\overline{\mathrm{p}}(T,\varphi)$.

(c) $\lim_{\varepsilon\to 0}\left\{\sup\left\{\mathrm{P}(T,\varphi,\mathcal{U}): \mathcal{U} \textit{ open cover with } \mathrm{diam}(\mathcal{U}) \le \varepsilon\right\}\right\}$.

(d) $\sup\left\{\overline{\mathrm{p}}(T,\varphi,\mathcal{U}): \mathcal{U} \textit{ open cover with } \mathrm{diam}(\mathcal{U}) \le \delta\right\}$ *for any* $\delta > 0$.

(e) $\lim_{\varepsilon\to 0}\mathrm{P}(T,\varphi,\mathcal{U}_\varepsilon)$ *for any family of open covers* $(\mathcal{U}_\varepsilon)_{\varepsilon\in(0,\infty)}$ *such that* $\lim_{\varepsilon\to 0}\mathrm{diam}(\mathcal{U}_\varepsilon) = 0$.

(f) $\lim_{\varepsilon\to 0}\overline{\mathrm{p}}(T,\varphi,\mathcal{U}_\varepsilon)$ *for any family of open covers* $(\mathcal{U}_\varepsilon)_{\varepsilon\in(0,\infty)}$ *such that* $\lim_{\varepsilon\to 0}\mathrm{diam}(\mathcal{U}_\varepsilon) = 0$.

(g) $\lim_{n\to\infty}\mathrm{P}(T,\varphi,\mathcal{U}_n)$ *for any sequence of open covers* $(\mathcal{U}_n)_{n=1}^{\infty}$ *such that* $\lim_{n\to\infty}\mathrm{diam}(\mathcal{U}_n) = 0$.

(h) $\lim_{n\to\infty}\overline{\mathrm{p}}(T,\varphi,\mathcal{U}_n)$ *for any sequence of open covers* $(\mathcal{U}_n)_{n=1}^{\infty}$ *such that* $\lim_{n\to\infty}\mathrm{diam}(\mathcal{U}_n) = 0$.

Note that $\overline{\mathrm{p}}$ can be replaced by $\underline{\mathrm{p}}$ in the statements above.

Proof We already know that (a) = (b) by Lemma 7.1.32. It is clear that (b) $\ge$ (d). It is also obvious that (d) $\ge$ (f) and (c) $\ge$ (e) for any family $(\mathcal{U}_\varepsilon)_{\varepsilon\in(0,\infty)}$ as described, and that (d) $\ge$ (h) and (c) $\ge$ (g) for any sequence $(\mathcal{U}_n)_{n=1}^{\infty}$ as specified. It, thus, suffices to prove that (f) $\ge$ (b), that (h) $\ge$ (b), that (e) $\ge$ (a), that (g) $\ge$ (a), and that (b) $\ge$ (c).

We will prove that (g) $\ge$ (a). The proofs of the other inequalities are similar. Let $\mathcal{V}$ be any open cover of X. Since $\lim_{n\to\infty}\mathrm{diam}(\mathcal{U}_n) = 0$, there exists $N \in \mathbb{N}$ such that

$$\mathcal{V} \prec \mathcal{U}_n$$

for all $n \ge N$. By Proposition 7.1.28(a), we obtain that, for all sufficiently large n,

$$\mathrm{P}(T,\varphi,\mathcal{U}_n) \ge \mathrm{P}(T,\varphi,\mathcal{V}) - \mathrm{osc}(\varphi,\mathcal{V}).$$

We immediately deduce that

$$\liminf_{n\to\infty}\mathrm{P}(T,\varphi,\mathcal{U}_n) \ge \mathrm{P}(T,\varphi,\mathcal{V}) - \mathrm{osc}(\varphi,\mathcal{V}).$$

As the open cover $\mathcal{V}$ was chosen arbitrarily, passing to the supremum over all open covers allows us to conclude that

$$\liminf_{n\to\infty} \mathrm{P}(T,\varphi,\mathcal{U}_n) \geq \mathrm{P}(T,\varphi).$$

But $\lim_{n\to\infty} \mathrm{osc}(\varphi,\mathcal{U}_n) = 0$ since $\lim_{n\to\infty} \mathrm{diam}(\mathcal{U}_n) = 0$ and φ is uniformly continuous. Therefore,

$$\begin{aligned}
\mathrm{P}(T,\varphi) &= \sup_{\mathcal{V}}\big[\mathrm{P}(T,\varphi,\mathcal{V}) - \mathrm{osc}(\varphi,\mathcal{V})\big] \geq \limsup_{n\to\infty}\big\{\mathrm{P}(T,\varphi,\mathcal{U}_n) - \mathrm{osc}(\varphi,\mathcal{U}_n)\big\} \\
&= \limsup_{n\to\infty} \mathrm{P}(T,\varphi,\mathcal{U}_n) - \lim_{n\to\infty} \mathrm{osc}(\varphi,\mathcal{U}_n) \\
&= \limsup_{n\to\infty} \mathrm{P}(T,\varphi,\mathcal{U}_n) \\
&\geq \liminf_{n\to\infty} \mathrm{P}(T,\varphi,\mathcal{U}_n) \\
&\geq \mathrm{P}(T,\varphi).
\end{aligned}$$

Hence, $\mathrm{P}(T,\varphi) = \lim_{n\to\infty} \mathrm{P}(T,\varphi,\mathcal{U}_n)$. ■

We can now obtain a slightly stronger estimate than Remark 7.1.33(b) for the difference between topological entropy and topological pressure when the underlying space is metrizable.

Corollary 7.1.37 *If $T: X \longrightarrow X$ is a topological dynamical system and $\varphi: X \longrightarrow \mathbb{R}$ is a continuous potential, then*

$$\mathrm{h_{top}}(T) + \inf(\varphi) \leq \mathrm{P}(T,\varphi) \leq \mathrm{h_{top}}(T) + \sup(\varphi).$$

Proof The upper bound was already mentioned in Remark 7.1.33(b). In order to derive the lower bound, we return to Remark 7.1.27(b). Let $(\mathcal{U}_n)_{n=1}^{\infty}$ be a sequence of open covers of X such that $\lim_{n\to\infty} \mathrm{diam}(\mathcal{U}_n) = 0$. According to Remark 7.1.27(b), for each $n \in \mathbb{N}$, we have that

$$\mathrm{h_{top}}(T,\mathcal{U}_n) + \inf(\varphi) \leq \mathrm{P}(T,\varphi,\mathcal{U}_n).$$

Passing to the limit $n \to \infty$ and using Proposition 7.1.36, we conclude that

$$\mathrm{h_{top}}(T) + \inf(\varphi) \leq \mathrm{P}(T,\varphi).$$ ■

The preceding lemma characterized the topological pressure of a potential as the limit of the topological pressure of the potential relative to a sequence of covers. An even better result would be the characterization of the topological pressure of a potential as the topological pressure of that potential with respect to a single cover. As might by now be expected, such a characterization exists when the system is expansive. We will, therefore, now devote a little bit of time to looking more closely at expansive maps.

First, the expansiveness of a system can be expressed in terms of the following "dynamical," also called Bowen, metrics.

Definition 7.1.38 Let $T\colon (X,d) \longrightarrow (X,d)$ be a topological dynamical system. For every $n \in \mathbb{N}$, let $d_n\colon X \times X \longrightarrow [0,\infty)$ be the metric defined by the following formula:

$$d_n(x,y) := \max\{d(T^j(x),T^j(y))\colon 0 \le j \le n-1\}.$$

Although this notation does not make explicit the dependence on T, it is crucial to remember that the metrics d_n arise from the dynamics of the system T. It is in this sense that they are dynamical metrics.

Observe that $d_1 = d$ and that, for each $x, y \in X$, we have that

$$d_n(x,y) \ge d_m(x,y)$$

whenever $n \ge m$. Moreover, it is worth noticing that all the metrics d_n, $n \in \mathbb{N}$, are topologically equivalent.

Indeed, given a sequence $(x_k)_{k=1}^{\infty}$ in X, the continuity of T ensures that

$$\begin{aligned}\lim_{k\to\infty} d(x_k,y) = 0 &\iff \lim_{k\to\infty} d(T^j(x_k),T^j(y)) = 0 \ \ \forall 0 \le j \le n-1 \ \ \forall n \in \mathbb{N}\\ &\iff \lim_{k\to\infty} d_n(x_k,y) = 0 \ \ \forall n \in \mathbb{N}.\end{aligned}$$

Furthermore, it is easy to see that a dynamical system $T\colon (X,d) \longrightarrow (X,d)$ is δ-expansive if and only if

$$\sup\{d_n(x,y)\colon n \in \mathbb{N}\} \le \delta \implies x = y.$$

Proposition 7.1.39 *A topological dynamical system $T\colon (X,d) \longrightarrow (X,d)$ is δ-expansive if and only if, for every $\eta > 0$, there exists $N = N(\eta) \in \mathbb{N}$ such that*

$$d(x,y) > \eta \implies d_N(x,y) > \delta.$$

Proof The implication ($\Rightarrow$) is obvious. For the converse one, suppose by way of contradiction that $T\colon (X,d) \longrightarrow (X,d)$ is a δ-expansive system (see Definition 6.4.16) and the assertion of our proposition fails. Then there exist $\eta > 0$ and two sequences

$$(x_n)_{n=0}^{\infty} \text{ and } (y_n)_{n=0}^{\infty}$$

in X such that

$$d(x_n,y_n) > \eta \text{ but } d_n(x_n,y_n) \le \delta.$$

Since X is compact, we may assume, by passing to subsequences if necessary, that the sequences $(x_n)_{n=0}^{\infty}$ and $(y_n)_{n=0}^{\infty}$ converge to, say, x and y, respectively. On one hand, this implies that

$$d(x, y) = \lim_{n\to\infty} d(x_n, y_n) \geq \eta > 0;$$

hence,

$$x \neq y. \tag{7.6}$$

On the other hand, if we fix momentarily any $N \in \mathbb{N}$, then, for all $n \geq N$, we have that

$$d_N(x_n, y_n) \leq d_n(x_n, y_n) \leq \delta.$$

Therefore,

$$d_N(x, y) = \lim_{n\to\infty} d_N(x_n, y_n) \leq \delta.$$

Since $d_N(x, y) \leq \delta$ for every $N \in \mathbb{N}$, the δ-expansiveness of the system implies that $x = y$, contrary to (7.6). We are done. ■

As an immediate consequence of this proposition, we get the following.

Corollary 7.1.40 *If (X,d) is a compact metric space and $T: X \longrightarrow X$ is an expansive topological dynamical system with an expansive constant δ, then*

$$\lim_{n\to\infty} \operatorname{diam}(\mathcal{U}^n) = 0$$

for every cover $\mathcal{U}$ with $\operatorname{diam}(\mathcal{U}) \leq \delta$.

We can now easily prove the following.

Theorem 7.1.41 *If (X,d) is a compact metric space and $T: X \longrightarrow X$ is an expansive topological dynamical system with an expansive constant δ, then*

$$\mathrm{P}(T,\varphi) = \underline{p}(T,\varphi,\mathcal{U}) = \overline{p}(T,\varphi,\mathcal{U})$$

for any finite open cover $\mathcal{U}$ of X with $\operatorname{diam}(\mathcal{U}) \leq \delta$. *Moreover,*

$$\mathrm{P}(T,\varphi) = \mathrm{P}(T,\varphi,\mathcal{U})$$

for any finite open partition $\mathcal{U}$ of X with $\operatorname{diam}(\mathcal{U}) \leq \delta$.

Proof It follows from Corollary 7.1.40 and Propositions 7.1.36 and 7.1.29 that

$$\mathrm{P}(T,\varphi) = \lim_{n\to\infty} \overline{p}(T,\varphi,\mathcal{U}^n) = \lim_{n\to\infty} \overline{p}(T,\varphi,\mathcal{U}) = \overline{p}(T,\varphi,\mathcal{U}).$$

A similar argument leads to the statements for $\underline{p}(T,\varphi,\mathcal{U})$ and $\mathcal{U}$ being a partition. ■

For the final result of this section, we study the behavior of topological pressure with respect to the iterates of the system.

Theorem 7.1.42 *If (X,d) is a compact metric space and $T\colon X \longrightarrow X$ is a topological dynamical system, then*

$$\mathrm{P}\big(T^n, S_n\varphi\big) = n\,\mathrm{P}(T,\varphi)$$

for every $n \in \mathbb{N}$.

Proof Fix $n \in \mathbb{N}$. Let $\mathcal{U}$ be an open cover of X. The action of the map T^n on $\mathcal{U}$ until time $j-1$ will be denoted by $\mathcal{U}^j_{T^n}$. Note that $\mathcal{U}^{mn} = (\mathcal{U}^n)^m_{T^n}$ for all $m \in \mathbb{N}$. Furthermore, for all $x \in X$,

$$S_{mn}\varphi(x) = \sum_{k=0}^{mn-1} \varphi \circ T^k(x) = \sum_{j=0}^{m-1} (S_n\varphi) \circ T^{jn}(x) = \sum_{j=0}^{m-1} (S_n\varphi) \circ (T^n)^j(x).$$

Hence,

$$\overline{S}_{mn}\varphi(Y) = S^{T^n}_m\big(\overline{S}_n\varphi\big)(Y)$$

for all subsets Y of X, and, in particular, for all $Y \in \mathcal{U}^{mn} = (\mathcal{U}^n)^m_{T^n}$, where

$$S^{T^n}_m\psi(x) = \sum_{j=0}^{m-1} \psi((T^n)^j(x)).$$

Thus,

$$Z_{mn}(T,\varphi,\mathcal{U}) = Z_m\big(T^n, S_n\varphi, \mathcal{U}^n\big)$$

for all $m \in \mathbb{N}$. Therefore,

$$\begin{aligned}\mathrm{P}(T,\varphi,\mathcal{U}) &= \lim_{m\to\infty} \frac{1}{mn} \log Z_{mn}(T,\varphi,\mathcal{U}) = \frac{1}{n} \lim_{m\to\infty} \frac{1}{m} \log Z_m\big(T^n, S_n\varphi, \mathcal{U}^n\big)\\ &= \frac{1}{n}\mathrm{P}\big(T^n, S_n\varphi, \mathcal{U}^n\big).\end{aligned}$$

Let $(\mathcal{U}_k)_{k=1}^\infty$ be a sequence of open covers such that

$$\lim_{k\to\infty} \mathrm{diam}(\mathcal{U}_k) = 0.$$

Then

$$\lim_{k\to\infty} \mathrm{diam}\big((\mathcal{U}_k)^n\big) = 0$$

since $\mathrm{diam}((\mathcal{U}_k)^n) \le \mathrm{diam}(\mathcal{U}_k)$ for all $k \in \mathbb{N}$. It, thus, follows from Proposition 7.1.36 that

$$\mathrm{P}\big(T^n, S_n\varphi\big) = \lim_{k\to\infty} \mathrm{P}\big(T^n, S_n\varphi, (\mathcal{U}_k)^n\big) = n \lim_{k\to\infty} \mathrm{P}(T,\varphi,\mathcal{U}_k) = n\,\mathrm{P}(T,\varphi).$$

■

7.2 Bowen's Definition of Topological Pressure

In this section, we provide the characterization of topological pressure, due to Bowen, by means of separated or spanning sets. Although this characterization does depend on a metric and is not immediately seen to be topologically invariant, it nevertheless has both theoretical and practical (calculation of pressures) advantages. We will define and discuss separated and spanning sets now and will then characterize topological pressure in terms of them.

Definition 7.2.1 A subset F of X is said to be *(n,ε)-separated* if F is ε-separated with respect to the metric d_n, which is to say that

$$d_n(x,y) \geq \varepsilon$$

for all $x, y \in F$ with $x \neq y$.

Remark 7.2.2

(1) If F is an (n,ε)-separated set and $m \geq n$, then F is also (m,ε)-separated.
(2) If F is an (n,ε)-separated set and $0 < \varepsilon' < \varepsilon$, then F is also (n,ε')-separated.
(3) Given that X is a compact metric space, any (n,ε)-separated set is finite. Indeed, let F be an (n,ε)-separated set and consider the family of balls

$$\{B_n(x,\varepsilon/2) \colon x \in F\}.$$

If the intersection of $B_n(x,\varepsilon/2)$ and $B_n(y,\varepsilon/2)$ is nonempty for some $x, y \in F$, then there exists

$$z \in B_n(x,\varepsilon/2) \cap B_n(y,\varepsilon/2),$$

and it follows that

$$d_n(x,y) \leq d_n(x,z) + d_n(z,y) < \varepsilon/2 + \varepsilon/2 = \varepsilon.$$

As F is an (n,ε)-separated set, this inequality implies that $x = y$. This means that the balls

$$\{B_n(x,\varepsilon/2) \colon x \in F\}$$

are mutually disjoint. Hence, as we are in a compact metric space, there can only be finitely many of them.

The largest separated sets will be especially useful in describing the complexity of the dynamics that the system exhibits.

Definition 7.2.3 A subset F of X is called a *maximal* (n,ε)*-separated set* if, for any (n,ε)-separated set F' with $F \subseteq F'$, we have that $F = F'$. In other words, no strict superset of F is (n,ε)-separated.

The counterpart of the notion of a separated set is the concept of the spanning set.

Definition 7.2.4 A subset E of X is said to be an (n,ε)*-spanning set* if

$$\bigcup_{x \in E} B_n(x,\varepsilon) = X.$$

That is, the orbit of every point in the space is ε-shadowed by the orbit of a point of E until at least time $n-1$.

The smallest spanning sets play a special role in describing the complexity of the dynamics that the system possesses.

Definition 7.2.5 A subset E of X is called a *minimal* (n,ε)*-spanning set* if, for any (n,ε)-spanning set E' with $E \supseteq E'$, we have that $E = E'$. In other words, no strict subset of E is (n,ε)-spanning.

Remark 7.2.6

(1) If E is an (n,ε)-spanning set and $m \le n$, then E is also (m,ε)-spanning.
(2) If E is an (n,ε)-spanning set and $\varepsilon' > \varepsilon$, then E is also (n,ε')-spanning.
(3) Given that we are in a compact metric space, any minimal (n,ε)-spanning set $E \subseteq X$ is finite since the open cover $\{B_n(x,\varepsilon) : x \in E\}$ of the compact metric space X admits a finite subcover.

The next lemma describes two useful relations between separated and spanning sets.

Lemma 7.2.7 *The following statements hold:*

(a) *Every maximal (n,ε)-separated set is an (n,ε)-spanning set.*
(b) *Every $(n,2\varepsilon)$-separated set can be embedded into any (n,ε)-spanning set.*

Proof (a). Let F be a maximal (n,ε)-separated set. By way of contradiction, suppose that F is not an (n,ε)-spanning set. Then there would exist a point

$$y \in X \backslash \bigcup_{x \in F} B_n(x,\varepsilon).$$

It is then easy to verify that the set

$$F \cup \{y\}$$

is (n,ε)-separated, hence contradicting the maximality of F. Therefore, if F is a maximal (n,ε)-separated set, then F is also (n,ε)-spanning.

(b). Let F be an $(n,2\varepsilon)$-separated set and E an (n,ε)-spanning set. For each $x \in F$, choose $i(x) \in E$ such that

$$x \in B_n(i(x),\varepsilon).$$

We claim that the map

$$i\colon F \longrightarrow E$$

is injective. In order to show this, let $x, y \in F$ be such that

$$i(x) = i(y) =: z.$$

Then $x, y \in B_n(z,\varepsilon)$. Therefore, $d_n(x,y) < 2\varepsilon$. Since F is $(n,2\varepsilon)$-separated, we deduce that $x = y$, i.e., the map i is injective and the proof is complete. ■

We are now ready to provide the announced characterization of topological pressure that is based on the concepts of separated and spanning sets. To allege notation, for any $n \in \mathbb{N}$ and $Y \subseteq X$, let

$$\Sigma_n(Y) = \sum_{x\in Y} e^{S_n\varphi(x)}.$$

Theorem 7.2.8 *Let (X,d) be a compact metric space and let $T\colon X \longrightarrow X$ be a topological dynamical system. For all $n \in \mathbb{N}$ and for all $\varepsilon > 0$, let $F_n(\varepsilon)$ be a maximal (n,ε)-separated set and $E_n(\varepsilon)$ be a minimal (n,ε)-spanning set. Then*

$$\begin{aligned}\mathrm{P}(T,\varphi) &= \lim_{\varepsilon\to 0}\liminf_{n\to\infty}\frac{1}{n}\log\Sigma_n(E_n(\varepsilon)) = \lim_{\varepsilon\to 0}\limsup_{n\to\infty}\frac{1}{n}\log\Sigma_n(E_n(\varepsilon))\\ &= \lim_{\varepsilon\to 0}\limsup_{n\to\infty}\frac{1}{n}\log\Sigma_n(F_n(\varepsilon)) = \lim_{\varepsilon\to 0}\liminf_{n\to\infty}\frac{1}{n}\log\Sigma_n(F_n(\varepsilon)).\end{aligned}$$

Proof Fix $\varepsilon > 0$. Let $\mathcal{U}_\varepsilon$ be an open cover of X consisting of balls of radius $\varepsilon/2$. Fix $n \in \mathbb{N}$. Let $\mathcal{U}$ be a subcover of $\mathcal{U}_\varepsilon^n$ such that

$$Z_n(\varphi,\mathcal{U}_\varepsilon) \geq e^{-1}\sum_{U\in\mathcal{U}}\exp(\overline{S}_n\varphi(U)).$$

For each $x \in F_n(\varepsilon)$, let $U(x)$ be an element of the cover $\mathcal{U}$ that contains x and define the function $i\colon F_n(\varepsilon) \to \mathcal{U}$ by setting

$$i(x) = U(x).$$

Since $E_n(\varepsilon)$ and $\mathcal{U}$ is a subcover of $\mathcal{U}_\varepsilon^n$, it follows that this function is injective. Therefore,

$$Z_n(\varphi,\mathcal{U}_\varepsilon) \geq e^{-1}\sum_{U\in\mathcal{U}} e^{\overline{S}_n\varphi(U)} \geq e^{-1}\sum_{x\in F_n(\varepsilon)} e^{\overline{S}_n\varphi(U(x))} \geq e^{-1}\sum_{x\in F_n(\varepsilon)} e^{S_n\varphi(x)}.$$

Since this is true for all $n \in \mathbb{N}$, we deduce that

$$\mathrm{P}(T,\varphi,\mathcal{U}_\varepsilon) = \lim_{n\to\infty}\frac{1}{n}\log Z_n(\varphi,\mathcal{U}_\varepsilon) \geq \limsup_{n\to\infty}\frac{1}{n}\log\sum_{x\in F_n(\varepsilon)} e^{S_n\varphi(x)}.$$

Letting $\varepsilon \to 0$ and using Proposition 7.1.36 yields that

$$\mathrm{P}(T,\varphi) \geq \limsup_{\varepsilon\to 0}\limsup_{n\to\infty}\frac{1}{n}\log\sum_{x\in E_n(\varepsilon)} e^{S_n\varphi(x)}. \tag{7.7}$$

On the other hand, if $\mathcal{V}$ is an arbitrary open cover of X, if $\delta(\mathcal{V}) > 0$ is a Lebesgue number for $\mathcal{V}$, if $0 < \varepsilon < \delta(\mathcal{V})/2$, and if $n \in \mathbb{N}$, then, for all integers $0 \leq k \leq n-1$ and for all $x \in F_n(\varepsilon)$, we have that

$$T^k(B_n(x,\varepsilon)) \subseteq B(T^k(x),\varepsilon).$$

Consequently,

$$\operatorname{diam}\big(T^k(B_n(x,\varepsilon))\big) \leq 2\varepsilon < \delta(\mathcal{V}).$$

Hence, for all integers $0 \leq k \leq n-1$, the set $T^k(B_n(x,\varepsilon))$ is contained in at least one element of the cover $\mathcal{V}$. Denote one of these elements by $V_k(x)$. It follows that

$$B_n(x,\varepsilon) \subseteq T^{-k}(V_k(x))$$

for each integer $0 \leq k \leq n-1$; in other words,

$$B_n(x,\varepsilon) \subseteq \bigcap_{k=0}^{n-1} T^{-k}(V_k(x)).$$

But this latter intersection is an element of $\mathcal{V}^n$. Let us denote it by

$$V(x).$$

Since $F_n(\varepsilon)$ is a maximal (n,ε)-separated set, by Lemma 7.2.7 it is also (n,ε)-spanning. So, the family

$$\{B_n(x,\varepsilon)\}_{x\in F_n(\varepsilon)}$$

is an open cover of X. Each one of these balls is contained in the corresponding set $V(x)$. Hence, the family $\{V(x)\}_{x\in F_n(\varepsilon)}$ is also an open cover of X. Therefore, it is a subcover of $\mathcal{V}^n$. Consequently,

$$Z_n(\varphi,\mathcal{V}) \le \sum_{x\in F_n(\varepsilon)} e^{\overline{S}_n\varphi(V(x))} \le e^{n\,\mathrm{osc}(\varphi,\mathcal{V})} \sum_{x\in F_n(\varepsilon)} e^{S_n\varphi(x)},$$

where the last inequality is due to Lemma 7.1.16. It follows that

$$\mathrm{P}(T,\varphi,\mathcal{V}) \le \mathrm{osc}(\varphi,\mathcal{V}) + \liminf_{n\to\infty}\frac{1}{n}\log\sum_{x\in E_n(\varepsilon)} e^{S_n\varphi(x)}.$$

Since $\mathcal{V}$ is independent of $\varepsilon > 0$, we deduce that

$$\mathrm{P}(T,\varphi,\mathcal{V}) - \mathrm{osc}(\varphi,\mathcal{V}) \le \liminf_{\varepsilon\to 0}\liminf_{n\to\infty}\frac{1}{n}\log\sum_{x\in F_n(\varepsilon)} e^{S_n\varphi(x)}.$$

Then, as $\mathcal{V}$ was chosen to be an arbitrary open cover of X, we conclude that

$$\mathrm{P}(T,\varphi) \le \liminf_{\varepsilon\to 0}\liminf_{n\to\infty}\frac{1}{n}\log\sum_{x\in E_n(\varepsilon)} e^{S_n\varphi(x)}. \tag{7.8}$$

The inequalities (7.7) and (7.8) taken together complete the proof of our theorem. ■

In Theorem 7.2.8, the topological pressure of the system is expressed in terms of a specific family of maximal separated (resp. minimal spanning) sets. However, to derive theoretical results, it is sometimes simpler to use the following quantities.

Definition 7.2.9 Let (X,d) be a compact metric space and let $T: X \longrightarrow X$ be a topological dynamical system. For all $n \in \mathbb{N}$ and $\varepsilon > 0$, let

$$\mathrm{P}_n(T,\varphi,\varepsilon) := \sup\Big\{\Sigma_n(E_n(\varepsilon)) \colon E_n(\varepsilon) \text{ maximal } (n,\varepsilon)\text{-separated set}\Big\}$$

and

$$Q_n(T,\varphi,\varepsilon) := \inf\Big\{\Sigma_n(F_n(\varepsilon)) : F_n(\varepsilon) \text{ minimal } (n,\varepsilon)\text{-spanning set}\Big\}.$$

Thereafter, let

$$\underline{\mathrm{P}}(T,\varphi,\varepsilon) := \liminf_{n\to\infty}\frac{1}{n}\log \mathrm{P}_n(T,\varphi,\varepsilon), \quad \overline{\mathrm{P}}(T,\varphi,\varepsilon) := \limsup_{n\to\infty}\frac{1}{n}\log \mathrm{P}_n(T,\varphi,\varepsilon)$$

and

$$\underline{Q}(T,\varphi,\varepsilon) := \liminf_{n\to\infty} \frac{1}{n} \log Q_n(T,\varphi,\varepsilon),$$

$$\overline{Q}(T,\varphi,\varepsilon) := \limsup_{n\to\infty} \frac{1}{n} \log Q_n(T,\varphi,\varepsilon).$$

The following are simple but key observations about these quantities.

Remark 7.2.10 Let (X,d) be a compact metric space and let $T\colon X \longrightarrow X$ be a topological dynamical system. Let $m \leq n \in \mathbb{N}$ and $0 < \varepsilon < \varepsilon'$. The following relations hold.

(a) By Remark 7.2.2*(a)*:

$$\mathrm{P}_m(T,\varphi,\varepsilon) \leq \mathrm{P}_n(T,\varphi,\varepsilon)e^{(n-m)\|\varphi\|_\infty}.$$

(b)

$$e^{-n\|\varphi\|_\infty} \leq \mathrm{P}_n(T,\varphi,\varepsilon) \leq r_n(\varepsilon)e^{n\|\varphi\|_\infty}$$

and

$$\mathrm{P}_n(T,0,\varepsilon) = r_n(\varepsilon).$$

(c) By Remark 7.2.6*(a)*:

$$Q_m(T,\varphi,\varepsilon) \leq Q_n(T,\varphi,\varepsilon)e^{(n-m)\|\varphi\|_\infty}.$$

(d)

$$e^{-n\|\varphi\|_\infty} \leq Q_n(T,\varphi,\varepsilon) \leq s_n(\varepsilon)e^{n\|\varphi\|_\infty}$$

and

$$Q_n(T,0,\varepsilon) = s_n(\varepsilon).$$

(e) By Remarks 7.2.2 and 7.2.6*(b)*:

$$\mathrm{P}_n(T,\varphi,\varepsilon) \geq \mathrm{P}_n(T,\varphi,\varepsilon')$$

and

$$Q_n(T,\varphi,\varepsilon) \geq Q_n(T,\varphi,\varepsilon').$$

(f) By Lemma 7.2.7:

$$0 < Q_n(T,\varphi,\varepsilon) \leq P_n(T,\varphi,\varepsilon) < \infty.$$

(g)

$$\underline{\mathrm{P}}(T,\varphi,\varepsilon) \le \overline{\mathrm{P}}(T,\varphi,\varepsilon)$$

and

$$\underline{Q}(T,\varphi,\varepsilon) \le \overline{Q}(T,\varphi,\varepsilon).$$

(h) By *(b)*:

$$-\|\varphi\|_\infty < \underline{\mathrm{P}}(T,\varphi,\varepsilon) \le \underline{r}(\varepsilon) + \|\varphi\|_\infty$$

and

$$-\|\varphi\|_\infty < \overline{\mathrm{P}}(T,\varphi,\varepsilon) \le \overline{r}(\varepsilon) + \|\varphi\|_\infty.$$

(i) By *(d)*:

$$-\|\varphi\|_\infty < \underline{Q}(T,\varphi,\varepsilon) \le \underline{s}(\varepsilon) + \|\varphi\|_\infty$$

and

$$-\|\varphi\|_\infty < \overline{Q}(T,\varphi,\varepsilon) \le \overline{s}(\varepsilon) + \|\varphi\|_\infty.$$

(j) By *(e)*:

$$\underline{\mathrm{P}}(T,\varphi,\varepsilon) \ge \underline{\mathrm{P}}(T,\varphi,\varepsilon')$$

and

$$\overline{\mathrm{P}}(T,\varphi,\varepsilon) \ge \overline{\mathrm{P}}(T,\varphi,\varepsilon').$$

(k) By *(e)*:

$$\underline{Q}(T,\varphi,\varepsilon) \ge \underline{Q}(T,\varphi,\varepsilon')$$

and

$$\overline{Q}(T,\varphi,\varepsilon) \ge \overline{Q}(T,\varphi,\varepsilon').$$

(l) By *(f)*:

$$-\|\varphi\|_\infty \le \overline{Q}(T,\varphi,\varepsilon) \le \overline{\mathrm{P}}(T,\varphi,\varepsilon) \le \infty$$

and

$$-\|\varphi\|_\infty \le \underline{Q}(T,\varphi,\varepsilon) \le \underline{\mathrm{P}}(T,\varphi,\varepsilon) \le \infty.$$

We will now describe a relationship between P_ns, Q_ns, and the cover-related quantities Z_ns and z_ns.

Lemma 7.2.11 *If (X,d) is a compact metric space and $T\colon X \longrightarrow X$ is a topological dynamical system, then the following relations hold.*

(a) *If $\mathcal{U}$ is an open cover of X with Lebesgue number 2δ, then*

$$z_n(T,\varphi,\mathcal{U}) \le Q_n(T,\varphi,\delta) \le \mathrm{P}_n(T,\varphi,\delta).$$

(b) *If $\varepsilon > 0$ and $\mathcal{V}$ is an open cover of X with $\operatorname{diam}(\mathcal{V}) \le \varepsilon$, then*

$$Q_n(T,\varphi,\varepsilon) \le \mathrm{P}_n(T,\varphi,\varepsilon) \le Z_n(T,\varphi,\mathcal{V}).$$

Proof We already know that

$$Q_n(T,\varphi,\delta) \le P_n(T,\varphi,\delta).$$

(a) Let $\mathcal{U}$ be an open cover with Lebesgue number 2δ and let F be an (n,δ)-spanning set. Then the dynamic balls $\{B_n(x,\delta)\colon x \in F\}$ form a cover of X. For every $0 \le i \le n-1$, the ball $B(T^i(x),\delta)$, which has diameter at most 2δ, is contained in an element of $\mathcal{U}$. Therefore

$$B_n(x,\delta) = \bigcap_{i=0}^{n-1} T^{-i}(B(T^i(x),\delta))$$

is contained in an element of

$$\mathcal{U}^n = \bigvee_{i=0}^{n-1} T^{-i}(\mathcal{U}).$$

That is,

$$\mathcal{U}^n \prec \{B_n(x,\delta)\colon x \in F\}.$$

Then there exists a map

$$i\colon \{B_n(x,\delta)\colon x \in F\} \to \mathcal{U}^n$$

such that $B_n(x,\delta) \subseteq i(B_n(x,\delta))$ for every $x \in F$. Let $\mathcal{W}$ be a subcover of $\{B_n(x,\delta)\colon x \in F\}$. Then

$$\begin{aligned}\Sigma_n(F) = \sum_{x\in F} e^{S_n\varphi(x)} &\ge \sum_{x\in F} e^{\underline{S}_n\varphi(B_n(x,\delta))} \ge \sum_{W\in\mathcal{W}} e^{\underline{S}_n\varphi(W)} \ge \sum_{W\in\mathcal{W}} e^{\underline{S}_n\varphi(i(W))}\\ &\ge \sum_{Z\in i(\mathcal{W})} e^{\underline{S}_n\varphi(Z)} \ge z_n(T,\varphi,\mathcal{U}).\end{aligned}$$

Since F is an arbitrary (n,δ)-spanning set, it follows that

$$Q_n(T,\varphi,\delta) \ge z_n(T,\varphi,\mathcal{U}).$$

(b) Let $\mathcal{V}$ be an open cover with $\operatorname{diam}(\mathcal{V}) \leq \varepsilon$ and let E be an (n,ε)-separated set. Then no element of the cover $\mathcal{V}^n$ can contain more than one element of E. Let $\mathcal{W}$ be a subcover of $\mathcal{V}^n$ and let a map $i\colon E \to \mathcal{W}$ be such that $x \in i(x)$ for all $x \in E$. Then

$$\Sigma_n(E) = \sum_{x\in E} e^{S_n\varphi(x)} \leq \sum_{x\in E} e^{\overline{S}_n\varphi(i(x))} = \sum_{W\in i(E)} e^{\overline{S}_n\varphi(W)} \leq \sum_{W\in\mathcal{W}} e^{\overline{S}_n\varphi(W)}.$$

As $\mathcal{W}$ is an arbitrary subcover of $\mathcal{V}^n$, it follows that $\Sigma_n(E) \leq Z_n(T,\varphi,\mathcal{V})$. Since E is an arbitrary (n,ε)-separated set, we deduce that

$$\mathrm{P}_n(T,\varphi,\varepsilon) \leq Z_n(T,\varphi,\mathcal{V}).$$ ■

These inequalities have the following immediate consequences.

Corollary 7.2.12 *If (X,d) is a compact metric space and $T\colon X \longrightarrow X$ is a topological dynamical system, then the following relations hold.*

(a) *If $\mathcal{U}$ is an open cover of X with Lebesgue number 2δ, then*

$$\underline{p}(T,\varphi,\mathcal{U}) \leq \underline{Q}(T,\varphi,\delta) \leq \underline{\mathrm{P}}(T,\varphi,\delta).$$

(b) *If $\varepsilon > 0$ and $\mathcal{V}$ is an open cover of X with* $\operatorname{diam}(\mathcal{V}) \leq \varepsilon$, *then*

$$\overline{Q}(T,\varphi,\varepsilon) \leq \overline{\mathrm{P}}(T,\varphi,\varepsilon) \leq \mathrm{P}(T,\varphi,\mathcal{V}).$$

We can then surmise new expressions for the topological pressure.

Corollary 7.2.13 *If (X,d) is a compact metric space and $T\colon X \longrightarrow X$ is a topological dynamical system, then the following equalities hold.*

$$\mathrm{P}(T,\varphi) = \lim_{\varepsilon\to 0} \underline{\mathrm{P}}(T,\varphi,\varepsilon) = \lim_{\varepsilon\to 0} \overline{\mathrm{P}}(T,\varphi,\varepsilon) = \lim_{\varepsilon\to 0} \underline{Q}(T,\varphi,\varepsilon) = \lim_{\varepsilon\to 0} \overline{Q}(T,\varphi,\varepsilon).$$

Proof Let $(\mathcal{U}_\varepsilon)_{\varepsilon\in(0,\infty)}$ be a family of open covers such that

$$\lim_{\varepsilon\to\infty} \operatorname{diam}(\mathcal{U}_\varepsilon) = 0.$$

Let $\delta_\varepsilon > 0$ be a Lebesgue number for $\mathcal{U}_\varepsilon$. Then

$$\lim_{\varepsilon\to\infty} \delta_\varepsilon = 0,$$

as $\delta_\varepsilon \leq \operatorname{diam}(\mathcal{U}_\varepsilon)$. Using Proposition 7.1.36 and Corollary 7.2.12(a), we deduce that

$$\mathrm{P}(T,\varphi) = \lim_{\varepsilon\to\infty} \underline{p}(T,\varphi,\mathcal{U}_\varepsilon) \leq \lim_{\varepsilon\to 0} \underline{Q}(T,\varphi,\varepsilon) \leq \lim_{\varepsilon\to 0} \underline{\mathrm{P}}(T,\varphi,\varepsilon). \tag{7.9}$$

On the other hand, using Proposition 7.1.36 and Corollary 7.2.12(b), we obtain

$$\lim_{\varepsilon\to 0} \overline{Q}(T,\varphi,\varepsilon) \le \lim_{\varepsilon\to 0} \overline{\mathrm{P}}(T,\varphi,\varepsilon)$$
$$\le \lim_{\varepsilon\to 0} \sup\left\{\mathrm{P}(T,\varphi,\mathcal{V}) : \operatorname{diam}(\mathcal{V}) \le \varepsilon\right\} = \mathrm{P}(T,\varphi). \quad (7.10)$$

Combining (7.9) and (7.10) allows us to conclude. ∎

Corollary 7.2.13 is useful to derive theoretical results. Nevertheless, in practice, Theorem 7.2.8 is simpler to use, as only one family of sets is needed. Sometimes, a single sequence of sets is enough.

Theorem 7.2.14 *Let (X,d) be a compact metric space and $T: X \longrightarrow X$ be an expansive topological dynamical system with an expansive constant δ. If $\mathcal{U}$ is an open cover of X whose Lebesgue number is 2η with some $\eta \in (0,\delta/2)$ (in particular* $\operatorname{diam}(\mathcal{U}) \le \delta$*), then the following statements hold for all* $0 < \varepsilon \le \eta$.

(a) *If $(E_n(\varepsilon))_{n=1}^{\infty}$ is a sequence of maximal (n,ε)-separated sets in X, then*

$$\mathrm{P}(T,\varphi) = \lim_{n\to\infty} \frac{1}{n} \log \Sigma_n(E_n(\varepsilon)).$$

(b) *If $(F_n(\varepsilon))_{n=1}^{\infty}$ is a sequence of minimal (n,ε)-spanning sets in X, then*

$$\mathrm{P}(T,\varphi) \le \liminf_{n\to\infty} \frac{1}{n} \log \Sigma_n(F_n(\varepsilon)).$$

(c) $\mathrm{P}(T,\varphi) = \lim_{n\to\infty} \frac{1}{n} \log \mathrm{P}_n(T,\varphi,\varepsilon).$

(d) $\mathrm{P}(T,\varphi) = \lim_{n\to\infty} \frac{1}{n} \log Q_n(T,\varphi,\varepsilon).$

Proof We will prove (a) and leave it to the reader as an exercise to show the other parts using similar arguments.

It follows from Theorem 7.1.41 that

$$\mathrm{P}(T,\varphi) = \underline{p}(T,\varphi,\mathcal{U}).$$

Fix $0 < \varepsilon \le \eta$. Observe that 2ε is also a Lebesgue number for $\mathcal{U}$. Choose any sequence $(E_n(\varepsilon))_{n=1}^{\infty}$ of maximal (n,ε)-separated sets. Since maximal (n,ε)-separated sets are (n,ε)-spanning sets, it follows from Lemma 7.2.11(a) that

$$z_n(T,\varphi,\mathcal{U}) \le Q_n(T,\varphi,\varepsilon) \le \Sigma_n(E_n(\varepsilon)).$$

Therefore,

$$\mathrm{P}(T,\varphi) = \underline{p}(T,\varphi,\mathcal{U}) = \liminf_{n\to\infty} \frac{1}{n} \log z_n(T,\varphi,\mathcal{U}) \le \liminf_{n\to\infty} \frac{1}{n} \log \Sigma_n(E_n(\varepsilon)). \quad (7.11)$$

On the other hand, since $\operatorname{diam}(\mathcal{U}) \le \delta$, it follows from Corollary 7.1.40 that there exists $N \in \mathbb{N}$ such that

$$\operatorname{diam}(\mathcal{U}^k) \le \varepsilon$$

for all integers $k \ge N$. It follows from Lemma 7.2.11(b) that

$$\Sigma_n(E_n(\varepsilon)) \le P_n(T,\varphi,\varepsilon) \le Z_n\left(T,\varphi,\mathcal{U}^k\right)$$

for all $k \ge N$. Consequently,

$$\limsup_{n\to\infty} \frac{1}{n} \log \Sigma_n(E_n(\varepsilon)) \le \lim_{n\to\infty} \frac{1}{n} \log Z_n\left(T,\varphi,\mathcal{U}^k\right) = \mathrm{P}\left(T,\varphi,\mathcal{U}^k\right)$$

for all $k \ge N$. It then follows from Proposition 7.1.36(g) that

$$\limsup_{n\to\infty} \frac{1}{n} \log \Sigma_n(E_n(\varepsilon)) \le \lim_{k\to\infty} \mathrm{P}\left(T,\varphi,\mathcal{U}^k\right) = \mathrm{P}(T,\varphi). \tag{7.12}$$

Combining (7.11) and (7.12) gives (a). ■

The ultimate result of this section is the following.

Theorem 7.2.15 *Let (X,d) be a compact metric space. If $T: X \to X$ is a δ-expansive topological dynamical system on a compact metric space (X,d), then items* (a)–(d) *of Theorem 7.2.14 hold for every* $0 < \varepsilon \le \delta/4$.

Proof In view of Theorem 7.2.14, it suffices to show that there exists $\mathcal{U}$, an open cover of X, whose Lebesgue number is $\delta/2$. So, since a Lebesgue number of the open cover

$$\mathcal{U} := \left\{B(x,\delta) \colon x \in X\right\}$$

is $\delta/2$, we are done. ■

7.3 Basic Properties of Topological Pressure

In this section, we give some of the most basic properties of topological pressure. First, we show that the addition or subtraction of a constant to the potential increases or decreases the pressure of the potential by that same constant.

Proposition 7.3.1 *If $T: X \longrightarrow X$ is a topological dynamical system and $\varphi: X \longrightarrow \mathbb{R}$ is a continuous function, then, for every constant $c \in \mathbb{R}$, we have that*

$$\mathrm{P}(T,\varphi + c) = \mathrm{P}(T,\varphi) + c.$$

Proof For each $n \in \mathbb{N}$ and each $\varepsilon > 0$, let $E_n(\varepsilon)$ be a maximal (n,ε)-separated set. Then

$$\begin{aligned}
\mathrm{P}(T,\varphi+c) &= \lim_{\varepsilon\to 0}\limsup_{n\to\infty}\frac{1}{n}\log\sum_{x\in E_n(\varepsilon)} e^{S_n(\varphi+c)(x)} \\
&= \lim_{\varepsilon\to 0}\limsup_{n\to\infty}\frac{1}{n}\log\sum_{x\in E_n(\varepsilon)} e^{S_n\varphi(x)}e^{nc} \\
&= \lim_{\varepsilon\to 0}\limsup_{n\to\infty}\frac{1}{n}\left[\log\left(\sum_{x\in E_n(\varepsilon)} e^{S_n\varphi(x)}\right)+nc\right] \\
&= \mathrm{P}(T,\varphi)+c.
\end{aligned}$$

■

Next, we shall show that topological pressure, as a function of the potential, is increasing.

Proposition 7.3.2 *If $T\colon X \longrightarrow X$ is a topological dynamical system and $\varphi,\psi\colon X \longrightarrow \mathbb{R}$ are continuous functions, such that $\varphi \le \psi$, then*

$$\mathrm{P}(T,\varphi) \le \mathrm{P}(T,\psi).$$

In particular,

$$\mathrm{h_{top}}(T)+\inf\varphi \le \mathrm{P}(T,\varphi) \le \mathrm{h_{top}}(T)+\sup\varphi.$$

Proof That $\mathrm{P}(T,\varphi) \le \mathrm{P}(T,\psi)$ whenever $\varphi \le \psi$ is clear from the characterization of pressure given in Theorem 7.2.8. The second statement was proved in Corollary 7.1.37. ■

In general, it is not the case that $\mathrm{P}(T,c\,\varphi) = c\mathrm{P}(T,\varphi)$. For example, suppose that $\mathrm{P}(T,0) \neq 0$. Then the equation $\mathrm{P}(T,c\cdot 0) = c\,\mathrm{P}(T,0)$ only holds when $c = 1$.

7.4 Examples

Example 7.4.1 Let E be a finite alphabet and let $\sigma\colon E^{\mathbb{N}} \longrightarrow E^{\mathbb{N}}$ be the corresponding shift map; see (3.2) for its definition and Section 11.1 its treatement at length. Let $\widetilde{\varphi}\colon E \longrightarrow \mathbb{R}$ be an arbitrary function. Then the function $\varphi\colon E^{\mathbb{N}} \longrightarrow \mathbb{R}$ defined by

$$\varphi(\omega) := \widetilde{\varphi}(\omega_1)$$

is a continuous function on E^∞ that depends only upon the first coordinate ω_1 of the word $\omega \in E^\infty$. We will show that

$$\mathrm{P}(\sigma,\varphi) = \log \sum_{e\in E} \exp(\widetilde{\varphi}(e)).$$

It can be seen immediately that the shift map $\sigma: E^{\mathbb{N}} \longrightarrow E^{\mathbb{N}}$ is expansive and any number $\delta \in (0,1)$ is an expansive constant when $E^{\mathbb{N}}$ is endowed with the metric given by (11.1). Note that

$$\mathcal{U} = \big\{[e]: e \in E\big\}$$

is a (finite) open cover of $E^{\mathbb{N}}$; furthermore, it is the partition of $E^{\mathbb{N}}$ into initial 1-cylinders. Since $\mathrm{diam}(\mathcal{U}) = s < 1$, in light of Theorem 7.1.41, we have that

$$\mathrm{P}(\sigma,\varphi) = \mathrm{P}(\sigma,\varphi,\mathcal{U}).$$

In order to compute $\mathrm{P}(\sigma,\varphi,\mathcal{U})$, observe that $\mathcal{U}^n = \{[\omega]: \omega \in E^n\}$ is the partition of E^∞ into initial cylinders of length n. Then

$$\begin{aligned}
\mathrm{P}(\sigma,\varphi) = \mathrm{P}(\sigma,\varphi,\mathcal{U}) &= \lim_{n\to\infty} \frac{1}{n} \log Z_n(\varphi,\mathcal{U}) = \lim_{n\to\infty} \frac{1}{n} \log \sum_{U\in\mathcal{U}^n} e^{\overline{S}_n\varphi(U)} \\
&= \lim_{n\to\infty} \frac{1}{n} \log \sum_{\omega\in E^n} e^{\overline{S}_n\varphi([\omega])} \\
&= \lim_{n\to\infty} \frac{1}{n} \log \sum_{\omega_1,\dots,\omega_n\in E^n} \exp\big(\widetilde{\varphi}(\omega_1) + \cdots + \widetilde{\varphi}(\omega_n)\big) \\
&= \lim_{n\to\infty} \frac{1}{n} \log \sum_{\omega_1\in E} \exp(\widetilde{\varphi}(\omega_1)) \cdots \sum_{\omega_n\in E} \exp(\widetilde{\varphi}(\omega_n)) \\
&= \lim_{n\to\infty} \frac{1}{n} \log \left(\sum_{e\in E} \exp(\widetilde{\varphi}(e))\right)^n \\
&= \log \sum_{e\in E} \exp(\widetilde{\varphi}(e)).
\end{aligned}$$

Example 7.4.2 Let E be a finite alphabet and $\sigma: E^{\mathbb{N}} \longrightarrow E^{\mathbb{N}}$ be the corresponding shift map; see, as in the previous example, (3.2) for its definition and Section 11.1 for its treatement at length. Let $\widetilde{\varphi}: E^2 \longrightarrow \mathbb{R}$ be an arbitrary function. Then the function $\varphi: E^{\mathbb{N}} \longrightarrow \mathbb{R}$ defined by

$$\varphi(\omega) := \widetilde{\varphi}(\omega_1,\omega_2)$$

is a continuous function on $E^{\mathbb{N}}$ that depends only upon the first two coordinates of the word $\omega \in E^{\mathbb{N}}$.

As in the previous example,

$$\mathrm{P}(\sigma,\varphi) = \mathrm{P}(\sigma,\varphi,\mathcal{U}),$$

where $\mathcal{U} = \{[e]\colon e \in E\}$ is the (finite) open partition of E^∞ into initial 1-cylinders and

$$\mathrm{P}(\sigma,\varphi) = \mathrm{P}(\sigma,\varphi,\mathcal{U}) = \lim_{n\to\infty} \frac{1}{n} \log \sum_{\omega\in E^n} e^{\overline{S}_n\varphi([\omega])}.$$

But, in this case,

$$\begin{aligned}\sum_{\omega\in E^n} e^{\overline{S}_n\varphi([\omega])} &= \sum_{\omega\in E^n} \exp\Big(\widetilde{\varphi}(\omega_1,\omega_2) + \widetilde{\varphi}(\omega_2,\omega_3) + \cdots + \widetilde{\varphi}(\omega_{n-1},\omega_n) \\ &\qquad + \max_{e\in E}\big\{\exp(\widetilde{\varphi}(\omega_n,e))\big\}\Big) \\ &= \sum_{\omega_1\in E}\sum_{\omega_2\in E} e^{\widetilde{\varphi}(\omega_1,\omega_2)} \sum_{\omega_3\in E} e^{\widetilde{\varphi}(\omega_2,\omega_3)} \\ &\qquad \cdots \sum_{\omega_n\in E} e^{\widetilde{\varphi}(\omega_{n-1},\omega_n)} \cdot \max_{e\in E}\big\{\exp(\widetilde{\varphi}(\omega_n,e))\big\}.\end{aligned}$$

Since

$$\begin{aligned} m := \min_{e,f\in E}\big\{\exp(\widetilde{\varphi}(f,e))\big\} &\le \max_{e\in E}\big\{\exp(\widetilde{\varphi}(\omega_n,e))\big\} \\ &\le \max_{e,f\in E}\big\{\exp(\widetilde{\varphi}(f,e))\big\} =: M\end{aligned}$$

for all $n \in \mathbb{N}$ and all $\omega_n \in E$, we have that

$$\sum_{\omega\in E^n} e^{\overline{S}_n\varphi([\omega])} \asymp \sum_{\omega_1\in E}\sum_{\omega_2\in E} e^{\widetilde{\varphi}(\omega_1,\omega_2)} \sum_{\omega_3\in E} e^{\widetilde{\varphi}(\omega_2,\omega_3)} \cdots \sum_{\omega_n\in E} e^{\widetilde{\varphi}(\omega_{n-1},\omega_n)}$$

for all n, with uniform constant of comparability $C := \max\{m^{-1}, M\}$.

Let $A\colon E^2 \longrightarrow (0,+\infty)$ be the positive matrix whose entries are

$$A_{ef} := \exp(\widetilde{\varphi}(e,f)).$$

Equip this matrix with the L^1 norm

$$\|A\| := \sum_{e\in E}\sum_{f\in E} A_{ef}.$$

It is then easy to prove by induction that

$$\|A^{n-1}\| = \sum_{\omega_1\in E}\sum_{\omega_2\in E} e^{\widetilde{\varphi}(\omega_1,\omega_2)} \sum_{\omega_3\in E} e^{\widetilde{\varphi}(\omega_2,\omega_3)} \cdots \sum_{\omega_n\in E} e^{\widetilde{\varphi}(\omega_{n-1},\omega_n)}$$

for all $n \ge 2$; hence,

$$\sum_{\omega\in E^n} e^{\overline{S}_n\varphi([\omega])} \asymp \|A^{n-1}\|.$$

Therefore,

$$
\begin{aligned}
\mathrm{P}(T,\varphi) &= \lim_{n\to\infty}\frac{1}{n}\log\sum_{\omega\in E^n} e^{\overline{S}_n\varphi([\omega])} \\
&= \lim_{n\to\infty}\frac{1}{n}\log\|A^{n-1}\| = \log\Big(\lim_{n\to\infty}\|A^n\|^{1/n}\Big) = \log r(A),
\end{aligned}
$$

where $r(A)$ is the spectral radius of A, i.e., the largest eigenvalue of A in modulus.

7.5 The Variational Principle and Equilibrium States

In this section, we will state and prove a fundamental result of thermodynamic formalism known as the *Variational Principle*. This deep result establishes a crucial relationship between topological dynamics and ergodic theory, by way of a formula linking topological pressure and measure-theoretic entropy. The Variational Principle in its classical form and full generality was proved in [**Wa1**] and [**Bow1**]. The proof we shall present follows that of Misiurewicz [**Mis**], which is particularly elegant, short, and simple. We further introduce the concept of equilibrium states and give some sufficient conditions for their existence, such as the upper semicontinuity of the metric entropy function, entailing its expansiveness. We single out the special class of equilibrium states, those of potentials that are identically equal to zero, and, following tradition, we call them measures of maximal entropy. We do not deal with the issue of uniqueness of equilibrium states in this section. We, however, provide an example of a topological dynamical system with positive and finite topological entropy that does not have any measure of maximal entropy.

7.5.1 The Variational Principle

For any topological dynamical system $T\colon X \longrightarrow X$, subject to a potential $\varphi\colon X \longrightarrow \mathbb{R}$ and equipped with a T-invariant Borel probability measure μ, the quantity

$$
\mathrm{h}_\mu(T) + \int_X \varphi\, d\mu
$$

is called the free energy of the system T with respect to μ. The Variational Principle states that the topological pressure of a system is the supremum of the free energies generated by that system.

Theorem 7.5.1 (Variational Principle) *If $T: X \longrightarrow X$ is a continuous map of a compact metrizable space X and $\varphi: X \longrightarrow \mathbb{R}$ is a continuous function, then*

$$\mathrm{P}(T,\varphi) = \sup\left\{\mathrm{h}_\mu(T) + \int_X \varphi \, d\mu : \mu \in M(T)\right\},$$

where $M(T) = M(T,\mathcal{B})$ is the set of all T-invariant Borel probability measures on X.

The proof will be given in two parts. In the first part, we will show that

$$\mathrm{P}(T,\varphi) \geq \mathrm{h}_\mu(T) + \int \varphi \, d\mu$$

for every measure $\mu \in M(T)$. The second part shall consist of the proof of the inequality

$$\sup\left\{\mathrm{h}_\mu(T) + \int_X \varphi \, d\mu : \mu \in M(T)\right\} \geq \mathrm{P}(T,\varphi).$$

The first part is relatively easier to prove than the second. For the proof of Part I, we will need Jensen's Inequality. Recall that a function $\psi : (a,b) \to \mathbb{R}$, where $-\infty \leq a < b \leq \infty$, is said to be *convex* if

$$\psi\big(tx + (1-t)y\big) \leq t\psi(x) + (1-t)\psi(y)$$

for all $t \in [0,1]$ and all $x, y \in (a,b)$.

Theorem 7.5.2 (Jensen's Inequality) *Let μ be a probability measure on a measurable space $(X,\mathfrak{F})$ and let $\psi: (a,b) \to \mathbb{R}$ be a convex function. If $f \in L^1(X,\mathfrak{F},\mu)$ and $f(X) \subseteq (a,b)$, then*

$$\psi\left(\int_X f \, d\mu\right) \leq \int_X \psi \circ f \, d\mu.$$

We shall also need the following lemma, which states that any finite Borel partition $\mathcal{A}$ of X can be, from a metric entropy viewpoint, approximated as closely as desired by a finite Borel partition $\mathcal{B}$ whose elements are all (but one) compact and contained in those of $\mathcal{A}$.

Lemma 7.5.3 *Let $\mu \in M(X)$, $\mathcal{A} := \{A_1, \ldots, A_s\}$ be a finite partition of X into Borel sets, and $\varepsilon > 0$. Then there exist compact sets $B_i \subseteq A_i$, $1 \leq i \leq s$, such that the partition $\mathcal{B} := \{B_1, \ldots, B_s, X \backslash (B_1 \cup \cdots \cup B_s)\}$ satisfies*

$$\mathrm{H}_\mu(\mathcal{A}|\mathcal{B}) \leq \varepsilon.$$

Proof Let the measure μ and the partition $\mathcal{A}$ be as stated and $\varepsilon > 0$. Recall from Definition 6.4.11 the continuous nonnegative function $k: [0,1] \to [0,1]$ defined by

$$k(t) := \begin{cases} -t \log t & \text{if } t \in (0,1] \\ 0 & \text{if } t = 0. \end{cases}$$

The continuity of k at 0 implies that there exists $\delta > 0$ such that $k(t) < \varepsilon/s$ when $0 \le t < \delta$. Since μ is regular and X is compact, for each $1 \le i \le s$, there exists a compact set $B_i \subseteq A_i$ such that $\mu(A_i \backslash B_i) < \delta$. Thus, $k(\mu(A_i \backslash B_i)) < \varepsilon/s$ for all $1 \le i \le s$. By Definition 6.3.2 of conditional entropy, it follows that

$$\begin{aligned}
\mathrm{H}_\mu(\mathcal{A}|\mathcal{B}) &= \sum_{j=1}^{s}\sum_{i=1}^{s} -\mu(A_i \cap B_j) \log \frac{\mu(A_i \cap B_j)}{\mu(B_j)} \\
&\quad + \sum_{i=1}^{s} -\mu\big(A_i \cap (X \backslash \cup_{j=1}^{s} B_j)\big) \log \frac{\mu\big(A_i \cap (X \backslash \cup_{j=1}^{s} B_j)\big)}{\mu(X \backslash \cup_{j=1}^{s} B_j)} \\
&= \sum_{j=1}^{s} -\mu(B_j) \log \frac{\mu(B_j)}{\mu(B_j)} \\
&\quad + \sum_{i=1}^{s} -\mu\big(A_i \cap (\cup_{j=1}^{s} A_j \backslash B_j)\big) \log \frac{\mu\big(A_i \cap (\cup_{j=1}^{s} A_j \backslash B_j)\big)}{\mu\big(\cup_{j=1}^{s} A_j \backslash B_j\big)} \\
&= 0 + \sum_{i=1}^{s} -\mu(A_i \backslash B_i) \log \frac{\mu(A_i \backslash B_i)}{\mu\big(\cup_{j=1}^{s} A_j \backslash B_j\big)} \\
&= \sum_{i=1}^{s} -\mu(A_i \backslash B_i)\Big[\log \mu(A_i \backslash B_i) - \log \mu\big(\cup_{j=1}^{s} A_j \backslash B_j\big)\Big] \\
&= \sum_{i=1}^{s} k(\mu(A_i \backslash B_i)) + \sum_{i=1}^{s} \mu(A_i \backslash B_i) \log \mu\big(\cup_{j=1}^{s} A_j \backslash B_j\big) \\
&\le \sum_{i=1}^{s} k(\mu(A_i \backslash B_i)) \le s \cdot \frac{\varepsilon}{s} = \varepsilon.
\end{aligned}$$

■

Proof of Part I Recall that our aim is to establish the inequality

$$\mathrm{P}(T,\varphi) \ge \mathrm{h}_\mu(T) + \int_X \varphi \, d\mu \tag{7.13}$$

for all $\mu \in M(T)$. To that end, let $\mu \in M(T)$ be arbitrary. We claim that it is sufficient to prove that

$$\mathrm{P}(T,\varphi) \ge \mathrm{h}_\mu(T) + \int_X \varphi \, d\mu - \log 2. \tag{7.14}$$

Indeed, suppose that (7.14) holds. Then, rather than considering directly the system (X,T) under the potential φ, we may alternatively consider the higher iterate system (X,T^n) under the potential

$$S_n\varphi = \sum_{k=0}^{n-1} \varphi \circ T^k.$$

Since the measure μ is T-invariant, it is also T^n-invariant for every integer $n \geq 1$. Using successively Theorem 7.1.42, inequality (7.14) with the quadruple $(X,T^n,S_n\varphi,\mu)$ instead of (X,T,φ,μ), and Theorem 6.4.9, we would then obtain that

$$\begin{aligned} n\mathrm{P}(T,\varphi) = \mathrm{P}(T^n,S_n\varphi) &\geq \mathrm{h}_\mu(T^n) + \int_X S_n\varphi\,d\mu + \log 2 \\ &= n\mathrm{H}_\mu(T) + n\int_X \varphi\,d\mu - \log 2. \end{aligned}$$

Dividing by n and letting n tend to infinity would then yield (7.13). Of course, to obtain (7.14) it suffices to show that

$$\mathrm{P}(T,\varphi) \geq \mathrm{h}_\mu(T,\mathcal{A}) + \int_X \varphi\,d\mu - \log 2 \tag{7.15}$$

for all finite Borel partitions $\mathcal{A}$ of X (see Definition 6.4.8). So let $\mathcal{A}$ be such a partition and $\varepsilon > 0$. To get (7.15), it is enough to prove that

$$\mathrm{P}(T,\varphi) \geq \mathrm{h}_\mu(T,\mathcal{A}) + \int_X \varphi\,d\mu - \log 2 - 2\varepsilon. \tag{7.16}$$

Because of Corollary 7.2.13, it suffices to demonstrate that

$$\underline{\mathrm{P}}(T,\varphi,\delta) \geq \mathrm{h}_\mu(T,\mathcal{A}) + \int_X \varphi\,d\mu - \log 2 - 2\varepsilon \tag{7.17}$$

for all sufficiently small $\delta > 0$. We will, in fact, show more; namely, that

$$\underline{\mathrm{P}}(T,\varphi,\delta) \geq \mathrm{h}_\mu(T,\mathcal{A}) + \int_X \varphi\,d\mu - \log 2 - 2\varepsilon \tag{7.18}$$

for all sufficiently small $\delta > 0$. In light of Definition 7.2.9 and in view of Definition 6.4.6 of the relative entropy $\mathrm{h}_\mu(T,\mathcal{A})$, it is sufficient to prove that

$$\frac{1}{n}\log \sum_{y\in F_n(\delta)} \exp(S_n\varphi(y)) \geq \frac{1}{n}\mathrm{H}_\mu(\mathcal{A}^n) + \frac{1}{n}\int_X S_n\varphi\,d\mu - \log 2 - 2\varepsilon \tag{7.19}$$

for all sufficiently small $\delta > 0$, all large enough $n \in \mathbb{N}$, and every (n,δ)-separated set $F_n(\delta)$. So, given a finite partition $\mathcal{A} := \{A_1, \ldots, A_s\}$ and $\varepsilon > 0$, let

$$\mathcal{B} := \{B_1, \ldots, B_s, X \backslash (B_1 \cup \cdots \cup B_s)\}$$

be the partition given by Lemma 7.5.3. Thus, $\mathrm{H}_\mu(\mathcal{A}|\mathcal{B}) \leq \varepsilon$. Fix $n \geq 1$. By Theorem 6.3.3(9) and Lemma 6.4.4(3), we know that

$$\mathrm{H}_\mu(\mathcal{A}^n) \leq \mathrm{H}_\mu(\mathcal{B}^n) + \mathrm{H}_\mu(\mathcal{A}^n|\mathcal{B}^n) \leq \mathrm{H}_\mu(\mathcal{B}^n) + n\mathrm{H}_\mu(\mathcal{A}|\mathcal{B}) \leq \mathrm{H}_\mu(\mathcal{B}^n) + n\varepsilon. \tag{7.20}$$

From (7.19) and (7.20), it, thus, suffices to establish that

$$\log \sum_{y \in F_n(\delta)} \exp(S_n\varphi(y)) \geq \mathrm{H}_\mu(\mathcal{B}^n) + \int_X S_n\varphi \, d\mu - (\log 2 - \varepsilon)n \tag{7.21}$$

for all sufficiently small $\delta > 0$, all large enough $n \in \mathbb{N}$, and every (n,δ)-separated set $F_n(\delta)$. In order to prove this inequality, we will estimate the term

$$\mathrm{H}_\mu(\mathcal{B}^n) + \int S_n\varphi \, d\mu$$

from above. Since the logarithm function is concave (so its negative is convex), Jensen's Inequality (Theorem 7.5.2) implies that

$$\begin{aligned}
\mathrm{H}_\mu(\mathcal{B}^n) + \int_X S_n\varphi \, d\mu &\leq \sum_{B \in \mathcal{B}^n} \mu(B)\Big[-\log \mu(B) + S_n\varphi(B)\Big] \\
&= \sum_{B \in \mathcal{B}^n} \mu(B) \log \frac{\exp(S_n\varphi(B))}{\mu(B)} \\
&= \int_X \log \frac{\exp\big(S_n\varphi(\mathcal{B}^n(x))\big)}{\mu(\mathcal{B}^n(x))} \, d\mu(x) \\
&\leq \log \int_X \frac{\exp\big(S_n\varphi(\mathcal{B}^n(x))\big)}{\mu(\mathcal{B}^n(x))} \, d\mu(x) \\
&= \log \sum_{B \in \mathcal{B}^n} \exp(S_n\varphi(B)).
\end{aligned} \tag{7.22}$$

Since each set $B_i \in \mathcal{B}$ is compact, it follows that $d(B_i, B_j) > 0$ for all $1 \leq i \neq j \leq s$. As φ is continuous, let $0 < \delta < \frac{1}{2}\min\{d(B_i, B_j) \colon 1 \leq i \neq j \leq s\}$ be such that

$$d(x,y) < \delta \quad \Longrightarrow \quad |\varphi(x) - \varphi(y)| < \varepsilon. \tag{7.23}$$

Now consider an arbitrary maximal (n,δ)-separated set $F_n(\delta)$ and fix temporarily $B \in \mathcal{B}$. According to Lemma 7.2.7, each maximal (n,δ)-separated set is an (n,δ)-spanning set. So, for every $x \in B$, there exists $y \in F_n(\delta)$ such that $x \in B_n(y,\delta)$; therefore, $|S_n\varphi(x) - S_n\varphi(y)| < n\varepsilon$ by (7.23). As the set $F_n(\delta)$ is finite, it follows that there exists $y_B \in F_n(\delta)$ such that

$$S_n\varphi(B) \le S_n\varphi(y_B) + n\varepsilon \qquad \text{and} \qquad B \cap B_n(y_B,\delta) \ne \emptyset. \tag{7.24}$$

Moreover, since $d(B_i,B_j) > 2\delta$ for each $1 \le i \ne j \le s$, any dynamic ball $B_k(z,\delta)$, $k \ge 1$, $z \in X$, intersects at most one B_i and perhaps $X \setminus \cup_{j=1}^{s} B_j$. Hence,

$$\#\{B \in \mathcal{B}\colon B \cap B_k(z,\delta) \ne \emptyset\} \le 2$$

for all $k \ge 1$ and $z \in X$. Thus,

$$\#\{B \in \mathcal{B}^n\colon B \cap B_n(z,\delta) \ne \emptyset\} \le 2^n$$

for all $z \in X$. So, the function $f\colon \mathcal{B}^n \longrightarrow F_n(\delta)$ defined by $f(B) := y_B$ is at most 2^n-to-one. Consequently, by (7.24), we obtain that

$$\begin{aligned}2^n \sum_{y \in F_n(\delta)} \exp(S_n\varphi(y)) &\ge \sum_{B \in \mathcal{B}^n} \exp(S_n\varphi(y_B)) \ge \sum_{B \in \mathcal{B}^n} \exp(S_n\varphi(B) - n\varepsilon)\\ &= \sum_{B \in \mathcal{B}^n} \exp(S_n\varphi(B)) \cdot e^{-n\varepsilon}.\end{aligned}$$

Multiplying both sides by 2^{-n} and then taking the logarithm of both sides and applying (7.22) yields

$$\begin{aligned}\log \sum_{y \in F_n(\delta)} \exp(S_n\varphi(y)) &\ge \log \sum_{B \in \mathcal{B}^n} \exp(S_n\varphi(B)) - n\varepsilon - n\log 2\\ &\ge \mathrm{H}_\mu(\mathcal{B}^n) + \int_X S_n\varphi\, d\mu + n(-\log 2 - \varepsilon).\end{aligned}$$

This inequality, which is nothing other than the sought inequality (7.21), holds for all

$$0 < \delta < \frac{1}{2}\min\{d(B_i,B_j)\colon 1 \le i \ne j \le s\},$$

all $n \in \mathbb{N}$, and all maximal (n,δ)-separated sets $F_n(\delta)$. This concludes the proof. ■

Let us now move on to the proof of Part II of the Variational Principle. For this, we shall need the following four lemmas. The first lemma states that, given any Borel probability measure μ, there exist finite Borel partitions of arbitrarily small diameters whose atoms have boundaries with zero μ-measure.

Lemma 7.5.4 *Let $\mu \in M(X)$. For every $\varepsilon > 0$, there exists a finite Borel partition $\mathcal{A}$ of X such that*

$$\operatorname{diam}(\mathcal{A}) < \varepsilon \quad \text{and} \quad \mu(\partial A) = 0$$

for each $A \in \mathcal{A}$.

Proof Let $\varepsilon > 0$ and let $E := \{x_1, \dots, x_s\}$ be an $(\varepsilon/4)$-spanning set of X. Since, for each fixed $1 \leq i \leq s$, the sets

$$\{x \in X : d(x, x_i) = r\},$$

where $\varepsilon/4 < r < \varepsilon/2$, are mutually disjoint, only countably many of them may have positive μ-measure. Hence, there exists $\varepsilon/4 < t < \varepsilon/2$ such that, for every $1 \leq i \leq s$, we have that

$$\mu\big(\{x \in X : d(x, x_i) = t\}\big) = 0. \tag{7.25}$$

Now let the set A_1 be defined by

$$A_1 := \{x \in X : d(x, x_1) \leq t\},$$

and for each $2 \leq i \leq s$ define the sets $A_2, \dots, A_s$ inductively by setting

$$A_i := \{x \in X : d(x, x_i) \leq t\} \backslash (A_1 \cup \cdots \cup A_{i-1}).$$

Since $t < \varepsilon/2$, the family $\mathcal{A} := \{A_1, \dots, A_s\}$ is a Borel partition of X with diameter smaller than ε. Noting that $\partial(A \backslash B) \subseteq \partial A \cup \partial B$ and $\partial(A \cup B) \subseteq \partial A \cup \partial B$, it follows from (7.25) that

$$\mu(\partial A_i) = 0$$

for each $1 \leq i \leq s$. ∎

The second lemma states that, given any finite Borel partition $\mathcal{A}$ whose atoms have boundaries with zero μ-measure, the entropy of $\mathcal{A}$ as a function of the underlying Borel probability measure is continuous at μ.

Lemma 7.5.5 *Let $\mu \in M(X)$. If $\mathcal{A}$ is a finite Borel partition of X such that $\mu(\partial A) = 0$ for every $A \in \mathcal{A}$, then the function $\mathrm{H}_{(\cdot)}(\mathcal{A}) \colon M(X) \longrightarrow [0, \infty]$, defined by*

$$\mathrm{H}_{(\cdot)}(\mathcal{A})(\nu) = \mathrm{H}_\nu(\mathcal{A}),$$

is continuous at μ.

Proof This follows directly from the fact that, according to the Portmanteau Theorem, a sequence of Borel probability measures $(\mu_n)_{n\geq 1}$ converges weakly to a measure μ if and only if $\lim_{n\to\infty} \mu_n(A) = \mu(A)$ for every Borel set A with $\mu(\partial A) = 0$. ■

In the third announced lemma, we show that the entropy of $\mathcal{A}$ as a function of the underlying Borel probability measure is a concave function.

Lemma 7.5.6 *For any finite Borel partition* $\mathcal{A}$ *of* X*, the function* $H_{(\cdot)}(\mathcal{A})$ *is concave.*

Proof Let $\mathcal{A}$ be a finite Borel partition of X, and μ and ν be Borel probability measures on X. Let also $t \in (0,1)$. Since the function $k(x) = -x\log x$ is concave, for any $A \in \mathcal{A}$, we have that

$$k\big(t\mu(A) + (1-t)\nu(A)\big) \geq tk(\mu(A)) + (1-t)k(\nu(A));$$

hence

$$\begin{aligned}
\mathrm{H}_{t\mu+(1-t)\nu}(\mathcal{A}) &= \sum_{A\in\mathcal{A}} k\big(t\mu(A) + (1-t)\nu(A)\big) \\
&\geq t\sum_{A\in\mathcal{A}} k(\mu(A)) + (1-t)\sum_{A\in\mathcal{A}} k(\nu(A)) \\
&= t\mathrm{H}_\mu(\mathcal{A}) + (1-t)\mathrm{H}_\nu(\mathcal{A}).
\end{aligned}$$

■

Finally, the fourth lemma is a generalization of the, already proven, Krylov–Bogolyubov Theorem asserting that any continuous self-map of a compact metrizable space has at least one Borel probability invariant measure.

Lemma 7.5.7 *Let* $T : X \longrightarrow X$ *be a topological dynamical system. If* $(\mu_n)_{n\geq 1}$ *is a sequence of measures in* $M(X)$*, then any weak* limit point of the sequence* $(m_n)_{n\geq 1}$*, where*

$$m_n := \frac{1}{n}\sum_{i=0}^{n-1} \mu_n \circ T^{-i},$$

is a T*-invariant measure.*

Proof By the compactness of $M(X)$ in the weak* topology, we know that the sequence $(m_n)_{n\geq 1}$ has accumulation points. So let $(m_{n_j})_{j\geq 1}$ be a subsequence which converges weakly* to, say, $m \in M(X)$. Let $f \in C(X)$ be arbitrary. We obtain that

$$
\begin{aligned}
\left|\int_X f\circ T\,dm-\int_X f\,dm\right| &= \lim_{j\to\infty}\left|\int_X f\circ T\,dm_{n_j}-\int_X f\,dm_{n_j}\right| \\
&= \lim_{j\to\infty}\left|\frac{1}{n_j}\int_X\sum_{i=0}^{n_j-1}(f\circ T^{i+1}-f\circ T^i)d\mu_{n_j}\right| \\
&= \lim_{j\to\infty}\left|\frac{1}{n_j}\int_X(f\circ T^{n_j}-f)d\mu_{n_j}\right| \\
&\le \lim_{j\to\infty}\frac{2\|f\|_\infty}{n_j}=0.
\end{aligned}
$$

Thus,

$$
\int_X f\circ T\,dm=\int_X f\,dm;
$$

therefore, the Borel probability measure m is T-invariant. ■

Proof of Part II Fix $\varepsilon>0$. Let $(F_n(\varepsilon))_{n\ge1}$ be a sequence of maximal (n,ε)-separated sets in X. For every $n\ge1$, define the measures μ_n and m_n by

$$
\mu_n=\frac{\sum_{x\in F_n(\varepsilon)}\delta_x\exp(S_n\varphi(x))}{\sum_{x\in F_n(\varepsilon)}\exp(S_n\varphi(x))}\quad\text{and}\quad m_n=\frac{1}{n}\sum_{k=0}^{n-1}\mu_n\circ T^{-k},
$$

where δ_x denotes the Dirac measure concentrated at the point x. For ease of exposition, define

$$
s_n:=\sum_{x\in F_n(\varepsilon)}\exp(S_n\varphi(x))
$$

and

$$
\mu(x):=\mu(\{x\}).
$$

Let $(n_i)_{i\ge1}$ be an increasing sequence of natural numbers such that $(m_{n_i})_{i\ge1}$ converges weakly to, say, m and

$$
\lim_{i\to\infty}\frac{1}{n_i}\log\sum_{x\in F_{n_i}(\varepsilon)}\exp(S_n\varphi(x))=\limsup_{n\to\infty}\sum_{x\in F_n(\varepsilon)}\exp(S_n\varphi(x)). \tag{7.26}
$$

By Lemma 7.5.7, the limit measure m belongs to $M(T)$. Also, in view of Lemma 7.5.4, there exists a finite Borel partition $\mathcal{A}$ such that

$$
\operatorname{diam}(\mathcal{A})<\varepsilon\quad\text{and}\quad m(\partial A)=0
$$

for all $A \in \mathcal{A}$. Since $\#(A \cap F_n(\varepsilon)) \leq 1$ for all $A \in \mathcal{A}^n$, we obtain that

$$\begin{aligned}
&\mathrm{H}_{\mu_n}(\mathcal{A}^n) + \int_X S_n\varphi \, d\mu_n \\
&= \sum_{x \in F_n(\varepsilon)} \mu_n(x)\Big[-\log \mu_n(x) + S_n\varphi(x)\Big] \\
&= \sum_{x \in F_n(\varepsilon)} \frac{\exp(S_n\varphi(x))}{s_n}\left[-\log \frac{\exp(S_n\varphi(x))}{s_n} + S_n\varphi(x)\right] \qquad (7.27) \\
&= \frac{1}{s_n} \sum_{x \in F_n(\varepsilon)} \exp(S_n\varphi(x))\Big[-S_n\varphi(x) + \log s_n + S_n\varphi(x)\Big] \\
&= \log s_n = \log \sum_{x \in F_n(\varepsilon)} \exp(S_n\varphi(x)).
\end{aligned}$$

Now fix $M \in \mathbb{N}$. Let $n \geq 2M$. For $j = 0, 1, \ldots, M-1$, define $s(j) := \lfloor \frac{n-j}{M} \rfloor - 1$, where $\lfloor r \rfloor$ denotes the integer part of r. Note that

$$\begin{aligned}
\bigvee_{k=0}^{s(j)} T^{-(kM+j)}(\mathcal{A}^M) &= T^{-j}(\mathcal{A}^M) \vee T^{-(M+j)}(\mathcal{A}^M) \vee \cdots \vee T^{-(s(j)M+j)}(\mathcal{A}^M) \\
&= T^{-j}(\mathcal{A}) \vee T^{-(j+1)}(\mathcal{A}) \vee \cdots \vee T^{-(s(j)M+j+M-1)}(\mathcal{A}) \\
&= T^{-j}(\mathcal{A}) \vee \cdots \vee T^{-((s(j)+1)M+j-1)}(\mathcal{A})
\end{aligned}$$

and

$$(s(j)+1)M + j - 1 = \left\lfloor \frac{n-j}{M} \right\rfloor M + j - 1 \leq n - j + j - 1 = n - 1.$$

Observe also that

$$\begin{aligned}
(n-1) - \big((s(j)+1)M + j\big) &\leq n - 1 - \left(\left\lfloor \frac{n-j}{M} \right\rfloor M + j\right) \\
&\leq n - j - \left(\frac{n-j}{M} - 1\right) M - 1 = M - 1.
\end{aligned}$$

Setting

$$R_j := \{0, 1, \ldots, j-1\} \cup \{(s(j)+1)M + j, \ldots, n-1\},$$

we have that $\#R_j \leq 2M$ and

$$\mathcal{A}^n = \bigvee_{k=0}^{s(j)} T^{-(kM+j)}(\mathcal{A}^M) \vee \bigvee_{i \in R_j} T^{-i}(\mathcal{A}).$$

Hence, using Theorem 6.3.3(7), we get that

$$
\begin{aligned}
\mathrm{H}_{\mu_n}(\mathcal{A}^n) &\le \sum_{k=0}^{s(j)} \mathrm{H}_{\mu_n}\big(T^{-(kM+j)}(\mathcal{A}^M)\big) + \mathrm{H}_{\mu_n}\left(\bigvee_{i\in R_j} T^{-i}(\mathcal{A})\right) \\
&\le \sum_{k=0}^{s(j)} \mathrm{H}_{\mu_n\circ T^{-(kM+j)}}\big(\mathcal{A}^M\big) + \log\#\left(\bigvee_{i\in R_j} T^{-i}(\mathcal{A})\right) \\
&\le \sum_{k=0}^{s(j)} \mathrm{H}_{\mu_n\circ T^{-(kM+j)}}\big(\mathcal{A}^M\big) + \log(\#\mathcal{A})^{\#R_j} \\
&\le \sum_{k=0}^{s(j)} \mathrm{H}_{\mu_n\circ T^{-(kM+j)}}\big(\mathcal{A}^M\big) + 2M\log\#\mathcal{A}.
\end{aligned}
$$

Summing over all $j = 0, 1, \ldots, M-1$ and using Lemma 7.5.6, we obtain that

$$
\begin{aligned}
M\mathrm{H}_{\mu_n}(\mathcal{A}^n) &\le \sum_{j=0}^{M-1}\sum_{k=0}^{s(j)} \mathrm{H}_{\mu_n\circ T^{-(kM+j)}}\big(\mathcal{A}^M\big) + 2M^2\log\#\mathcal{A} \\
&\le \sum_{l=0}^{n-1} \mathrm{H}_{\mu_n\circ T^{-l}}\big(\mathcal{A}^M\big) + 2M^2\log\#\mathcal{A} \\
&\le n\mathrm{H}_{\frac{1}{n}\sum_{l=0}^{n-1}\mu_n\circ T^{-l}}\big(\mathcal{A}^M\big) + 2M^2\log\#\mathcal{A} \\
&= n\mathrm{H}_{m_n}\big(\mathcal{A}^M\big) + 2M^2\log\#\mathcal{A}.
\end{aligned}
$$

Adding $M\int S_n\varphi\,d\mu_n$ to both sides and applying (7.27) yields

$$
M\log\sum_{x\in F_n(\varepsilon)}\exp(S_n\varphi(x)) \le n\mathrm{H}_{m_n}\big(\mathcal{A}^M\big) + M\int_X S_n\varphi\,d\mu_n + 2M^2\log\#\mathcal{A}.
$$

Dividing both sides by Mn and since $\frac{1}{n}\int S_n\varphi\,d\mu_n = \int\varphi\,dm_n$, it follows that

$$
\frac{1}{n}\log\sum_{x\in F_n(\varepsilon)}\exp(S_n\varphi(x)) \le \frac{1}{M}\mathrm{H}_{m_n}\big(\mathcal{A}^M\big) + \int_X\varphi\,dm_n + \frac{2M}{n}\log\#\mathcal{A}.
$$

Since $\partial T^{-1}(A) \subseteq T^{-1}(\partial A)$ for every set $A \subseteq X$, the m-measure of the boundaries of the partition $\mathcal{A}^M$ is equal to zero. Therefore, remembering that each maximal (n,ε)-separated set is (n,ε)-spanning, on letting n tend to infinity along the subsequence $(n_i)_{i\ge 1}$, we conclude from the above inequality and from Lemma 7.5.5 that

$$
\varlimsup_{n\to\infty}\frac{1}{n}\log\sum_{x\in F_n(\varepsilon)}\exp(S_n\varphi(x)) \le \frac{1}{M}\mathrm{H}_m\big(\mathcal{A}^M\big) + \int_X\varphi\,dm.
$$

Letting M tend to infinity, we obtain that

$$\varlimsup_{n\to\infty}\frac{1}{n}\log\sum_{x\in F_n(\varepsilon)}\exp(S_n\varphi(x))\le \mathrm{h}_m(T,\mathcal{A})+\int_X\varphi\,dm$$
$$\le\sup\left\{\mathrm{h}_\mu(T)+\int_X\varphi\,d\mu\colon\mu\in M(T)\right\}.$$

Finally, letting $\varepsilon>0$ tend to zero and applying Theorem 7.2.8 yields the desired inequality. This finishes the proof of Part II and consequently completes the proof of the Variational Principle. ■

Let us now state some immediate consequences of the Variational Principle. The first consequence concerns topological entropy of a topological dynamical system. Namely, the topological entropy of such a system is the supremum of its all measure-theoretic entropies.

Corollary 7.5.8 *If $T\colon X\longrightarrow X$ is a continuous map of a compact metrizable space, then*

$$\mathrm{h}_{\mathrm{top}}(T)=\sup\{\mathrm{h}_\mu(T)\colon\mu\in M(T)\}.$$

Proof This follows directly from Theorem 7.5.1 on letting $\varphi\equiv 0$. ■

Furthermore, the topological pressure of a topological dynamical system is determined by the supremum of the free energy of the system with respect to its ergodic measures.

Corollary 7.5.9 *If $T\colon X\longrightarrow X$ is a continuous map of a compact metrizable space, then*

$$\mathrm{P}(T,\varphi)=\sup\left\{\mathrm{h}_\mu(T)+\int_X\varphi\,d\mu\colon\mu\in E(T)\right\},$$

where $E(T)$ denotes the subset of all ergodic measures in $M(T)$.

Proof Let $\mu\in M(T)$. Since $M(T)$ is a compact convex metrizable space, the Choquet Representation Theorem (see [**Ph**]) implies that μ has a decomposition in terms of the extreme points of $M(T)$, which, according to Theorem 3.1.3, are exactly the ergodic measures $E(T)$ in M(T). This means that there exists a unique probability measure m on the Borel σ-algebra of $M(T)$ such that $m(E(T))=1$ and

$$\int_X f\,d\mu=\int_{E(T)}\left(\int_X f\,d\nu\right)dm(\nu)$$

for all $f \in C(X)$. In particular,

$$\int_X \varphi \, d\mu = \int_{E(T)} \left(\int_X \varphi \, d\nu \right) dm(\nu).$$

Moreover, we have that

$$\mathrm{h}_\mu(T) = \mathrm{h}_{\int_{E(T)} \nu \, dm(\nu)}(T) = \int_{E(T)} \mathrm{h}_\nu(T) dm(\nu).$$

It, therefore, follows that

$$\mathrm{h}_\mu(T) + \int_X \varphi \, d\mu = \int_{E(T)} \left[\mathrm{h}_\nu(T) + \int_X \varphi \, d\nu \right] dm(\nu).$$

Suppose, by way of contradiction, that

$$\mathrm{h}_\nu(T) + \int_X \varphi \, d\nu < \mathrm{h}_\mu(T) + \int_X \varphi \, d\mu$$

for every $\nu \in E(T)$. Then we would have that

$$\begin{aligned} \mathrm{h}_\mu(T) + \int_X \varphi \, d\mu &= \int_{E(T)} \left[\mathrm{h}_\nu(T) + \int_X \varphi \, d\nu \right] dm(\nu) \\ &< \int_{E(T)} \left[\mathrm{h}_\mu(T) + \int_X \varphi \, d\mu \right] dm(\nu) \\ &= \mathrm{h}_\mu(T) + \int_X \varphi \, d\mu. \end{aligned}$$

This is a contradiction; consequently, there exists $\nu \in E(T)$ such that

$$\mathrm{h}_\nu(T) + \int_X \varphi \, d\nu \geq \mathrm{h}_\mu(T) + \int_X \varphi \, d\mu.$$

So, an application of the Variational Principle, i.e., Theorem 7.5.1, finishes the proof. ∎

Finally, we will show that the pressure of any subsystem is at most the pressure of the entire system.

Corollary 7.5.10 *If $T: X \longrightarrow X$ is a topological dynamical system, $\varphi: X \longrightarrow \mathbb{R}$ is a continuous potential, and Y is a closed subset of X such that $T(Y) \subseteq Y$, then*

$$\mathrm{P}\big(T|_Y, \varphi|_Y\big) \leq \mathrm{P}(T, \varphi).$$

Proof Each Borel probability $T|_Y$-invariant measure ν on Y can be extended to the Borel probability T-invariant measure on X by the following formula:

$$\nu^*(B) = \nu(B \cap Y).$$

Then

$$\mathrm{h}_{\nu^*}(T) = \mathrm{h}_\nu(T) \quad \text{and} \quad \int_X \varphi \, d\nu^* = \int_Y \varphi \, d\nu.$$

Therefore, by virtue of the Variational Principle, i.e., Theorem 7.5.1, we get that

$$\mathrm{P}\big(T|_Y, \varphi|_Y\big) \le \mathrm{P}(T, \varphi).$$

■

7.5.2 Equilibrium States

In light of the Variational Principle, the measures that maximize the free energy of the system, i.e., the measures with respect to which free energy of the system coincides with its topological pressure, are given a special name.

Definition 7.5.11 If $T\colon X \longrightarrow X$ is a topological dynamical system and $\varphi\colon X \longrightarrow \mathbb{R}$ is a continuous potential, then a measure $\mu \in M(T)$ is called an *equilibrium state* for the potential φ if and only if

$$\mathrm{P}(T, \varphi) = \mathrm{h}_\mu(T) + \int_X \varphi \, d\mu.$$

Notice that if a potential φ has an equilibrium state, then, after invoking Theorem 3.2.4, the same reasoning as that of Corollary 7.5.9 shows that φ has an ergodic equilibrium state.

When $\varphi = 0$, the equilibrium states are also called *measures of maximal entropy*, i.e., measures for which

$$\mathrm{h}_\mu(T) = \mathrm{h}_{\mathrm{top}}(T).$$

In particular, if $\mathrm{h}_{\mathrm{top}}(T) = 0$, then every invariant measure is a measure of maximal entropy for T. Note that this is the case for homeomorphisms of the unit circle among other examples.

It is natural at this point to wonder whether equilibrium states exist for all topological dynamical systems. The answer, as the following example demonstrates, is that they do not.

Example 7.5.12 We will describe a topological dynamical system with finite topological entropy and with no measure of maximal entropy. Let

$$(T_n\colon X_n \to X_n)_{n=1}^{\infty}$$

be a sequence of topological dynamical systems with the property that

$$\mathrm{h}_{\mathrm{top}}(T_n) < \mathrm{h}_{\mathrm{top}}(T_{n+1}) \qquad \text{and} \qquad \sup_{n\ge 1}\big\{\mathrm{h}_{\mathrm{top}}(T_n)\big\} < +\infty.$$

Let $\bigoplus_{n=1}^{\infty} X_n$ denote the topological disjoint union of the spaces X_n and

$$X := \{\omega\} \cup \bigoplus_{n=1}^{\infty} X_n$$

be the one-point compactification of $\bigoplus_{n=1}^{\infty} X_n$. Define the map $T: X \to X$ by

$$T(x) = \begin{cases} T_n(x) & \text{if } x \in X_n \\ \omega & \text{if } x = \omega. \end{cases}$$

Then $T: X \longrightarrow X$ is continuous. Suppose that μ is an ergodic measure of maximal entropy for T. Then

$$\mu(\{\omega\}) \in \{0, 1\}$$

since $T^{-1}(\{\omega\}) = \{\omega\}$. But if $\mu(\{\omega\}) = 1$, then we would have that

$$\mu\left(\bigoplus_{n=1}^{\infty} X_n\right) = 0.$$

Hence, on the one hand, we would have that $\mathrm{h}_\mu(T) = 0$, while, on the other hand,

$$\mathrm{h}_\mu(T) = \mathrm{h}_{\mathrm{top}}(T) \geq \sup_{n \geq 1} \left\{\mathrm{h}_{\mathrm{top}}(T_n)\right\} > 0.$$

This contradiction implies that

$$\mu(\{\omega\}) = 0.$$

Similarly,

$$\mu(X_n) \in \{0, 1\}$$

for all $n \geq 1$ since $T^{-1}(X_n) = X_n$. Therefore, there exists a unique $N \geq 1$ such that $\mu(X_N) = 1$. It then follows that

$$\mathrm{h}_{\mathrm{top}}(T) = \mathrm{h}_\mu(T) = \mathrm{h}_\mu(T_N) \leq \mathrm{h}_{\mathrm{top}}(T_N) < \sup_{n \geq 1} \left\{\mathrm{h}_{\mathrm{top}}(T_n)\right\} \leq \mathrm{h}_{\mathrm{top}}(T).$$

This contradiction shows that there is no measure of maximal entropy for the system T.

Given that equilibrium states do not always exist, we would like to find conditions under which they do exist. But, since the function

$$\mu \longmapsto \int_X \varphi \, d\mu$$

is continuous in the weak* topology on the compact space $M(T)$, the function

$$\mu \longmapsto \mathrm{h}_\mu(T)$$

cannot be continuous in general. Otherwise, the sum of these last two functions would be continuous and would hence attain a maximum on the compact space $M(T)$, i.e., equilibrium states would always exist. Nevertheless, the function

$$\mu \longmapsto \mathrm{h}_\mu(T)$$

is sometimes upper semicontinuous and this is sufficient to ensure the existence of an equilibrium state. Let us first recall the notion of upper (and lower) semicontinuity.

Definition 7.5.13 Let X be a topological space. A function $f\colon X \longrightarrow [-\infty, +\infty]$ is *upper semicontinuous* if and only if, for all $x \in X$, it holds that

$$\limsup_{y \to x} f(y) \le f(x).$$

Equivalently, f is upper semicontinuous if and only if the set

$$\{x \in X\colon f(x) < r\}$$

is open in X for all $r \in \mathbb{R}$.

A function $f\colon X \to [-\infty, +\infty]$ is called *lower semicontinuous* if and only if $-f$ is upper semicontinuous.

Evidently, a function $f\colon X \to [-\infty, +\infty]$ is continuous if and only if it is both upper and lower semicontinuous. A perhaps most significant property of upper semicontinuous functions is the following well-known fact.

Lemma 7.5.14 *If $f\colon X \longrightarrow [-\infty, +\infty]$ is an upper semicontinuous function on a compact topological space X, then f attains its upper bound on X.*

One class of dynamical systems for which the function $\mu \mapsto \mathrm{h}_\mu(T)$ is upper semicontinuous are all expansive maps.

Theorem 7.5.15 *If $T\colon X \longrightarrow X$ is a (positively) expansive topological dynamical system, then the function*

$$M(T) \ni \mu \longmapsto \mathrm{h}_\mu(T)$$

is upper semicontinuous. Hence, each continuous potential $\varphi\colon X \longrightarrow \mathbb{R}$ has an equilibrium state.

Proof Fix $\delta > 0$ as an expansive constant for T. Let $\mu \in M(T)$. According to Lemma 7.5.4, there exists a finite Borel partition $\mathcal{A}$ of X with the property that

$$\mathrm{diam}(\mathcal{A}) < \delta \quad \text{and} \quad \mu(\partial A) = 0$$

for each $A \in \mathcal{A}$. Fix $\varepsilon > 0$. Notice that, since

$$\mathrm{h}_\mu(T) \geq \mathrm{h}_\mu(T, \mathcal{A}) = \inf_{n \to \infty} \left\{ \frac{1}{n} \mathrm{H}_\mu(\mathcal{A}^n) \right\}$$

by Definitions 6.4.8 and 6.4.6, so there exists $m \geq 1$ such that

$$\frac{1}{m} \mathrm{H}_\mu(\mathcal{A}^m) \leq \mathrm{H}_\mu(T) + \frac{\varepsilon}{2}.$$

Now let $(\mu_n)_{n \geq 1}$ be a sequence of measures in $M(T)$ converging weakly to μ. Since $\mathrm{diam}(\mathcal{A}) < \delta$, it follows from Theorem 6.4.19 that

$$\mathrm{h}_{\mu_n}(T) = \mathrm{h}_{\mu_n}(T, \mathcal{A})$$

for all $n \geq 1$. Moreover, by Lemma 7.5.5 (with $\mathcal{A}$ replaced by $\mathcal{A}^m$), we have that

$$\lim_{n \to \infty} \mathrm{H}_{\mu_n}(\mathcal{A}^m) = \mathrm{H}_\mu(\mathcal{A}^m).$$

Therefore, there exists $N \geq 1$ such that, for all $n \geq N$, we have that

$$\frac{1}{m} \left| \mathrm{H}_{\mu_n}(\mathcal{A}^m) - \mathrm{H}_\mu(\mathcal{A}^m) \right| \leq \frac{\varepsilon}{2}.$$

Hence, for all $n \geq N$, we deduce that

$$\mathrm{h}_{\mu_n}(T) = \mathrm{h}_{\mu_n}(T, \mathcal{A}) \leq \frac{1}{m} \mathrm{H}_{\mu_n}(\mathcal{A}^m) \leq \frac{1}{m} \mathrm{H}_\mu(\mathcal{A}^m) + \frac{\varepsilon}{2} \leq \mathrm{h}_\mu(T) + \varepsilon.$$

Consequently,

$$\limsup_{n \to \infty} \mathrm{h}_{\mu_n}(T) \leq \mathrm{h}_\mu(T)$$

for any sequence $(\mu_n)_{n \geq 1}$ in $M(T)$ converging weakly to μ. Thus,

$$\limsup_{\nu \to \mu} \mathrm{h}_\nu(T) \leq \mathrm{h}_\mu(T);$$

in other words, the function

$$M(T) \ni \mu \longmapsto \mathrm{h}_\mu(T)$$

is upper semicontinuous. Since the function $\mu \mapsto \int_X \varphi \, d\mu$ is continuous in the weak topology on the compact space $M(T)$, it follows that the function

$$\mu \longmapsto \mathrm{h}_\mu(T) + \int_X \varphi \, d\mu$$

is upper semicontinuous. Lemma 7.5.14 then yields that each continuous potential φ admits an equilibrium state. ■

Finally, we show that the pressure function is Lipschitz continuous.

Theorem 7.5.16 *If $T: X \longrightarrow X$ is a topological dynamical, then the pressure function* $\mathrm{P}: C(X) \to \mathbb{R}$ *is Lipschitz continuous with Lipschitz constant* 1.

Proof Let $\psi, \varphi \in C(X)$. Let also $\varepsilon > 0$. By the Variational Principle, there exists $\mu \in M(T)$ such that

$$\mathrm{P}(T, \psi) \le \mathrm{h}_\mu(T) + \int_X \psi \, d\mu + \varepsilon.$$

Then

$$\begin{aligned}
\mathrm{P}(T, \psi) &\le \mathrm{h}_\mu(T) + \int_X \varphi \, d\mu + \int_X (\psi - \varphi) \, d\mu + \varepsilon \\
&\le \mathrm{P}(T, \varphi) + \|\psi - \varphi\|_\infty + \varepsilon.
\end{aligned}$$

Since this is true for all $\varepsilon > 0$, we conclude that

$$\mathrm{P}(T, \psi) - \mathrm{P}(T, \varphi) \le \|\psi - \varphi\|_\infty.$$ ■

PART II

Complex Analysis, Conformal Measures, and Graph Directed Markov Systems

8

Selected Topics from Complex Analysis

From now on, throughout the book, most of our considerations will take place in the (extended) complex plane and the maps will be assumed to be analytic. We collect in the present chapter some selected concepts and theorems from complex analysis which will form for us indispensable tools in subsequent chapters, especially when we will deal with the dynamics of transcendental meromorphic and, more specifically, of elliptic functions.

The fundamental, indispensable concept and tool in the theory of iteration of meromorphic, either transcendental or not, function is the one of normal families due to Montel. Indeed, the seminal work of Montel (see the references in [**Al**, **AIR**]) on normal families of analytic functions was what made the modern, comprehensive, and systematic approach to iteration of meromorphic, rational, and transcendental functions both local and global, possible. We deal with normal families and prove Montel's Theorem in the first section of this chapter. We then use it very frequently throughout the rest of the book.

The most important for us, particularly for the needs of fractal geometry of Julia sets and of measurable dynamics induced by elliptic functions, are (various versions of) the Koebe Distortion Theorems. These will be used most frequently in the book. Section 8.2 about extremal lengths and moduli of topological annuli is a preparation for formulating the Koebe Distortion Theorems in the context of where "the Koebe space/collar" is an arbitrary topological annulus, and its results will be used throughout the book. All versions of the Koebe Distortion Theorems will be treated in detail with proofs in Section 8.3.

The Riemann–Hurwitz Formula, which we treat at length in the last section of this chapter, is an elegant and probably the best tool to control the topological structure of connected components of inverse images of open connected sets under meromorphic maps, especially to be sure that such connected components are simply connected.

Definition 8.0.1 If $H: D \to \mathbb{C}$ is an analytic map, $z \in \mathbb{C}$, and $r > 0$, then, by

$$\mathrm{Comp}(z, H, r),$$

we denote the only connected component of $H^{-1}(B_e(H(z), r))$ that contains z.

8.1 Riemann Surfaces, Normal Families, and Montel's Theorem

8.1.1 Riemann Surfaces

A Riemann surface is a connected one-dimensional complex manifold. This means that a set S is a Riemann surface if and only if there exists an atlas $\mathcal{A}$ on S whose elements, called charts, consist of injective maps $\phi: U_\phi \longrightarrow \mathbb{C}$ such that $\phi(U_\phi)$ is an open subset of $\mathbb{C}$ and all compositions

$$\psi \circ \phi^{-1}: \phi\big(U_\phi \cap U_\psi\big) \longrightarrow \mathbb{C},$$

$\phi, \psi \in \mathcal{A}$, are holomorphic. The topology on S, i.e., the one generated by the atlas $\mathcal{A}$, is the topology generated by the base consisting of all sets of the form $\phi^{-1}(V)$, where $\phi \in \mathcal{A}$ and V is an open subset of $\mathbb{C}$ (or $\phi(U_\phi)$). So, being precise, we should say that a Riemann surface is a pair $(S, \mathcal{A})$. We will sometimes refer to $\mathcal{A}$ as a holomorphic atlas generating S. Such an atlas is always contained in a unique largest holomorphic atlas generating S. The atlas $\mathcal{A}$ or its largest extension is frequently referred to as a conformal or complex structure on $\mathcal{S}$.

Each Riemann surface S can be naturally viewed as a two-dimensional real manifold and it is orientable. The latter follows immediately from the fact that, for all charts ϕ, ψ as above, the Jacobian of the map $\psi^{-1} \circ \phi$ is at every point $z \in U_\phi \cap \phi^{-1}\big(\psi(U_\psi)\big)$ equal to $\left|\big(\psi^{-1} \circ \phi\big)'(z)\right|^2 > 0$.

Given two Riemann surfaces R and S, respectively, generated by two atlases $\mathcal{A}$ and $\mathcal{B}$, we say that a map $f: R \to S$ is holomorphic if and only if the map

$$\psi \circ f \circ \phi^{-1}: \phi\big(U_\phi \cap f^{-1}(U_\psi)\big) \longrightarrow \mathbb{C}$$

is holomorphic for all $\phi \in \mathcal{A}$ and $\psi \in \mathcal{B}$. It is straightforward to check that a composition of two holomorphic maps is holomorphic.

Two Riemann surfaces R and S are called conformally equivalent or isomorphic if and only if there exists a holomorphic bijection $f: R \to S$. Note that then the inverse map $f^{-1}: S \to R$ is holomorphic, whence the conformal equivalence is an equivalence relation.

All simply connected Riemann surfaces are completely classified up to conformal equivalence; there are just three of them. Indeed, we have the following uniformization theorem.

Theorem 8.1.1 (Koebe–Poincaré Uniformization Theorem) *Every simply connected Riemann surface is conformally equivalent either to the unit disk* $\mathbb{D} := \{z \in \mathbb{C}: |z| < 1\}$*, to the complex plane* $\mathbb{C}$*, or to the Riemann sphere* $\widehat{\mathbb{C}} = \mathbb{C} \cup \{\infty\}$.

This theorem will be frequently used throughout our book, primarily in Volume 2, but its first important application is given here, in the present section, as an ingredient for the proof of the full version of Montel's Theorem. The Uniformization Theorem was proved by Poincaré and Koebe in 1907. Its proof can be found in several books on complex analysis and Riemann surfaces; for example, Theorem 7.4 in Kodaira's book [**Kod**] or Theorem 27.9 in Forster's book [**For**]. As an immediate consequence of this theorem and Picard's Little Theorem, we get the following.

Theorem 8.1.2 (Riemann Mapping Theorem) *Every open connected, simply connected subset of the Riemann sphere whose complement in* $\widehat{\mathbb{C}}$ *contains at least two points is conformally equivalent to the unit disk* $\mathbb{D}$*. Any conformal homeomorphism establishing such equivalence is commonly referred to either as the Riemann mapping or as a Riemann conformal homeomorphism. Such a homeomorphism is unique up to pre-composition with a conformal automorphism of the unit disk.*

Remark 8.1.3 The above hypothesis for a set $G \subseteq \widehat{\mathbb{C}}$ to be open connected, simply connected with the complement having at least two points is equivalent to G being open with connected complement having at least two points.

In fact, the proof of the Riemann Mapping Theorem, i.e., Theorem 8.1.2, is much more elementary than the proof of the Koebe–Poincaré Uniformization Theorem and is included in many textbooks on complex analysis.

If S is a Riemann surface, $\tilde{S}$ denotes its universal cover space, and $\pi : \tilde{S} \to S$ is the corresponding covering projection, then $\tilde{S}$ is via π canonically endowed with a structure of a Riemann surface. Indeed, if C is an open cover of S such that, for every $V \in C$, the inverse image $\pi^{-1}(V)$ is a disjoint union of open subsets U of $\tilde{S}$ such that the maps $\pi|_U : U \to V$ are a homeomorphism, and $\mathcal{A}$ is a holomorphic atlas generating S, then the maps $\phi \circ \pi : U \cap \pi^{-1}(U_\phi) \longrightarrow \mathbb{C}$, $\phi \in \mathcal{A}$ form a holomorphic atlas on $\tilde{S}$ since then

$$(\psi \circ \pi) \circ (\phi \circ \pi)^{-1} = \psi \circ \phi^{-1}$$

on appropriate domains. This conformal structure on $\tilde{S}$ is frequently referred to as the pull-back of the conformal structure of S via π.

Thus, given the Koebe–Poincaré Uniformization Theorem, Theorem 8.1.1, we geometrically classify each Riemann surface S as elliptic, parabolic, or hyperbolic, respectively, if its universal cover space $\tilde{S}$ is conformally equivalent to $\widehat{\mathbb{C}}$, $\mathbb{C}$, or the unit disk $\mathbb{D}$.

There is only one (up to conformal equivalence) elliptic Riemann surface, namely $\widehat{\mathbb{C}}$, since every deck transformation of $\widehat{\mathbb{C}}$ of any elliptic surface S is a Möbius transformation of $\widehat{\mathbb{C}}$ and each Möbius transformation has a fixed point. Therefore, the group of deck transformations of S is the trivial (identity) one, whence the projection $\pi : \widehat{\mathbb{C}} \to S$ is a conformal homeomorphism.

If S is a parabolic Riemann surface, then each of its deck transformations is a holomorphic homeomorphism of $\mathbb{C}$ onto itself, thus a translation $\mathbb{C} \ni z \mapsto z + w \in \mathbb{C}$ for some $w \in \mathbb{C}$. Since the only properly discontinuous (covering space action in the terminology of [**Ha**]) subgroups of the group of all translations of $\mathbb{C}$ are finitely generated with either zero, one, or two generators, each parabolic Riemann surface S is conformally equivalent to either

$$\mathbb{C}, \quad \mathbb{C}/\mathbb{Z}, \quad \text{or} \quad \mathbb{C}/(\mathbb{Z} + \tau\mathbb{Z}),$$

where $\tau \in \mathbb{C}$ with (normalization) $\mathrm{Im}(\tau) > 0$. We refer to the group $\mathbb{Z} + \tau\mathbb{Z}$ as a lattice, commonly denote it by Λ, and refer to the Riemann surface

$$\mathbb{T}_\Lambda := \mathbb{C}/(\mathbb{Z} + \tau\mathbb{Z})$$

as the complex torus generated by the lattice Λ. All such tori are conformally nonequivalent. We will encounter more of them in the second volume of the book when dealing with elliptic functions. We call the parabolic Riemann surface $\mathbb{C}/\mathbb{Z}$ the complex infinite cylinder.

Each Riemannian surface is endowed with a unique canonical Riemannian metric. For the Riemann sphere, this is the spherical metric

$$\frac{2}{1 + |z|^2}|dz|.$$

The distance between any two points w and z in $\widehat{\mathbb{C}}$ generated by this metric is equal to the length of the arc on the great circle in the unit two-dimensional sphere S^2 in $\mathbb{R}^3$ (centered at the point $(0, 0, 1)$ joining the images of w and z under stereographic projection).

For a parabolic surface S this is the pull-back of the Euclidean metric on $\mathbb{C}$ by any holomorphic covering map $\Pi_S \colon \mathbb{C} \to S$. Since, for any two such covering maps Π_1 and Π_2, the composition $\Pi_1^{-1} \circ \Pi_2 \colon \mathbb{C} \longrightarrow \mathbb{C}$ is a translation (thus, Euclidean isometry), this definition is independent of the chosen covering map. Let us record the following immediate observation.

Observation 8.1.4 If S is a parabolic Riemann surface, then the canonical projection $\Pi_S \colon \mathbb{C} \to S$ is Lipschitz continuous with a Lipschitz constant equal to 1 and it is a uniform local isometry. The latter more precisely means that there exists a largest radius $u_S \in (0, +\infty]$ such that, for every $w \in \mathbb{C}$, the map

$$\Pi_S\big|_{B(w,8u_S)} \colon B\big(w, 8u_S\big) \longrightarrow S$$

is an isometry. Note that $u_{\mathbb{C}} = +\infty$ and we also set $u_{\widehat{\mathbb{C}}} := +\infty$.

The above radius $8u_{\mathbb{T}}$ is called the injectivity radius of the surface S.

For a hyperbolic surface S, this is the pull-back of the Poincaré (hyperbolic) metric

$$\frac{2}{1-|z|^2}|dz|$$

on $\mathbb{D}$ by any holomorphic covering map $\pi \colon \mathbb{D} \to S$. Since, similarly to the case of parabolic surfaces, for any two such covering maps π_1 and π_2, the composition $\pi_1^{-1} \circ \pi_2 \colon \mathbb{C} \to \mathbb{C}$ is a conformal automorphism of the disk $\mathbb{D}$ (thus, isometric with respect to the Poincaré (hyperbolic) metric; the well-known fact proven in many books on complex analysis and all books on hyperbolic geometry), this definition is independent of the chosen covering map.

Although we will not need the concept of curvature in this book, we would like to add that the sectional spherical curvature on $\widehat{\mathbb{C}}$ is equal to 1, on each parabolic surface is equal to 0, and on each hyperbolic surface is equal to -1. Throughout the book, we will sometimes ignore the factor 2 appearing in the definitions of spherical and Poincaré metrics.

Dealing with a single Riemann surface, we will consider it to be, unless explicitly stated otherwise, endowed with its canonical Riemannian metric described above. If we deal with two or more Riemann surfaces, which is what will most often happen when one of them is contained in the other or one of them is mapped to the other, we may and frequently will consider Riemannian metrics of all Riemann surfaces involved. More specifically, this will be most often the case of Riemann surfaces $\mathbb{C}$ and $\widehat{\mathbb{C}}$, and then usually both Riemannian metrics, Euclidean and spherical, will be in simultaneous use.

Given a set $A \subseteq \mathbb{C}$ and $r > 0$, the symbol $B_e(A,r)$ denotes the Euclidean open ball (r-neighborhood) of the set A. If A is a singleton, say $A = \{w\}$, then we will write $B(w,r)$ instead. We denote

$$\operatorname{diam}_e(A) := \sup\{|y - x| : x, y \in A\}$$

and call $\operatorname{diam}_e(A)$ the Euclidean diameter of the set A. Similarly, if $A \subseteq \widehat{\mathbb{C}}$, we, respectively, denote by $B_s(A,r)$ and $\operatorname{diam}_s(A)$ the spherical open ball and the spherical diameter of the set A. If it is clear from the context or if it does not matter which metric, Euclidean or spherical, we use, we will occasionally write just $B(A,r)$ or $\operatorname{diam}(A)$. The spherical distance between any two points x and y in $\widehat{\mathbb{C}}$ will be denoted by $|x - y|_s$, although we may occasionally write $|x - y|$ instead. On parabolic Riemann surfaces, different from $\mathbb{C}$, we will use only the Euclidean metric throughout the book, and balls and diameters of sets will always be considered with respect to this metric. The distance between points x and y in $\mathbb{T}_\Lambda$ will be denoted, as for the complex plane $\mathbb{C}$, by $|y - x|$.

For the Riemann surfaces $\mathbb{C}$ and $\widehat{\mathbb{C}}$, all tangent planes at all points except ∞ in $\widehat{\mathbb{C}}$ are naturally identified with $\mathbb{C}$. If R and S are two Riemann surfaces, U is an open subset of R, and $f : U \to S$ is a holomorphic map, then, for every point $z \in S$, $D_z f : T_z R \longrightarrow T_{f(z)} S$, we will denote the derivative (linear) map from the tangent plane of R at z to the tangent plane of S at $f(z)$. If $R = \mathbb{C}$, $S = \widehat{\mathbb{C}}$, $z \in \mathbb{C}$, and $f(z) \in \mathbb{C}$, then there exists a complex $f'(z)$ (ordinary derivative) such that

$$D_z f(v) = f'(z)v$$

for every vector $v \in T_z\mathbb{C} = \mathbb{C}$. Treating then both $\mathbb{C}$ and $\widehat{\mathbb{C}}$ as subsets of the Riemann sphere $\widehat{\mathbb{C}}$ endowed with the spherical metric, the norm $\|D_z f\|_s$ of the linear map $D_z f : T_z\mathbb{C} \longrightarrow T_{f(z)}\widehat{\mathbb{C}}$ is equal to

$$\frac{1 + |z|^2}{1 + |f(z)|^2}|f'(z)|.$$

For ease and simplicity of notation, this will be denoted by

$$|f'(z)|_s.$$

For parabolic Riemann surfaces S, all tangent planes are canonically identified with $\mathbb{C}$ via the projection map

$$\Pi_S : \mathbb{C} \longrightarrow S$$

and, given a holomorphic map f from an open subset U of a parabolic or elliptic Riemann surface to any of such Riemann surfaces, if $z \in U\backslash\{\infty\}$ and $f(z) \neq \infty$, then there also exists a complex number $f'(z)$ such that

$$D_z f(v) = f'(z)v$$

for every vector v in the appropriate tangent plane. We will then denote the norm of the linear map $D_z f$ by $|f'(z)|$.

Furthermore, if R and S are arbitrary Riemann surfaces, U is an open subset of R, and $f: U \to S$ is a holomorphic map, then, for every $z \in R$, we will frequently denote

$$|f'(z)| := \|D_z f\|.$$

We may even sometimes write $f'(z)$ for $D_z f$.

8.1.2 Normal Families and Montel's Theorems

We now discuss at length normal families of analytic/meromorphic functions.

Definition 8.1.5 Let X and Y be topological spaces. We say that sequence $(f_n)_{n=1}^{\infty}$ of functions from X to Y diverges off compact sets of Y if and only if, for every compact set $F \subseteq X$ and every compact set $L \subseteq Y$, there exists a number $N \geq 1$ such that

$$f_n(F) \cap L = \emptyset$$

for every integer $n \geq N$.

Let us record the following immediate observation

Observation 8.1.6 If X and Y are topological spaces, then a sequence $(f_n)_{n=1}^{\infty}$ of functions from X to Y diverges off compact sets of Y if and only if every subsequence of this sequence contains a subsequence diverging off compact sets of Y.

Definition 8.1.7 Let X be a topological space and Y be a metric space. A family $\mathcal{F}$ of Y-valued functions defined on X is called normal if and only if, for every sequence $(f_n)_{n=1}^{\infty}$ of functions in $\mathcal{F}$, there exists a subsequence $(f_{n_k})_{k=1}^{\infty}$ of the sequence $(f_n)_{n=1}^{\infty}$ such that either this subsequence diverges off compact sets of Y or for every compact set $F \subseteq X$ the sequence $\left(f_{n_k}|_F\right)_{k=1}^{\infty}$ converges uniformly.

We record the following three immediate observations.

Observation 8.1.8 If X is a topological space and Y is a compact metric space, then a family $\mathcal{F}$ of Y-valued functions defined on X is normal if and only if, for every sequence $(f_n)_{n=1}^{\infty}$ of functions in $\mathcal{F}$, there exists a subsequence $(f_{n_k})_{k=1}^{\infty}$ of the sequence $(f_n)_{n=1}^{\infty}$ such that, for every compact set $F \subseteq X$, the sequence $\left(f_{n_k}|_F\right)_{k=1}^{\infty}$ converges uniformly.

Observation 8.1.9 If X is a topological space and Y is a locally compact metric space, then a family $\mathcal{F}$ of Y-valued functions defined on X is normal if and only if this family is normal if treated as a family of $\hat{Y}$-valued functions, where $\hat{Y}$ is the Alexandrov (one-point) compactification.

Since $\widehat{\mathbb{C}}$ is the Alexandrov compactification of $\mathbb{C}$, as an immediate consequence of this observation, we get the following.

Observation 8.1.10 If X is a topological space then a family $\mathcal{F}$ of $\mathbb{C}$-valued functions defined on X is normal if and only if this family is normal if treated as a family of $\widehat{\mathbb{C}}$-valued functions.

When talking about normality, the domain (the topological space X) and the co-domain (the metric spaces Y) of the families of considered functions will always be clear from the context or explicitly indicated. A standard and most common way to establish normality of a family of functions is to invoke the Ascoli–Arzelà Theorem. In this book, in particular in this section, we will be primarily interested in families of holomorphic/meromorphic functions taking values in elliptic or parabolic Riemann surfaces, all of which are also defined on Riemann surfaces; note that open subsets of Riemann surfaces are Riemann surfaces. Then the limits of corresponding subsequences are meromorphic functions, possibly constant. The concept of normal families, in the context of complex analysis, was introduced at the beginning of the twentieth century and is a beautiful and very powerful tool in complex analysis and complex dynamics as well. Its treatment can be found in many textbooks on complex analysis. The whole book *Normal Families* [**Sc**] by Schiff is devoted to its systematic exposition.

We will now state and prove, beginning with the help of the above-mentioned Ascoli–Arzelà Theorem, several versions of Montel's Theorem, which form an extremely powerful tool for establishing the normality of families of meromorphic functions and which make normality so important and so useful in complex analysis and complex dynamics. Indeed, it is an absolutely indispensable and very frequently used tool in the theory of iteration of all meromorphic functions, particularly elliptic ones, which is the ultimate subject of our book.

To begin with, we do need to show that some distinguished Riemann surfaces are hyperbolic. Indeed, given the full classification of elliptic and parabolic Riemann surfaces, the following three results look almost obvious but actually require proofs. We will obtain them as a consequence of the Koebe–Poincaré Uniformization Theorem and Picard's Little Theorem. We will first prove the following.

Theorem 8.1.11 *If a, b, c are three distinct points in $\widehat{\mathbb{C}}$, then $\widehat{\mathbb{C}} \setminus \{a, b, c\}$ is a hyperbolic Riemann surface, i.e., the unit disk $\mathbb{D}$ is a universal covering space of $\widehat{\mathbb{C}} \setminus \{a, b, c\}$ and there exists a meromorphic covering map $\pi : \mathbb{D} \longrightarrow \widehat{\mathbb{C}} \setminus \{a, b, c\}$.*

Proof Let D be a (topological) universal, i.e., simply connected, covering space of $\widehat{\mathbb{C}} \setminus \{a, b, c\}$ and $P \colon D \longrightarrow \widehat{\mathbb{C}} \setminus \{a, b, c\}$ be a covering map. This map uniquely induces the structure of a two-dimensional holomorphic manifold (Riemann surface) on D. In particular, the map P becomes meromorphic. The surface D cannot be conformally equivalent to $\widehat{\mathbb{C}}$ since then $\widehat{\mathbb{C}} \setminus \{a, b, c\}$ would be compact, and it cannot be conformally equivalent to $\mathbb{C}$ because of Picard's Little Theorem. So, because of the Koebe–Poincaré Uniformization Theorem above, D is conformally equivalent to the unit disk $\mathbb{D}$. This means that there exists a conformal homeomorphism $h \colon \mathbb{D} \to D$. Then $\pi := P \circ h \colon \mathbb{D} \longrightarrow \widehat{\mathbb{C}} \setminus \{a, b, c\}$ is the required holomorphic map. ■

As a special case of this theorem, we get the following.

Corollary 8.1.12 *If a and b are two distinct points in $\mathbb{C}$, then $\mathbb{C} \setminus \{a, b\}$ is a hyperbolic Riemann surface.*

Theorem 8.1.13 *If S is a parabolic Riemann surface different from $\mathbb{C}$, then, for every point $a \in S$, the Riemann surface $S \setminus \{a\}$ is hyperbolic.*

Proof Since $S \setminus \{a\}$ is not compact, it is not elliptic. Seeking contradiction, suppose that $S \setminus \{a\}$ is parabolic. Denote the corresponding projection map $\Pi_{S\setminus\{a\}} \colon \mathbb{C} \longrightarrow S \setminus \{a\}$ by Π_a. Since the complex plane $\mathbb{C}$ is simply connected, there is a continuous map $\tilde{\Pi}_a \colon \mathbb{C} \longrightarrow \mathbb{C}$ such that $\mathrm{P}_S \circ \tilde{\Pi}_a = \Pi_a$, i.e., the following diagram commutes.

$$\begin{array}{ccc} \mathbb{C} & \xrightarrow{\tilde{\Pi}_a} & \mathbb{C} \\ & \underset{\Pi_a}{\searrow} & \downarrow \Pi_S \\ & & S \end{array}$$

So, the map $\tilde{\Pi}_a \colon \mathbb{C} \longrightarrow \mathbb{C}$ is holomorphic as both Π_S and Π_a are locally invertible holomorphic maps. Thus, the map Π_a is entire. Also,

$$\Pi_S \circ \tilde{\Pi}_a(\mathbb{C}) = \Pi_a(\mathbb{C}) = S \setminus \{a\}.$$

Hence,

$$\tilde{\Pi}_a(\mathbb{C}) \subseteq \Pi_S^{-1}(a).$$

So, since the set $\Pi_S^{-1}(a)$ is (countable) infinite, it follows from Picard's Little Theorem that the map $\widetilde{\Pi}_a : \mathbb{C} \longrightarrow \mathbb{C}$ is constant. Thus, the covering map $\Pi_{S\setminus\{a\}} : \mathbb{C} \longrightarrow S\setminus\{a\}$ is also constant, and this contradiction, because of the Koebe–Poincaré Uniformization Theorem, finishes the proof. ∎

Now we move to actual Montel's Theorems. We gradually prove their more and more general versions. First, the simplest one.

Lemma 8.1.14 *Each family $\mathcal{F}$ of holomorphic functions from the unit disk $\mathbb{D}$ into $\mathbb{D}$ is normal.*

Proof It is enough to show that if $w \in \mathbb{D}$ and $R > 0$ is taken such that

$$B(w, 4R) \subseteq \mathbb{D},$$

then the family $\mathcal{F}_w := \{f|_{B(w,R)} : f \in \mathcal{F}\}$ is equicontinuous (with respect to the Euclidean metric on $\mathbb{D}$). Let $\gamma : [0,1] \longrightarrow B(w, 4R)$ be the simple, closed, smooth (and rectifiable) curve given by

$$\gamma(t) := w + 3Re^{2\pi i t}.$$

The Cauchy Integral Formulas then give, for any function $f \in \mathcal{F}$ and for every $z \in \overline{B(w, 2R)}$, that

$$\begin{aligned} |f'(z)| &= \left| \frac{1}{2\pi i} \int_\gamma \frac{f(z)}{(z-w)^2} \, dz \right| \\ &\le \frac{1}{2\pi} \left| \int_\gamma \frac{|f(z)|}{|z-w|^2} \, |dz| \right| \le \frac{1}{2\pi R^2} \int_\gamma |dz| = \frac{3}{R} < \infty. \end{aligned}$$

All functions $f|_{\overline{B(w,2R)}}$, $f \in \mathcal{F}$, are Lipschitz continuous with a Lipschitz constant $3/R$. So the family $\{f|_{\overline{B(w,2R)}} : f \in \mathcal{F}\}$ is equicontinuous. Since it is uniformly bounded (by 1) and since the sets $\overline{B(w,R)}$ are compact, this family is relatively compact by the Arzelà–Ascoli Theorem. Therefore, by virtue of this theorem, each sequence $(f_n)_{n=1}^\infty$ of functions in $\mathcal{F}$ has a subsequence $(f_{n_k})_{k=1}^\infty$ such that, for every compact set $F \subseteq U$, the sequence $\left(f_{n_k}|_F\right)_{k=1}^\infty$ converges uniformly. Denote the limit of the sequence $(f_{n_k})_{k=1}^\infty$ by g. Then $g : \mathbb{D} \to \overline{\mathbb{D}}$ is an analytic function. So, if $g(\mathbb{D}) \subseteq \mathbb{D}$, we are done. If, on the other hand, $g(\mathbb{D}) \cap \partial\mathbb{D} \neq \emptyset$, then, by the Open Mapping Theorem, the map g is constant and its only value, denoted by ξ, belongs to $\partial\mathbb{D}$. So, if F and L are two compact subsets of $\mathbb{D}$, then $\overline{\mathbb{D}}\setminus L$ is an open (in $\overline{\mathbb{D}}$) neighborhood of ξ; therefore, $f_{n_k}(F) \subseteq \overline{\mathbb{D}}\setminus L$ for all $k \ge 1$ large enough. The proof of Lemma 8.1.14 is complete. ∎

Remember that each open subset of a Riemann surface is a Riemann surface. As a fairly direct consequence of Lemma 8.1.14, we shall prove the following.

Theorem 8.1.15 (Montel's Theorem I) *If R is a Riemann surface and S is a hyperbolic Riemann surface, then every family $\mathcal{F}$ of holomorphic functions from R to S is normal.*

Proof Since normality is a local property, we may and we do assume without loss of generality that R is the unit disk $\mathbb{D}$. Fix an arbitrary sequence $(f_n)_{n=1}^{\infty}$ of functions in $\mathcal{F}$. If this sequence diverges off compact sets in S, we are done. So, suppose that it does not. Then there exist a compact set $F \subseteq \mathbb{D}$, a compact set $L \subseteq S$, and a subsequence $(f_{n_k})_{k=1}^{\infty}$ of the sequence $(f_n)_{n=1}^{\infty}$ such that

$$f_{n_k}(F) \cap L \neq \emptyset$$

for all $k \geq 1$. Then, for every $k \geq 1$, there exists a point $x_k \in F$ such that

$$f_{n_k}(x_k) \in L.$$

Since the set F is compact, by passing to a subsequence, we may assume that the sequence $(x_k)_{k=1}^{\infty}$ converges. Denote its limit by x. Then $x \in F$. Since L is compact, by passing to yet another subsequence, we may assume that the sequence $\left(f_{n_k}(x_k)\right)_{k=1}^{\infty}$ converges. Denote its limit by y. Then $y \in L$. Fix a lift $\tilde{y}$ of y by Π_S, i.e., a point $\tilde{y}$ in $\Pi_S^{-1}(y)$. Since $\Pi_S \colon \mathbb{D} \longrightarrow S$ is a covering map, there exists an open set $U \subseteq \mathbb{D}$ such that $\overline{U}$ is a compact subset of $\mathbb{D}$, $\tilde{y} \in U$, and the map $\Pi_S|_U$ is one-to-one. Since $\Pi_S(U)$ is an open subset containing y, by disregarding finitely many terms, we may assume that

$$f_{n_k}(x_k) \in \Pi_S(U)$$

for all $k \geq 1$. Thus, for every $k \geq 1$, there exists a point $\widetilde{f_{n_k}(x_k)} \in U$ such that

$$\Pi_S\left(\widetilde{f_{n_k}(x_k)}\right) = f_{n_k}(x_k).$$

Therefore, since the unit disk $\mathbb{D}$ is simply connected, for every $k \geq 1$ there exists a continuous, thus holomorphic, map $\widetilde{f}_{n_k} \colon \mathbb{D} \longrightarrow \mathbb{D}$, a lift of f_{n_k}, such that $f_{n_k} = \Pi_S \circ \widetilde{f}_{n_k}$, i.e., the following diagram commutes:

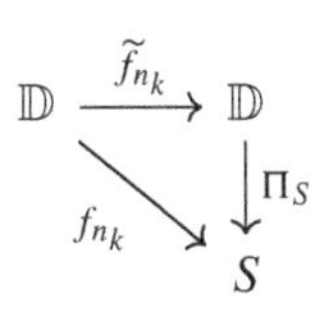

$$\begin{array}{ccc} \mathbb{D} & \xrightarrow{\widetilde{f}_{n_k}} & \mathbb{D} \\ & \underset{f_{n_k}}{\searrow} & \downarrow \Pi_S \\ & & S \end{array}$$

and

$$\widetilde{f}_{n_k}(x_k) = \widetilde{f_{n_k}(x_k)}.$$

Since the set $\overline{U}$ is compact, by passing to yet another subsequence, we may assume that the sequence $\left(\widetilde{f_{n_k}(x_k)}\right)_{k=1}^{\infty}$ converges. Denoting its limit by $\xi \in \overline{U} \subseteq \mathbb{D}$, we have that both sets

$$A := \{x\} \cup \{x_k : k \geq 1\} \quad \text{and} \quad B := \{\xi\} \cup \left\{\widetilde{f_{n_k}(x_k)} : k \geq 1\right\}$$

are contained in $\mathbb{D}$, compact, and $\widetilde{f}_{n_k}(A) \cap B \neq \emptyset$ (this intersection contains the point $\widetilde{f_{n_k}(x_k)}$) for all $k \geq 1$. Thus, the sequence $\left(\widetilde{f}_{n_k}\right)_{k=1}^{\infty}$ does not diverge off compact sets of $\mathbb{D}$. So, by Lemma 8.1.14, it has a subsequence, denoted by $\left(\widetilde{f}_{n_{k_j}}\right)_{j=1}^{\infty}$, converging uniformly on all compact subsets of $\mathbb{D}$. Then, denoting the limit of this subsequence by $g : \mathbb{D} \to \mathbb{D}$, and by invoking Observation 8.1.4, we see that the sequence

$$\left(f_{n_{k_j}} = \Pi_S \circ \widetilde{f}_{n_{k_j}}\right)_{k=1}^{\infty}$$

converges to the holomorphic function $\Pi_S \circ g : \mathbb{D} \longrightarrow S$ uniformly on all compact subsets of $\mathbb{D}$. The proof of Theorem 8.1.15 is complete. ■

Now we can prove the ultimate theorem of this section.

Theorem 8.1.16 (Montel's Theorem II) *If R is a Riemann surface and S is a parabolic or elliptic Riemann surface, then every family $\mathcal{F}$ of holomorphic functions from R to S that omits one point if S is either an infinite complex cylinder or a complex torus, two points if $S = \mathbb{C}$, and three points if $S = \widehat{\mathbb{C}}$, then the family $\mathcal{F}$ is normal. Furthermore, because of Observations 8.1.8–*8.1.10, *in the case of surfaces S being complex tori or $\widehat{\mathbb{C}}$, all sequences in $\mathcal{F}$ have subsequences converging uniformly on compact subsets of R. Likewise if $S = \mathbb{C}$, but then the limit functions are possibly $\widehat{\mathbb{C}}$-valued.*

Proof Denote by $a \in S$ a missing point if S is either an infinite complex cylinder or a complex torus, by $a, b \in S$ two distinct missing points if $S = \mathbb{C}$, and by $a, b, c \in S$ three, mutually distinct, missing points if $S = \widehat{\mathbb{C}}$. Denote then by D the respective set $\{a\}$, $\{a, b\}$, or $\{a, b, c\}$. Fix an arbitrary sequence $(f_n)_{n=1}^{\infty}$ of functions in $\mathcal{F}$. If this sequence contains a subsequence converging uniformly (with respect to the canonical metric on the Riemann surface $S \setminus D$) on all compact subsets of R to a function from R to $S \setminus D$, then this subsequence also converges uniformly (on all compact subsets of R) with respect to the canonical metric on the Riemann surface S. We are then done. Otherwise, by Theorem 8.1.13, Theorem 8.1.15, and Observation 8.1.6, the sequence $(f_n)_{n=1}^{\infty}$

diverges off compact sets of $S\backslash D$. If it diverges off compact sets of S, we are done. Otherwise, by passing to a subsequence, we may assume that there are two compact sets $A \subseteq R$ and $B \subseteq S$ such that

$$f_n(A) \cap B \neq \emptyset$$

for all integers $n \geq 1$. Now let F be an arbitrary compact subset of R. Since R is a Riemann surface, there exists a compact connected set Γ contained in R and containing $A \cup F$. Now let $\eta > 0$ be so small that the open balls of radii η centered at elements of A have mutually disjoint closures. Then let L be such a large compact subset of S that

$$B \subseteq L \quad \text{and} \quad (S\backslash L) \cap B(A,\eta) = \emptyset.$$

Of course, if either $S = \widehat{\mathbb{C}}$ or S is a complex torus, we can just take $L = S$. Since $L \cap (S\backslash B(A,\eta))$ is a compact subset of $S\backslash A$, there exists $N \geq 1$ such that $f_n(\Gamma) \cap L \cap \big(S\backslash B(A,\eta)\big) = \emptyset$ for all $n \geq N$. This means that

$$f_n(\Gamma) \subseteq (S\backslash L) \cup B(A,\eta). \tag{8.1}$$

So, since all the sets $f_n(\Gamma)$, $n \geq N$, are connected and since $f_n(\Gamma)$ is not contained in $S\backslash L$, we conclude that, for every $n \geq N$, there exists exactly one element $d_n \in A$ such that $f_n(\Gamma) \subseteq B(d_n \eta)$. Therefore, since the set A has at most three elements, there exist $d \in A$ and a strictly increasing sequence $(n_k)_{k=1}^{\infty}$ of positive integers such that

$$f_{n_k}(\Gamma) \subseteq B(d,\eta) \tag{8.2}$$

for all $k \geq 1$. Now fix any $\varepsilon \in (0,\eta]$. By the same argument as the one leading to (8.1), there exists $M \geq 1$ such that

$$f_{n_k}(\Gamma) \subseteq B(A,\varepsilon)$$

for all $k \geq M$. Along with (8.2) and the choices of η and ε, this implies that

$$f_{n_k}(\Gamma) \subseteq B(d,\varepsilon)$$

for all $k \geq M$. Thus, the sequence $f_{n_k}|_\Gamma$ converges uniformly to d. So, since $F \subseteq \Gamma$, we also have that the sequence $f_{n_k}|_F$ converges uniformly to d. The proof of Theorem 8.1.16 is complete. ■

Invoking again the fact that normality is a local property, as a direct consequence of Theorem 8.1.16, we obtain the following, even more general, theorem.

Theorem 8.1.17 (Montel's Theorem III) *If U is an open subset of a Riemann surface S and $\mathcal{F}$ is a family of meromorphic functions from U to $\widehat{\mathbb{C}}$ such* *that, for every $w \in U$, there exists a ball $B(w,r) \subseteq U$ for which the family $\{f|_{B(w,r)}: f \in \mathcal{F}\}$ omits one point if S is either an infinite complex cylinder or a complex torus, two points if $S = \mathbb{C}$, and three points if $S = \widehat{\mathbb{C}}$, then the family $\mathcal{F}$ is normal. Furthermore, because of Observations 8.1.8–8.1.10, in the case of surfaces S being complex tori or $\widehat{\mathbb{C}}$, all sequences in $\mathcal{F}$ have subsequences converging uniformly on compact subsets of R. Likewise if $S = \mathbb{C}$, but then the limit functions are possibly $\widehat{\mathbb{C}}$-valued.*

8.1.3 Fatou and Julia Sets

In this subsection, we take the first fruits of the concept of normality and Montel's Theorems. We will define and touch on the study of Fatou and Julia sets of a holomorphic map from one Riemann surface into another which contains it. Fatou and Julia sets will be the objects of our interest in all but one (Chapter 11) of the subsequent chapters of this volume and, even more substantially, in Volume 2 when dealing with the iteration of meromorphic and, more specifically, elliptic functions.

Let S be a Riemann surface. Let R be a nonempty open subset of S. Let

$$f: R \longrightarrow S$$

be an analytic map. As in the case of meromorphic functions treated in Volume 2, we say that $y \in S$ is a regular point of f^{-1} if and only if, for every $r > 0$ small enough and every connected component C of $f^{-1}(B(y,r))$, the restriction

$$f|_C: C \longrightarrow B(y,r)$$

is a (conformal) homeomorphism from C onto $B(y,r)$. Otherwise, we say that y is a singular point of f^{-1} and we denote by $\text{Sing}(f^{-1})$ the set of all such singular points. We also set

$$\text{PS}(f) := \bigcup_{n=0}^{\infty} f^n(\text{Sing}(f^{-1})).$$

We will use it only rarely but to be formally correct we take the convention that

$$f(\{A\}) = \emptyset$$

if $A \subseteq S \backslash R$. We call $\text{PS}(f)$ the post-singular set of f.

We say that a point $z \in R$ belongs to the Fatou set $F(f)$ of f if and only if there is an open neighborhood U of z such that all the iterates

$$f|_U^n : U \longrightarrow S, \quad n \geq 1,$$

are well defined and contain a subsequence forming a normal family. The Julia set $J(f)$ is defined as $S \backslash F(f)$. Clearly, $J(f)$ is a closed subset of S and

$$f^{-1}(J(f)) \subseteq J(f), \quad J(f) \cap R \subseteq f^{-1}(J(f)), \quad J(f) \cap R \subseteq J(f),$$

and

$$J(f) \cap f(R) \subseteq f(J(f)).$$

We would like to note that quite often – and, most notably, this will be the case in Volume 2, where we deal with the iteration of meromorphic functions, particularly of elliptic functions – we get the same concept of Fatou and Julia sets if in the definition of the former set all iterates $f|_U^n$, $n \geq 1$, are required to form a normal family.

When dealing with Fatou and Julia sets, we will very frequently, essentially almost always, use the following property, which will always be satisfied in our applications.

In order to formulate them, we will need the following property and some results about it.

Property 8.1.18 (Standard Property) Let S be either a parabolic or elliptic ($\widehat{\mathbb{C}}$) Riemann surface. Let R be an open subset of S, which, moreover, is a subset of $\mathbb{C}$ if $S = \widehat{\mathbb{C}}$. An analytic map

$$f : R \longrightarrow S$$

is said to have the Standard Property if and only if the set $(S \backslash R) \cup \mathrm{Per}(f)$ contains at least one point if S is either a complex torus or a conformal infinite cylinder, two points if $S = \mathbb{C}$, and three points if $S = \widehat{\mathbb{C}}$, where, as usual, $\mathrm{Per}(f)$ denotes the set of all periodic points of f.

Given a point $w \in S$, a radius $R > 0$ if $S = \mathbb{C}$ or $\widehat{\mathbb{C}}$, $R \in (0, u_S]$ if S is a parabolic Riemann surface, and W, an open neighborhood of w compactly contained in $B(w,r)$, for every integer $n \geq 0$, we denote by

$$\mathrm{Comp}_n^*(w,r)$$

the collection of all connected components Γ of $f^{-n}(B(w,r))$ such that the map

$$f^n|_{\tilde{\Gamma}} : \tilde{\Gamma} \longrightarrow B(w,2r)$$

is a conformal homeomorphism from $\tilde{\Gamma}$ onto $B(w,2r)$, where $\tilde{\Gamma}$ is the (only) connected component of $f^{-n}(B(w,2r))$ containing Γ. We denote by

$$f_{\Gamma}^{-n}: B(w,2r) \longrightarrow \tilde{\Gamma}$$

its inverse $(f^n|_{\tilde{\Gamma}})^{-1}$. We then set $n(\Gamma)$ to be n. Note that if $w \notin \overline{\mathrm{PS(f)}}$, $r \leq \frac{1}{2}\mathrm{dist}(w,\overline{\mathrm{PS(f)}})$, and for every $\Gamma \in \mathrm{Comp}_n(w,r)$ the map $f^n|_{\tilde{\Gamma}}: \tilde{\Gamma} \longrightarrow B(w,2r)$ is proper, then, by virtue of Corollary 8.6.20 and Remark 8.6.21, we get that

$$\mathrm{Comp}_n^*(w,r) = \mathrm{Comp}_n(w,r),$$

the latter being the collection of all connected components of $f^{-n}(B(w,r))$. One notation more. Given a set $F \subseteq R$ we denote by $\mathrm{Comp}_n(w,r;F)$ and $\mathrm{Comp}_n^*(w,r;F)$ the collection of all elements in $\mathrm{Comp}_n(w,r)$ or, respectively, in $\mathrm{Comp}_n^*(w,r)$ that intersect F.

We start with the following general result, which will be used now and frequently from now on in both volumes.

Lemma 8.1.19 *Let S be either a parabolic or elliptic ($\widehat{\mathbb{C}}$) Riemann surface. Let R be an open subset of S, which, moreover, is a subset of $\mathbb{C}$ if $S = \widehat{\mathbb{C}}$. Let*

$$f: R \longrightarrow S$$

be an analytic map with the Standard Property. Let $Q \subseteq S$ be a set witnessing this property. Fix

1. *A radius $r \in (0, u_S)$.*
2. *A radius $u \in \big(0, \min\{r, \mathrm{dist}(w,Q)/2\}\big)$.*
3. *A point $w \in J(f) \backslash O_+(Q)$, where, we recall, $O_+(Q) = \bigcup_{n=0}^{\infty} f^n(Q)$.*
4. *A compact set $L \subseteq R$.*

Then

$$\liminf_{n\to\infty}\big\{\inf\{|(f^n)'(z)|:\ z \in V\}: V \in \mathrm{Comp}_n^*(w,u;L)\big\} = +\infty. \tag{8.3}$$

Proof For every $n \geq 0$ and every $V \in \mathrm{Comp}_n^*(w,u)$, recall that

$$f_V^{-n}: B(w,2u) \longrightarrow \tilde{V}$$

is the unique holomorphic inverse branch of f^{-n} defined on $B(w,2u)$ and mapping it onto V and $\tilde{V}$ denotes the only connected component of $f_V^{-n}(B(w,2u))$ containing V. By our choice of Q and w, the family

$$\mathfrak{F} := \big\{f_V^{-n}: B(w,2u) \longrightarrow R \subseteq S \,|\, n \geq 1,\ V \in \mathrm{Comp}_n^*(w,u;L)\big\}$$

omits the set Q, which consists of at least one point if S is a complex torus, two points if $S = \mathbb{C}$, and three points if $S = \widehat{\mathbb{C}}$. Thus, by Montel's Theorem II, Theorem 8.1.16, the family $\mathfrak{F}$ of S-valued functions is normal.

Seeking contradiction, suppose that (8.3) fails. This means that there exist $\gamma > 0$ and an infinite sequence $(n_k)_1^\infty$ such that, for every $k \geq 1$, there exist $V_k \in \mathrm{Comp}_{n_k}^*(w, u; L)$ and a point $\xi_k \in V_k$ such that $|(f^{n_k})'(\xi_k)| \leq \gamma$. It then follows from Theorem 8.3.9 that

$$|(f^{n_k})'(z)| \leq K^2 \gamma \tag{8.4}$$

for all $z \in V_k$. Since, by item (4), no sequence of elements of $\mathfrak{F}$ diverges off compact sets of S, by passing to a subsequence we may assume without loss of generality that the sequence $\left(f_{V_k}^{-n_k} : B(w, 2u) \longrightarrow S\right)_{k=1}^\infty$ converges to an analytic function

$$g : B(w, 2u) \longrightarrow S$$

uniformly on compact subsets of $B(w, 2u)$. But, since all functions $f_{V_k}^{-n_k}$ are S-valued and since, again by item (4), $g : B(w, 2u) \cap R \neq \emptyset$, we conclude that

$$g(B(w, 2u)) \subseteq R.$$

It, therefore, follows from (8.4) that

$$|g'(z)| \geq (K^2 \gamma)^{-1}$$

for all $z \in B(w, u)$. In particular, $g : B(w, 2u) \longrightarrow R$ is not a constant function, and, so, $g(B(w, u/2))$ is an open neighborhood of $g(w)$. Since $g(\overline{B(w, u/2)})$ is a compact subset of the open set $g(B(w, u))$, we, thus, conclude that

$$f^{n_k}(g(B(w, u/2))) \subseteq B(w, u)$$

for all $k \geq 1$ large enough. Hence, the family

$$\left(f^{n_k}|_{g(B(w, u/2))}\right)_{k=1}^\infty$$

is normal; therefore, the point $g(w)$ belongs to $F(f)$, the Fatou set of f. But, on the other hand,

$$g(w) = \lim_{n \to \infty} f_{V_k}^{-n_k}(w) \in \overline{J(f)} = J(f).$$

This contradiction finishes the proof. ■

8.2 Extremal Lengths and Moduli of Topological Annuli

The concepts of extremal length and moduli of topological annuli are intimately connected to the theory of quasiconformal maps. We will present a concise but self-contained account of these concepts without using quasiconformal maps. The results of this section will be used in the next one, on the Koebe Distortion Theorems, and throughout the entire book. More complete and comprehensive exposition of these topics can be found in many books on, the already mentioned, theory of quasiconformal maps. Among them are [**Ah**], [**Le**], [**LV**], [**AIM**], [**Gar**], [**Hub**], [**FdM**], [**FM**], and [**BF**].

Let $D \subseteq \mathbb{C}$ be a domain, i.e., an open connected set. Let Γ be a family of rectifiable curves in D, i.e., of such continuous almost everywhere differentiable maps $\gamma: I \to D$ where I is a closed and bounded interval in $\mathbb{R}$ that the length of γ

$$\ell(\gamma) := \int_\gamma |dz| = \int_I |\gamma'(t)|dt < +\infty. \tag{8.5}$$

A Borel measurable function $\rho: D \longrightarrow [0, +\infty]$ is said to be admissible if and only if

$$A(\rho) := \int_D \rho^2 dS \in (0, +\infty),$$

where, we recall, S denotes the planar Lebesgue measure on $\mathbb{C}$.

The expression $\rho(z)|dz|$ should be thought of as a measurable, though conformal, change of the standard Riemannian metric $|dz|$. More generally than (8.5), for every curve $\gamma: I \to D$ in Γ, we define the length of γ with respect to the metric ρ as

$$L_\rho(\gamma) := \int_\gamma \rho|dz| = \int_I \rho(\gamma(t))|\gamma'(t)|dt.$$

Then

$$L(\rho) = L_\Gamma(\rho) := \inf\{L_\rho(\gamma): \gamma \in \Gamma\}.$$

The inverse extremal length of Γ is defined as

$$\lambda^{-1}(\Gamma) := \inf\left\{\frac{A(\rho)}{L^2(\rho)}\right\}, \tag{8.6}$$

where the infimum is taken over all admissible metrics $\rho: D \longrightarrow [0, +\infty]$. The quantity

$$\lambda(\Gamma) := \frac{1}{\lambda^{-1}(\Gamma)}$$

is then called the extremal length of Γ. Note that $\lambda^{-1}(\Gamma)$ is a conformal invariant. More precisely, we have the following.

Observation 8.2.1 If $D \subseteq \mathbb{C}$ is a domain, Γ is a family of rectifiable curves in D, and $\varphi\colon D \longrightarrow \mathbb{C}$ is a conformal isomorphism onto $\varphi(D)$, then

$$\lambda^{-1}(\varphi(\Gamma)) := \lambda^{-1}(\Gamma),$$

where $\varphi(\Gamma) = \{\varphi(\gamma)\colon \gamma \in \Gamma\}$.

If $I = [a,b] \subseteq \mathbb{R}$ with $a < b$, then we say that a curve, i.e., a continuous function, $\gamma\colon I \to \mathbb{C}$ is closed if and only if

$$\gamma(a) = \gamma(b).$$

Then there exists a unique continuous function $\hat{\gamma}\colon S^1 \to \mathbb{C}$ such that

$$\gamma(t) = \hat{\gamma}\left(\exp\left(2\pi i \frac{t-a}{b-a}\right)\right)$$

for every $t \in [a,b]$

Given a domain D in $\mathbb{C}$, we say that a closed curve $\gamma\colon [a,b] \longrightarrow D$ is essential if and only if it is not homotopic to a constant curve in D relative to the set $\{a,b\}$. Equivalently, the function $\hat{\gamma}\colon S^1 \longrightarrow D$ is not homotopic to a constant curve in D relative to the point 1 in S^1.

We then denote by Γ_D the family of all closed rectifiable essential curves $\gamma\colon [0,2\pi] \longrightarrow D$.

For every $R > 0$, we denote by S_R the circle in $\mathbb{C}$ with center 0 and radius R , i.e.,

$$S_R := \{z \in \mathbb{C}\colon |z| = R\}.$$

However, following the commonly adopted notation, we write S^1 for S_1.

We say that an open subset of the complex plane $\mathbb{C}$ is a topological annulus if its complement consists of exactly two connected components, one of which is bounded. It is well known from topology that any two annuli are homeomorphic and their fundamental group is isomorphic to $\mathbb{Z}$, the additive group of all integers. In fact, more is true.

Theorem 8.2.2 *An open connected subset of the Riemann sphere $\widehat{\mathbb{C}}$ is a topological annulus if and only if its fundamental group is isomorphic to $\mathbb{Z}$, the additive group of integers.*

We will now deal at length with annuli. Our nearest goal is to classify them up to conformal equivalence.

Given an annulus $A \subseteq \mathbb{C}$, we define its modulus of A to be

$$\mathrm{Mod}(A) := 2\pi\lambda^{-1}(\Gamma_A). \tag{8.7}$$

First, we shall easily prove two monotonicity relations between moduli of annuli. We say that an annulus A is essentially contained in an annulus B if and only if A separates the (two) boundary components of $\mathbb{C}\backslash B$. Equivalently, the inclusion $i\colon A \to B$ induces an isomorphism of fundamental groups, both of which, as we already know, are isomorphic to $\mathbb{Z}$.

Theorem 8.2.3 *If $\{A_n\}_{n=1}^{\infty}$ are mutually disjoint annuli, all essentially contained in an annulus $A \subseteq \mathbb{C}$, then*

$$\mathrm{Mod}(A) \geq \sum_{n=1}^{\infty} \mathrm{Mod}(A_n).$$

Proof Let ρ be an admissible metric for A. Since all annuli A_n are mutually disjoint, we have that $A(\rho) \geq \sum_{n=1}^{\infty} A(\rho|A_n)$. Also, since all annuli A_n, $n \geq 1$, are essentially contained in A, we have that $L(\rho) \leq L(\rho|_{A_n})$ for all $n \geq 1$. Hence,

$$\frac{A(\rho)}{L^2(\rho)} \geq \sum_{n=1}^{\infty} \frac{A(\rho|_{A_n})}{L^2(\rho|_{A_n})} \geq \sum_{n=1}^{\infty} \mathrm{Mod}(A_n).$$

Thus,

$$\mathrm{Mod}(A) \geq \sum_{n=1}^{\infty} \mathrm{Mod}(A_n).$$

The proof is complete. ■

As an immediate consequence of this theorem (also note that its independent proof would be a sub-proof of this theorem), we get the following.

Corollary 8.2.4 *If an annulus A is essentially contained in an annulus $B \subseteq \mathbb{C}$, then*

$$\mathrm{Mod}(A) \leq \mathrm{Mod}(B).$$

Given $w \in \mathbb{C}$ and $0 \leq R_1 \leq R_2 \leq +\infty$, we define the geometric annulus as

$$A(w; R_1, R_2) := \{z \in \mathbb{C}\colon R_1 < |z - w| < R_2\}.$$

In order to ease notation, denote $\Gamma_{A(w;\,R_1,\,R_2)}$ just by Γ. Note that a closed curve $\gamma\colon [0, 2\pi] \longrightarrow A(w; R_1, R_2)$ is essential if and only if its index, or the winding number, with respect to w does not vanish, i.e.,

$$\operatorname{ind}_w(\gamma) = \int_0^{2\pi} \frac{\gamma'(\theta)}{\gamma(\theta) - w} d\theta \neq 0.$$

Note also that the range $\gamma([0, 2\pi])$ of each curve $\gamma \in \Gamma$ separates S_{R_1} from S_{R_2} but the converse is not in general true.

We shall prove the following.

Theorem 8.2.5 *If $w \in \mathbb{C}$ and $0 \le R_1 \le R_2 \le +\infty$, then*

$$\operatorname{Mod}(A(w; R_1, R_2)) = \begin{cases} \log(R_2/R_1) & \text{if } 0 < R_1 < R_2 < +\infty \\ +\infty & \text{otherwise.} \end{cases}$$

Proof Assume without loss of generality that $w = 0$. Also, assume first that $0 < R_1 < R_2 < +\infty$. Denote

$$A := A(0; R_1, R_2).$$

Let ρ be an admissible Riemannian metric on A. Working in polar coordinates (θ, r), we have, for every $r \in (R_1, R_2)$, that

$$L(\rho) \le \int_0^{2\pi} r\rho(re^{it}) d\theta.$$

Hence,

$$\begin{aligned} \int_{R_1}^{R_2} \int_0^{2\pi} \rho(re^{it}) d\theta dr &= \int_{R_1}^{R_2} \int_0^{2\pi} r\rho(re^{it}) d\theta \frac{dr}{r} \ge \int_{R_1}^{R_2} L(\rho) \frac{dr}{r} \\ &= L(\rho) \log(R_2/R_1). \end{aligned} \tag{8.8}$$

On the other hand, using the Cauchy–Schwarz inequality, we get that

$$\begin{aligned} \left(\int_{R_1}^{R_2} \int_0^{2\pi} \rho(re^{it}) d\theta dr \right)^2 &= \left(\int_{R_1}^{R_2} \int_0^{2\pi} \sqrt{r}\rho(re^{it}) \frac{1}{\sqrt{r}} d\theta dr \right)^2 \\ &\le \int_{R_1}^{R_2} \int_0^{2\pi} r\rho^2(re^{it}) d\theta dr \int_{R_1}^{R_2} \int_0^{2\pi} \frac{1}{r} d\theta dr \\ &= 2\pi \log(R_2/R_1) A(\rho). \end{aligned}$$

Combining this with (8.8), we get that $A(\rho)/L^2(\rho) \ge \log(R_2/R_1)$, whence

$$\lambda^{-1}(\Gamma) \ge \frac{1}{2\pi} \log(R_2/R_1). \tag{8.9}$$

To prove the opposite inequality, consider the Riemannian metric τ on A that is defined as

$$\tau(re^{it}) = 1/r. \tag{8.10}$$

So,

$$A(\tau) = \int_{R_1}^{R_2} \int_0^{2\pi} r\tau^2(re^{it})d\theta dr = \int_{R_1}^{R_2} \int_0^{2\pi} \frac{1}{r} d\theta dr = 2\pi \log(R_2/R_1). \tag{8.11}$$

But, for every curve $\gamma \in \Gamma$, we have that

$$\begin{aligned} L_\tau(\gamma) &= \int_\gamma \tau|dz| = \int_0^{2\pi} \frac{1}{|\gamma(\theta)|}|\gamma'(\theta)|d\theta \\ &= \int_0^{2\pi} \left|\frac{\gamma'(\theta)}{\gamma(\theta)}\right| d\theta \geq \left|\int_0^{2\pi} \frac{\gamma'(\theta)}{\gamma(\theta)} d\theta\right| \\ &\geq 2\pi |\text{ind}_0(\gamma)| \geq 2\pi. \end{aligned} \tag{8.12}$$

Thus,

$$L(\tau) \geq 2\pi. \tag{8.13}$$

Hence, invoking (8.11), we get that

$$\lambda^{-1}(\Gamma) \leq \frac{A(\tau)}{L^2(\tau)} \leq \frac{1}{2\pi} \log(R_2/R_1).$$

Combining this with (8.9), we, thus, get that

$$\lambda^{-1}(\Gamma) = \frac{1}{2\pi} \log(R_2/R_1).$$

So, finally,

$$\text{Mod}(A) = 2\pi\lambda^{-1}(\Gamma) = \log(R_2/R_1).$$

The second case follows immediately from the first one and Corollary 8.2.4. The proof is complete. ■

Remark 8.2.6 It follows from the proof of this theorem that $L(\tau) = 2\pi$ and

$$\frac{A(\tau)}{L^2(\tau)} = \text{Mod}(A(w; R_1, R_2)) = \log(R_2/R_1).$$

In the case when $R_1 = 0$ and $R_2 = +\infty$, we actually, for the ease of exposition, redefine the modulus of $\text{Mod}(A(w; R_1, R_2))$ to be $+2\infty$ (and we consider in this context $+2\infty$ to be different than $+\infty$).

Since if $R_1, R_1 > 0$ and $R_2, R_2' < +\infty$, then, for all $z \in \mathbb{C}$, all the annuli $A(z; 0, \mathbb{R}_2)$, $A(z; 0, \mathbb{R}_2')$, $A(z; R_1, +\infty)$, and $A(z; R_1', +\infty)$ are conformally equivalent, while the annuli $A(z; 0, +\infty)$ are not conformally equivalent to any of them, as an immediate consequence of this theorem we get the following.

Corollary 8.2.7 *Two geometric annuli $A(w; R_1, R_2)$ and $A(z; R_1', R_2')$ are conformally equivalent if and only if their moduli are the same, meaning, in the case when all radii are positive and finite, that $R_2'/R_1' = R_2/R_1$.*

This corollary could have been proved without using the concept of extremal length; a particularly appealing short argument is given at the beginning of Chapter VII of Nehari's book [**Ne**]. We will primarily need the concepts of moduli and extremal length for other purposes.

We shall prove the following.

Theorem 8.2.8 *Two annuli are conformally equivalent if and only if their moduli are equal.*

In addition, if $A \subseteq \mathbb{C}$ is an annulus, then $\mathrm{Mod}(A) < +\infty$ *if and only if both connected components of $\mathbb{C}\backslash A$ contain at least two (so continuum many) points.*

Furthermore, $\mathrm{Mod}(A) = +\infty$ *if and only if exactly one connected component of $\mathbb{C}\backslash A$ is a singleton, while* $\mathrm{Mod}(A) = +2\infty$ *if and only if the two connected components of $\mathbb{C}\backslash A$ are singletons.*

Let $\mathbb{H}$ be the open upper half-plane, i.e.,

$$\mathbb{H} := \{z \in \mathbb{C} \colon \mathrm{Im}(z) > 0\}.$$

In order to prove Theorem 8.2.8, we will need the following.

Proposition 8.2.9 *For every $\kappa > 0$, the holomorphic map $\Pi_\kappa \colon \mathbb{H} \longrightarrow \mathbb{C}$ given by the formula*

$$\Pi_\kappa(z) := \exp\left(\frac{\kappa}{\pi} i \log z\right)$$

is a holomorphic covering map from $\mathbb{H}$ onto $A(0; e^{-\kappa}, 1)$, where

$$\log \colon \mathbb{H} \longrightarrow \mathbb{C}$$

is the restriction to $\mathbb{H}$ of the principal branch of logarithm from $\mathbb{C}\backslash(-\infty, 0]$, to $\mathbb{C}$, i.e., the one sending 1 to 0.

Furthermore, the group of deck transformations of Π_κ coincides with $\langle g \rangle$, the group generated by g, where $g \colon \mathbb{H} \longrightarrow \mathbb{H}$ is given by the formula

$$g(z) := \exp\left(\frac{2\pi^2}{\kappa}\right) z.$$

In particular, the annulus $A(0; e^{-\kappa}, 1)$ and the quotient space $\mathbb{H}/\langle g \rangle$ are conformally isomorphic

Proof First observe that $\log\colon \mathbb{H} \longrightarrow \mathbb{C}$ is a conformal covering of $\mathbb{H}$ onto its image $\{w \in \mathbb{C}\colon 0 < \mathrm{Im}(w) < \pi\}$. So,

$$i\frac{\kappa}{\pi}\{w \in \mathbb{C}\colon 0 < \mathrm{Im}(w) < \pi\} = \{\xi \in \mathbb{C}\colon -\kappa < \mathrm{Re}(\xi) < 0\}.$$

Furthermore, $\exp(\{\xi \in \mathbb{C}\colon -\kappa < \mathrm{Re}\xi < 0\}) = A(0; e^{-\kappa}, 1)$. So, Π_κ, as a composition of conformal covering maps, is a conformal covering map onto its image $A(0; e^{-\kappa}, 1)$. The first part of the proposition is, thus, proved. In order to prove the second part, we note that

$$\begin{aligned}
\Pi_\kappa(g(z)) &= \exp\left(\frac{\kappa}{\pi} i \log\left(\exp\left(\frac{2\pi^2}{\kappa}\right) z\right)\right) = \exp\left(i\frac{\kappa}{\pi}\left(\frac{2\pi^2}{\kappa} + \log z\right)\right) \\
&= \exp\left(\frac{\kappa}{\pi} i \log z\right)\exp(2\pi i) = \exp\left(\frac{\kappa}{\pi} i \log z\right) \\
&= \Pi_\kappa(z).
\end{aligned}$$

So,

$$\Pi_\kappa \circ g = \Pi_\kappa;$$

therefore, g belongs to the deck group of the covering map Π_κ. Now if $\Pi_\kappa(z) = \Pi_\kappa(w)$, then

$$i\frac{\kappa}{\pi}\log z - i\frac{\kappa}{\pi}\log w = 2\pi i n$$

for some integer $n \in \mathbb{Z}$. This means that

$$\log(w/z) = \frac{2\pi^2}{\kappa} n.$$

So,

$$w = \exp\left(\frac{2\pi^2}{\kappa} n\right) z = g^n(z).$$

Thus, the group of deck transformations of Π_κ coincides with $\langle g \rangle$, the group generated by g. Since $A(0; e^{-\kappa}, 1)$, as the range of Π_κ, is conformally isomorphic to the quotient of $\mathbb{H}$ by the deck group of Π_κ, we are, thus, done. ■

Proposition 8.2.10 *The function $\Pi_\infty\colon \mathbb{H} \longrightarrow \mathbb{C}$, given by the formula $\Pi_\infty(z) = e^{iz}$, is a holomorphic covering map from $\mathbb{H}$ onto $A(0; 0, 1)$. In particular, $A(0; 0, 1)$ and $\mathbb{H}/\langle g \rangle$ are conformally equivalent, where $g\colon \mathbb{H} \longrightarrow \mathbb{H}$ is the translation given by the formula*

$$g(z) = z + 2\pi.$$

Proof Note that $i\mathbb{H} = \{w \in \mathbb{C}\colon \mathrm{Re}(w) < 0\}$. So,

$$\Pi_\infty(\mathbb{H}) = e^{i\mathbb{H}} = \{\xi \in \mathbb{C}\colon 0 < |\xi| < 1\} = A(0; 0, 1).$$

Since Π_∞ is clearly a conformal covering map, we are, thus, done with the first part.

In order to prove the second part, note that

$$\Pi_\infty(g(z)) = e^{i(z+2\pi)} = e^{iz} = \Pi_\infty(z).$$

Hence, Π_∞ is a member of the deck group of Π_∞. If

$$\Pi_\infty(w) = \Pi_\infty(z),$$

then $iz = iw + 2\pi in$ for some $n \in \mathbb{Z}$. This means that $w = z + 2\pi n = g^n(z)$. Thus, the group of the deck transformation of Π_∞ is equal to $\langle g \rangle$. So, since $A(0; 0, 1) = \Pi_\infty(\mathbb{H})$ and the quotient of $\mathbb{H}$ by the deck group of Π_∞ are conformally equivalent, we are, thus, done. ■

Proof of Theorem 8.2.8 Proving the first assertion of this theorem we see that, in view of Corollary 8.2.7, it is sufficient to prove that any annulus A is conformally equivalent either to a geometric annulus $A(0; R_1, 1)$, $R_1 \geq 0$, or to the annulus $\mathbb{C}\backslash\{0\}$.

If $\mathbb{C}\backslash A$ is a singleton, say ξ, then the translation $z \mapsto z + \xi$ establishes a conformal equivalence between $A(0; 0, \infty)$ and A. So, we may assume that $\mathbb{C}\backslash A$ contains at least two points. The Koebe–Poincaré Uniformization Theorem, i.e., Theorem 8.1.1, asserts that there are exactly three conformally distinct simply connected Riemann surfaces; namely, $\widehat{\mathbb{C}}$, $\mathbb{C}$, and the upper half-plane $\mathbb{H}$. The universal cover of A cannot be $\widehat{\mathbb{C}}$ since A is not compact (purely topological obstacle), and it cannot be $\mathbb{C}$ because of Picard's Little Theorem. So, the universal cover of A must be the half-plane $\mathbb{H}$. Let

$$\Pi\colon \mathbb{H} \longrightarrow A$$

be a conformal universal cover of A. The deck group Γ of Π, being isomorphic to the fundamental group of A, must be isomorphic to $\mathbb{Z}$, the group of integers. Let g be a generator of Γ. Being one element of a deck group, g cannot have fixed points, but g extends continuously to $\overline{\mathbb{H}}$ (even biholomorphically to $\widehat{\mathbb{C}}$ because of the Schwartz Reflection Principle), mapping it univalently onto itself. Keep for this extension the same symbol g. Then $g\colon \overline{\mathbb{H}} \to \overline{\mathbb{H}}$ has either exactly one fixed point or exactly two distinct fixed points. In the former case, applying conformal conjugacy, we may assume without loss of generality that g fixes ∞ and is of the form

$$g(z) = z + 2\pi.$$

In the latter case, also because of conformal conjugacy, we may assume that $g(0) = 0$ and $g(\infty) = \infty$, and then $g : \overline{\mathbb{H}} \longrightarrow \overline{\mathbb{H}}$ must be of the form

$$g(z) = \lambda z$$

with some $\lambda > 0$. Let us deal first with the latter case. Writing

$$\lambda = \exp\bigl(2\pi^2/\kappa\bigr),$$

we conclude from Proposition 8.2.9 that the annulus $A(0; e^{-\kappa}, 1)$ is conformally isomorphic to $\mathbb{H}/\langle g\rangle$ which is conformally isomorphic to A as $\langle g\rangle$ is the deck group of $\Pi : \mathbb{H} \to A$. The same argument, but based on Proposition 8.2.10, works also for the case when $g(z) = z + 2\pi$.

The second and the third assertions of Theorem 8.2.8 are now obtained by a standard straightforward application of the Schwarz Reflection Principle. The proof is complete. ■

Now, given $r \in (0,1)$, let

$$B(r) := \mathbb{D}\backslash[0,r].$$

Clearly, $B(r)$ is an annulus and, following tradition, we denote its modulus by $\mu(r)$. We call the map

$$(0,1) \ni r \longmapsto \mu(r) \in [0,+\infty]$$

the Grötzsch modulus function. We immediately have the following from Corollary 8.2.4.

Observation 8.2.11 The Grötzsch modulus function $(0,1) \ni r \longmapsto \mu(r) \in [0,+\infty]$ is monotone decreasing.

We shall now show that $B(r)$ is in a sense extremal among of all annuli that separate 0 and r from $S^1 = \partial\mathbb{D}$ (see [**LV**]; comp. [**Gr**]).

Theorem 8.2.12 (Grötzsch Module Theorem) *If $r \in (0,1)$ and an annulus $A \subseteq \mathbb{C}$ separates 0 and r from $\partial\mathbb{D}$, then*

$$\mathrm{Mod}(A) \le \mu(r).$$

Proof By virtue of Theorem 8.2.8, there exists a conformal homeomorphism $\varphi : B(r) \longrightarrow A\bigl(0; 1, e^{\mu(r)}\bigr)$. Let

$$B_+(r) := \{z \in B(r) : \mathrm{Im} z > 0\}.$$

Since $\partial B_+(r)$ is a Jordan curve, by virtue of Carathéodory's Theorem, the map $\varphi_+ := \varphi_{|B_+(r)}$ extends homeomorphically to a map from $\overline{B_+(r)}$ to $\mathbb{C}$. By the

Schwarz Reflection Principle, we can then extend the map φ_+ to a holomorphic map $\varphi^*: \mathbb{D} \to \mathbb{C}$ by setting

$$\varphi^*(z) := \overline{\varphi(\overline{z})}$$

for all $z \in \{w \in \mathbb{D}\colon \operatorname{Im}(w) < 0\}$. But φ and φ^* are holomorphic and $\varphi^*|_{B_+(r)} = \varphi$, whence

$$\varphi^*|_{B(r)} = \varphi.$$

So, we can now get rid of the notation φ^* and can consider only a holomorphic map

$$\varphi\colon \mathbb{D} \longrightarrow \overline{A\big(0; 1, e^{\mu(r)}\big)}$$

such that $\varphi|_{B(r)}$ is a conformal homeomorphism onto $A\big(0; 1, e^{\mu(r)}\big)$ and φ is symmetric with respect to the x-axis, meaning that

$$\varphi(\overline{z}) = \overline{\varphi(z)}.$$

Now consider the metric τ on $A\big(0; 1, e^{\mu(r)}\big)$ given by the formula (8.10). Define the pull-back metric on $B(r)$ by the formula

$$\varphi^*(\tau|dz|) := \tau(\varphi(z))|\varphi'(z)||dz| = \frac{|\varphi'(z)|}{|\varphi(z)|}|dz|,$$

i.e., put

$$\rho(z) := \frac{|\varphi'(z)|}{|\varphi(z)|}.$$

Since the map $\mathbb{C} \ni z \longmapsto \overline{z}$ is a Euclidean isometry, we conclude from the Chain Rule that $|\varphi'(\overline{z})| = |\varphi'(z)|$; therefore,

$$\rho(\overline{z}) = \rho(z), \tag{8.14}$$

which just means that the metric ρ is symmetric with respect to the x-axis. Consider ρ as a metric defined on B. Since the area of the segment $[0, r]$ with respect to the metric ρ is equal to zero, we have that

$$A_\rho(B) \le A_\rho(B(r)) = A_\tau\big(A\big(0; 1, e^{\mu(r)}\big)\big) =: A(\tau).$$

It, therefore, follows from (8.11) that

$$\operatorname{Mod}(A) \le 2\pi \frac{A(\rho)}{L^2(\rho)} \le 4\pi^2 \frac{\mu(r)}{L^2(\rho)}. \tag{8.15}$$

Now consider an arbitrary curve $\gamma \in \Gamma_A$. Since γ separates the points 0 and r from the unit circle S^1, it can be divided into two subarcs γ_1 and γ_2 with disjoint interiors, both having one common endpoint on the segment

$(-1,0)$ and one on the segment $(r,1)$. Let $\hat{\gamma}_1$ be the curve resulting from γ_1 by replacing each point z on γ_1 with $\operatorname{Im} z < 0$ by its conjugate $\overline{z}$. Let $\hat{\gamma}_2$ be the curve resulting from γ_2 by replacing each point z on γ_2 with $\operatorname{Im} z > 0$ by its conjugate $\overline{z}$. Since the metric ρ is symmetric with respect to the x-axis (see (8.14)), we have that $L_\rho(\hat{\gamma}_i) = L_\rho(\gamma_i)$, $i = 1,2$. Obviously, $\varphi(\hat{\gamma}_1 \cup \hat{\gamma}_2) \in \Gamma_{\mathbb{C}\setminus\{0\}}$. It then follows from (8.12) that

$$\begin{aligned} L(\rho) &\geq L_\rho(\gamma) = L_\rho(\gamma_1 \cup \gamma_2) = L_\rho(\gamma_1) + L_\rho(\gamma_2) = L_\rho(\hat{\gamma}_1) + L_\rho(\hat{\gamma}_2) \\ &= L_\rho(\hat{\gamma}_1 \cup \hat{\gamma}_2) = L_\tau(\varphi(\hat{\gamma}_1 \cup \hat{\gamma}_2)) \\ &\geq 2\pi. \end{aligned}$$

Inserting this into (8.15), we finally get that $\operatorname{Mod}(A) \leq \mu(r)$. The proof is complete. ■

Now we shall prove the following estimates for $\mu(r)$.

Proposition 8.2.13 *For all $r \in (0,1)$, we have that $\mu(r) < \log 4 - \log r$.*

Proof Let $R > 0$. Let $M\colon \widehat{\mathbb{C}} \longrightarrow \widehat{\mathbb{C}}$ be the unique linear affine map (Möbius transformation) sending the points 0, r, and 1, respectively, to $-1, 1$, and R. Since all Möbius transformations preserve cross-ratios, we get, for all $z \in \widehat{\mathbb{C}}$, that

$$\frac{1}{r} \cdot \frac{z-r}{z-1} = \frac{1+R}{2} \cdot \frac{M(z)-1}{M(z)-R}.$$

We now require R to be such that $M(-1) = -R$. Then we get the following quadratic equation for R:

$$\begin{aligned} \frac{1+R}{2} \cdot \frac{1+R}{2R} = \frac{1}{r} \cdot \frac{1+r}{2} \quad &\Leftrightarrow \quad r(1+R)^2 \\ = 2(1+r)R \quad &\Leftrightarrow \quad rR^2 - 2R + r = 0. \end{aligned} \tag{8.16}$$

With $R \in \mathbb{R}$ being a solution to this equation, the map M maps the unit circle $\{z \in \mathbb{C}\colon |z| = 1\}$ onto the circle $\{z \in \mathbb{C}\colon |z| = R\}$ as $M(S^1)$ is a circle, $M(\overline{\mathbb{R}}) = \overline{\mathbb{R}}$, $M(S^1)$ intersects $\overline{\mathbb{R}}$ at right angles (as a Möbius map preserve angles), $M(-1) = -R$, and $M(1) = R$. Note that (8.16) has two solutions

$$R = r^{-1}(1 - \sqrt{1-r}) \text{ and } R = r^{-1}(1 + \sqrt{1-r}) \in (0,1).$$

We fix the later one. Then $M(B(0,1)) = B(0,R)$, as can be easily checked by evaluating $M(\infty)$. As also $M(0) = -1$ and $M(r) = 1$, we get that $M([0,r]) = [-1,1]$. Therefore, $\hat{M}$, the restriction of M to $B(r)$, establishes a conformal isomorphism between $B(r)$ and

$$G(r) := M(B(r)) = B(0,R)\setminus[-1,1].$$

Now consider the rational function $H: \widehat{\mathbb{C}} \longrightarrow \widehat{\mathbb{C}}$ given by the formula

$$H(z) = \frac{1}{2}\left(z + \frac{1}{z}\right).$$

This is a rational function of degree 2 and, as $H(z) = H(1/z)$, it is injective on $\{z \in \widehat{\mathbb{C}}: |z| > 1\}$. If $|z| = \rho$, then writing $z = \rho e^{i\theta}$, $0 \leq \theta < 2\pi$, we get that

$$H(z) = \rho e^{it} + \frac{1}{\rho}e^{-i\theta} = \left(\rho + \frac{1}{\rho}\right)\cos\theta + i\left(\rho - \frac{1}{\rho}\right)\sin\theta.$$

Hence, if $\rho = 1$, then

$$H(\{z \in \mathbb{C}: |z| = 1\}) = [-1, 1];$$

if $\rho > 1$, then

$$H(\{z \in \mathbb{C}: |z| = \rho\}) = E_\rho,$$

the ellipse with axis points $\pm(\rho + \frac{1}{\rho})$ and $\pm(\rho - \frac{1}{\rho})$. This is so because

$$\frac{\left((\rho + \frac{1}{\rho})\cos\theta\right)^2}{\left(\rho + \frac{1}{\rho}\right)^2} + \frac{\left((\rho - \frac{1}{\rho})\sin\theta\right)^2}{\left(\rho - \frac{1}{\rho}\right)^2} = 1.$$

Thus, $\hat{H}$, the map H restricted to the annulus $A(0; 1, \rho)$, establishes a conformal isomorphism between the annulus $A(0; 1, \rho)$ and $E_\rho \backslash [-1, 1]$. Now if $\rho = \frac{4}{r}$, then

$$\rho - \frac{1}{\rho} = \frac{4}{r} - \frac{r}{4} = \frac{1}{r}\left(4 - \frac{r^2}{4}\right).$$

We check that $4 - \frac{r^2}{4} \geq 2\left(1 + \sqrt{1 - r^2}\right)$. This inequality means that

$$2 - \frac{r^2}{4} \geq \sqrt{1 - r^2} \iff 4 - r^2 + \frac{r^4}{16} \geq 1 - r^2,$$

which is true. Hence, $\rho - \frac{1}{\rho} \geq 2R$; therefore, $G(r) \subsetneq E_{4/r} \backslash [-1, 1]$. Applying Corollary 8.2.4, we, thus, obtain that

$$\begin{aligned}\mu(r) = \operatorname{Mod}(B(r)) = \operatorname{Mod}(G(r)) &\leq \operatorname{Mod}(E_{4/r} \backslash [-1, 1]) \\ = \operatorname{Mod}(A(0; 1, 4/r)) &= \log 4 - \log r.\end{aligned}$$

We are done. ■

We are now in a position to easily prove the following.

Theorem 8.2.14 *If B is a compact connected subset of the complex plane $\mathbb{C}$, $x \in B$, and $R > \text{Dist}(x, B)$, then*

$$\text{Dist}_e(x, B) \leq 4R\exp\big(-\text{Mod}((B(x, R)\backslash B)_*)\big),$$

where $(B(x, R)\backslash B)_*$ is the connected component (an annulus) separating x from $\partial B(x, R)$.

Proof By Theorem 8.2.12 and Proposition 8.2.13, we get that

$$\begin{aligned}\text{Mod}((B(x, R)\backslash B)_*) &\leq \mu\left(\frac{\text{Dist}_e(x, B)}{R}\right) < \log 4 - \log\left(\frac{\text{Dist}_e(x, B)}{R}\right) \\ &= \log(4R) - \log(\text{Dist}_e(x, B)).\end{aligned}$$

Exponentiating this inequality, we get that

$$\text{Dist}_e(x, B) \leq 4R\exp\big(-\text{Mod}((B(x, R)\backslash B)_*)\big).$$

We are done. ■

We know, by Observation 8.2.11, that the limit $\lim_{r\nearrow 1}\mu(r)$ exists. We need to know that the value of this limit is equal to zero. We will prove it now. In order to do this, we shall first provide an upper estimate of the modulus of an annulus, which is also interesting on its own.

Theorem 8.2.15 *Let $A \subseteq \mathbb{C}$ be an annulus, δ be the minimum of spherical diameters of boundary components of A, and ε be the spherical distance between these (two) components. If $0 < \varepsilon < \delta$, then*

$$\text{Mod}(A) \leq \frac{\pi^2}{\log\left(\frac{\tan(\delta/2)}{\tan(\varepsilon/2)}\right)}. \tag{8.17}$$

Proof Let A_0^c be the bounded connected component of $\mathbb{C}\backslash A$ and A_*^c be the unbounded connected component of $\mathbb{C}\backslash A$. Since $\varepsilon < \delta$, there exist two points $x \in A_0^c$ and $y \in A_*^c$ such that $|y - x|_s < \delta$. Since

$$\big|\tan(|y - x|_s/2) - (-\tan(|y - x|_s/2))\big|_s = |y - x|_s,$$

by an affine change of coordinates, which corresponds to a rotation of the Riemann sphere $\widehat{\mathbb{C}}$, we may assume without loss of generality that

$$x = \tan(|y - x|_s/2) \text{ and } y = -\tan(|y - x|_s/2).$$

Both components of the complement of the annulus

$$\tilde{A} := A\big(0; \tan(|y - x|_s/2), \tan(\delta/2)\big)$$

then contain points of both A_0^c and A_*^c. It follows that if $\gamma \in \Gamma_A$, then $\gamma([0,2\pi])$ intersects both connected components of $\mathbb{C}\backslash\tilde{A}$. Hence, the intersection $\gamma([0,2\pi])\cap\tilde{A}$ has at least two connected components, each of which joins the circles $S_{\tan(|y-x|_s/2)}$ and $S_{\tan(\delta/2)}$. More precisely, there exist two closed intervals $[a_1,b_1]\subseteq[0,2\pi]$ and $[a_1,b_1]\subseteq[0,2\pi]$ such that $(a_1,b_1)\cap(a_2,b_2)=\emptyset$, $\gamma([a_1,b_1]\cup[a_2,b_2])\subseteq\tilde{A}$, $\gamma(a_1),\gamma(a_2)\in S_{\tan(\delta/2)}$ and $\gamma(b_1),\gamma(b_2)\in S_{\tan(\delta/2)}$. Set

$$\tau(z) := \begin{cases} 1/|z| & \text{if } z \in \tilde{A} \\ 0 & \text{if } z \in A\backslash\tilde{A}, \end{cases}$$

i.e., $\tau\colon A \longrightarrow [0,+\infty)$ is a measurable Riemannian metric similar to the one defined in (8.10). Calculating similarly as in (8.12), we get that

$$\begin{aligned}\int_{\gamma|_{[a_1,b_1]}} \tau|dz| = \int_{a_1}^{b_1} \frac{|\gamma'(\theta)|}{|\gamma(\theta)|}d\theta &\geq \left|\int_{a_1}^{b_1} \frac{\gamma'(\theta)}{\gamma(\theta)}d\theta\right| \\ &= \left|\log\big(\gamma(b_1)\big) - \log\big(\gamma(a_1)\big)\right| \\ &\geq \left|\log\big(|\gamma(b_1)|\big) - \log\big(\gamma(a_1)\big)\right| \\ &= \left|\log\big(\tan(|y-x|_s/2)\big) - \log\big(\tan(\delta/2)\big)\right| \\ &= \log\left(\frac{\tan(|y-x|_s/2)}{\tan(\delta/2)}\right),\end{aligned}$$

where $\log\big(\gamma(b_1)\big)$ and $\log\big(\gamma(a_1)\big)$ are some appropriate choices of logarithms, respectively, of b_1 and a_1. The same calculation with a_1 replaced by a_2 and b_1 replaced by b_2 gives the same estimate for the integral of τ over $\gamma|_{[a_2,b_2]}$. Hence,

$$L_\tau(\gamma) = \int_\gamma \tau|dz| \geq \int_{\gamma_1} \tau|dz| + \int_{\gamma_2} \tau|dz| \geq 2\log\left(\frac{\tan(|y-x|_s/2)}{\tan(\delta/2)}\right).$$

Thus,

$$L(\tau) \geq 2\log\left(\frac{\tan(|y-x|_s/2)}{\tan(\delta/2)}\right). \tag{8.18}$$

On the other hand, similarly as in (8.11), with $R_1 := \tan(|y-x|_s/2)$ and $R_2 := \tan(\delta/2)$, we get that

$$\begin{aligned}A(\tau) \leq \int_{\tilde{A}} |z|^{-2}dS(z) &\leq \int_{R_1}^{R_2}\int_0^{2\pi} r\tau^2(re^{it})d\theta dr \\ &= \int_{R_1}^{R_2}\int_0^{2\pi} \frac{1}{r}d\theta dr = 2\pi\log(R_2/R_1) \\ &= 2\pi\log\left(\frac{\tan(|y-x|_s/2)}{\tan(\delta/2)}\right).\end{aligned}$$

It directly follows from (8.18), along with (8.7) and (8.6), that

$$\mathrm{Mod}(A) \leq 2\pi \frac{A(\tau)}{L^2(\tau)} \leq \frac{\pi^2}{\log\left(\frac{\tan(|y-x|_s/2)}{\tan(\delta/2)}\right)}.$$

So, letting $|y - x|_s$ converge to ε, we get that

$$\mathrm{Mod}(A) \leq \frac{\pi^2}{\log\left(\frac{\tan(\delta/2)}{\tan(\varepsilon/2)}\right)}$$

and the proof is complete. ∎

Now, as an immediate consequence of this theorem, we get the result we were after.

Proposition 8.2.16 *If* $(0,1) \ni r \longmapsto \mu(r) \in [0, +\infty]$ *is the Grötzsch modulus function, then*

$$\lim_{r \nearrow 1} \mu(r) = 0.$$

We end this section by examining the behavior of moduli of annuli under covering maps.

We recall that a continuous surjective map $f : X \to Y$ from a topological space X to a topological space Y is called covering if and only if, for every point $y \in Y$, there exists an open set $V_y \subseteq Y$ containing y such that, for every $x \in f^{-1}(y)$, there exists an open set $U_x(y) \subseteq X$ such that the map

$$f|_{U_x(y)} : U_x \longrightarrow V_y$$

is a homeomorphism.

Note that any open subset of the above set V_y containing y also supports the definition of f being a covering map. Furthermore, any open subset of V_y supports the definition of f being a covering map for all of its elements. Let us record the following straightforward fact.

Theorem 8.2.17 *If X and Y are connected, locally connected topological spaces and $f : X \to Y$ is a covering map, then each above open set V_y can be taken to be connected and then the corresponding sets $U_x(y)$, $x \in f^{-1}(y)$, are connected components of $f^{-1}(V_y)$.*

Since, for covering maps, the cardinality function

$$Y \ni y \longmapsto \# f^{-1}(y)$$

is locally constant (thus, also continuous), we get, in the setting of Theorem 8.2.17, the following.

Theorem 8.2.18 *If X and Y are connected, locally connected topological spaces and $f: X \to Y$ is a covering map, then the function $Y \ni y \longmapsto \# f^{-1}(y)$ constant, we denote its only value by* $\deg(f)$*, and we call it the degree of the covering map $f: X \to Y$.*

Given a pathwise connected topological space X and a point $\xi \in X$, we denote by $\pi_1(X,\xi)$ the fundamental group of X with the base point ξ. Since, for all points $\xi \in X$, all such groups are isomorphic, if we do not really care about the base point, we simply write $\pi_1(X)$ for the fundamental group of X.

Here, we raise two classical basic facts about covering maps, the first one is, for example, Theorem 4.1 in [**Mas**], while the second one is, for example, Proposition 1.32 in [**Ha**].

Theorem 8.2.19 *If X and Y are pathwise connected topological spaces and $f: X \to Y$ is a covering map, then the induced homomorphisms*

$$f_*: \pi_1(X) \longrightarrow \pi_1(Y)$$

is a monomorphism, i.e., it is one-to-one.

Theorem 8.2.20 *If X and Y are pathwise connected topological spaces and $f: X \to Y$ is a covering map, then* $\deg(f)$*, the degree of the covering map f, is equal to the index of the group $f_*(\pi_1(X))$ in $\pi_1(Y)$.*

Since, as it directly follows from its definition, every covering map is open, as an immediate consequence of Theorem 8.2.20, we get the following.

Theorem 8.2.21 *If X and Y are pathwise connected topological spaces, Y is simply connected (meaning that $\pi_1(Y) = \{0\}$), and $f: X \to Y$ is a covering map, then this map is a homeomorphism and, consequently, the space X is also simply connected.*

Since any subgroup of the additive group of integers $\mathbb{Z}$ is either a zero group or is isomorphic to $\mathbb{Z}$ with finite index, as an immediate consequence of Theorems 8.2.2, 8.2.20, and 8.1.1, we get the following.

Theorem 8.2.22 *If U is an open connected subset of $\widehat{\mathbb{C}}$, $A \subseteq \widehat{\mathbb{C}}$ is a topological annulus, and $f: U \to A$ is a covering map, then U is conformally equivalent to either the complex plane $\mathbb{C}$, the unit disk $\mathbb{D}$, or a topological annulus. In the last case, the degree of f is finite, i.e.,* $\deg(f) < +\infty$.

Theorem 8.2.23 *If A and B are two annuli contained in the complex plane $\mathbb{C}$, and $f: A \to B$ is a conformal covering map, then*

$$\mathrm{Mod}(B) = \deg(f)\mathrm{Mod}(A),$$

where $\deg(f)$, *the degree of f, is by Theorem 8.2.22 finite.*

Proof Because of Theorems 8.2.8 and 8.2.5, we may assume without loss of generality that $B = A(0; 1, R)$. Let $d = \deg(f)$. Consider the map $E_d: A(0; 1, R^{1/d}) \longrightarrow B$, given by the formula

$$E_d(z) := z^d.$$

This is a covering map of degree d. Fix a point $w \in B$ and $w_1 \in A$, $w_2 \in A(0; 1, R^{1/d})$ such that

$$f(w_1) = w \text{ and } E_d(w_2) = w.$$

Consider the fundamental groups

$$\pi_1(A, w_1),\ \pi_1\left(A\left(0; 1, R^{1/d}\right), w_1\right), \text{ and } \pi_1(B, w).$$

Since both maps f and E_d are coverings with degree d, we get that

$$f_*(\pi_1(A, w_1)) = d\pi_1(B, w) = (E_d)_*\left(\pi_1\left(A\left(0; 1, R^{1/d}\right), w_1\right)\right).$$

So, there exists a continuous map $H: A \longrightarrow A\left(0; 1, R^{1/d}\right)$ such that $E_d \circ H = f$ and $H(w_1) = w_2$. Since both maps f and E_d are covering, so is H. In particular, the degree, $\deg(H)$, is well defined and

$$d = \deg(f) = \deg(E_d \circ H) = \deg(E_d)\deg(H) = d\deg(H).$$

So, $\deg(H) = 1$ and, as $H: A \longrightarrow A(0; 1, R^{1/d})$ is covering, it must be a homeomorphism. Thus, $H: A \to A(0; 1, R^{1/d})$ is a conformal isomorphism. So,

$$\mathrm{Mod}(A) = \mathrm{Mod}\left(A\left(0; 1, R^{1/d}\right)\right) = \frac{1}{d}\log R = \frac{1}{\deg(f)}\mathrm{Mod}(B).$$

Equivalently, $\mathrm{Mod}(B) = \deg(f)\mathrm{Mod}(A)$ and we are done. ■

8.3 Koebe Distortion Theorems

This section is entirely devoted to formulating and proving various versions of the Koebe Distortion Theorems. They are truly amazing features of univalent holomorphic functions and form one of the main indispensable tools when dealing with dynamical and, especially, geometric aspects of meromorphic maps in the complex plane. The Koebe Distortion Theorems will be one of the most frequently invoked theorems in Volume 2 of this book, where we will be dealing with elliptic functions. It will also be quite heavily used in subsequent chapters of the current volume. The Koebe $\frac{1}{4}$-Theorem, i.e., Theorem 8.3.3, was conjectured by Koebe in 1907 and was proved by Bieberbach in 1916 [**Bie**]. The proof is a fairly easy consequence of the Bieberbach Coefficient Inequality, i.e., Theorem 8.3.2, obtained in [**Bie**]. Likewise, the analytic Koebe Distortion Theorems, such as Theorems 8.3.5 and 8.3.6, follow from the Bieberbach Coefficient Inequality. Our proofs are standard and closely follow the exposition in [**Hi**].

All other distortion theorems proved in this section are relatively straightforward consequences of those mentioned above and some results of Section 8.2.

We start by proving Theorems 8.3.1 and 8.3.2. The latter will form a crucial ingredient in the proof of the full version of the first Koebe Distortion Theorem, i.e., Theorem 8.3.3, following it. Let $\mathcal{S}$ denote the class of all univalent holomorphic functions $f\colon B(0,1) \longrightarrow \mathbb{C}$ such that

$$f(0) = 0 \ \text{ and } \ f'(0) = 1.$$

Theorem 8.3.1 (Area Theorem) *Let $g\colon B_e(0,1) \longrightarrow \mathbb{C}$ be a univalent meromorphic function with a simple pole at 0. Assume that the residue of g at 0 is equal to 1, so that the function g can be represented in a form*

$$g(z) = 1/z + b_0 + b_1 z + \cdots .$$

Then

$$\sum_{n=1}^{\infty} n|b_n|^2 \le 1.$$

Proof For every $0 < r < 1$, put $D_r = \mathbb{C}\backslash g(B_e(0,r))$. By Green's Theorem, we get that

$$S(D_r) = \int\int_{B_e(0,r)} dxdy = \frac{1}{2i}\int_{\partial B_e(0,r)} \overline{z}dz = -\frac{1}{2i}\int_{\partial B_e(0,r)} \overline{g}dg. \quad (8.19)$$

Recall that

$$\frac{1}{2\pi i}\int_{\partial B_e(0,r)} z^k \overline{z}^l dz = \delta_{kl}.$$

So, substituting the power series expansions for g and g' into (8.19), and performing the integration, we obtain that

$$S(B_e(0,r)) = \pi\left(\frac{1}{r^2} - \sum_{n=1}^{\infty} n|b_n|^2 r^{2n}\right).$$

Since $S(B_e(0,r)) \geq 0$, taking the limit as $r \nearrow 1$ yields the desired result. ■

Theorem 8.3.2 (Bieberbach Coefficient Inequality) *If $f(z) = z + \sum_{n=2}^{\infty} a_n z^n \in \mathcal{S}$, then $|a_2| \leq 2$.*

Proof Note that the formula

$$h(z) := \frac{f(z^2)}{z^2} = 1 + \sum_{n=2}^{\infty} a_n z^{2n-2} = 1 + \sum_{n=2}^{\infty} a_n z^{2(n-1)}$$

defines a holomorphic function from $B_e(0,1)$ to $\mathbb{C}\backslash\{0\}$ such that $h(0) = 1$. Let

$$\sqrt{h}\colon B_e(0,1) \longrightarrow \mathbb{C}\backslash\{0\}$$

be the square root of h uniquely determined by the requirement that $0 \mapsto 1$. Let

$$g(z) = \frac{1}{z\sqrt{h(z)}} = \frac{1}{z} - \frac{1}{2}a_2 z + \cdots,$$

and this series contains only odd powers of z so that the function g is odd. If now $g(z_1) = g(z_2)$, then $f(z_1^2) = f(z_2^2)$, so $z_1^2 = z_2^2$ and $z_1 = \pm z_2$. But since g is odd, we get that $z_1 = z_2$. Thus, g is univalent, whence Theorem 8.3.1 gives that $|a_2| \leq 2$. The proof is complete. ■

Theorem 8.3.3 (Koebe $\frac{1}{4}$-Theorem). *If $w \in \mathbb{C}$, $r > 0$, and $H\colon B_e(w,r) \longrightarrow \mathbb{C}$ is an arbitrary univalent analytic function, then*

$$H(B_e(w,r)) \supseteq B_e(H(w), 4^{-1}|H'(w)|r).$$

Proof Precomposing H with the scaled (by factor $1/r$) translation moving 0 to w and $B_e(0,r)$ onto $B_e(w,r)$, and postcomposing with a scaled (by factor r) translation moving $H(w)$ to 0, we may assume without loss of generality that $H \in \mathcal{S}$. Fix now a point $c \in \mathbb{C}\backslash H(B(0,1))$. Then an immediate inspection shows that the function

$$B_e(0,r) \ni z \longmapsto \frac{cH(z)}{c - H(z)} \in \mathbb{C}$$

belongs to $\mathcal{S}$ and takes on the form

$$B_e(0,r) \ni z \longmapsto z + \left(a_2 + \frac{1}{c}\right)z^2 + \cdots, \tag{8.20}$$

where, as usual, a_2 is the coefficient of z^2 in the Taylor series expansion of H about 0. Applying now Theorem 8.3.2 twice, to the function in (8.20) and to H, we obtain that

$$\frac{1}{|c|} \leq |a_2| + \left|a_2 + \frac{1}{c}\right| \leq 2 + 2 = 4.$$

The proof is complete. ■

The Koebe function

$$B_e(0,1) \ni z \longmapsto f(z) := \frac{z}{1-z^2} = \sum_{n=1}^{\infty} nz^n \in \mathbb{C}$$

univalently maps the ball unit $B_e(0,1)$ to the slit plane $\mathbb{C}\backslash(-\infty, 1/4]$. This shows that the number $1/4$ is optimal in the above theorem.

We will now prove a series of Koebe Distortion Theorems. First, we establish Lemma 8.3.4, which will form a crucial ingredient in the proof of the full version of Theorem 8.3.6, following it.

Lemma 8.3.4 *If $f \in \mathcal{S}$ and $|z| < 1$, then*

$$\left|\frac{1}{2}(1-|z|^2)\frac{f''(z)}{f'(z)} - \overline{z}\right| \leq 2 \tag{8.21}$$

for all $z \in B_e(0,1)$.

Proof We fix $z \in B_e(0,1)$. The parameter $w \in B_e(0,1)$ will be a variable throughout the proof. The Möbius transformation

$$w \longmapsto \frac{w+z}{1+\overline{z}w}$$

maps the ball $B_e(0,1)$ onto itself, sending 0 to z. It follows that, for any choice of constants C_1 and C_2, the function

$$g(w) = C_1 + C_2 f\left(\frac{w+z}{1+\overline{z}w}\right)$$

is univalent in $B_e(0,1)$. Now we specify C_1 and C_2 so that $g(0) = 0$ and $g'(0) = 1$, i.e., so that $g \in \mathcal{S}$. These conditions give

$$C_1 = -\frac{f(z_0)}{f'(z)(1-|z|^2)}, \quad C_2 = \frac{1}{f'(z)(1-|z_0|^2)}.$$

Direct calculations give

$$g'(z) = \frac{1}{f'(z)}\frac{1}{(1+\overline{z}w)^2}f'\left(\frac{w+z}{1+\overline{z}w}\right)$$

and

$$g''(w) = \frac{1}{f'(z)}\frac{2\overline{z}}{(1+\overline{z}w)^3}f'\left(\frac{w+z}{1+\overline{z}w}\right) + \frac{1}{f'(z)}\frac{1-|z|^2}{(1+\overline{z}w)^4}f''\left(\frac{w+z}{1+\overline{z}w}\right).$$

It, thus, follows that

$$g(w) = w + \left(\frac{1}{2}(1-|z|^2)\frac{f''(z)}{f'(z)} - \overline{z}\right)w^2 + \cdots.$$

Since $g \in \mathcal{S}$, the coefficient of z^2 is in modulus ≤ 2 because of Theorem 8.3.2. And this is the assertion of our lemma. ■

Now we formulate and prove the following theorem, which is the central one among the Koebe Distortion Theorems.

Theorem 8.3.5 (Koebe Distortion Theorem, Analytic Version I). *If $f \in \mathcal{S}$, then*

$$\frac{1-r}{(1+r)^3} \le |f'(z)| \le \frac{1+r}{(1-r)^3}, \tag{8.22}$$

and there is a choice of argument of $f'(z)$ such that

$$|\arg f'(z)| \le 2\log\left(\frac{1+r}{1-r}\right) \tag{8.23}$$

for all $r \in [0,1)$ and all $z \in \overline{B}(0,r)$.

Proof By Lemma 8.3.4, for every $t \in B_e(0,1)$, there exists $\eta(t) \in \overline{B}_e(0,1)$ such that

$$\frac{f''(t)}{f'(t)} - \frac{2\overline{t}}{1-|t|^2} = 4\frac{\eta(t)}{1-|t|^2}.$$

We integrate this expression along the line segment from 0 to z in $B(0,1)$ to obtain that

$$\log f'(z) + \log(1-|z|^2) = 4\int_0^z \frac{\eta(t)}{1-|t|^2}dt, \tag{8.24}$$

where the modulus of the right-hand side expression does not exceed

$$4\int_0^r \frac{ds}{1-|s|^2} = 2\log\frac{1+r}{1-r} \tag{8.25}$$

for all $z \in \overline{B}(0,r)$. Then taking the real parts of both sides of (8.24), we get the inequality

$$2\log\frac{1-r}{1+r} \le \log\big((1-|z|^2)|f'(z)|\big) \le 2\log\frac{1+r}{1-r}.$$

Since the function $\overline{B}(0,r) \ni z \longmapsto \log|f'(z)|$ is harmonic, it assumes its maximum and minimum values on $\partial B(0,r)$. This yields (8.22). Formula (8.23) is what we get by equating imaginary parts in (8.24) and using (8.25). ■

Part (8.23) is called the Rotation Theorem. It was discovered by Bieberbach in 1919. The estimates in (8.22) are the best possible; equality is reached for the functions

$$f_\beta(z) = \frac{z}{\left(1+e^{i\beta}z\right)^2}, \tag{8.26}$$

where β is an arbitrary real number.

Let $\mathcal{U}$ denote the class of all univalent and holomorphic functions from the unit disk $B_e(0,1)$ into $\mathbb{C}$. With obvious translations and rescalings, as an immediate consequence of Theorem 8.3.5, we get the following.

Theorem 8.3.6 (Koebe Distortion Theorem, Analytic Version II) *If $f \in \mathcal{U}$, then*

$$\frac{1-r}{(1+r)^3} \le \frac{|f'(z)|}{|f'(0)|} \le \frac{1+r}{(1-r)^3}, \tag{8.27}$$

and there is a holomorphic branch of argument of $f'(z)$ such that

$$\left|\arg\left(\frac{f'(z)}{f'(0)}\right)\right| \le 2\log\left(\frac{1+r}{1-r}\right) \tag{8.28}$$

for all $r \in [0,1)$ and all $z \in \overline{B}_e(0,r)$.

As an immediate consequence of Theorem 8.3.6, we get the following two facts.

Theorem 8.3.7 (Koebe Distortion Theorem, Analytic Version III) *There exists a monotone increasing continuous function*

$$K\colon [0,1) \longrightarrow [1,+\infty)$$

such that $K(0) = 1$ and with the following property. If $w \in \mathbb{C}$, $R > 0$, and $H\colon B_e(w,R) \longrightarrow \mathbb{C}$ is an arbitrary univalent analytic function, then

$$\left|\frac{|H'(z)|}{|H'(w)|} - 1\right| \le K(r/R)|z-w| \tag{8.29}$$

for every $r \in [0,R]$ and all $z \in \overline{B}_e(w,r)$.

Theorem 8.3.8 (Koebe Distortion Theorem I, Euclidean Version) *There exists a monotone increasing continuous function*

$$k\colon [0,1) \to [1,\infty)$$

such that $k(0) = 1$ and, for any $w \in \mathbb{C}$, any $r > 0$, all $t \in [0,1)$, and any univalent analytic function $H\colon B_e(w,r) \longrightarrow \mathbb{C}$, we have that

$$\sup\left\{|H'(z)|\colon z \in B_e(w,tr)\right\} \le k(t) \inf\left\{|H'(z)|\colon z \in B_e(w,tr)\right\}.$$

We put $K := k(1/2)$.

As a fairly simple consequence of this theorem, we get its following extension.

Theorem 8.3.9 (Koebe Distortion Theorem I for Parabolic Surfaces) *If S is a parabolic Riemann surface, then, for every $w \in S$, every $r \in (0,4u_S)$, all $t \in [0,1)$, and every univalent analytic function $H\colon B(w,r) \longrightarrow S$, we have that*

$$\sup\left\{|H'(z)|\colon z \in B(w,tr)\right\} \le k(t) \inf\left\{|H'(z)|\colon z \in B(w,tr)\right\},$$

where $k\colon [0,1) \to [1,\infty)$ is the function produced in Theorem 8.3.8. We recall that $k(1/2)$ is denoted by K.

Proof Fix a point $\tilde{w} \in \Pi_S^{-1}(w)$. Since both spaces $B_e(\tilde{w},r)$ and $\mathbb{C}$ are simply connected, there exists a lift $G\colon B_e(\tilde{w},r) \longrightarrow \mathbb{C}$ of the map $H \circ \Pi_S\colon B_e(\tilde{w},r) \longrightarrow S$ via the covering map $\Pi_S\colon \mathbb{C} \longrightarrow S$. So, G is analytic and the following diagram commutes:

$$\begin{array}{ccc} B_e(\tilde{w},r) & \xrightarrow{G} & \mathbb{C} \\ & \searrow^{H\circ\Pi_S} & \downarrow \Pi_S \\ & & \mathbb{T} \end{array}$$

Since the composition $H \circ \Pi_S\colon B_e(\tilde{w},r) \longrightarrow S$ is univalent, so is the holomorphic map G. Since the projection Π_S is a local isometry, it follows that $\left|H'\big(\Pi_S(z)\big)\right| = |G'(z)|$ for all $z \in B_e(\tilde{w},r)$. Thus, a direct application of Theorem 8.3.8 completes the proof. ■

With the same approach to that in the above proof, as a respective consequence of the Koebe $\frac{1}{4}$-Theorem (Theorem 8.3.3) and the Koebe Distortion Theorem, Analytic Version II (Theorem 8.3.6), we can prove the following two theorems.

Theorem 8.3.10 (Koebe $\frac{1}{4}$-Theorem for Parabolic Surfaces). *If S is a parabolic Riemann surface, then, for every $w \in S$, every $r \in (0,4u_S)$, and every univalent analytic function $H\colon B(w,r) \longrightarrow S$, we have that*

$$H(B(w,r)) \supseteq B\big(H(w),\, \min\{4u_{\mathbb{T}}, 4^{-1}|H'(w)|r\}\big).$$

Theorem 8.3.11 (Koebe Distortion Theorem, Analytic Version for Parabolic Surfaces) *If S is a parabolic Riemann surface, then, for every $w \in S$, every $r \in (0, 4u_S)$, every $t \in (0,1)$, and every univalent analytic function $H\colon B(w,r) \longrightarrow S$, we have that*

$$\frac{1-t}{(1+t)^3} \le \frac{|H'(z)|}{|H'(0)|} \le \frac{1+t}{(1-t)^3} \tag{8.30}$$

and there is a holomorphic branch of argument of $H'(z)$ such that

$$\left|\arg\left(\frac{H'(z)}{H'(0)}\right)\right| \le 2\log\left(\frac{1+t}{1-t}\right) \tag{8.31}$$

for all $z \in \overline{B}(w,tr)$.

Now we shall pass for a moment to the Riemann sphere $\widehat{\mathbb{C}}$. We shall prove the following.

Theorem 8.3.12 (Koebe Distortion Theorem I, Spherical Version). *Given two numbers $R, s \in (0,\pi)$, there exists a monotone increasing continuous function $k_{R,s}\colon [0,1) \longrightarrow [1,\infty)$ such that, for every $w \in \widehat{\mathbb{C}}$, every $r \in (0,R]$, all $t \in [0,1)$, and any univalent analytic function $H\colon B_s(w,r) \longrightarrow \widehat{\mathbb{C}}$ such that the complement $\widehat{\mathbb{C}} \backslash H(B_s(w,r))$ contains a spherical ball of radius s,*

$$\sup\left\{|H'(z)|_s : z \in B_s(w,tr)\right\} \le k_{R,s}(t) \inf\left\{|H'(z)|_s : z \in B_s(w,tr)\right\}.$$

Proof Let

$$M_w : \widehat{\mathbb{C}} \to \widehat{\mathbb{C}}$$

be the Möbius transformation corresponding via the stereographic projection to the rotation of sphere $S^2 = \{x \in \mathbb{R}^3 : \|x\| = 1\}$ from the corresponding point of 0 to the corresponding point of w. Then

$$M_w^{-1}(B_s(w,R)) = B_s(0,R) = B_e(0,R_1) \quad \text{and}$$
$$M_w^{-1}(B_s(w,r)) = B_s(0,r) = B_e(0,r_1)$$

with some appropriate radius $R_1 \in (0,+\infty)$ depending only on R and radius $r_1 \in (0,R_1)$ depending only on r. Note also that, for every $t \in [0,1)$, there exists $t' \in (0,1)$ such that

$$M_w^{-1}(B_s(w,tr)) \subseteq B_e(0,t'r_1). \tag{8.32}$$

Let $B_s(\xi,s)$ be a spherical ball of radius s disjoint from $H(B_s(w,r))$. Let

$$M_\xi : \widehat{\mathbb{C}} \to \widehat{\mathbb{C}}$$

be the Möbius transformation corresponding via the stereographic projection to the rotation of sphere S^2 from the corresponding point of ξ to the corresponding point of ∞. Then

$$M_\xi\left(\widehat{\mathbb{C}}\backslash B_s(\xi,s)\right) = \widehat{\mathbb{C}}\backslash B_s(\infty,s) = B_e(0,s_2)$$

with some appropriate radius $s_2 \in (0,+\infty)$ depending only on s. Then

$$M_\xi \circ H \circ M_w\colon B_e(0,r_1) \longrightarrow B_e(0,s_2)$$

is a univalent analytic function. Noting then that the moduli of Euclidean and spherical derivatives of $|(M_2 \circ H \circ M_1)|'$ and $|(M_2 \circ H \circ M_1)'|_s$ are uniformly comparable on $B_e(0,r_1)$ with (multiplicative) comparability constants depending only on R and s, and observing also that the moduli of spherical derivatives of M_w and M_ξ are everywhere equal to 1 as these are isometries with respect to the spherical metric on $\widehat{\mathbb{C}}$, our theorem follows from Theorem 8.3.8 by taking into account (8.32). ■

Employing the Mean Value Inequality, the following two lemmas are straightforward respective consequences of the last four theorems and the Koebe $\frac{1}{4}$-Theorem (Theorem 8.3.3).

Lemma 8.3.13 *If S is a parabolic Riemann surface, then, for every $w \in S$, every $r \in (0,4u_S)$, and every univalent analytic function $H\colon B(w,2r) \longrightarrow S$, we have that*

$$B\left(H(w),\frac{1}{4}r|H'(w)|\right) \subseteq H(B(w,r)) \subseteq B(H(w),Kr|H'(w)|).$$

Lemma 8.3.14 *If $w \in \widehat{\mathbb{C}}$, $R \in 0,(0,\pi/2)$, and $H\colon B(w,2R) \longrightarrow \widehat{\mathbb{C}}$ is a univalent analytic function whose range $H(B_s(w,2r))$ is disjoint from some spherical ball of some radius $s \in (0,\pi)$, then, for every $0 \le r \le R$, we have that*

$$\begin{aligned} B_s\left(H(w),k_{2R,s}^{-1}(1/2)r|H'(w)|_s\right) &\subseteq H(B(w,r)) \\ &\subseteq B_s\left(H(w),k_{2R,s}(1/2)r|H'(w)|_s\right). \end{aligned}$$

Since conformal homeomorphisms preserve moduli of annuli, employing the Riemann Mapping Theorem, i.e., Theorem 8.1.2, we fairly easily obtain from Theorem 8.3.8 a more geometric version of the Koebe Distortion Theorems, the one that involves moduli of annuli.

Theorem 8.3.15 (Koebe Distortion Theorem II, Euclidean Version). *There exists a function $w\colon (0,+\infty) \longrightarrow [1,\infty)$ such that, for any two open* *topological disks* $Q_1 \subseteq Q_2$ *with* $\mathrm{Mod}(Q_2 \setminus Q_1) \geq t$ *and any univalent analytic function* $H\colon Q_2 \to \widehat{\mathbb{C}}$, *we have that*

$$\sup\{|H'(\xi)|\colon \xi \in Q_1\} \leq w(t) \inf\{|H'(\xi)|\colon \xi \in Q_1\}.$$

Proof Let $R\colon B(0,1) \longrightarrow Q_2$ be a Riemann mapping (conformal homeomorphism), produced in Theorem 8.1.2, such that $R(0) \in Q_1$. Then, by Theorem 8.2.23, $\mathrm{Mod}(R^{-1}(Q_2 \setminus \overline{Q}_1)) \geq t$. Let

$$r := \sup\left\{|R^{-1}(z)|\colon z \in Q_1\right\} < 1.$$

Then, by Theorem 8.2.12 (Grötzsch Module Theorem), $\mathrm{Mod}(R^{-1}(Q_2 \setminus \overline{Q}_1)) \leq \mu(r)$. Hence, $t \leq \mu(r)$. Therefore, by virtue of Proposition 8.2.16, there exists $s(t) < 1$ such that

$$r < s(t).$$

Thus, applying Theorem 8.3.8, Koebe Distortion Theorem I, Euclidean Version (in fact, Theorem 8.3.6 would directly apply), we get, for all $\xi, z \in Q_1$, that

$$\begin{aligned}\frac{|H'(\xi)|}{|H'(z)|} &= \frac{\left|\left((H \circ R) \circ R^{-1}\right)'(\xi)\right|}{\left|\left((H \circ R) \circ R^{-1}\right)'(z)\right|} \\ &= \frac{\left|(H \circ R)'(R^{-1}(\xi))\right|}{\left|(H \circ R)'(R^{-1}(z))\right|} \cdot \frac{\left|R'\left(R^{-1}(z)\right)\right|}{\left|R'\left(R^{-1}(z)\right)\right|} \leq k^2(s(t)).\end{aligned}$$

So, setting $w(t) := k^2(s(t))$ finishes the proof. ∎

8.4 Local Properties of Critical Points of Holomorphic Functions

This short technical section provides a short, not fully commonly known, description of what is going on near critical points of holomorphic maps.

Given an open set $D \subseteq \mathbb{C}$ and an analytic function $H\colon D \to \mathbb{C}$, we put

$$\mathrm{Crit}(H) := \{z \in D\colon H'(z) = 0\}.$$

Its image, $H(\mathrm{Crit}(H))$, is called the set of critical values of H. Suppose now that $c \in \mathrm{Crit}(H)$. Then Taylor's Theorem asserts that there exists a unique integer $p_c := p(H,c) \geq 2$ such that

$$H(z) = H(c) + \sum_{n=p_c}^{\infty} a_n (z-c)^n \tag{8.33}$$

on some sufficiently small open neighborhood of c. The number p_c is called the order, or multiplicity, of H at the critical point c. It is also just called the order, or multiplicity, of the critical point c. Finally, it is also frequently denoted by $\deg_c(H)$ and called the local degree of H at c. It follows from (8.33) that there exist two numbers $R := R(H,c) > 0$ and $A := A(H,c) \geq 1$ such that

$$A^{-1}|z-c|^{p_c} \leq |H(z) - H(c)| \leq A|z-c|^{p_c} \tag{8.34}$$

and

$$A^{-1}|z-c|^{p_c-1} \leq |H'(z)| \leq A|z-c|^{p_c-1}, \tag{8.35}$$

for every $z \in \mathrm{Comp}(c, H, R)$, and that

$$H(\mathrm{Comp}(c, H, R)) = B_e(H(c), R).$$

In particular,

$$\mathrm{Comp}(c, H, R) \subseteq B_e\big(c, (AR)^{1/p_c}\big).$$

Moreover, by taking $R > 0$ sufficiently small, we can ensure that the two above inequalities hold for every $z \in B_e\big(c, (AR)^{1/p_c}\big)$.

Observation 8.4.1 The ball $B\big(c, (AR)^{1/p_c}\big)$ can be expressed as a union of p_c closed topological disks with smooth boundaries and mutually disjoint interiors such that the map H restricted to each of these interiors is injective.

Of course, all of the above is also true if $p_c = 1$, i.e., if c is an ordinary noncritical point.

Remark 8.4.2 By working with local holomorphic charts, we see that (8.34) and (8.35) along with Observation 8.4.1 remain true if D is an open subset of a Riemann surface S and $H: D \to S$ is a holomorphic map. In particular, this is true if S is a Riemann sphere, complex plane, or a complex torus.

In Volume 2, we will need the following technical lemma proven in [**U3**] as Lemma 2.11.

Lemma 8.4.3 *Suppose that a parabolic Riemann surface S, an open subset D of S, a composition $Q \circ H: D \longrightarrow S$ of analytic maps, a radius $R \in (0, 4u_S)$, and a point $z \in D$ are given such that*

(a)

$$\mathrm{Comp}(H(z), Q, 2R) \cap \mathrm{Crit}(Q) = \emptyset;$$

moreover, the map $Q|_{\mathrm{Comp}(H(z),Q,2R)}: \mathrm{Comp}(H(z), Q, 2R) \longrightarrow B(Q \circ H(z), 2R)$ is bijective; and

(b)

$$\text{Comp}(z, Q \circ H, R) \cap \text{Crit}(H) \neq \emptyset.$$

If c belongs to the last intersection and

$$\text{diam}\big(\text{Comp}(z, Q \circ H, R)\big) \leq (AR(H,c))^{1/p_c},$$

then

$$|z - c| \leq KA^2 |(Q \circ H)'(z)|^{-1} R.$$

Proof In view of Theorem 8.3.13, we get that

$$\text{Comp}(H(z), Q, R) \subseteq B(H(z), KR|Q'(H(z))|^{-1}).$$

So, since $H(c) \in \text{Comp}(H(z), Q, R)$, we get that

$$H(c) \in B(H(z), KR|Q'(H(z))|^{-1}).$$

Thus, using this, (8.34), Remark 8.4.2, and hypothesis (b), we obtain that

$$\begin{aligned} A^{-1}|z - c|^{p_c} &\leq |H(z) - H(c)| \leq KR|Q'(H(z))|^{-1} \\ &= KR|(Q \circ H)'(z)|^{-1}|H'(z)| \\ &\leq KR|(Q \circ H)'(z)|^{-1} A|z - c|^{p_c - 1}. \end{aligned}$$

So, $|z - c| \leq KA^2|(Q \circ H)'(z)|^{-1}R$. ■

8.5 Proper Analytic Maps and Their Degree

In this short section, we bring up, with proofs, the concept of proper maps, in particular holomorphic proper maps, and the degree of proper holomorphic maps. We follow closely Section 4.2 of Forster's book [**For**].

Let X and Y be topological spaces. A continuous map $f: X \longrightarrow Y$ is called discrete if and only if the inverse image $f^{-1}(w)$ is a discrete subset of X for every $w \in Y$; this means that $f^{-1}(w)$ viewed as a topological subspace of X with relative topology is a discrete topological space or, equivalently, for every point $z \in f^{-1}(w)$ there exists an open set $U \subseteq X$ such that $U \cap f^{-1}(w) = \{z\}$.

Let X and Y be topological spaces. A continuous map $f: X \longrightarrow Y$ is said to be proper if and only if $f^{-1}(K)$ is compact for every compact subset K of Y. We shall prove the following well-known result.

Proposition 8.5.1 *Every proper map between locally compact Hausdorff topological spaces is closed, meaning that images of all closed sets are closed.*

Proof Let X and Y be locally compact Hausdorff topological spaces and let $f: X \to Y$ be a proper map. Let $F \subseteq X$ be a closed set. Let $w \in \overline{f(F)}$. Since Y is a T_2 locally compact space, there exists $W \subseteq Y$, an open neighborhood of w, such that $\overline{W}$ is compact. Now if V is an arbitrary open neighborhood of w, then so is $W \cap V$, whence

$$(f(F) \cap W) \cap V = f(F) \cap (W \cap V) \neq \emptyset.$$

Thus,

$$w \in \overline{f(F) \cap W} \subseteq \overline{f(F) \cap \overline{W}}.$$

Since $f(F) \cap \overline{W} = f(F \cap f^{-1}(\overline{W}))$ and $F \cap f^{-1}(W)$ is also a compact set, we, therefore, get that

$$w \in \overline{f(F \cap f^{-1}(\overline{W}))} = f(F \cap f^{-1}(\overline{W})) \subseteq f(F).$$

The proof is complete. ■

Lemma 8.5.2 *Suppose that X and Y are locally compact Hausdorff topological spaces and $p: X \to Y$ is a proper discrete map. Then the following hold.*

(a) *For every point $y \in Y$, the set $p^{-1}(y)$ is finite.*
(b) *If $y \in Y$ and V is a neighborhood of $p^{-1}(y)$, then there exists a neighborhood U of y with $p^{-1}(U) \subseteq V$.*
(c) *If $D \subseteq X$ is a closed discrete set, then $p(D)$ is also discrete.*

Proof Item (a) follows from the fact that $p^{-1}(y)$ is a compact discrete subset of Y.

For item (b), note that $p(X \backslash V)$ is closed by applying Proposition 8.5.1 and as the set $X \backslash V$ is closed. Also, $y \notin p(Y \backslash V)$. Thus, $U := Y \backslash p(X \backslash V)$ is an open neighborhood of x such that $p^{-1}(U) \subseteq V$.

Proving (c), fix $z \in D$. Since D is closed and discrete and since, by (a), the set $p^{-1}(p(z))$ is finite, there exists U, an open neighborhood of $p^{-1}(p(z))$, such that

$$U \cap D = D \cap p^{-1}(p(z)).$$

Then, by Proposition 8.5.1, $p(X \backslash U)$ is a closed set and $p(z) \notin p(X \backslash U)$. So, $Y \backslash p(X \backslash U)$ is an open neighborhood of $p(z)$ and

$$p(z) \in \big(Y \backslash p(X \backslash U)\big) \cap p(D) \subseteq p(U \cap D) \subseteq p(D) \cap \{p(z)\} = \{p(z)\}.$$

Thus,

$$\big(Y \backslash p(X \backslash U)\big) \cap p(D) = \{p(z)\},$$

proving that $p(D)$ is discrete. ■

Theorem 8.5.3 *Suppose that X and Y are Riemann surfaces and $f: X \longrightarrow Y$ is a proper nonconstant holomorphic map. Then there exists a natural number $n \geq 1$ such that f takes every value $w \in Y$, counting multiplicities, n times, i.e.,*

$$\sum_{z \in f^{-1}(w)} \deg_z(f) = n,$$

where $\deg_z(f)$ is the order of f at z; it is equal to 1 if z is not a critical point of f. In particular, the map $f: X \longrightarrow Y$ is surjective. The above number n is denoted by $\deg(f)$ and is called the (topological) degree of the map f.

Proof The set $\mathrm{Crit}(f)$ of branch (critical) points of the map $f: X \to Y$ is closed and discrete. Since the map f is proper and discrete, $B := f(A)$ is also closed and discrete, respectively, by Proposition 8.5.1 and Lemma 8.5.2(c). Let

$$Y' := Y \backslash B,$$

$$X' := X \backslash f^{-1}(B) \subseteq X \backslash A.$$

Since

$$f|_{X'}: X' \to Y'$$

is a proper discrete (as holomorphic) covering (as a proper local homeomorphism between locally compact Hausdorff spaces) map and the space X' is pathwise connected, it follows from Lemma 8.5.2(a) that there exists a natural number $n \geq 1$ such that

$$\#\left(f|_{X'}\right)^{-1}(y) = n \tag{8.36}$$

for all $y \in Y'$.

We are, thus, only left to deal with points $b \in B$. Then there exist mutually disjoint neighborhoods U_x of respective points $x \in f^{-1}(b)$, and neighborhoods V_x of b, $x \in f^{-1}(b)$, such that $V_x \backslash \{b\} \subseteq Y'$ and for every $w \in V_x \backslash \{b\}$ the set

$$\#\left(f^{-1}(w) \cap U_x\right) = \deg_x(f).$$

By Lemma 8.5.2(b), we can find a neighborhood

$$V \subseteq \bigcap_{x \in f^{-1}(b)} V_x$$

of b such that

$$f^{-1}(V) \subseteq \bigcup_{x \in f^{-1}(b)} U_x.$$

Then, for every point $w \in V \setminus \{b\} \subseteq Y'$, we have that

$$\#\big(f^{-1}(w)\big) = \sum_{x \in f^{-1}(b)} \deg_x(f).$$

Combining this with (8.36), we, thus, get that

$$\sum_{x \in f^{-1}(b)} \deg_x(f) = n$$

and the proof is complete. ■

8.6 Riemann–Hurwitz Formula

In this section, we present a relation between holomorphic maps, their degrees, and critical points and the topological structure of images and preimages. This is known as the Riemann–Hurwitz Formula. This formula has a long history and is treated in many textbooks on Riemann surfaces and algebraic geometry. It is usually formulated for compact Riemann surfaces. However, we need larger generality. We, therefore, provide here a complete self-contained exposition which suffices for our needs throughout the book.

In particular, the topological structure, mentioned above, enters our considerations in terms of the Euler characteristic. We make a brief introduction to it in a restricted scope and, probably, in the most elementary way; in particular, we speak of triangulations only rather than dealing with a more general and flexible concept of CW-complexes.

The results of this section, i.e., various versions of the Riemann–Hurwitz Formula and their corollaries, will be instrumental when dealing with elliptic functions, particularly the nonrecurrent ones. These formulas provide an elegant and probably the best tool to control the topological structure of connected components of inverse images of open connected sets under meromorphic maps, especially to know that such connected components are simply connected.

Our exposition in this section stems from that in Beardon's book [**Bea**], developing it in many respects.

8.6.1 The Euler Characteristic of Plane Bordered Surfaces

We devote this section to a brief, restricted in scope, introduction to the Euler characteristic. More about this concept and its much deeper treatment can be found in virtually any book on algebraic topology.

We say that a domain (open connected set) in $\widehat{\mathbb{C}}$ is an open bordered surface if and only if its boundary (in $\widehat{\mathbb{C}}$) consists of a finite number (possibly 0) of mutually disjoint simple closed (Jordan) curves. Note that the complex plane $\mathbb{C}$ is not an open bordered surface but the Riemann sphere $\widehat{\mathbb{C}}$ is. We call any open bordered surface proper if and only if it is different from $\widehat{\mathbb{C}}$.

We call a set $S \subseteq \widehat{\mathbb{C}}$ a closed bordered surface if and only if it is the closure of an open bordered surface. Note again that the boundary (border) of S can be empty and then $S = \widehat{\mathbb{C}}$. In fact, $\widehat{\mathbb{C}}$ is the only set that is simultaneously an open and closed bordered surface. A triangulation T of S is a finite partition of S into mutually disjoint subsets called vertices, edges, and faces, respectively denoted by V, E, and F, with the following properties.

(1) Each vertex $v \in V$ is a point of S.

(2) For each edge $e \in E$, there is a compact interval $[a,b] \subseteq \mathbb{R}$ and a homeomorphism $\varphi\colon [a,b] \to \overline{e}$ such that

$$\varphi\big((a,b)\big) = e \quad \text{and} \quad \varphi(\{a,b\}) \subseteq V.$$

(3) For each face $f \in F$, there is a closed triangle $\Delta \subseteq \mathbb{C}$ and a homeomorphism $\varphi\colon \Delta \to \overline{f}$ which maps the edges and vertices of Δ (in the usual sense) to E and V, respectively, and such that

$$\varphi(\operatorname{Int}\Delta) = f.$$

Of course, if (2) holds for some interval (resp. if (3) holds for some triangle) then (2) holds for any closed interval (resp. (3) holds for any triangle). We stress once again that T partitions S into mutually disjoint subsets of S. Each such subset is either a vertex, an edge, or a face and we call each of these a simplex of T of dimension 0, 1, and 2, respectively. For any simplex s of dimension m, the Euler characteristic $\chi(s)$ of s is defined to be

$$(-1)^m.$$

More generally, if S' is any subset of S comprising a union of simplices, say $s_1, \ldots, s_k$, where s_j has dimension m_j, then S' is called a subcomplex of S relative to T, and we define

$$\chi(S', T) := \sum_{j=1}^{k} \chi(s_j) = \sum_{j=1}^{k} (-1)^{m_j}. \tag{8.37}$$

In particular, if a triangulation T of S consists of faces F, edges E, and vertices V, then the Euler characteristic $\chi(S)$ is, by definition,

$$\chi(S) = \chi(S,T) := \#F - \#E + \#V.$$

The crucial fact that is well known in algebraic topology (which we accept here without proof) is that $\chi(S,T)$ is independent of the particular triangulation T used. In particular, it is a topological invariant, meaning that two homeomorphic closed bordered surfaces have the same Euler characteristic. So, we can compute $\chi(S)$ using the triangulation of our choice. Of course, we need to know for this, and we indeed do by obvious geometry, that each such surface admits a triangulation.

For each edge $e \in E$, the closure $\overline{e}$ is a subcomplex of S relative to T and

$$\chi(\overline{e},T) = 1. \tag{8.38}$$

The boundary ∂S of S is its subcomplex relative to any triangulation of S and it is again evident that

$$\chi(\partial S,T) = 0,$$

as every connected component of ∂S is a Jordan curve which is a subcomplex of any triangulation of S and whose Euler characteristic is clearly equal to zero. The above formula permits us to get rid of the triangulation T and to express it in the following form:

$$\chi(\partial S) = 0. \tag{8.39}$$

We say that the Euler characteristic of ∂S is zero.

Calculations of χ can often be simplified by making use of the above and the following simple idea extending it. If T is a triangulation of S and S_1 and S_2 are two subcomplexes of S relative to T, then $S_1 \cup S_2$ and $S_1 \cap S_2$ are subcomplexes of S relative to T as well. Then we obtain from (8.37) that

$$\chi(S_1 \cup S_2,T) + \chi(S_1 \cap S_2,T) = \chi(S_1,T) + \chi(S_2,T). \tag{8.40}$$

Let us illustrate this idea with simple examples. Let $\mathrm{Int}S = S\backslash\partial S$ denote the interior of S. The interior $\mathrm{Int}S$ of S is its subcomplex relative to any triangulation of S. Then, by (8.40) and (8.39),

$$\chi(S) = \chi(\mathrm{Int}S,T) + \chi(\partial S) = \chi(\mathrm{Int}S,T).$$

In particular, $\chi(\mathrm{Int}S,T)$ is again independent of T and we can speak of $\chi(\mathrm{Int}S)$, calling it the Euler characteristic of $\mathrm{Int}S$. We, thus, have that

$$\chi(\mathrm{Int}S) = \chi(S). \tag{8.41}$$

For open bordered surfaces V in $\widehat{\mathbb{C}}$, we define

$$\chi(V) := \chi(\overline{V}). \tag{8.42}$$

However, a warning. We have not established yet that $\chi(\text{Int}S)$ is a topological invariant, i.e., we do not know that if $\widehat{S}$ is another bordered surface in $\widehat{\mathbb{C}}$, and $\text{Int}\widehat{S}$ is homeomorphic to $\text{Int}S$, then $\chi(\text{Int}\widehat{S}) = \chi(\text{Int}S)$ (we know this though if $\widehat{S}$ and S are homeomorphic). We will address this issue later in this section.

By constructing explicit triangulations, it is immediate that $\chi(\widehat{\mathbb{C}}) = 2$ and $\chi(\overline{D}) = 1$ for any closed topological disk $\overline{D}$; these computations are indeed simple but important.

By its very definition, each closed bordered surface S in $\widehat{\mathbb{C}}$ is the complement in $\widehat{\mathbb{C}}$ of some finite number, say $k \geq 0$ mutually disjoint open topological disks $D_1, \ldots, D_k$, whose boundaries are Jordan curves, so D is of connectivity k, meaning that its complement has k connected components. We can triangulate the sphere such that each of the sets S, $D_1, \ldots, D_k$ is its subcomplex. Then (8.40) yields

$$2 = \chi(\widehat{\mathbb{C}}) = \chi(S) + \sum_{j=1}^{k} \chi(D_j) = \chi(S) + k.$$

So,

$$\chi(S) = 2 - k. \tag{8.43}$$

Note that, for any such closed bordered surface S of the sphere $\widehat{\mathbb{C}}$, we have the following.

(a) $\chi(S) = 2$ if and only if S is the sphere $\widehat{\mathbb{C}}$ $(k = 0)$.
(b) $\chi(S) = 1$ if and only if S is simply connected, but is not $\widehat{\mathbb{C}}$, i.e., S is a closed topological disk; equivalently, if $k = 1$.
(c) $\chi(S) = 0$ if and only if S is doubly connected, i.e., if $k = 2$.

In all other cases, i.e., if $k \geq 3$,

$$\chi(S) < 0.$$

8.6.2 The Riemann–Hurwitz Formula for Bordered Surfaces in $\widehat{\mathbb{C}}$

Our first main goal is to obtain and to prove the formula, commonly called the Riemann–Hurwitz Formula, in the setting we start to describe now. Our ambient space will always be the Riemann sphere $\widehat{\mathbb{C}}$ and the boundary ∂A of a subset A of $\widehat{\mathbb{C}}$ will always be understood with respect to the topological space $\widehat{\mathbb{C}}$, i.e., it will be equal to $\overline{A} \cap \overline{\widehat{\mathbb{C}} \setminus A}$. If $A \subseteq B \subseteq \widehat{\mathbb{C}}$, then the boundary of A with respect to the topological space B will be denoted by $\partial_B A$. Note then that

$$\partial_B A \subseteq \partial A.$$

Now let S be a domain in $\widehat{\mathbb{C}}$, i.e., an open connected subset of $\widehat{\mathbb{C}}$. We record the following obvious observation.

Observation 8.6.1 If S is a domain in $\widehat{\mathbb{C}}$ and K is a compact subset of S, then

$$\text{dist}_s(K, \partial S) > 0.$$

Now let S_1 and S_2 be two domains in $\widehat{\mathbb{C}}$ and let

$$R\colon S_1 \to S_2$$

be an analytic map which has a (unique) continuous extension from $\overline{S_1}$ to $\overline{S_2}$. Keep the same symbol R for this extension. We shall prove the following.

Theorem 8.6.2 *If S_1 and S_2 are two domains in $\widehat{\mathbb{C}}$, then an analytic map $R\colon S_1 \to S_2$ with a (unique) continuous extension from $\overline{S_1}$ to $\overline{S_2}$ is proper if and only if*

$$R(\partial S_1) \subseteq \partial S_2. \tag{8.44}$$

Proof For the sake of this proof, we reserve the symbol R for the map from S_1 to S_2 and we denote the unique continuous extension of R to $\overline{S_1}$ by $\overline{R}$. By way of contradiction, suppose that the map $R\colon S_1 \to S_2$ is proper but that (8.44) does not hold. This means that there exists a point $z \in \partial S_1$ such that $\overline{R}(z) \in S_2$. As S_2 is open, there exists $r > 0$ such that

$$\overline{B(R(z), r)} \subseteq S_2.$$

Since R is proper, the preimage $R^{-1}(\overline{B(R(z), r)})$ is a compact subset of S_1. By Observation 8.6.1, there then exists $\varepsilon > 0$ such that

$$B(z, \varepsilon) \cap R^{-1}\left(\overline{B(R(z), r)}\right) = \emptyset. \tag{8.45}$$

Given that $\overline{R}\colon \overline{S_1} \to \overline{S_2}$ is continuous, there exists $\delta \in (0, \varepsilon)$ such that

$$\overline{R}(B(z, \delta)) \subseteq B(R(z), r). \tag{8.46}$$

But, since $z \in \overline{S_1}$, there exists a point $w \in S_1 \cap B(z, \delta)$. It then follows from (8.45) that $R(w) \notin B(R(z), r)$, while it ensues from (8.46) that $R(w) \in B(R(z), r)$. This contradiction finishes the proof of the first implication.

Now suppose that (8.44) holds but the map R is not proper. Then there exists a compact set $K \subseteq S_1$ such that $R^{-1}(K)$ is not compact. But since $\overline{R}^{-1}(K)$ is compact, we get that $\overline{R}^{-1}(K)\backslash R^{-1}(K) \neq \emptyset$. However, $\overline{R}^{-1}(K)\backslash R^{-1}(K) \subseteq \partial S_1$, whence $\overline{R}(\partial S_1) \supseteq \overline{R}(\overline{R}^{-1}(K)\backslash R^{-1}(K)) \subseteq K \subseteq S_1$. This contradicts (8.44), completing the proof of the second implication and of the whole theorem. ■

Theorem 8.6.3 *Let S_1 and S_2 be two domains in $\widehat{\mathbb{C}}$ and $R\colon S_1 \to S_2$ be a proper analytic map which has a (unique) continuous extension from $\overline{S}_1$ to $\overline{S}_2$. If V is an open connected subset of S_2 and U is a connected component of $f^{-1}(V)$, then $R|_U\colon U \to V$ is a proper analytic map. Hence (see Theorem 8.5.3), $R(U) = V$. In addition,*

$$R(\partial U) = \partial V \text{ and } \overline{U} \cap R^{-1}(\partial V) = \partial U. \tag{8.47}$$

Proof We first prove that the map $R|_U\colon U \to V$ is proper. Let $K \subseteq V$ be a compact set. Let $\mathcal{G}$ be the collection of all connected components of $R^{-1}(V)$. Then

$$R|_U^{-1}(K) = R^{-1}(K) \cap \left(\widehat{\mathbb{C}} \backslash \bigcup_{G \in \mathcal{G}\backslash\{U\}} G \right).$$

Since the set $R^{-1}(K)$ is compact (as the map $R\colon S_1 \to S_2$ is proper) and each element of $\mathcal{G}$ is an open subset of $\widehat{\mathbb{C}}$, we conclude that the set $R|_U^{-1}(K)$ is compact. Thus, the map $R|_U\colon U \to V$ is proper.

So, it follows from Theorem 8.6.2 that

$$R(\partial U) \subseteq \partial V$$

and from Theorem 8.5.3 that

$$R(U) = V.$$

This immediately implies that $U \cap R^{-1}(\partial V) = \emptyset$. Hence,

$$\overline{U} \cap R^{-1}(\partial V) \subseteq \partial U. \tag{8.48}$$

Thus, in order to complete the proof, it suffices to show that

$$\partial V \subseteq R(\partial U). \tag{8.49}$$

Let $w \in \partial V$. There exists a sequence $(w_n)_{n=1}^{\infty}$ of points in V converging to w. For every $n \in \mathbb{N}$, choose a point $z_n \in U \cap R^{-1}(w_n)$. Passing to a subsequence, we may assume that the sequence $(z_n)_{n=1}^{\infty}$ converges in $\widehat{\mathbb{C}}$. Denote its limit by z. Then $z \in \overline{U}$ and

$$R(z) = \lim_{n\to\infty} R(z_n) = \lim_{n\to\infty} w_n = w.$$

So, $z \in \overline{U} \cap R^{-1}(\partial V)$. Hence, $z \in \partial U$ by (8.48). Thus, $w = R(z) \in R(\partial U)$ and (8.49) is proved. The proof of Theorem 8.6.3 is complete. ■

From now on, we assume throughout the whole section that S_1 and S_2 are two domains in $\widehat{\mathbb{C}}$ and $R\colon S_1 \to S_2$ is a proper analytic map having a (unique) continuous extension. We also assume throughout the whole section that V is an open connected subset of S_2 such that $\overline{V} \subseteq S_2$ and that

$$\partial V \cap R(\text{Crit}(R)) = \emptyset. \tag{8.50}$$

Furthermore, we let U be a connected component of $R^{-1}(V)$. For the purpose of dealing later in this chapter with Euler characteristics and the Riemann–Hurwitz Formula, we will establish some properties of the map R in relation to V and U that will turn out to also be useful in further stages of the proof of the Riemann–Hurwitz Formula.

Having proved Theorem 8.6.3, we may, by virtue of Theorem 8.5.3, speak about $\deg\big(R|_U\big) \geq 1$, the degree of the map $R|_U\colon U \to V$. Now we shall prove the following.

Lemma 8.6.4 *Let S_1 and S_2 be two domains in $\widehat{\mathbb{C}}$ and $R\colon S_1 \to S_2$ be a proper analytic map which has a (unique) continuous extension from $\overline{S}_1$ to $\overline{S}_2$. If $V \subseteq S_2$ is an open bordered surface (in particular, $\overline{V}$ is a closed bordered surface in $\widehat{\mathbb{C}}$ and* $\text{Int}(\overline{V}) = V$*), then U is also an open bordered surface. In particular, $\overline{U}$ is a closed bordered surface in $\widehat{\mathbb{C}}$ and* $\text{Int}(\overline{U}) = U$.

In addition, $R|_{\partial U}\colon \partial U \to \partial V$ is a proper covering map of finite degree.

Proof Because of Theorem 8.6.3 and formula (8.54), the map

$$R|_{\partial U}\colon \partial U \to \partial V$$

is a locally homeomorphic surjection. Since ∂U is a compact set, $R|_{\partial U}\colon \partial U \to \partial V$ is, thus, a covering map of finite degree whose value we denote by q. If Γ is a connected component of ∂V (we know then that Γ is a Jordan curve), then $R\colon R^{-1}(\Gamma) \to \Gamma$ is also a covering map of degree $\leq q$. Let L be a connected component of $R^{-1}(\Gamma)$. We claim that $R(L) = \Gamma$. Otherwise, there would exist a point $x \in L$ such that $R(x) \in \partial_\Gamma R(L)$. But, because of (8.50), there exists an open topological arc α containing x such that $R(\alpha) \subseteq \Gamma$ and $R(\alpha)$ is an open topological arc containining $R(x)$. But then $L \cup \alpha$ is connected and $R(L \cup \alpha) \subseteq \Gamma$. Hence, $L \cup \alpha = L$; therefore, $R(x) \in R(L) \cup R(\alpha)$, yielding $R(x) \in \text{Int}_\Gamma R(L)$. This contradiction gives that

$$R(L) = \Gamma.$$

Thus, the map $R\colon L \to \Gamma$ is a covering surjection. In consequence, it is at most q-to-one and $R^{-1}(\Gamma)$ has at most q connected components. It is, thus, left for us to show that L is a Jordan curve. But L is compact and $R\colon L \to \Gamma$ is locally homeomorphic, so L is a compact connected one-dimensional topological manifold. It is well known then that L is a Jordan curve. Another argument, less topological but more analytic, would be this. Fix a point $\xi \in \Gamma$. Then the set $L \cap R^{-1}(\xi)$ has, at most, q elements and each element $z \in R^{-1}(\xi)$ gives rise to a local inverse map R_z^{-1} from a sufficiently small neighborhood of ξ onto a neighborhood of z; in particular, $R_z^{-1}(\xi) = z$. Continuing R_z^{-1} analytically from ξ to ξ along Γ, we traverse a path Γ_z in L whose other endpoint $\widehat{z}$ belongs to $R^{-1}(\xi)$. The path Γ_z is either a closed topological arc or a Jordan curve depending on whether $\widehat{z} \neq z$ or $\widehat{z} = z$. Furthermore, the map

$$R^{-1}(\xi) \ni z \longmapsto \widehat{z} \in R^{-1}(\xi)$$

is a bijection. We then continue analytically getting points $z, \widehat{z}, \widehat{\widehat{z}}, \ldots$, until, with at most q iterates, we reach z again. Then the consecutive closed topological arcs $\Gamma_z, \Gamma_{\widehat{z}}, \Gamma_{\widehat{\widehat{z}}}, \ldots$, have exactly one common endpoint, $\widehat{z}, \widehat{\widehat{z}}, \ldots$, respectively. So, their union, up to the second to last element, is a closed topological arc again. Its union with the last closed topological arc in the sequence $\Gamma_z, \Gamma_{\widehat{z}}, \Gamma_{\widehat{\widehat{z}}}, \ldots$, is a Jordan curve since these two arcs have two common endpoints and no other endpoints. The proof is complete ■

Lemma 8.6.5 *With the hypotheses of Lemma 8.6.4 (in particular V is an open bordered surface in $\widehat{\mathbb{C}}$), we have that*

$$\deg\big(R|_{\partial U}\big) = \deg(R|_U).$$

Proof Fix $\xi \in \partial V$. Since ∂V contains no critical values of R, there exists $\varepsilon > 0$ such that all analytic branches of R^{-1} are defined on $B(\xi, \varepsilon)$. Since, by Lemma 8.6.4, ∂U is a finite disjoint union of Jordan curves, and since, for every $z \in \partial U \cap R^{-1}(\xi)$, the map $R\colon R_z^{-1}(B(z,\varepsilon)) \to B(z,\varepsilon)$ is a homeomorphism, there exists $\Gamma_z \subseteq R_z^{-1}(B(z,\varepsilon)) \cap \overline{U}$, an open, relative to $\overline{U}$, neighbourhood of z in $\overline{U}$ such that $R(\Gamma_z)$ is an open, relative to $\overline{V}$, neighbourhood of ξ in $\overline{V}$. Hence,

$$V_\xi := \bigcap \{R(\Gamma_z) \colon z \in \partial U \cap R^{-1}(\xi)\}$$

is also an open, relative to $\overline{V}$, neighborhood of ξ in $\overline{V}$. Fix $y \in V_\xi \cap V$. Since both families

$$\mathcal{F}_\xi := \{R_z^{-1}(B(\xi,\varepsilon)) \colon z \in \partial U \cap R^{-1}(\xi)\}$$

and

$$\mathcal{F}_y := \{R_x^{-1}(B(\xi,\varepsilon)) : x \in U \cap R^{-1}(y)\}$$

consist of mutually disjoint sets, we have that

$$\deg(R|_{\partial U}) = \#\mathcal{F}_\xi \quad \text{and} \quad \deg(R|_U) = \#\mathcal{F}_y.$$

Thus, in order to complete the proof, it suffices to show that

$$\mathcal{F}_\xi = \mathcal{F}_y. \tag{8.51}$$

Indeed, if $z \in \partial U \cap R^{-1}(\xi)$, then there exists a (unique) point $x \in \Gamma_z$ such that $R(x) = y$. Furthermore, $x \in \overline{U} \cap R^{-1}(V) = U$, so $x \in U \cap R^{-1}(y)$ and $R_x^{-1}(y) = x = R_z^{-1}(y)$. Thus, $R_z^{-1}(B(\xi,\varepsilon)) = R_x^{-1}(B(\xi,\varepsilon)) \in \mathcal{F}_y$ and the inclusion

$$\mathcal{F}_\xi \subseteq \mathcal{F}_y \tag{8.52}$$

is proved. For the proof of the opposite inclusion, suppose that $x \in U \cap R^{-1}(y)$. Then $R_x^{-1}(V_\xi \cap V) \subseteq U$, whence

$$z := R_x^{-1}(\xi) \in R_x^{-1}(\overline{V_\xi \cap V}) = \overline{R_x^{-1}(V_\xi \cap V)} \subseteq \overline{U}.$$

Therefore, $z \in \partial U \cap R^{-1}(\xi)$ and $R_z^{-1}(\xi) = z = R_x^{-1}(\xi)$. Thus, $R_z^{-1}(B(\xi,\varepsilon)) = R_x^{-1}(B(\xi,\varepsilon))$. So, $R_x^{-1}(B(\xi,\varepsilon)) \in \mathcal{F}_\xi$ and the inclusion $\mathcal{F}_y \subseteq \mathcal{F}_\xi$ is proved. Along with (8.52), this completes the proof of formula (8.51) and, simultaneously, the proof of Lemma 8.6.5. ■

Now we turn to preparing the appropriate data from R. This will involve a contribution from the critical points of R. In order to quantify this, we introduce the *deficiency* of R at a point z belonging to S_1 as

$$\delta_R(z) := p_z - 1, \tag{8.53}$$

where, we recall, p_z is the order of z with respect to R; if z is not a critical point of R, then $p_z = 1$ and $\delta_R(z) = 0$. For any set $A \subseteq S_1$, we define the *total deficiency* of R *over* A as

$$\delta_R(A) := \sum_{z \in A} \delta_R(z).$$

The function $A \longmapsto \delta_R(A)$ is additive, meaning that, for disjoint sets A and B in S_1, we have that

$$\delta_R(A \cup B) = \delta_R(A) + \delta_R(B).$$

Frequently, but only when R is understood, we omit the suffix R and use δ instead of δ_R.

We are now ready to relate the quantities $\chi(U)$, $\chi(V)$, $\deg\big(R|_U\big)$, and $\delta_R(U)$. We shall prove the following.

Theorem 8.6.6 (Riemann–Hurwitz Formula for Bordered Surfaces in $\widehat{\mathbb{C}}$) *Let S_1 and S_2 be two domains in $\widehat{\mathbb{C}}$, i.e., open connected subsets of $\widehat{\mathbb{C}}$. Let*

$$R\colon S_1 \to S_2$$

be a proper analytic map which has a (unique) continuous extension from $\overline{S}_1$ to $\overline{S}_2$. Let $V \subseteq S_2$ be an open bordered surface. Assume also that

$$\partial V \cap R(\mathrm{Crit}(R)) = \emptyset. \tag{8.54}$$

If U is a connected component of $R^{-1}(V)$, then

$$\chi(U) + \delta_R(U) = \deg\big(R|_U\big)\chi(V). \tag{8.55}$$

Proof We triangulate the closure of V, ensuring (as we may) that all critical values of R in V are vertices of the triangulation. Indeed, since each connected component on ∂V is a Jordan curve, given any triangulation, we can always connect with a closed topological arc each critical value of R with at least two distinct vertices of T, bounding a one-dimensional simplex on one connected component of ∂V, and forming, in this way, a larger triangulation with the required property. Denote such a triangulation by T, and, as always, its vertices, edges, and faces, respectively, by V, E, and F.

Now we construct a triangulation T_U of ∂U in the following way. Denote $\deg\big(R|_U\big)$ by m. Let $\Delta \in F$. Since Δ is a topological closed disk, there exists an open topological disk Δ' (whose closure is a topological closed disk) containg $\overline{\Delta}\backslash V$ and disjoint from the union of critical values of R and vertices of Δ. Then, using the Monodromy Theorem, we see that there are exactly m distinct analytic branches $R_j^{-1}\colon \Delta' \to S_1$ (so, $j = 1,2,\dots,m$) of R^{-1} such that $R_j^{-1}(\Delta) \subseteq U$. We define

$$F_U := \big\{R_j^{-1}(\Delta)\colon \Delta \in F,\ j = 1,2,\dots,m\big\}.$$

These are to be the faces of the ultimate triangulation T_U. The edges of every face $R_j^{-1}(\Delta)$ are to be the sets $R_j^{-1}(e)$, where e is one of the three edges of Δ. So, we define

$$E_U := \big\{R_j^{-1}(e)\colon \Delta \in F, e \in \partial\Delta,\ j = 1,2,\dots,m\big\}.$$

Finally, the set V_U consists of the endpoints of the closures of all elements of E_U. Note that the endpoints of the closure of each element e of E_U are distinct since these are mapped by R onto two different points; namely, the endpoints of $R(e) in E$. It follows immediately from this construction that the sets F_U labeled as faces, E_U labeled as edges, and V_U labeled as vertices form a triangulation of $\overline{U}$. We denote this by T_U.

It is immediate from this construction that

$$\#F_U = m\#F, \quad \#E_U = m\#E$$

and

$$\#V_U = m\#V - \sum_{c \in V_U \cap \mathrm{Crit}(R)} (p_c - 1) = m\#V - \delta_R(U).$$

Therefore,

$$\begin{aligned}
\chi(U) = \chi(U, T_U) &= \#F_U - \#E_U + \#V_U = m\#F - m\#F + (m\#V - \delta_R(U)) \\
&= m(\#F - \#E + \#V) - \delta_R(U) \\
&= m\chi(V) - \delta_R(U).
\end{aligned}$$

The proof is complete. ■

Now we shall derive some fairly straightforward consequences of this theorem.

Corollary 8.6.7 *With the hypotheses of Theorem 8.6.6, if V is simply connected and different from $\widehat{\mathbb{C}}$ (knowing that then because of the Koebe–Poincaré Uniformization Theorem, Theorem 8.1.1, and since $V \neq \mathbb{C}$, V is conformally equivalent to the unit disk $\mathbb{D}$), then $U \neq \widehat{\mathbb{C}}$, so that $\chi(U) \leq 1$, and*

$$\delta_U(R) \geq \deg(R|_U) - 1,$$

with equality holding if and only if U is simply connected, i.e., conformally equivalent to the unit disk $\mathbb{D}$. Then, i.e., if U is simply connected, also

$$\#\big(\mathrm{Crit}(R|_U)\big) \leq \deg(R|_U) - 1. \tag{8.56}$$

Corollary 8.6.8 *With the hypotheses of Theorem 8.6.6, if V is simply connected, i.e., (being different from $\mathbb{C}$) either equal to $\widehat{\mathbb{C}}$ or conformally equivalent to the unit disk $\mathbb{D}$, and U contains no critical points of R, then the map $R|_U : U \to V$ is a conformal homeomorphism. In consequence, in the former case $U = \widehat{\mathbb{C}}$, while in the latter case U is conformally equivalent to the unit disk $\mathbb{D}$.*

Proof If $V = \widehat{\mathbb{C}}$, then $\chi(V) = 2$. Since also $\chi(U) \leq 2$, we conclude from Theorem 8.6.6 that $\deg(R|_U) \leq 1$. Thus, $\deg(R|_U) = 1$, whence $R|_U : U \to V$ is a conformal homeomorphism. In consequence, $U = \widehat{\mathbb{C}}$.

If $V \neq \widehat{\mathbb{C}}$, then $U \neq \widehat{\mathbb{C}}$; otherwise, V would be compact, thus equal to $\widehat{\mathbb{C}}$. Hence, $\chi(U) \leq 1$. We, therefore, conclude from Theorem 8.6.6 that

$\deg(R|_U) \le 1$. Thus, $\deg(R|_U) = 1$, whence $R|_U : U \to V$ is a conformal homeomorphism. This, in turn, immediately yields that U is conformally equivalent to the unit disk $\mathbb{D}$. ■

Corollary 8.6.9 *With the hypotheses of Theorem 8.6.6, if V is conformally equivalent to the unit disk $\mathbb{D}$ and the map $R|_U : U \to V$ has only one critical point, which we denote by c, then*

(a) *the set U is conformally equivalent to the unit disk $\mathbb{D}$, and*

(b)

$$\deg\big(R|_U\big) = p_c.$$

Proof Obviously, $\deg\big(R|_U\big) \ge p_c$. Since $U \ne \widehat{\mathbb{C}}$, as otherwise V would be compact, thus equal to $\widehat{\mathbb{C}}$, we get that $\chi(U) \le 1$. It, thus, follows from Theorem 8.6.6 that $\deg\big(R|_U\big) \le p_c$. Hence, $\deg\big(R|_U\big) = p_c$, proving item (b). Applying Theorem 8.6.6 once more, we now conclude that $\chi(U) = 1$, and item (a) is also proved. ■

8.6.3 Euler Characteristic: The General $\widehat{\mathbb{C}}$ Case

The Euler characteristic of a domain D in the complex sphere has been defined in Subsection 8.6.1 whenever its boundary ∂D consists of a finite number of Jordan curves. In general, however, the boundary of a domain D is much more complicated than this; we may not be able to triangulate the closure of D and, in these circumstances, $\chi(D)$ is as yet undefined. As this is likely to be so in the case of major interest to us (when D is a component of a Fatou set), this presents us with a problem which we must now address. We propose to show that, given any domain D, we can define $\chi(D)$ as the limiting value of the Euler characteristic of smooth subdomains which exhaust D; once this has been done, we can then use the Euler characteristic as a tool to study the way in which R maps one component of the Fatou set onto another. We shall restrict our discussion to subdomains of $\widehat{\mathbb{C}}$, nevertheless. The following development is closely related to the construction of the ideal boundary components of a Riemann surface.

Let $D \subseteq \widehat{\mathbb{C}}$ be a domain, i.e., an open connected set. A subdomain Δ of D is said to be a regular subdomain of D if and only if:

(1) Δ is an open bordered surface whose closure $\overline{\Delta}$ is contained in D, i.e., $\overline{\Delta} \subseteq D$, and $\partial\Delta$ is a finite, possibly zero, union of mutually disjoint Jordan curves, say $\gamma_1, \dots, \gamma_n$,

and

(2) if $\Gamma_1, \ldots, \Gamma_n$ denote all connected components of $\widehat{\mathbb{C}} \backslash \Delta$ enumerated so that their boundaries are, respectively, of $\gamma_1, \ldots, \gamma_n$, then $\Gamma_j \cap (\widehat{\mathbb{C}} \backslash D) \neq \emptyset$ for every $j = 1, 2, \ldots, n$.

For example, the unit open disk $\mathbb{D} = \{z \in \mathbb{C}: |z| < 1\}$ is a regular subdomain of $\mathbb{C}$, whereas the annulus $\{z \in \mathbb{C}: 1 < |z| < 2\}$ is not. For ease of future reference, we would like to record the following immediate observation.

Observation 8.6.10 The only regular subdomain of the Riemann sphere $\widehat{\mathbb{C}}$ is $\widehat{\mathbb{C}}$ itself.

A crucial fact is that, according to the previous section, the Euler characteristic $\chi(\Delta)$ is defined for each regular subdomain Δ of D.

Of course, if a subdomain Δ of D satisfies (1) but not (2), we can adjoin to Δ those sets Γ_j which do not intersect the complement of D to form a regular subdomain of D, which we denote by Δ_*. Obviously, $\chi(\Delta_*) \geq \chi(\Delta)$; in fact, $\chi(\Delta_*) = \chi(\Delta) + k$, where $k \geq 0$ is the number of adjoined sets Γ_j.

We want to consider D as a limit of regular subdomains; as no canonical sequence of subdomains of D presents itself, it is best to reject the idea of a sequential limit and to consider instead convergence with respect to the directed set (or net) of regular subdomains. There is no need for great generality here and the details are quite simple and explicit. First, we shall prove the following.

Lemma 8.6.11 *If D is a subdomain of $\widehat{\mathbb{C}}$, then*

(a) *any compact subset of D lies in some regular subdomain of D*
and
(b) *if Δ_1 and Δ_2 are regular subdomains of D, then there is a regular subdomain Δ of D which contains $\Delta_1 \cup \Delta_2$.*

Proof If ξ is an arbitrarily chosen point in D, then the map $\widehat{\mathbb{C}} \ni z \mapsto (z - \xi)^{-1} \in \widehat{\mathbb{C}}$ sends ξ to ∞, so we may assume without loss of generality that $\infty \in D$. Let n be a positive integer and cover the plane with a square grid (including the axes), each square having diameter $1/n$.

If $D = \widehat{\mathbb{C}}$, the lemma is immediate by virtue of Observation 8.6.10. So, suppose that D is a proper subset of $\widehat{\mathbb{C}}$. Let K_n be the union of those closed squares that intersect $\widehat{\mathbb{C}} \backslash D$. If $n \geq 1$ is large enough, say $n \geq q$, then $\infty \notin K_n$. Let D_n be the connected component of $\widehat{\mathbb{C}} \backslash K_n$ that contains ∞. Then it is easy to see that $(D_n)_{n=q}^{\infty}$ is an ascending sequence of regular subdomains of D whose union $\bigcup_{n=q}^{\infty} D_n$ is D.

Having this, the rest of the proof is straightforward. Given any compact subset K of D, the family $\{D_n\}_{n=q}^{\infty}$ is an open cover of K and so is covered by a finite collection of the sets D_n. As the sequence $\left(D_n\right)_{n=q}^{\infty}$ is ascending, this finite collection contains a largest domain, say D_m, and then D_m contains K. This proves (a).

Finally, (b) follows from (a) for if Δ_1 and Δ_2 are regular subdomains of D, then $\overline{\Delta}_1 \cup \overline{\Delta}_2$ is a compact subset of D and so, by (a), it lies in some regular subdomain of D. ■

Lemma 8.6.11 states that the class $\mathcal{R}(D)$ of all regular subdomains of D is a net with the order given by the (direct) inclusion relation.

Our next task is to show that the function

$$\mathcal{R}(D) \ni \Delta \longmapsto \chi(\Delta)$$

is monotone decreasing, and, so, tends to a limit, which may be $-\infty$. We shall prove the following.

Lemma 8.6.12 *The Euler characteristic function*

$$\mathcal{R}(D) \ni \Delta \longmapsto \chi(\Delta)$$

is decreasing. Explicitly, if Δ_1 and Δ_2 are regular subdomains of D such that $\Delta_1 \subseteq \Delta_2$, then

$$\chi(\Delta_2) \leq \chi(\Delta_1).$$

Proof Let $V_1, \ldots, V_m$ be the connected components of $\widehat{\mathbb{C}} \backslash \Delta_1$ and $W_1, \ldots, W_n$ be the connected components of $\widehat{\mathbb{C}} \backslash \Delta_2$. Since $\Delta_1 \subseteq \Delta_2$, we have that

$$W_1 \cup \cdots \cup W_n \subseteq V_1 \cup \cdots \cup V_m.$$

For each $j \in \{1, \ldots, m\}$, fix a point $z_j \in V_j \backslash D$. As z_j is not in D, it lies in some set W_k, $1 \leq k \leq n$, and so W_k (being connected) lies in V_j. It follows that each set V_j, $j \in \{1, \ldots, m\}$, contains some set W_k with $k \in \{1, \ldots, n\}$. Hence, $m \leq n$. The given inequality now follows since, using (8.43), we get that

$$\chi(\Delta_2) = 2 - n \leq 2 - m = \chi(\Delta_1).$$

The proof is complete. ■

Lemma 8.6.12 states that the function

$$\chi : \mathcal{R}(D) \longrightarrow \{2, 1, 0, -1, -2, \ldots, -\infty\},$$

defined on the net $\mathcal{R}(D)$, is monotone decreasing. Thus, we have the following.

Proposition 8.6.13 *For any subdomain D of $\widehat{\mathbb{C}}$, the limit*

$$\chi(D) := \lim_{\Delta \in \mathcal{R}(D)} \chi(\Delta)$$

exists and

$$\chi(D) = \inf\{\chi(\Delta) : \Delta \in \mathcal{R}(D)\}.$$

Quite explicitly, either:

(1) $\chi(D) = -\infty$, and there are regular subdomains Δ_n of D with $\chi(\Omega_n) \to -\infty$ or, equivalently, with the connectivity of Δ_n diverging to $+\infty$; or else
(2) there is some regular subdomain D_* of D such that

$$\chi(D_*) = \chi(D) > -\infty;$$

then (from Lemma 8.6.12) $\chi(\Delta) = \chi(D)$ whenever Δ is a regular subdomain which contains D_*.

If D is a simply connected domain, then ∂D is connected and each regular subdomain Δ can only have at most one complementary connected component, regardless of the nature of ∂D: no component if $D = \widehat{\mathbb{C}}$ and always one if $D \neq \widehat{\mathbb{C}}$. So, using also Theorem 8.1.2, the Riemann Mapping Theorem, we have the following.

Theorem 8.6.14 *A domain D in $\widehat{\mathbb{C}}$ is simply connected if and only if either $\chi(D) = 2$ or $\chi(D) = 1$. In the former case, $D = \widehat{\mathbb{C}}$, while in the latter it is either $\mathbb{C}$ or it is conformally equivalent to the unit disk $\mathbb{D}$.*

More generally, if D has connectivity $k \geq 0$, then $\chi(D) = 2 - k$ for all sufficiently large regular subdomains Δ of D; so,

$$\chi(D) = 2 - k, \tag{8.57}$$

again irrespective of the complexity of ∂D.

Furthermore, the current definition of the Euler characteristic of an open domain in $\widehat{\mathbb{C}}$ coincides with that of (8.41) in the case that D is a bordered surface. Also, we now know that the Euler characteristic of open domains is a topological invariant.

8.6.4 Riemann–Hurwitz Formula: The General $\widehat{\mathbb{C}}$ Case

The main theorem of this subsection and the entire Section 8.6 is the following.

Theorem 8.6.15 (General Riemann–Hurwitz Formula) *Let S_1 and S_2 be two domains in $\widehat{\mathbb{C}}$, i.e., open connected subsets of $\widehat{\mathbb{C}}$. Let*

$$R\colon S_1 \to S_2$$

be a proper analytic function which has a (unique) continuous extension from $\overline{S}_1$ to $\overline{S}_2$. Let V be an open connected subset of S_2. If U is a connected component of $R^{-1}(V)$, then

$$\chi(U) + \delta_R(U) = \deg\bigl(R|_U\bigr)\chi(V). \tag{8.58}$$

Proof Fix a point $w \in V\backslash R(\mathrm{Crit}(R))$. Because of Lemma 8.6.11(a) there exists Δ_0, a regular subdomain of U such that

(1) $\mathrm{Crit}(R) \cap U \subseteq \Delta_0$ and
(2) $\overline{U} \cap R^{-1}(w) \subseteq \Delta_0$.

Again because of Lemma 8.6.11(a) and since $R(\overline{\Delta}_0)$ is a compact subset of V, there exists a regular subdomain Δ_1 of V such that

$$R(\overline{\Delta}_0) \subseteq \Delta_1. \tag{8.59}$$

We shall prove the following

Claim 1°.

$$\Delta_2 := U \cap R^{-1}(\Delta_1)$$

is a regular subdomain of U and $\Delta_0 \subseteq \Delta_2$.

Proof Since Δ_0 is connected and since, by (8.59), $R(\Delta_0) \subseteq \Delta_1$, there exists $\Delta_2^{'}$, a unique connected component of $R^{-1}(\Delta_1)$ such that

$$\Delta_0 \subseteq \Delta_2'. \tag{8.60}$$

Since every connected component of $R^{-1}(\Delta_1)$ is either disjoint from U or is contained in U, and since any such component $\Delta_2^{''}$ contained in U intersects $R^{-1}(w)$ (as, by Theorem 8.6.3, $R(\Delta_2^{'}) = \Delta_1$), it follows from (2) that $\Delta_2^{''} \cap \Delta_0 \neq \emptyset$. But then, by (8.60), $\Delta_2^{''} \cap \Delta_2^{'} \neq \emptyset$. Consequently, $\Delta_2^{''} = \Delta_2^{'}$. We have, thus, proved that

$$\Delta_2^{'} = U \cap R^{-1}(\Delta_1).$$

So, Δ_2 is a subdomain of U and, invoking (8.60), $\Delta_0 \subseteq \Delta_2$. Furthermore, since, by Lemma 8.6.4, $\Delta_2^{'}$ is an open bordered surface, and since $\Delta_2 = \Delta_2^{'}$, we get that Δ_2 is a bordered surface. We are, thus, left to show that the subdomain Δ_2 of U is regular. Proving this, let W be a connected component of $\widehat{\mathbb{C}}\backslash\Delta_2$. Seeking contradiction, suppose that $W \cap (\widehat{\mathbb{C}}\backslash U) = \emptyset$. This means that

$$W \subseteq U. \tag{8.61}$$

We know that $R(W)$ is a connected set contained in $V\backslash\Delta_1$. Let W_1 be the connected component of $V\backslash\Delta_1$ containing $R(W)$ and W_2 be the connected component of $\widehat{\mathbb{C}}\backslash\Delta_1$ containing W_1. In particular, W_2 is a connected component of $\widehat{\mathbb{C}}\backslash\Delta_1$; therefore, since Δ_1 is a regular subdomain of V, we have that

$$W_2 \cap (\widehat{\mathbb{C}}\backslash V) \neq \emptyset. \tag{8.62}$$

Since $W_1 \subseteq V \cap W_2$, we have that

$$\partial W_1 \subseteq \partial V \cup \partial W_2.$$

Seeking contradiction, suppose that

$$\partial W_1 \cap \partial V = \emptyset.$$

Thus, then

$$\partial W_1 \subseteq \partial W_2.$$

Hence, $W_2 = W_1 \cup (W_2\backslash\overline{W_1})$. Since W_2 is connected and $W_1 \neq \emptyset$, this implies that $W_2\backslash\overline{W}_1 = \emptyset$. Equivalently, $W_2 \subseteq \overline{W}_1$. Since both W_2 and W_1 are open, this yields $W_2 \subseteq W_1$. Thus, $W_2 = W_1$, whence $W_2 \subseteq V$, contrary to (8.62). We have, thus, proved that

$$\partial W_1 \cap \partial V \neq \emptyset. \tag{8.63}$$

Being still in the process of proving Claim 1°, we shall now prove the following.

Claim 2°. W is a connected component of $R^{-1}(W_1)$ contained in U.

Proof $W \subseteq U$ by (8.61). Furthermore,

$$W \subseteq U \cap R^{-1}(W_1) \subseteq U \cap R^{-1}(V\backslash\Delta_1) = U\backslash\Delta_2 \subseteq \widehat{\mathbb{C}}\backslash\Delta_2.$$

Therefore, since W is a connected component of $\widehat{\mathbb{C}}\backslash\Delta_2$, it is also a connected component of $U \cap R^{-1}(W_1)$. Since each connected component of $R^{-1}(W_1)$ is either contained in U or is disjoint from U, W is, thus, a connected component of $R^{-1}(W_1)$ contained in U. Claim 2° is proved. ■

It follows from this claim and Theorem 8.6.3 that $R(\partial W) = \partial W_1$. Thus, invoking (8.63), we get that $R(\partial W) \cap \partial V \neq \emptyset$. Hence,

$$R(\partial W) \cap (\widehat{\mathbb{C}} \backslash V) \neq \emptyset. \tag{8.64}$$

But $\partial W \subseteq \partial \Delta_2 \subseteq \overline{U} \cap \overline{\Delta_2} \subseteq \overline{U} \cap R^{-1}(\overline{\Delta_1}) \subseteq \overline{U} \cap R^{-1}(V) \subseteq U$. So, $R(\partial W) \subseteq R(U) = V$, contrary to (8.64). The proof of Claim 1° is complete. ■

Now, since, by Claim 1°, $\Delta_0 \subseteq \Delta_2 \subseteq U$, it follows from Lemma 8.6.12 that

$$\chi(\Delta_0) \geq \chi(\Delta_2) \geq \chi(U), \tag{8.65}$$

and from (1) of this proof that

$$\delta_R(\Delta_0) = \delta_R(\Delta_2) = \delta_R(U). \tag{8.66}$$

Because of Claim 1°, Theorem 8.6.6 is applicable to give, with the use of items (2) and (1) of this proof, along with (8.65), that

$$\begin{aligned} \deg(R|_U)\chi(\Delta_1) &= \deg(R|_{\Delta_2})\chi(\Delta_1) = \chi(\Delta_2) + \delta_R(\Delta_2) \\ &= \chi(\Delta_2) + \delta_R(U) \geq \chi(U) + \delta_R(U). \end{aligned} \tag{8.67}$$

Now the only requirement on Δ_1 is (8.59), so, because of Lemmas 8.6.11 and 8.6.12, the infinimum of $\chi(\Delta_1)$ over all regular subdomains Δ_1 of V is $\chi(V)$. Therefore,

$$\deg\big(R|_U\big)\chi(V) \geq \chi(U) + \delta_R(U). \tag{8.68}$$

On the other hand, (8.67) and (8.65) along with Lemma 8.6.12 give

$$\deg\big(R|_U\big)\chi(V) \leq \deg\big(R|_U\big)\chi(\Delta_1) = \chi(\Delta_2) + \delta_R(U) \leq \chi(\Delta_0) + \delta_R(U). \tag{8.69}$$

Since the only requirements on Δ_0 are (1) and (2), because of Lemmas 8.6.11 and 8.6.12, we see that the supremum of $\chi(\Delta_0)$ over all such domains Δ_0 is $\chi(U)$. Therefore, (8.69) yields

$$\deg\big(R|_U\big)\chi(V) \leq \chi(U) + \delta_R(U).$$

Along with (8.68), this completes the proof of Theorem 8.6.15. ■

Remark 8.6.16 An important straightforward observation is that all the corollaries, Corollary 8.6.7 through Corollary 8.6.9, hold true under the weaker hypotheses of Theorem 8.6.15, rather than those of Theorem 8.6.6. The proofs are essentially the same. We list all these corollaries below for the convenience of the reader and ease of reference.

Corollary 8.6.17 *With the hypotheses of Theorem 8.6.15, if V is simply connected and different from $\widehat{\mathbb{C}}$, then*

$$\delta_U(R) \geq \deg(R|_U) - 1,$$

with equality holding if and only if U is simply connected. Then, i.e., if U is simply connected, also

$$\#\big(\mathrm{Crit}(R|_U)\big) \leq \deg(R|_U) - 1. \tag{8.70}$$

Corollary 8.6.18 *With the hypotheses of Theorem 8.6.15, if V is simply connected and U contains no critical points of R, then the map $R|_U : U \to V$ is a conformal homeomorphism. In particular, U is simply connected.*

Corollary 8.6.19 *With the hypotheses of Theorem 8.6.15, if V is simply connected and different from $\widehat{\mathbb{C}}$ and if the map $R|_U : U \to V$ has only one critical point, which we denote by c, then*

(a) *U is simply connected.*
(a′) *If V is conformally equivalent to the unit disk $\mathbb{D}$, then U is also conformally equivalent to $\mathbb{D}$.*
(b)

$$\deg\big(R|_U\big) = p_c.$$

Item (a′) of this corollary immediately follows from item (a) since the possibility of $U = \mathbb{C}$ is ruled out by Liouville's Theorem and the possibility of $U = \widehat{\mathbb{C}}$ is excluded by the noncompactness of V.

We will frequently use, often without explicit invoking, the following reformulation of Corollary 8.6.18.

Corollary 8.6.20 *With the hypotheses of Theorem 8.6.15, if V is simply connected, $\xi \in S_1$, $R(\xi) \in V$, and the (unique) connected component of $R^{-1}(V)$ containing ξ contains no critical points of R, then there exists a unique holomorphic map $\psi : V \longrightarrow S_1$ such that*

$$R \circ \psi = \mathrm{Id}_V \quad \text{and} \quad \psi(R(\xi)) = \xi.$$

Moreover, $\psi(V)$ is the connected component of $R^{-1}(V)$ containing ξ and both maps $\psi : V \longrightarrow \psi(V)$ and $R|_{\psi(V)} : \psi(V) \longrightarrow V$ are conformal homeomorphisms.

The map $\psi : V \longrightarrow \psi(V)$ will be referred to throughout the whole book as the unique holomorphic branch of R^{-1} defined on V and sending $R(\xi)$ to ξ. It will be denoted as R_ξ^{-1}.

Remark 8.6.21 Since, for every complex torus $\mathbb{T}$, the projection $\Pi_\mathbb{T} : \mathbb{C} \longrightarrow \mathbb{T}$ is an isometry from any ball $B(w, 8u_\mathbb{T})$, $w \in \mathbb{C}$, onto $B(\Pi(w), 8u_\mathbb{T})$, we immediately conclude that all results of this section continue to be true if S_1 and S_2 are any domains in $\mathbb{T}$ with diameters smaller than $8u_\mathbb{T}$.

9

Invariant Measures for Holomorphic Maps f in $\mathcal{A}(X)$ or in $\mathcal{A}^w(X)$

Let Y be a parabolic Riemann surface. Let X be a compact subset of Y.

Definition 9.0.1 We say that $f \in \mathcal{A}(X)$ if and only if

(a) $f: X \to X$ is a continuous map.
(b) $f: X \to X$ can be extended to a holomorphic map from $U(f) = U(f,X)$ to Y, where $U(f)$ is some open neighborhood of X in Y.
(c) $X_*(f) \cap \mathrm{Crit}(f) = \emptyset$.
(d) X is a subset of the Julia sets $J(f)$ of the map $f: U(f) \longrightarrow Y$. We recall that this means that, whenever $V \subseteq Y$ is an open set intersecting X and all iterates $f^n: V \to Y$ are well defined, no subsequence of the sequence $\left(f^n|_V\right)_{n=1}^{\infty}$ forms a normal family.
(e) Considering the analytic map $f: U(f) \longrightarrow Y$, there exists $\iota_f \in (0, u_Y)$ such that if $x \in X$ and $\mathrm{Comp}(x, f, 2\iota_f)$ is the connected component of $f^{-1}(B(f(x), 2\iota_f))$ containing x, then the map $f|_{\mathrm{Comp}(x,f,2\iota_f)}: \mathrm{Comp}(x, f, 2\iota_f) \longrightarrow B(f(x), 2\iota_f)$ is proper.

If, in addition, the map $f: X \to X$ has no critical points, then we say that it belongs to $\mathcal{A}_0(X)$. Then, by the previous paragraph, the map $f: U(f) \longrightarrow Y$ has no critical points. If $X_*(f) = \emptyset$, then we say that f belongs to $\mathcal{A}_*(X)$.

Finally, we define

$$\mathcal{A}_{0,*}(X) = \mathcal{A}_0(X) \cap \mathcal{A}_*(X).$$

Sometimes, instead of (d), we will merely assume that the following condition, weaker than (d), holds.

(d') If $V \subseteq Y$ is an open set intersecting X and all iterates $f^n: V \to Y$ are well defined, then they do not form a normal family.

We will then say that $f \in \mathcal{A}^w(X)$. We then naturally define the sets $\mathcal{A}_0^w(X)$, $\mathcal{A}_*^w(X)$, and $\mathcal{A}_{0,*}^w(X)$ similarly to be respective sets $\mathcal{A}_0(X)$, $\mathcal{A}_*(X)$, and $\mathcal{A}_{0,*}(X)$.

Remark 9.0.2 Shrinking then the set $U(f)$ if necessary, we may and we always will assume that

1. all critical points of the map $f : U(f) \longrightarrow Y$ belong to X. In particular, the set $\mathrm{Crit}(f)$ is finite; and
2. the closure $\overline{U(f)}$ is a compact subset of Y.

9.1 Preliminaries

The first technical result of this section, which is a consequence of condition (d) or (d′), is the following.

Lemma 9.1.1 *Let Y be a parabolic Riemann surface. Let X be a compact subset of Y and $f \in \mathcal{A}^w(X)$. If $V \subseteq Y$ is an open set intersecting X and all iterates $f^n : V \to Y$ are well defined, then*

$$\limsup_{n\to\infty} \mathrm{diam}(f^n(V)) \geq \mathrm{diam}(Y).$$

Furthermore, if $f \in \mathcal{A}(X)$, then

$$\liminf_{n\to\infty} \mathrm{diam}(f^n(V)) \geq \mathrm{diam}(Y).$$

Proof Proving the first assertion, suppose for a contradiction that

$$\limsup_{n\to\infty} \mathrm{diam}(f^n(V)) < \mathrm{diam}(Y).$$

Disregarding finitely many terms of this sequence, we may assume without loss of generality that

$$D := \sup\left\{\mathrm{diam}(f^n(V)) : n \geq 0\right\} < \mathrm{diam}(Y).$$

Then

$$\bigcup_{n=0}^{\infty} f^n(V) \subseteq B(X, D).$$

Let $\left(f^{n_k}\right)_{k=1}^{\infty}$ be an arbitrary sequence of iterates of f. Fix any point $w \in X$. Since X is compact, passing to a subsequence, we may assume without loss of generality that the sequence $\left(f^{n_k}(w)\right)_{k=1}^{\infty}$ converges. Denote its limit by ξ. Fix any number $D' \in (D, \mathrm{diam}(Y))$. Then, disregarding finitely many terms of the sequence $\left(f^{n_k}\right)_{k=1}^{\infty}$, we will have that

$$f^{n_k}(V) \subseteq B(\xi, D') \subseteq \overline{B(\xi, D')} \subsetneq Y$$

for all $k \geq 1$. Hence, it follows from Montel's Theorem II, Theorem 8.1.16, that the sequence $\left(f^{n_k}|_V\right)_{k=1}^{\infty}$ is normal. Thus, it either has a convergent subsequence or it diverges off compact sets of Y; therefore, the family $\left(f^n|_V\right)_{n=1}^{\infty}$ is normal, contrary to item (d′). The proof of the first assertion is complete.

The proof of the second assertion is analogous and left to the reader as an exercise. ■

Our next technical result, interesting on its own, is the following.

Lemma 9.1.2 *Let Y be a parabolic Riemann surface. If $X \subseteq Y$ is a compact set and $f \in \mathcal{A}_0^w(X)$, then the series*

$$\sum_{n=1}^{\infty} |(f^n)'(z)|^{\frac{1}{3}}$$

diverges for all $z \in X$.

Proof By our hypothesis, there exists $\varepsilon \in \left(0, \min\{\iota_f, \mathrm{diam}(Y)/2\}\right)$ such that, for every $w \in X$, we have that $B(w,\varepsilon) \subseteq U(f)$ and the map f restricted to the ball $B(w,\varepsilon)$ is one-to-one. Since, by decreasing the set $U(f)$ if necessary, the map $f\colon U(f) \longrightarrow Y$ is uniformly continuous, there exists $0 < \alpha < 1$ such that, for every $x \in X$, we have that

$$f(B(x,\alpha\varepsilon)) \subseteq B(f(x),\varepsilon). \tag{9.1}$$

Suppose, on the contrary, that the series $\sum_{n=1}^{\infty} |(f^n)'(z)|^{\frac{1}{3}}$ converges for some $z \in X$. Then there exists $q \geq 1$ such that

$$\sup\left\{\left(2|(f^n)'(z)|\right)^{\frac{1}{3}} : n \geq q\right\} < 1.$$

Choose $0 < \varepsilon_1 = \varepsilon_2 = \cdots = \varepsilon_q < \alpha\varepsilon$ so small that, for every $n = 1, 2, \ldots, q$,

$$\text{the map } f^n \text{ restricted to the ball } B(z,\varepsilon_n) \text{ is one-to-one.} \tag{9.2}$$

and

$$f^n(B(z,\varepsilon_n)) \subseteq B(f^n(z),\varepsilon). \tag{9.3}$$

For every $n \geq q$, define ε_{n+1} inductively by

$$\varepsilon_{n+1} = \left(1 - \left(2|(f^n)'(z)|\right)^{\frac{1}{3}}\right)\varepsilon_n. \tag{9.4}$$

Then $0 < \varepsilon_n < \alpha\varepsilon$ for every $n \geq 1$. Assume that (9.2) and (9.3) are satisfied for some $n \geq q$. Then, by virtue of the Koebe Distortion Theorem, Analytic Version for Parabolic Surfaces (Theorem 8.3.11) and (9.4), we have that

$$f^n(B(z,\varepsilon_{n+1})) \subseteq B(f^n(z),r_n), \tag{9.5}$$

where

$$r_n = \varepsilon_{n+1}|(f^n)'(z)|\frac{2}{(1-\varepsilon_{n+1}/\varepsilon_n)^3} = \frac{2\varepsilon_{n+1}|(f^n)'(z)|}{2|(f^n)'(z)|} = \varepsilon_{n+1} < \alpha\varepsilon. \tag{9.6}$$

Then, since f is injective on $B(f^n(z),\varepsilon)$, (9.2) is satisfied for $n+1$. Also using (9.1), we get that

$$f^{n+1}\big(B(z,\varepsilon_{n+1})\big) = f\big(f^n(B(z,\varepsilon_{n+1}))\big) \subseteq f\big(B(f^n(z),\alpha\varepsilon)\big) \subseteq B\big(f^{n+1}(z),\varepsilon\big).$$

Thus, (9.3) is satisfied for $n+1$. Since the sequence $(\varepsilon_n)_{n=1}^{\infty}$ is monotone decreasing and is bounded below by 0, it converges. Denote its limit by ε_∞. Then

$$\varepsilon_\infty \leq \varepsilon < \mathrm{diam}(Y)/2. \tag{9.7}$$

Since the series $\sum_{k=1}^{\infty} |(f^k)'(z)|^{\frac{1}{3}}$ converges, it also follows from (9.4) that

$$\varepsilon_\infty > 0.$$

We then directly derive from (9.3), (9.5), (9.6), and (9.7) that

$$\sup\big\{\mathrm{diam}\big(f^n(B(z,\varepsilon_\infty))\big) : n \geq 1\big\} \leq 2\varepsilon < \mathrm{diam}(Y).$$

This, however, contradicts Lemma 9.1.1 and finishes the proof of Lemma 9.1.2. ■

Let μ be a Borel probability f-invariant ergodic measure on X. The characteristic Lyapunov exponent $\chi_\mu(f)$ of the map f with respect to the measure μ is

$$\chi_\mu(f) := \int_X \log|f'(x)|\,d\mu(x) < +\infty.$$

As an immediate consequence of Lemma 9.1.2 and of the ergodic part of the Birkhoff Ergodic Theorem (Theorem 2.3.2), we get the following.

Corollary 9.1.3 *If Y is a parabolic Riemann surface, $X \subseteq Y$ is compact, and $f \in \mathcal{A}_0^w(X)$, then*

$$\chi_\mu(f) \geq 0$$

for every Borel probability f-invariant ergodic measure μ on X.

We would like to remark that such an inequality was proved by Przytycki in [**P2**] for all rational functions $f: \widehat{\mathbb{C}} \longrightarrow \widehat{\mathbb{C}}$ and all f-invariant measures supported on the Julia set of f.

We would now like to derive one technical-looking consequence of Corollary 9.1.3. It will, however, be an indispensable ingredient in the conclusion of the proof of Theorem 17.6.7 in Volumne 2, asserting the existence of conformal measures (with some additional properties) for general elliptic functions.

Corollary 9.1.4 *If Y is a parabolic Riemann surface, $X \subseteq Y$ is compact, $f \in \mathcal{A}_0^w(X)$, and μ is a Borel probability f-invariant ergodic measure on X, then*

$$\limsup_{n\to\infty} \left|(f^n)'(x)\right| \geq 1$$

for μ-a.e. $x \in X$.

Proof If the measure μ is supported on a single periodic orbit of f, then, by virtue of Corollary 9.1.3, the modulus of the multiplier of this orbit must be ≥ 1 and we are done. So, suppose that μ is not supported on a single periodic orbit of f. Seeking contradiction, assume that there exists a Borel measurable subset Z_1 of X such that $\mu(Z_1) > 0$ and

$$\limsup_{n\to\infty} \left|(f^n)'(z)\right| < 1$$

for all $z \in Z_1$. Hence, there exist a number $\eta \in (0, 1)$ and a Borel measurable subset Z_2 of Z_1 such that $\mu(Z_2) > 0$ and

$$\limsup_{n\to\infty} \left|(f^n)'(z)\right| < \eta$$

for all $z \in Z_2$. We conclude from this that there exist an integer $N \geq 1$ and a Borel measurable subset Z_3 of Z_2 such that $\mu(Z_3) > 0$ and

$$\left|(f^n)'(z)\right| < \eta \tag{9.8}$$

for all $n \geq N$ and all $z \in Z_3$. Since the measure μ is ergodic and it is not supported on a single periodic orbit of f, it does not charge any periodic orbit of f. Since also, because the map $f: U(f) \longrightarrow Y$ is holomorphic, the set of periodic orbits of f is countable, it follows that $\mu(Z_4) = \mu(Z_3) > 0$, where Z_4 results from Z_3 by removing all periodic points of f from Z_3. Let $w \in Z_3$ be a Lebesgue density point with respect to the measure μ produced in Lebesgue Density Theorem (Theorem 1.3.7). Since w is not a periodic point of f, there exists $r_1 > 0$ such that the first return of w to $B(w, r_1)$, under forward iterates of f, is $\geq N + 1$ (possibly $+\infty$). So, since the map f is continuous, there exists $r_2 \in (0, r_1]$ such that, for every $z \in B(w, r_2)$, the first return time $\tau(z)$ of

z to $B(w,r_2)$ is $\geq N$. So, if $Z_5 := Z_4 \cap B(w,r_2)$, then, with the notation and terminology of Section 2.2, it follows from (9.8) that

$$\log|f'|_{Z_5} = \log\left|f'_{Z_5}(z)\right| < \log\eta < 0$$

for all $z \in Z_5$. So, since also, by the choice of w and the Lebesgue Density Theorem (Theorem 1.3.7), $\mu(Z_5) > 0$, we conclude from Proposition 2.2.9 that

$$\chi_\mu = \int_X \log|f'|\,d\mu = \mu(Z_5)\int_{Z_5}\log|f'|_{Z_5}\,d\mu_{Z_5} \leq \mu(Z_5)\log\eta < 0.$$

This, however, contradicts Corollary 9.1.3 and finishes the proof of Corollary 9.1.4. ■

9.2 Three Auxiliary Partitions

For the purpose of this section, for every Lebesgue measurable subset A of $\mathbb{R}$, denote by $|A|$ the Lebesgue measure of A. We start with the following technical fact, which is, however, interesting on its own. The results of the first two subsections are adapted from [**PU2**], while the third subsection basically follows Mañé's book [**M3**].

9.2.1 Boundary Partition

Lemma 9.2.1 *Every monotone increasing function $k\colon I \to \mathbb{R}$ defined on a bounded closed interval $I \subset \mathbb{R}$ is Lipschitz continuous at Lebesgue almost every point in I. More precisely, for every $\varepsilon > 0$, there exist a number $L > 0$ and a Lebesgue measurable set $A \subset I$ such that $|I\backslash A| < \varepsilon$ and the function $k\colon I \to \mathbb{R}$ is Lipschitz continuous at each point of A with the Lipschitz constant not exceeding L.*

Proof For every $y \in I$, let

$$L(y) := \sup\left\{z \in I\backslash\{y\}\colon \frac{|k(z)-k(y)|}{|z-y|}\right\}.$$

Of course, the function $I \ni y \longmapsto L(y) \in [0,+\infty]$ is Lebesgue measurable. Seeking a contradiction, suppose that the set

$$B := \{y \in I\colon L(y) = +\infty\}$$

has positive Lebesgue measure. Write $I := [a,b]$. Replacing B by its subset if necessary, we may assume without loss of generality that the set B is compact and contains neither a nor b. For every $y \in B$, choose $y' \in I\backslash\{y\}$ such that

$$\frac{|k(y) - k(y')|}{|y - y'|} > 2\frac{k(b) - k(a)}{|B|}. \tag{9.9}$$

Now replace every such pair y, y' by a pair x, x' such that $I \supseteq (x, x') \supset [y, y']$, and the points x, x' are so close to the respective points y, y' that (9.9) still holds with x, x'. However, if (at least) one of the points y or y' happens to be an endpoint of I, we make no replacement. Now, from the family of intervals (x, x'), choose a finite family $\mathcal{I}$ covering our compact set B. From this family, it is, in turn, possible to choose a subfamily that is a union of two subfamilies $\mathcal{I}_1$ and $\mathcal{I}_2$, each of which consists of mutually disjoint open intervals. Also using monotonicity of the function $k\colon I \to \mathbb{R}$ and (9.9), we get, for $i = 1, 2$, that

$$k(b) - k(a) \geq \sum_{(w,w') \in \mathcal{I}^i} (k(w') - k(w)) > 2\frac{k(b) - k(a)}{|B|} \sum_{(w,w') \in \mathcal{I}^i} (w' - w).$$

Hence, taking into account that $\mathcal{I}^1 \cup \mathcal{I}^2$ covers B, we get that

$$\begin{aligned} 2(k(b) - k(a)) &> 2\frac{k(b) - k(a)}{|B|} \sum_{(w,w') \in \mathcal{I}^1 \cup \mathcal{I}^2} (w' - w) \geq 2\frac{k(b) - k(a)}{|B|} |B| \\ &= 2(k(b) - k(a)), \end{aligned}$$

which is a contradiction. Thus,

$$|B| = 0.$$

Hence,

$$I \backslash B = \bigcup_{n=1}^{\infty} L^{-1}([0, n]).$$

Since also the measurable sets $L^{-1}([0, n])$, $n \geq 1$, form an ascending sequence, we conclude that, for every $\varepsilon > 0$, there exists $n_\varepsilon \geq 1$ such that

$$|L^{-1}([0, n_\varepsilon]|) \geq |I \backslash B| - \varepsilon = |I| - \varepsilon.$$

So, taking $A := L^{-1}([0, n_\varepsilon])$ finishes the proof. ■

Corollary 9.2.2 *For every Borel probability measure ν on a compact metric space (X, ρ) and for every $r > 0$, there exists a finite partition $\mathcal{P} = \{P_j\}_{j=1}^M$ of X consisting of Borel sets with the following three properties.*

(1)

$$\nu(P_j) > 0 \tag{9.10}$$

for all $j = 1, 2, \ldots, M$.

(2)

$$\operatorname{diam}(\mathcal{P}) < r. \tag{9.11}$$

(3) *There exists a constant $C > 0$ such that, for every $\eta > 0$,*

$$\nu(\partial_\eta \mathcal{P}) \leq C\eta, \tag{9.12}$$

where

$$\partial_\eta \mathcal{P} := \bigcap_{j=1}^{M} \left(\bigcup_{k \neq j} B(P_s, \eta) \right).$$

Proof Let $\{x_1, \dots, x_N\}$ be a finite $r/4$-spanning set in X. Fix $\varepsilon \in (0, r/4N)$. For each monotone increasing function,

$$I := [r/4, r/2] \ni t \longmapsto k_i(t) := \nu(B(x_i, t)),$$

$i = 1, 2, \dots, N$, apply Lemma 9.2.1, and consider respective constants L_i and sets A_i. Let

$$L := \max\{L_i, i = 1, \dots, N\}$$

and

$$A = \bigcap_{i=1,\dots,N} A_i.$$

The set A has positive Lebesgue measure by the choice of ε. So, we can choose its point s different from $r/4$ and $r/2$. Therefore, for all $0 < \eta < \eta_0 := \min\{s - r/4, r/2 - s\}$ and for all $i \in \{1, 2, \dots, N\}$, we have that

$$\nu(B(x_i, r_0 + \eta) \backslash B(x_i, s - \eta)) \leq 2L\eta.$$

Hence, putting

$$\Delta(a) := \bigcup_{i=1}^{N} \big(B(x_i, s + \eta) \backslash B(x_i, s - \eta) \big),$$

we get that $\nu(\Delta(\eta)) \leq 2LN\eta$. Put

$$B^+(x_i, s) := B(x_i, s) \text{ and } B^-(x_i, s) := X \backslash B(x_i, s).$$

Let Ω be the set of all functions $\kappa : \{1, \dots, N\} \to \{+, -\}$. Define

$$\mathcal{P} = \left\{ \bigcap_{i=1}^{N} B^{\kappa(i)}(x_i, s) : \kappa \in \Omega \right\}.$$

After removing from X a set of measure 0, the partition $\mathcal{P}$ covers X. Since $s \geq r/4$, the balls $B(x_i, s)$, $i - 1, 2, \dots, N$, cover X. Hence, for each nonempty $P_j \in \mathcal{P}$, at least one value of κ is equal to $+$. Thus, $\mathrm{diam}(P_j) \leq 2s < r$. We shall now show that

$$\partial_\eta \mathcal{P} \subset \Delta(a). \tag{9.13}$$

Indeed, let $x \in \partial_\eta \mathcal{P}$. Since $\mathcal{P}$ covers X, there exists j_0 such that $x \in P_{j_0}$; so, $x \notin P_j$ for all $j \neq j_0$. However, since $x \in \bigcup_{j \neq j_0} B(P_t, \eta)$, there exists $j_1 \neq j_0$ such that $\mathrm{dist}(x, P_{j_1}) < a$. Let $B := B(x_i, s)$ be such that

$$P_{j_0} \subset B^+ \quad \text{and} \quad P_{tj1} \subset B^-,$$

or vice versa.

In the case when

$$x \in P_{t_0} \subset B^+,$$

we get by the triangle inequality that $\rho(x, x_i) > s - \eta$ and, since $\rho(x, x_i) < s$, that

$$x \in \Delta(\eta).$$

In the case when

$$x \in P_{j_0} \subset B^-,$$

we have that

$$x \in B(x_i, s + \eta) \backslash B(x_i, s) \subset \Delta(\eta).$$

Formula (9.13) is therefore proved.

We, thus, conclude that

$$\nu(\partial_\eta \mathcal{P}) \leq \nu(\Delta(a)) \leq 2LNa$$

for all $\eta < \eta_0$. For $\eta \geq \eta_0$, it suffices to take $C \geq 1/\eta_0$. So, the corollary is proved, with $C := \max\{2LN, 1/\eta_0\}$. ■

9.2.2 Exponentially Large Partition

In this subsection, using the fruits of the previous one, we shall prove the following.

Corollary 9.2.3 *Let ν be a Borel probability measure on a compact metric space (X, ρ) and $T\colon X \longrightarrow X$ be a Borel endomorphism on X preserving the measure ν.*

Then, for every $r > 0$, there exists a finite partition $\mathcal{P} = \{P_j\}_{j=1}^M$ of X into Borel sets of positive measure ν with $\mathrm{diam}(\mathcal{P}) < r$ *such that, for every $\delta > 0$ and ν-a.e. $x \in X$, there exists an integer $N = N(x)$ such that, for every $n \geq n_0$,*

$$B\big(T^n(x), \exp(-n\delta)\big) \subset \mathcal{P}(T^n(x)). \tag{9.14}$$

Proof Let $\mathcal{P}$ be the partition from Corollary 9.2.2. Fix an arbitrary $\delta > 0$. Then, by Corollary 9.2.2,

$$\sum_{n=0}^{\infty} \nu\big(\partial_{\exp(-n\delta)}\mathcal{P}\big) \leq \sum_{n=0}^{\infty} C \exp(-n\delta) < \infty.$$

Hence, by the T-invariance of ν, we obtain that

$$\sum_{n=0}^{\infty} \nu\big(T^{-n}(\partial_{\exp(-n\delta)}\mathcal{P})\big) < \infty.$$

Applying now the Borel–Cantelli Lemma for the family $\big\{T^{-n}\big(\partial_{\exp(-n\delta)}\mathcal{P}\big)\big\}_{n=1}^{\infty}$, we conclude that, for ν-a.e. $x \in X$, there exists $N = N(x)$ such that, for every $n \geq N$, we have that $x \notin T^{-n}\big(\partial_{\exp(-n\delta)}\mathcal{P}\big)$. This means that $T^n(x) \notin \partial_{\exp(-n\delta)}\mathcal{P}$. Hence, by the definition of $\partial_{\exp(-n\delta)}\mathcal{P}$, if $T^n(x) \in P_j$ for some $j = 1, 2, \ldots, M$, then

$$T^n(x) \notin \bigcup_{k \neq j} B\big(P_k, \exp(-n\delta)\big).$$

Thus,

$$B\big(T^n(x), \exp -n\delta\big) \subset \mathcal{P}(T^n(x)).$$

The proof is finished. ■

9.2.3 Mañé Partition

In this subsection, basically following Mañé's book [**M3**], we construct the so-called Mañé partition, which will play an important role in the proof of a part of the Volume Lemma given in the next section. We begin with the following elementary fact.

Lemma 9.2.4 *If $x_n \in (0,1)$ for every $n \geq 1$ and $\sum_{n=1}^{\infty} nx_n < +\infty$, then*

$$\sum_{n=1}^{\infty} -x_n \log x_n < +\infty.$$

Proof Let $S := \{n \geq 1 \colon -\log x_n \geq n\}$. Then

$$\sum_{n=1}^{\infty} -x_n \log x_n = \sum_{n \notin S} -x_n \log x_n + \sum_{n \in S} -x_n \log x_n \leq \sum_{n=1}^{\infty} n x_n + \sum_{n \in S} -x_n \log x_n.$$

Since $n \in S$ means that $x_n \leq e^{-n}$ and since $\log t \leq 2\sqrt{t}$ for all $t \geq 1$, we have that

$$\sum_{n \in S} x_n \log \frac{1}{x_n} \leq 2 \sum_{n=1}^{\infty} x_n \sqrt{\frac{1}{x_n}} \leq 2 \sum_{n=1}^{\infty} e^{-\frac{1}{2}n} < \infty.$$

The proof is finished. ■

The next lemma is the main, and simultaneously the last, result of this subsection.

Lemma 9.2.5 *If μ is a Borel probability measure supported on a compact subset M of a finitely dimensional Riemannian manifold and $\rho \colon M \longrightarrow (0,1]$ is a measurable function such that* $\log \rho$ *belongs to $L^1(\mu)$, then there exists a countable measurable partition, called the Mañé partition, $\mathcal{P}$, of M such that $\mathrm{H}_\mu(\mathcal{P}) < +\infty$ and*

$$\operatorname{diam}(\mathcal{P}(x)) \leq \rho(x)$$

for μ-a.e. $x \in M$.

Proof Let q be the dimension of the Riemannian manifold containing M. Since M is compact, there exists a constant $C > 0$ such that, for every $0 < r < 1$, there exists a partition $\mathcal{P}_r$ of M of diameter $\leq r$ that consists of at most Cr^{-q} elements. For every $n \geq 0$, put

$$U_n := \left\{x \in M \colon e^{-(n+1)} < \rho(x) \leq e^{-n}\right\}.$$

Since $\log \rho$ is a nonpositive integrable function, we have that

$$\sum_{n=1}^{\infty} -n\mu(U_n) \geq \sum_{n=1}^{\infty} \int_{U_n} \log \rho \, d\mu = \int_M \log \rho \, d\mu > -\infty, \tag{9.15}$$

so that

$$\sum_{n=1}^{\infty} n\mu(U_n) < +\infty. \tag{9.16}$$

Define now $\mathcal{P}$ as the partition whose atoms are of the form $Q \cap U_n$, where $n \geq 0$ and $Q \in \mathcal{P}_{r_n}$, $r_n = e^{-(n+1)}$. Then

$$\mathrm{H}_\mu(\mathcal{P}) = \sum_{n=0}^{\infty} \left(- \sum_{U_n \supseteq P \in \mathcal{P}} \mu(P) \log \mu(P) \right).$$

But, for every $n \geq 0$,

$$\begin{aligned} - \sum_{U_n \supseteq P \in \mathcal{P}} \mu(P) \log \mu(P) &= \mu(U_n) \sum_{U_n \supseteq P \in \mathcal{P}} - \frac{\mu(P)}{\mu(U_n)} \log \left(\frac{\mu(P)}{\mu(U_n)} \right) \\ &\quad - \mu(U_n) \sum_{U_n \supseteq P \in \mathcal{P}} \frac{\mu(P)}{\mu(U_n)} \log(\mu(U_n)) \\ &\leq \mu(U_n)(\log C - q \log r_n) - \mu(U_n) \log \mu(U_n) \\ &\leq \mu(U_n) \log C + q(n+1)\mu(U_n) - \mu(U_n) \log \mu(U_n). \end{aligned}$$

Thus, summing over all $n \geq 0$, we obtain that

$$\mathrm{H}_\mu(\mathcal{P}) \leq \log C + q + q \sum_{n=0}^{\infty} n\mu(U_n) + \sum_{n=0}^{\infty} -\mu(U_n) \log \mu(U_n).$$

Therefore, looking at (9.16) and Lemma 9.2.4, we conclude that $\mathrm{H}_\mu(\mathcal{P})$ is finite. Also, if $x \in U_n$, then the atom $\mathcal{P}(x)$ is contained in some atom of P_{r_n}; therefore,

$$\mathrm{diam}(\mathcal{P}(x)) \leq r_n = e^{-(n+1)} < \rho(x).$$

Now the remark that the union of all the sets U_n is of measure 1 completes the proof. ■

9.3 Pesin's Theory

In this section, we work in the same general setting of this chapter and we follow the same notation as in the previous sections. We present here a (very special, for the maps in $\mathcal{A}^w(X)$) version of Pesin's Theory, whose foundations were laid down in the fundamental works of Pesin [**Pe1**] and [**Pe2**]. Since then, Pesin's Theory has been extended, generalized, refined, and cited in numerous articles, research books, and textbooks. Our exposition here follows Section 11.2 of the book [**PU2**].

We begin with the following.

Lemma 9.3.1 *If μ is a Borel finite measure on a metric space (M,ρ), a is an arbitrary point in M, and the function $M \ni z \longmapsto \log\big(\rho(z,a)\big)$ restricted to every sufficiently small open neighborhood of a is μ-integrable, then, for every $C > 0$ and every $0 < t < 1$,*

$$\sum_{n\geq 1} \mu(B(a, Ct^n)) < +\infty.$$

Proof By our hypotheses, there exists $s \in (0,1)$ such that the function $B(a,s) \ni z \longmapsto \log\big(\rho(z,a)\big)$ is μ-integrable. So, since μ is finite and since, given any $t \in (0,1)$, there exists $q \geq 1$ such that $Ct^n \leq s^n$ for all $n \geq q$, we may assume without loss of generality that $t \leq s$ and $C = 1$. Recall that, given $b \in \mathbb{R}^d$ and two numbers $0 \leq r < R$,

$$A(b;r,R) = \{z \in \mathbb{R}^d : r \leq \rho(z,b) < R\}.$$

Since $-\log(t^n) \leq -\log\big(\rho(z,a)\big)$ for every $z \in B(a,t^n)$, we get that

$$\begin{aligned}\sum_{n\geq 1} \mu\big(B(a,t^n)\big) &= \sum_{n\geq 1} n\mu\big(A\big(a;t^{n+1},t^n\big)\big) = \frac{-1}{\log t}\sum_{n\geq 1} -\log(t^n)\mu\big(A\big(a;t^{n+1},t^n\big)\big)\\ &\leq \frac{-1}{\log t}\int_{B(a,t)} -\log\big(\rho(z,a)\big)\,d\mu(z)\\ &< +\infty.\end{aligned}$$

The proof is finished. ■

Lemma 9.3.2 *Let Y be a parabolic Riemann surface, $X \subseteq Y$ be a compact set, and $f \in \mathcal{A}^w(X)$. If μ is a Borel finite measure on X and the function $\log|f'|$ belongs to $L^1(\mu)$, then, for every critical point c of f, the function*

$$X \ni z \longmapsto \log|z-c| \in \mathbb{R}$$

belongs to $L^1(\mu)$. If, in addition, μ is f-invariant, then, for every $w \in f(\mathrm{Crit}(f))$, the function

$$X \ni z \longmapsto \log|z-c| \in \mathbb{R}$$

also belongs to $L^1(\mu)$.

Proof The first assertion follows immediately from (8.35) and Remark 8.4.2.
Proving the second assertion, let

$$R := \min\{R(f,c) : c \in \mathrm{Crit}(f)\} \quad \text{and} \quad A := \max\{A(f,c) : c \in \mathrm{Crit}(f)\},$$

where the numbers $R(f,c)$ and $A(f,c)$ come from Section 8.4, and the minimum and maximum exist since the set $\mathrm{Crit}(f)$ is finite. For every

$c \in \mathrm{Crit}(f)$, let $B_c := B\big(c,(AR)^{1/p_c}\big)$. Decreasing $R > 0$ if necessary, we may, and we do, assume that the balls B_c, $c \in \mathrm{Crit}(f)$, are mutually disjoint. Since the set X is compact, the function $X \ni z \longmapsto \log|z-w|$ is bounded. Since μ is also finite, we only need to show that

$$\int_X \log|z-w|\,d\mu(z) > -\infty.$$

By invoking (8.34) and f-invariance of the measure μ, we get that

$$\begin{aligned}
&\int_X \log|z-w|\,d\mu(z)\\
&\quad= \int_X \log|T(z)-w|\,d\mu(z)\\
&\quad= \int_{X\setminus\bigcup_{c\in\mathrm{Crit}(f)\cap T^{-1}(w)} B_c} \log|T(z)-w|\,d\mu(z)\\
&\qquad+ \sum_{c\in\mathrm{Crit}(f)\cap T^{-1}(w)} \int_{B_c} \log|T(z)-T(c)|\,d\mu(z)\\
&\quad\geq -\log A\mu(X) + \int_{X\setminus\bigcup_{c\in\mathrm{Crit}(f)\cap T^{-1}(w)} B_c} \log\|T(z)-w|\,d\mu(z)\\
&\qquad+ \sum_{c\in\mathrm{Crit}(f)\cap T^{-1}(w)} p_c\int_{B_c} \log|z-c|\,d\mu(z);
\end{aligned}$$

this sum is $> -\infty$ since the integrand in the second summand is bounded below while each term in the sum over all $c \in \mathrm{Crit}(f)\cap T^{-1}(w)$ is bounded below by the already proven first assertion of our lemma. ■

Theorem 9.3.3 *Let $(Z,\mathfrak{F},\nu)$ be a measure space with an ergodic measure-preserving automorphism $T\colon Z \to Z$. Let Y be a parabolic Riemann surface, $X \subseteq Y$ be a compact set, and $f \in \mathcal{A}^w(X)$.*

Suppose that μ is an f-invariant ergodic measure on X with positive Lyapunov exponent $\chi_\mu(f)$. Suppose also that $h\colon Z \to X$ is a measurable mapping such that

$$\nu\circ h^{-1} = \mu \text{ and } h\circ T = f\circ h$$

ν-a.e.

Then, for ν-a.e. $z \in Z$, there exists $r(z) \in (0,\iota_f]$ such that the function $z \mapsto r(z)$ is measurable and the following is satisfied.

For every $n \geq 1$, there exists $f_{x_n}^{-n}\colon B(x,r(z)) \longrightarrow Y$, a holomorphic inverse branch of f^n sending $x := h(z)$ to $x_n := h(T^{-n}(z))$. Given, in addition, an *arbitrary* $\varepsilon > 0$, *there exists a constant* $K_\varepsilon(z) \in (0, +\infty)$ *independent of* n *and* z *such that*

$$K_\varepsilon^{-1}(z)\exp\big(-(\chi_\mu+\varepsilon)n\big) \le |\big(f_{x_n}^{-n}\big)'(y)| \le K_\varepsilon(z)\exp\big(-(\chi_\mu-\varepsilon)n\big)$$

and

$$\frac{\big|\big(f_{x_n}^{-n}\big)'(w)\big|}{\big|\big(f_{x_n}^{-n}\big)'(y)\big|} \le K$$

for all $y, w \in B(x, r(z))$. *K here is the Koebe constant of Theorem 8.3.9 corresponding to the scale* $1/2$.

Proof Suppose, first, that

$$\mu\left(\bigcup_{n\ge1} f^n(\mathrm{Crit}(f))\right) > 0.$$

Since the measure μ is ergodic, this implies that μ must be concentrated on a periodic orbit of an element $w \in \bigcup_{n\ge1} f^n(\mathrm{Crit}(f))$. This means that $w = f^q(c) = f^{q+k}(c)$ for some $q, k \ge 1$ and $c \in \mathrm{Crit}(f)$ and that

$$\mu\big(\big\{f^q(c), f^{q+1}(c), \dots, f^{q+k-1}(c)\big\}\big) = 1.$$

Since $\int \log|f'|\,d\mu > 0$, we have that $|(f^k)'(f^q(c))| > 1$. Thus, the theorem is obviously true for the set

$$h^{-1}\big(\big\{f^q(c), f^{q+1}(c), \dots, f^{q+k-1}(c)\big\}\big)$$

of ν measure 1. So, suppose that

$$\mu\left(\bigcup_{n\ge1} f^n(\mathrm{Crit}(f))\right) = 0.$$

Fix $\lambda \in \left(\exp\left(-\frac{1}{4}(\chi_\mu-\varepsilon)\right), 1\right)$. Consider $z \in Z$ such that $x = h(z) \notin \bigcup_{n\ge1} f^n(\mathrm{Crit}(f))$,

$$\lim_{n\to\infty}\frac{1}{n}\log\big|(f^n)'(h(T^{-n}(z)))\big| = \chi_\mu(f),$$

and $x_n = h(T^{-n}(z))$ belongs to $B(f(\mathrm{Crit}(f)), \iota_f\lambda^n)$ for finitely many ns only, where the number $\iota_f > 0$ comes from item (e) of Definition 9.0.1. We shall first demonstrate that the set of points satisfying these properties is of full measure ν. Indeed, the first requirement is satisfied by our hypothesis and the second is due to the ergodic part of the Birkhoff Ergodic Theorem (Theorem 2.3.2). In order to prove that the set of points satisfying the third condition has ν measure 1, notice that

$$\begin{aligned}\sum_{n\ge1}\nu\big(T^n(h^{-1}(B(f(\mathrm{Crit}(f)),\iota_f\lambda^n)))\big) &= \sum_{n\ge1}\nu\big(h^{-1}(B(f(\mathrm{Crit}(f)),\iota_f\lambda^n))\big)\\ &= \sum_{n\ge1}\mu(B(f(\mathrm{Crit}(f)),\iota_f\lambda^n))\\ &< +\infty,\end{aligned}$$

where the last inequality is due to Lemmas 9.3.2 and 9.3.1. The application of the Borel–Cantelli Lemma finishes the demonstration. Fix now an integer $n_1 = n_1(z) \ge 0$ so large that

$$x_n = h(T^{-n}(z)) \notin B\big(f(\mathrm{Crit}(f)),\iota_f\lambda^n\big)$$

for all $n \ge n_1$. Notice that, because of our choices, there exists $n_2 \ge n_1$ such that

$$\big|(f^n)'(x_n)\big|^{-1/4} < \lambda^n$$

for all $n \ge n_2$. Finally, set

$$S := \sum_{n\ge1}\big|(f^n)'(x_n)\big|^{-1/4},$$

$$b_n := \frac{1}{2}S^{-1}\big|\big(f^{n+1}\big)'(x_{n+1})\big|^{-\frac14},$$

and

$$\Pi := \Pi_{n=1}^{\infty}(1-b_n)^{-1}.$$

This infinite product converges because the series $\sum_{n\ge1} b_n$ does. Choose now $r = r(z) > 0$ so small that $16r(z)\Pi K S^3 \le \iota_f$, all the inverse branches $f_{x_n}^{-n}\colon B\big(x_0,\Pi r(z)\big) \longrightarrow Y$ are well defined (because of the Inverse Function Theorem) for all $n = 1,2,\dots,n_2$, and

$$\mathrm{diam}\big(f_{x_{n_2}}^{-n_2}\big(B\big(x_0, r\Pi_{k\ge n_2}(1-b_k)^{-1}\big)\big)\big) \le \lambda^{n_2}\iota_f.$$

We shall show by induction that, for every $n \ge n_2$, there exists an analytic inverse branch $f_{x_n}^{-n}\colon B\big(x_0,r\Pi_{k\ge n}(1-b_k)^{-1}\big) \longrightarrow Y$, sending x_0 to x_n and such that

$$\mathrm{diam}\big(f_{x_n}^{-n}\big(B\big(x_0, r\Pi_{k\ge n}(1-b_k)^{-1}\big)\big)\big) \le \lambda^{n}\iota_f.$$

Indeed, for $n = n_2$, this immediately follows from our requirements imposed on $r(z)$. So, suppose that the claim is true for some $n \geq n_2$. Since

$$x_n = f_{x_n}^{-n}(x_0) \notin B\big(f(\mathrm{Crit}(f)), \iota_f \lambda^n\big),$$

and since $\lambda^n \iota_f \leq \iota_f$, it follows from both item (e) of Definition 9.0.1 and Corollary 8.6.20 that there exists a holomorphic inverse branch $f_{x_{n+1}}^{-1} : B(x_n, \lambda^n \iota_f) \longrightarrow Y$ of f sending x_n to x_{n+1}. Since

$$\mathrm{diam}\big(f_{x_n}^{-n}\big(B\big((x_0, r\Pi_{k\geq n}(1-b_k)^{-1}\big)\big)\big) \leq \lambda^n \iota_f,$$

the composition

$$f_{x_{n+1}}^{-1} \circ f_{x_n}^{-n} B\big(x_0, r\Pi_{k\geq n}(1-b_k)^{-1}\big) \longrightarrow Y$$

is well defined and forms the inverse branch of f^{n+1} that sends x_0 to x_{n+1}. By Theorem 8.3.9, we now estimate that

$$\begin{aligned}
&\mathrm{diam}\big(f_{x_{n+1}}^{-(n+1)}\big(B(x_0, r\Pi_{k\geq n+1}(1-b_k)^{-1})\big)\big) \\
&\leq 2r\Pi_{k\geq n+1}(1-b_k)^{-1}|\big(f^{n+1}\big)'(x_{n+1})|^{-1}Kb_n^{-3} \\
&\leq 16r\Pi K S^3|\big(f^{n+1}\big)'(x_{n+1})|^{-1}|\big(f^{n+1}\big)'(x_{n+1})|^{\frac{3}{4}} \\
&= 16r\Pi K S^3|\big(f^{n+1}\big)'(x_{n+1})|^{-\frac{1}{4}} \\
&\leq \iota_f \lambda^{n+1},
\end{aligned}$$

where the last inequality sign is due to our choice of r and the number n_2. Putting $r(z) = r/2$, the second part of this theorem follows as a combined application of the equality $\lim_{n\to\infty} \frac{1}{n} \log|(f^n)'(x_n)| = \chi_\mu(f)$ and Theorem 8.3.9. ■

As an immediate consequence of Theorem 9.3.3, we get the following.

Corollary 9.3.4 *Assume the same notation and hypotheses as in Theorem 9.3.3. Fix $\varepsilon > 0$. Then there exist a measurable set $Z(\varepsilon) \subseteq Z$, the numbers $r(\varepsilon) \in (0, \iota_f)$ and $K(\varepsilon) \geq 1$ such that $\mu(Z(\varepsilon)) > 1 - \varepsilon$, $r(z) \geq r(\varepsilon)$ for all $z \in Z(\varepsilon)$ and denoting $x_n = h(T^{-n}(z))$, we have that*

$$K^{-1}(\varepsilon) \leq \exp(-(\chi_\mu + \varepsilon)n) \leq |\big(f_{x_n}^{-n}\big)'(y)| \leq K(\varepsilon)\exp(-(\chi_\mu - \varepsilon)n)$$

and

$$K^{-1} \leq \frac{|\big(f_{x_n}^{-n}\big)'(w)|}{|\big(f_{x_n}^{-n}\big)'(y)|} \leq K$$

for all $n \geq 1$, all $z \in Z(\varepsilon)$, and all $y, w \in B(x_0, r(\varepsilon))$. K here is the Koebe constant of Theorem 8.3.9 corresponding to the scale $1/2$.

In our future applications, the system (Z, f, ν) will usually be given by the Rokhlin natural extension (see Theorem 4.3.7) of the holomorphic system (f, μ). In this context, Theorem 9.3.3 and Corollary 9.3.4 give the following.

Theorem 9.3.5 *Let Y be a parabolic Riemann surface, $X \subseteq Y$ be a compact set, and $f \in \mathcal{A}^w(X)$. If μ is an f-invariant ergodic measure on X with positive Lyapunov exponent $\chi_\mu(f)$, then, for $\tilde{\mu}$-a.e. $x \in \tilde{X}$, there exists $r(x) \in (0, \iota_f)$ such that the function $x \mapsto r(x)$ is measurable and the following is satisfied.*

For every $n \geq 1$, there exists $f_{x_n}^{-n}\colon B(x, r(z)) \longrightarrow Y$, a holomorphic inverse branch of f^n sending x_0 to x_n. Given, in addition, an arbitrary $\varepsilon > 0$, there exists a constant $K_\varepsilon(z) \in (0, +\infty)$ independent of n and z such that

$$K_\varepsilon^{-1}(z) \exp\big(-(\chi_\mu + \varepsilon)n\big) \leq \big|\big(f_{x_n}^{-n}\big)'(y)\big| \leq K_\varepsilon(z) \exp\big(-(\chi_\mu - \varepsilon)n\big)$$

and

$$\frac{\big|\big(f_{x_n}^{-n}\big)'(w)\big|}{\big|\big(f_{x_n}^{-n}\big)'(y)\big|} \leq K$$

for all $y, w \in B(x_0, r(x))$. K here, as usual, is the Koebe constant corresponding to the scale $1/2$.

Consequently, for every $\varepsilon > 0$, there exist a closed set $\tilde{X}(\varepsilon) \subseteq \tilde{X}$, the numbers $r(\varepsilon) \in (0,1)$, and $K(\varepsilon) \geq 1$ such that $\tilde{\mu}\big(\tilde{X}(\varepsilon)\big) > 1 - \varepsilon$, $r(x) \geq r(\varepsilon)$ for all $x \in \tilde{X}(\varepsilon)$. We have that

$$K^{-1}(\varepsilon) \leq \exp(-(\chi_\mu + \varepsilon)n) \leq \big|\big(f_{x_n}^{-n}\big)'(y)\big| \leq K(\varepsilon) \exp(-(\chi_\mu - \varepsilon)n)$$

and

$$K^{-1} \leq \frac{\big|\big(f_{x_n}^{-n}\big)'(w)\big|}{\big|\big(f_{x_n}^{-n}\big)'(y)\big|} \leq K$$

for all $n \geq 1$, all $x \in \tilde{X}(\varepsilon)$, and all $y, w \in B(x_0, r(\varepsilon))$.

9.4 Ruelle's Inequality

The first, famous, relation between entropy and the Lyapunov exponent is the following version of Ruelle's Inequality proven by Ruelle in [**Ru2**] in the context of smooth diffeomorphisms of multi-dimensional smooth manifolds.

Theorem 9.4.1 (Ruelle's Inequality) *If Y is a parabolic Riemann surface, $X \subseteq Y$ is a compact set, $f \in \mathcal{A}^w(X)$, and μ is a Borel probability f-invariant ergodic measure on X, then*

$$\mathrm{h}_\mu(f) \leq 2 \max\{0, \chi_\mu(f)\}.$$

If, in addition, $f \in \mathcal{A}_0^w(X)$, *then*

$$\mathrm{h}_\mu(f) \le 2\chi_\mu(f).$$

Proof Since the set X is compact, there exists $\kappa \in [1, +\infty)$ such that

$$\kappa^{-1} r^2 \le V(B(x,r)) \le \kappa r^2 \tag{9.17}$$

for all $r \in [0,2]$ and all $x \in X$, where V is the two-dimensional volume measure (area) on Y generated by the Riemann metric of Y. For every $r \in (0,1]$, let $F(r) \subseteq X$ be a maximal $2r$-separated subset of X. Since X is compact, the set $F(r)$ is finite. Denote its elements by $x_1, x_2, \dots, x_{n_r}$. Construct inductively the following family $\mathcal{P}(r) = \{P_1, P_2, \dots, P_{n_r}\}$ of subsets of Y:

$$P_1 := B(x_1, 2r) \setminus \bigcup_{i=2}^{n_r} B(x_i, r)$$

and, for $1 \le j \le n_r - 1$,

$$P_{j+1} := B\big(x_{j+1}, 2r\big) \setminus \left(\bigcup_{i=1}^{j} P_i \cup \bigcup_{i=j+2}^{n_r} B(x_i, r) \right).$$

The elements of $\mathcal{P}(r)$ are, of course, mutually disjoint. Furthermore, since $F(r)$ is a maximal $2r$-separated set,

$$\bigcup_{i=1}^{n_r} B(x_i, 2r) \supseteq X,$$

and the balls $B(x_i, r)$, $i = 1, 2, \dots, n_r$, are mutually disjoint. Therefore,

$$\bigcup_{j=1}^{n_r} P_j \supseteq X$$

and

$$P_j \supseteq B(x_j, r)$$

for every $j = 1, 2, \dots, n_r$. For every integer $k \ge 1$, let

$$\mathcal{P}_k := \mathcal{P}(1/k).$$

Then

$$\mathrm{diam}\big(\mathcal{P}_k\big) \le \frac{2}{k} \quad \text{and} \quad V\big(\mathcal{P}_k\big) \ge \frac{1}{\kappa k^2}. \tag{9.18}$$

For every $z \in X$ and every $k \ge 1$, let

$$N(f,z,k) := \#\{P \in \mathcal{P}_k : f(\mathcal{P}_k(z)) \cap P \neq \emptyset\}.$$

We shall first show that, for every $k \geq 1$ large enough, say $k \geq k(f)$, we have that

$$N(f,z,k) \leq 2\kappa^2(|f'(z)|+2)^2. \tag{9.19}$$

Indeed, let $\geq k(f) \geq 1$ be so large that, for every $k \geq k(f)$, the Lipschitz constant of $f|_{\mathcal{P}_k(z)}$ does not exceed $|f'(z)| + 1$. Then the set $T(\mathcal{P}_k(z))$ is contained in the ball $B\big(f(z), 2(|f'(z)| + 1)k^{-1}\big)$. Thus, $f(\mathcal{P}_k(z)) \cap P \neq \emptyset$ implies that

$$P \subseteq B\big(f(z), 2(|f'(z)|+1)k^{-1} + 2k^{-1}\big) = B\big(f(z), 2(|f'(z)|+2)k^{-1}\big).$$

Since also, by (9.17),

$$V\big(B(f(z), 2(|f'(z)|+2)k^{-1})\big) \leq 2\kappa\big((|f'(z)|+2)k^{-1}\big)^2,$$

we, therefore, get, with the help of (9.18), that

$$N(f,z,k) \leq \frac{2\kappa\big((|f'(z)|+2)k^{-1}\big)^2}{(\kappa k^2)^{-1}} = 2\kappa^2(|f'(z)|+2)^2,$$

and (9.19) is proved. Let

$$N(f,z) := \sup\{N(f,z,k) : k \geq k(f)\}.$$

In view of (9.19), we get that

$$N(f,z) \leq 2\kappa^2(|f'(z)|+2)^2. \tag{9.20}$$

Observe that (see (6.1))

$$\begin{aligned}
\mathrm{H}_{\mu_{\mathcal{P}_k(z)}}\big(f^{-1}(\mathcal{P}_k)|\mathcal{P}_k(z)\big) &\leq \log \#\{P \in \mathcal{P}_k : f^{-1}(P) \cap \mathcal{P}_k(z) \neq \emptyset\} \\
&= \log N(f,z,k).
\end{aligned}$$

Using Theorems 6.4.14 and 6.4.7(b), we then get that

$$\begin{aligned}
\mathrm{h}_\mu(f) = \lim_{k\to\infty} \mathrm{h}_\mu(f,\mathcal{P}_k) &\leq \liminf_{k\to\infty} \mathrm{H}_\mu(f^{-1}(\mathcal{P}_k)|\mathcal{P}_k) \\
&= \liminf_{k\to\infty} \int_X \mathrm{H}_{\mu_{\mathcal{P}_k(z)}}\big(f^{-1}(\mathcal{P}_k)|\mathcal{P}_k(z)\big)\,d\mu(z) \\
&\leq \liminf_{k\to\infty} \int_X \log N(f,z,k)\,d\mu(z) \\
&\leq \int_X \log N(f,z)\,d\mu(z).
\end{aligned}$$

Applying this inequality and (9.20) to f^n and employing Theorem 6.4.9 and (9.20), we obtain that

$$\begin{aligned} \mathrm{h}_\mu(f) = \frac{1}{n}\mathrm{h}_\mu(f^n) &\le \frac{1}{n}\int_X \log N(f^n,z)\,d\mu(z) \\ &\le \int_X \frac{1}{n}\log\left(2\kappa^2(|(f^n)'(z)|+2)^2\right)d\mu(z). \end{aligned} \tag{9.21}$$

Obviously,

$$0 \le \frac{1}{n}\log\left(2\kappa^2(\|(f^n)'|_X\|_\infty+2)^2\right) \le \left(\log\|f'|_X\|_\infty+2\right)$$

for all $n \ge 1$, say $n \ge M$. Moreover, the ergodic part of the Birkhoff Ergodic Theorem (Theorem 2.3.2) implies that

$$\lim_{n\to\infty}\frac{1}{n}\log|(f^n)'(z)| = \int_X \log|f'|\,d\mu = \chi_\mu(f)$$

for μ-a.e. $z \in X$. Therefore,

$$\lim_{n\to\infty}\frac{1}{n}\log\left(2\kappa^2(|(f^n)'(z)|+2)\right) = \max\{0,\chi_\mu(f)\}$$

for μ-a.e. $z \in X$. So, invoking (9.21) and applying the Lebesgue Dominated Convergence Theorem, we conclude that

$$\mathrm{h}_\mu(f) \le \lim_{n\to\infty}\int_X \frac{1}{n}\log\left(2\kappa^2(|(f^n)'(z)|+2)^2\right)d\mu(z) = 2\max\{0,\chi_\mu(f)\}.$$

Ruelle's Inequality is, thus, proved. ■

Remark 9.4.2 This proof could be simplified if we took more substantial advantage of the fact that Y is a parabolic surface. This would give us a natural grid on a neighborhood of X and the use of maximal separated sets would not be needed. But we did this on purpose, having in mind possible future extensions of the theory of holomorphic maps in $\mathcal{A}(X)$ and $\mathcal{A}^w(X)$ to the case of hyperbolic Riemann surfaces Y.

9.5 Volume Lemmas and Hausdorff Dimensions of Invariant Measures

This section is entirely devoted to proving a closed formula for the Hausdorff and packing dimensions of Borel probability measures invariant under a map from $\mathcal{A}^w(X)$. For historical reasons, which are partly justified, this formula is frequently referred to as a Volume Lemma. Its first forms can be traced back to the works of Eggleston and Billingsley, [**Eg**] and [**Bil1**] respectively. From a dynamical point of view, a kind of breakthrough was given in a paper by Young [**LSY1**]. Since then, a multitude of papers have appeared. We would like to mention only some of the early ones: [**MM**], [**P1**], [**M2**]. Also, early papers [**Bow2**] and [**PUZI**] shed some light on the nature of dimensions of measures.

The main result of this subsection and of the entire section is the following.

Theorem 9.5.1 (Volume Lemma) *Let Y be a parabolic Riemann surface and $X \subseteq Y$ be a compact set. If $f \in \mathcal{A}^w(X)$ and μ is a Borel probability f-invariant ergodic measure on X with $\chi_\mu(f) > 0$, then*

$$\lim_{r \searrow 0} \frac{\log \mu(B(x,r))}{\log r} = \frac{\mathrm{h}_\mu(f)}{\chi_\mu(f)}$$

for μ-a.e. $x \in X$. In particular, the measure μ is dimensional exact and, by Proposition 1.7.8,

$$\mathrm{HD}(\mu) = \mathrm{PD}(\mu) = \frac{\mathrm{h}_\mu(f)}{\chi_\mu(f)}.$$

Proof In view of Theorem 1.7.6, the second formula follows from the first one; therefore, we only need to prove the first one. Let us prove, first, that

$$\liminf_{r \to 0} \frac{\log(\mu(B(x,r)))}{\log r} \geq \frac{\mathrm{h}_\mu(f)}{\chi_\mu(f)} \tag{9.22}$$

for μ-a.e. $x \in X$. By Corollary 9.2.3, there exists a finite partition $\mathcal{P}$ such that, for an arbitrary $\varepsilon > 0$ and every x in a set X_o of full measure μ, there exists $n(x) \geq 0$ such that, for all $n \geq n(x)$,

$$B(f^n(x), e^{-\varepsilon n}) \subseteq \mathcal{P}(f^n(x)). \tag{9.23}$$

Let us work, from now on, in the Rokhlin natural extension (see Theorem 4.3.7) $(\tilde{X}, \tilde{f}, \tilde{\mu})$. Let $\tilde{X}(\varepsilon)$ and $r(\varepsilon) > 0$ be given by Theorem 9.3.5. In view of the ergodic part of the Birkhoff Ergodic Theorem (Theorem 2.3.2), there exists a measurable set $\tilde{F}(\varepsilon) \subseteq \tilde{X}(\varepsilon)$ such that $\tilde{\mu}(\tilde{F}(\varepsilon)) = \tilde{\mu}(\tilde{X}(\varepsilon))$ and

$$\lim_{n \to \infty} \frac{1}{n} \sum_{j=1}^{n-1} \chi_{\tilde{X}(\varepsilon)} \circ \tilde{f}^n(\tilde{x}) = \tilde{\mu}(\tilde{X}(\varepsilon))$$

for every $\tilde{x} \in \tilde{F}(\varepsilon)$. Let $F(\varepsilon) = \pi(\tilde{F}(\varepsilon))$. Then

$$\mu(F(\varepsilon)) = \tilde{\mu}(\pi^{-1}(F(\varepsilon))) \geq \tilde{\mu}(\tilde{F}(\varepsilon)) = \tilde{\mu}(\tilde{X}(\varepsilon))$$

converges to 1 if $\varepsilon \searrow 0$. Consider now an arbitrary point

$$x \in F(\varepsilon) \cap X_o$$

and take $\tilde{x} \in \tilde{F}(\varepsilon)$ such that $x = \pi(\tilde{x})$. Then, by the above, there exists an increasing sequence $\{n_k = n_k(x) : k \geq 1\}$ such that $\tilde{f}^{n_k}(\tilde{x}) \in \tilde{X}(\varepsilon)$ and

$$\frac{n_{k+1} - n_k}{n_k} \leq \varepsilon \tag{9.24}$$

for every $k \geq 1$. Moreover, we can assume that $n_1 \geq n(x)$. Consider now an integer $n \geq n_1$ and the ball

$$B\Big(x, Cr(\varepsilon)\exp\big(-(\chi_\mu + (2 + \log\|f'\|_\infty)\varepsilon)n\big)\Big),$$

where $0 < C < K(\varepsilon)^{-1}$ is a constant (possibly depending on x) so small that

$$f^q\left(B\left(x, Cr(\varepsilon)\exp\left(-(\chi_\mu + (2 + \log\|f'\|_\infty)\varepsilon)n\right)\right)\right) \subseteq P(f^q(x)) \tag{9.25}$$

for every $q \leq n_1$ and $K(\varepsilon) \geq 1$ is the constant appearing in Theorem 9.3.5. Take now any q, $n_1 \leq q \leq n$, and associate k such that $n_k \leq q \leq n_{k+1}$. Since $\tilde{f}^{n_k}(\tilde{x}) \in \tilde{X}(\varepsilon)$ and since $\pi(\tilde{f}^{n_k}(\tilde{x})) = f^{n_k}(x)$, Theorem 9.3.5 produces a holomorphic inverse branch

$$f_x^{-n_k} : B\big(f^{n_k}(x), r(\varepsilon)\big) \longrightarrow Y$$

of f^{n_k} such that $f_x^{-n_k}(f^{n_k}(x)) = x$. This corollary also yields

$$f_x^{-n_k}\big(B(f^{n_k}(x), r(\varepsilon))\big) \supseteq B\Big(x, K(\varepsilon)^{-1}r(\varepsilon)\exp\big(-(\chi_\mu + \varepsilon)n_k\big)\Big).$$

Since

$$\begin{aligned} B\left(x, Cr(\varepsilon)\exp\left(-(\chi_\mu + (2 + \log\|f'\|)\varepsilon)n\right)\right) \\ \subseteq B\Big(x, K(\varepsilon)^{-1}r(\varepsilon)\exp\big(-(\chi_\mu + \varepsilon)n_k\big)\Big), \end{aligned}$$

it follows from Theorem 9.3.5 that

$$\begin{aligned} &f^{n_k}\Big(B\Big(x, Cr(\varepsilon)\exp\big(-(\chi_\mu + (2 + \log\|f'\|)\varepsilon)n\big)\Big)\Big) \\ &\quad \subseteq B\Big(f^{n_k}(x), CK(\varepsilon)r(\varepsilon)e^{-\chi_\mu(n-n_k)}\exp\big(\varepsilon(n_k - (2 + \log\|f'\|_\infty)n)\big)\Big). \end{aligned}$$

Since $n \geq n_k$ and since, by (9.24), $q - n_k \leq \varepsilon n_k$, we, therefore, obtain that

$$\begin{aligned}
f^q\Big(B\Big(x, Cr(\varepsilon)\exp(-(\chi_\mu + (2+\log\|f'\|)\varepsilon)n)\Big)\Big) \\
\subseteq B\Big(f^q(x), CK(\varepsilon)r(\varepsilon)e^{-\chi_\mu(n-n_k)}\exp(\varepsilon(n_k - (2+\log\|f'\|)n)) \\
\times \exp((q-n_k)\log\|f'\|_\infty)\Big) \\
\subseteq B\Big(f^q(x), CK(\varepsilon)r(\varepsilon)\exp(\varepsilon(n_k\log\|f'\|_\infty + n_k - 2n - n\log\|f'\|))\Big) \\
\subseteq B\big(f^q(x), CK(\varepsilon)r(\varepsilon)e^{-\varepsilon n}\big) \\
\subseteq B\big(f^q(x), e^{-\varepsilon q}\big).
\end{aligned}$$

Combining this, (9.23), and (9.25), we get that

$$B\Big(x, Cr(\varepsilon)\exp\Big(-\big(\chi_\mu + (2+\log\|f'\|)\varepsilon\big)n\Big)\Big) \subseteq \bigvee_{j=0}^{n} f^{-j}(\mathcal{P})(x).$$

Therefore, applying Theorem 6.5.5 (the ergodic case of the Shannon–McMillan–Breiman Theorem), we have that

$$\begin{aligned}
\varliminf_{n\to\infty} -\frac{1}{n}\log\mu\Big(B\big((x, Cr(\varepsilon)\exp(-\big(\chi_\mu + (2+\log\|f'\|)\varepsilon\big)n))\big)\Big) \\
\geq \mathrm{h}_\mu(f,\mathcal{P}) \geq \mathrm{h}_\mu(f) - \varepsilon.
\end{aligned}$$

This means that, denoting the number $Cr(\varepsilon)\exp\big(-(\chi_\mu + (2+\log\|f'\|)\varepsilon)n\big)$ by r_n, we have that

$$\liminf_{n\to\infty}\frac{\log\mu(B(x,r_n))}{\log r_n} \geq \frac{\mathrm{h}_\mu(f)-\varepsilon}{\chi_\mu(f) + (2+\log\|f'\|)\varepsilon}.$$

Now, since $\{r_n\}$ is a geometric sequence and since $\varepsilon > 0$ can be taken arbitrarily small, we conclude that, for μ-a.e. $x \in X$,

$$\liminf_{n\to\infty}\frac{\log\mu(B(x,r))}{\log r} \geq \frac{\mathrm{h}_\mu(f)}{\chi_\mu(f)}.$$

This completes the proof of (9.22).

Now let us prove that

$$\limsup_{r\to 0}\frac{\log(\mu(B(x,r)))}{\log r} \leq \frac{\mathrm{h}_\mu(f)}{\chi_\mu(f)} \tag{9.26}$$

for μ-a.e. $x \in X$.

In order to prove this formula, we again work in the Rokhlin natural extension $(\tilde{X}, \tilde{f}, \tilde{\mu})$ and we apply Pesin's Theory. In particular, the sets $\tilde{X}(\varepsilon)$ and $\tilde{F}(\varepsilon) \subseteq \tilde{X}(\varepsilon)$ and the radius $r(\varepsilon) > 0$ have the same meaning as in the current proof of (9.22); compare this with Theorem 9.3.5.

To begin with, notice that there exist two numbers $R > 0$ and $0 < Q < \min\{1, r(\varepsilon)/2\}$ such that the following two conditions are satisfied.

$$\text{If } z \notin B(\mathrm{Crit}(f), R), \text{ then } f|_{B(z,Q)} \text{ is injective.} \tag{9.27}$$

$$\text{If } z \in B(\mathrm{Crit}(f), R), \text{ then } f|_{B(z,Q\mathrm{dist}(z,\mathrm{Crit}(f)))} \text{ is injective.} \tag{9.28}$$

Observe also that if z is sufficiently close to a critical point c, then $f'(z)$ is of order $(z-c)^{q-1}$, where $q \geq 2$ is the order of critical point c. In particular, the quotient of $f'(z)$ and $(z-c)^{q-1}$ remains bounded away from 0 and ∞; therefore, there exists a constant number $B > 1$ such that

$$|f'(z)| \leq B\mathrm{dist}(z, \mathrm{Crit}(f)).$$

So, in view of Lemma 9.3.2, the logarithm of the function

$$\rho(z) := Q\min\{1, \mathrm{dist}(z, \mathrm{Crit}(f))\}$$

is integrable and consequently Lemma 9.2.5 applies. Let $\mathcal{P}$ be the Mañé partition produced by this lemma. Then $B(x, \rho(x)) \supseteq \mathcal{P}(x)$ for μ-a.e. $x \in X$, say for a subset X_ρ of X of measure 1. Consequently,

$$B_n(x,\rho) = \bigcap_{j=0}^{n-1} f^{-j}\big(B(f^j(x), \rho(f^j(x)))\big) \supseteq \mathcal{P}_0^n(x) \tag{9.29}$$

for every $n \geq 1$ and every $x \in X_\rho$. By our choice of Q and the definition of ρ, the function f is injective on all balls $B(f^j(x), \rho(f^j(x)))$, $j \geq 0$; therefore, f^k is injective on the set $B_n(x,\rho)$ for every $0 \leq k \leq n-1$. Now let

$$x \in F(\varepsilon) \cap X_\rho$$

and k be the greatest subscript such that $q = n_k(x) \leq n-1$. Denote by f_x^{-q} the unique holomorphic inverse branch of f^q produced by Theorem 9.3.5, which sends $f^q(x)$ to x. Of course,

$$B_n(x,\rho) \subseteq f^{-q}(B(f^q(x), \rho(f^q(x)))).$$

Having this and knowing that f^q is injective on $B_n(x,\rho)$, we even have that

$$B_n(x,\rho) \subseteq f_x^{-q}(B(f^q(x), \rho(f^q(x)))).$$

By Corollary 9.3.4,

$$\mathrm{diam}\big(f_x^{-q}(B(f^q(x), \rho(f^q(x))))\big) \leq K\exp(-q(\chi_\mu - \varepsilon)).$$

Since, by (9.24), $n \le q(1+\varepsilon)$, we finally deduce that

$$B_n(x,\rho) \subseteq B\left(x, K\exp\left(-n\frac{\chi_\mu - \varepsilon}{1+\varepsilon}\right)\right).$$

Thus, in view of (9.29),

$$B\left(x, K\exp\left(-n\frac{\chi_\mu - \varepsilon}{1+\varepsilon}\right)\right) \supseteq \mathcal{P}_0^n(x).$$

Therefore, denoting by r_n the radius of the ball above, it follows from Theorem 6.5.5 (ergodic case of the Shannon–McMillan–Breiman Theorem) that, for μ-a.e $x \in X$,

$$\limsup_{n\to\infty} -\frac{1}{n}\log\mu(B(x,r_n)) \le \mathrm{h}_\mu(f,\mathcal{P}) \le \mathrm{h}_\mu(f).$$

So,

$$\limsup_{n\to\infty} \frac{\log\mu(B(x,r_n))}{\log r_n} \le \frac{\mathrm{h}_\mu(f)}{\chi_\mu(f)-\varepsilon}(1+\varepsilon).$$

Now, since $(r_n)_{n=1}^\infty$ is a geometric sequence and since ε can be taken arbitrarily small, we conclude that, for μ-a.e. $x \in X$, we have that

$$\limsup_{n\to\infty} \frac{\log\mu(B(x,r))}{\log r} \le \frac{\mathrm{h}_\mu(f)}{\chi_\mu(f)}.$$

This completes the proof of (9.26) and, because of (9.22), also the proof of the first part of Theorem 9.5.1. The second part is now an immediate consequence of the first part and Proposition 1.7.8. ■

10

Sullivan Conformal Measures for Holomorphic Maps f in $\mathcal{A}(X)$ and in $\mathcal{A}^w(X)$

As we have already written in the Preface and in the Introduction to this volume, conformal measures were first defined and introduced by Patterson in his seminal paper [**Pat1**] (see also [**Pat2**]) in the context of Fuchsian groups. Sullivan extended this concept to all Kleinian groups in [**Su2**]–[**Su4**]. He then, in the papers [**Su5**] –[**Su7**], defined conformal measures for all rational functions of the Riemann sphere $\widehat{\mathbb{C}}$; he also proved their existence therein. Both Patterson and Sullivan came up with conformal measures to get an understanding of geometric measures, i.e., Hausdorff and packing ones. Although Sullivan had already noticed that there are conformal measures for Kleinian groups that are not equal, nor even equivalent, to any Hausdorff or packing (generalized) measure, the main purpose of dealing with them is still to understand Hausdorff and packing measures. Chapter 11, Section 17.6 in Volume 2, and, especially, three chapters, Chapters 20–22, all in Volume 2, provide good evidence.

Conformal measures, in the sense of Sullivan, have been studied in the context of rational functions in greater detail in [**DU3**], where, in particular, the structure of the set of their exponents was examined and fairly clarified.

Since then, conformal measures in the context of rational functions have been studied in numerous research works. We list here only a very few of them that appear in the early stages of the development of their theory: [**DU2**], [**DU5**], [**DU6**]. Subsequently, the concept of conformal measures, in the sense of Sullivan, has been extended to countable alphabet iterated functions systems in [**MU1**] and to conformal graph directed Markov systems in [**MU2**]. These were treated at length in Chapter 11. The concept was furthermore extended to some transcendental meromorphic dynamics in [**KU2**], [**UZ1**], and [**MyU2**]; see also [**UZ2**], [**MyU3**], [**BKZ1**], and [**BKZ2**]. Our current construction fits well in this line of development.

Last, the concept of conformal measures also found its place in random dynamics; we cite only [**MSU**].

Describing briefly our approach in this chapter, we would like to note that there could be essentially three ways of constructing the Sullivan conformal measures for maps f in $\mathcal{A}(X)$ and in $\mathcal{A}^w(X)$. One way would be to apply the Schauder–Tichonov Fixed Point Theorem for the dual Perron–Frobenius operator acting on the space of bounded linear functionals from $C(X)$ to $\mathbb{R}$. This method is, however, inapplicable in the current context since, because of either of the two reasons, the existence of critical points of f or $f: X \to X$, not being open, the Perron–Frobenius operator itself does not preserve the space of continuous functions on X. This method would be applicable if we restricted ourselves to open expanding (at least expansive) maps $f: X \to X$. But we cannot and we do not. The second method would be the original Sullivan construction. This would, however, essentially mean to choose the point ξ out of the set X (but in $U(f)$), and then we would not have good means to equate the minimal exponent t for which a t-Sullivan conformal measure exists with other fractal characteristics (dimensions) of the set X. But this is an important part of the theory of conformal measures and Section 10.3 in this chapter is devoted to such a task. It will be even more important in the elliptic function chapters of Volume 2 mentioned above. We, therefore, choose the third method, the one stemming from [**DU1**]. It is motivated by the Patterson–Sullivan approach but the point ξ belongs to the set X in this method and the limit construction of candidates for conformal measures is different. Furthermore, this method works within a very general setting of continuous maps of compact metrizable spaces (in fact, local compactness and separability suffices) and nonnegative continuous functions serving as Jacobians in the definition of conformal measures. The very first section of this chapter, Section 10.1, is devoted to this task. Then, in order to get genuine Sullivan conformal measures for holomorphic maps in $\mathcal{A}(X)$, $\mathcal{A}(X)$ and for elliptic functions and to understand the exponents of these measures, a substantial amount of additional work is needed. This is done in the second and the third sections of this chapter, and later, in Volume 2, for elliptic functions.

10.1 General Concept of Conformal Measures

In this section, we deal at length with the concept, properties, and construction of general conformal measures in the sense of [**DU1**]. Our approach here stems from and develops the one of [**DU1**]. It will be used in the next two sections to produce genuine Sullivan conformal or semi-conformal measures for holomorphic maps in $\mathcal{A}(X)$.

10.1.1 Motivation for and Definition of General Conformal Measures

Consider first an arbitrary, quasi-invariant Borel measure m for a Borel measurable map $T\colon X \to X$, where X is a completely metrizable space. Assume that T is (at most) countable-to-one, i.e.,

$$X = \bigcup_{j\in I} X_j,$$

where all the sets X_j, $j \in I$, are Borel, pairwise disjoint, and, for each $j \in I$, the map

$$T|_{X_j}\colon X_j \longrightarrow T(X_j)$$

is one-to-one. Note then that (see Theorem 4.5.4 in Srivastava [**Sr**]) all the sets $T(X_j)$, $j \in I$, are also Borel. Recall that

$$\mathcal{L}_m\colon L^1(m) \longrightarrow L^1(m)$$

is the transfer operator corresponding to the quasi-invariant measure m, set up in (2.8). We now introduce the measure $\mathcal{L}_m^* m$ by the formula

$$\mathcal{L}_m^* m(\psi) = m(\mathcal{L}_m \psi). \tag{10.1}$$

Putting in the formula $g := \mathbb{1}_X$ and $f = \psi \in L^1(m)$, we get from this formula that

$$\int_X \psi \, d\mathcal{L}_m^* m = \int_X \psi \, dm. \tag{10.2}$$

This precisely means that

$$\mathcal{L}_m^* m = m. \tag{10.3}$$

Denote

$$J_m^{-1}(T) := \frac{d\big(m \circ \big(T|_{X_j}\big)^{-1}\big)}{dm}. \tag{10.4}$$

Obviously, for all $u \in L^1(m)$, we have that

$$\mathcal{L}_m(u)(x) = \mathcal{L}_{\log J_m^{-1}(T)}(x) := \sum_{y\in T^{-1}(x)} u(y) J_m^{-1}(T)(y). \tag{10.5}$$

We say that $A \subseteq X$ is a special set if A is a Borel set and the restriction $T|_A$ is injective. Then, see the above-mentioned Theorem 4.5.4 in [**Sr**], the set $T(A)$ is also Borel. Assume now, in addition, that

$$T\colon X \to X \text{ is nonsingular with respect to } m,$$

meaning that

$$m\big(T\big((J_m(T)^{-1}(+\infty))\big)\big) = 0.$$

Hence, if $A \subseteq X$ is a special set, then this formula along with (10.2) and (10.5) yields

$$\begin{aligned} \int_A J_m(T)dm &= \int_X 1\!\!1_A J_m(T)dm = \int_X 1\!\!1_A J_m(T)d\mathcal{L}_m^* m \\ &= \int_X \mathcal{L}_m\big(1\!\!1_A J_m(T)\big)dm \\ &= \int_X \left(\sum_{y \in T^{-1}(x)} 1\!\!1_A(y) J_m(T)(y) J_m^{-1}(T)(y) \right) dm(x) \\ &= \int_{T(A)} \left(\sum_{y \in T^{-1}(x)} 1\!\!1_A(y) \right) dm(x) \\ &= \int_{T(A)} 1\!\!1 dm = m(T(A)). \end{aligned} \tag{10.6}$$

We want to single out this property by saying that the nonsingular measure m is $J_m(T)$-conformal. This prompts us to introduce the following.

Definition 10.1.1 Let $T\colon X \to X$ be a Borel measurable map of a completely metrizable topological space X and $g\colon X \to [0, +\infty)$ be a Borel measurable function. A Borel probability measure m on X is said to be g-conformal for $T\colon X \to X$ if and only if

$$m(T(A)) = \int_A g\,dm \tag{10.7}$$

for every special set $A \subseteq X$.

Observation 10.1.2 Observe that, on the other hand, if, in the above definition, $g > 0$, then the g-conformal measure m is nonsingular with respect to the map T and

$$J_m^{-1}(T) = 1/g.$$

Notice that, even if T is continuous, the operator $\mathcal{L}_m$ need not in general map $C(X)$ into $C(X)$. But this is the case if, in addition to being continuous, the map T is also open. Nevertheless, if we just assume that $\mathcal{L}_m(C(X)) \subseteq C(X)$, then the linear operator $\mathcal{L}_m : C(X) \to C(X)$ is bounded, and so is its dual $\mathcal{L}_m^* : C^*(X) \to C^*(X)$. Then, under the above constraints, T being countable-to-one and g being positive, we get the following.

Proposition 10.1.3 *A probability measure m is g-conformal if and only if*

$$\mathcal{L}_{-\log g}^*(m) = m.$$

Therefore, since we can have trouble with the operator $\mathcal{L}^*$ for the maps T that are not open, rather than looking for theorems that would assure the existence of fixed points of $\mathcal{L}^*$, we shall, as we have already said, provide another general method of constructing conformal measures; it stems from the Patterson–Sullivan method [**Pat1**] (see also [**Pat2**]) in the context of Fuchsian groups, [**Su5**]–[**Su7**] for Kleinian groups, and [**Su2**] and [**Su3**] for all rational functions.

10.1.2 Selected Properties of General Conformal Measures

Definition 10.1.4 Let $T : X \to Y$ be a map from a topological space X to a topological space Y. A point $x \in X$ is said to be singular for T if and only if at least one of the following two conditions is satisfied.

(a) There is no open neighborhood U of x such that the map $T|_U$ is one-to-one.
(b) For every open neighborhood V of x, there exists an open set $U \subseteq V$ such that the set $T(U)$ is not open.

The set of all singular points will be denoted by $\mathrm{Sing}(T)$ and will be referred to as the set of singular points of T. The set of all points satisfying condition *(a)* will be denoted by $\mathrm{Crit}(T)$ and will be referred to as the set of critical points of T. The set of all points satisfying condition *(b)* will be denoted by $X_*(T)$.

First, let us record the following immediate observation.

Observation 10.1.5 Let $T : X \to Y$ be a map from a topological space X to a topological space Y. If $z \in X$, then

(a) $\mathrm{Sing}(T) = \mathrm{Crit}(T) \cup X_*(T)$.
(b) All three sets $\mathrm{Crit}(T)$, $X_*(T)$, and $\mathrm{Sing}(T)$ are closed.
(c) $X \backslash X_*(T)$ is the largest, in the sense of inclusion, open subset of X, restricted to which map T is open.

It is easy to give examples of continuous maps T where $X_*(T) \cap \mathrm{Crit}(T) \neq \emptyset$. We will need the following, kind of obvious, topological fact whose proof we provided for the sake of completeness and convenience for the reader.

Lemma 10.1.6 *If X and Y are metrizable (regularity would suffice) topological spaces and X $T: X \to Y$ is a continuous map, then, for every $x \in X \backslash \mathrm{Sing}(T)$, there exists an open neighborhood $U(x)$ of x with the following properties:*

(a) $U(x) \subseteq X \backslash \mathrm{Sing}(T)$.
(b) *The map $T|_{U(x)}$ is one-to-one.*
(c) *The set $T(U(x))$ is open and the map $T|_{U(x)}: U(x) \to Y$ is open.*
(d) *If $G \subseteq U(x)$ is an open set, then $\partial(T(G)) = T(\partial G)$.*

Proof Since $x \in X \backslash \mathrm{Sing}(T)$ and $\mathrm{Sing}(T)$ is closed, there exists $W \subseteq X \backslash \mathrm{Sing}(T)$, an open neighborhood of x, such that the map $T|_W: W \to Y$ is one-to-one and open. In particular, $T(W)$ is an open neighborhood of $T(x)$. Since the space Y is regular, there exists an open neighborhood V of $T(x)$ such that

$$\overline{V} \subseteq T(W). \tag{10.8}$$

Since the map T is continuous, there exists an open neighborhood $U'(x)$ of x such that $U'(x) \subseteq W$ and

$$T(U'(x)) \subseteq V. \tag{10.9}$$

We shall prove the following.

Claim 1°. If $F \subseteq U'(x)$ is a closed subset of X, then $T(F) \subseteq Y$ is also closed.

Proof Since $W \backslash F \subseteq X$ is open, the set $T(W \backslash F)$ is open. As $T|_W$ is one-to-one, it is also clear that $T(W \backslash F) = T(W) \backslash T(F)$. So $T(F)$ is a closed subspace of $T(W)$. Thus, there exists a closed subset H of X such that $T(F) = H \cap T(W)$. By virtue of (10.8) and (10.9), we get that

$$T(F) = H \cap T(U'(x)) \cap T(W) = H \cap \overline{V} \cap T(W) = H \cap \overline{V}.$$

So, $T(F)$ is closed in Y and the claim holds. ■

By regularity of the space X, there exists an open neighborhood $U(x)$ of x such that

$$\overline{U(x)} \subseteq U'(x). \tag{10.10}$$

As a fairly immediate consequence of Claim 1°, we obtain the following.

Claim 2°. If $A \subseteq U(x)$, then $\overline{T(A)} = T(\overline{A})$.

Proof By continuity of T, we have that

$$T(\overline{A}) \subseteq \overline{T(A)}. \tag{10.11}$$

By Claim 1° and (10.10), the set $T(\overline{A})$ is closed in X. Therefore, as $T(A) \subseteq T(\overline{A})$, we get that $\overline{T(A)} \subseteq T(\overline{A})$. Along with (10.11), this shows that $\overline{T(A)} = T(\overline{A})$, proving Claim 2°. ■

To conclude the proof of the lemma, let $G \subseteq U(x) \subseteq U'(x) \subseteq W \subseteq X\backslash\text{Sing}(T)$ be an open set. Then $T(G)$ is also open and, using Claim 2°, we deduce that

$$\partial(T(G)) = \overline{T(G)}\backslash T(G) = T(\overline{G})\backslash T(G) = T(\overline{G}\backslash G) = T(\partial G).$$

The proof of Lemma 10.1.6 is complete. ■

Lemma 10.1.7 *Let $T\colon X \to Y$ be a continuous map from a Polish, i.e, completely metrizable and separable, topological space X to a Polish space Y and m be a Borel probability measure on Y. Let $\Gamma \subseteq X$ be a closed set containing* Sing(T). *If $g\colon X \to [0,+\infty)$ is a Borel measurable function integrable with respect to the measure m and* (10.7) *holds for every special set A whose closure is disjoint from Γ and for which $m(\partial A) = m(\partial T(A)) = 0$, then* (10.7) *continues to hold for every special set A disjoint from Γ.*

Proof Fix a complete metric ρ on X compatible with the topology on X. Let A be a special set disjoint from Γ. Fix $\varepsilon > 0$. Since $A \subseteq X\backslash\Gamma$, since the set $X\backslash\Gamma$ is open and since the measure m is outer regular, there exists an open set $V \subseteq X\backslash\Gamma$ such that $A \subseteq V$ and

$$\int_V g\,dm < \int_A g\,dm + \varepsilon. \tag{10.12}$$

By virtue of Lemma 10.1.6 for every $x \in V$, there exists $r(x) > 0$ such that

$$B(x,2r(x)) \subseteq U(x) \cap V, \tag{10.13}$$

where $U(x)$ is the open neighborhood of x produced in Lemma 10.1.6. Since $\partial B(x,r) \subseteq \{z \in X\colon \rho(z,x) = r\}$ for every $r > 0$ and since the sets $\{z \in X\colon \rho(z,x) = r\}$, $r > 0$, are mutually disjoint, the set

$$R_1(x) := \{r \in (0,r(x)]\colon m(\partial B(x,r)) > 0\}$$

is countable. Also, by items (b) and (d) of Lemma 10.1.6, the sets $\partial(T(B(x,r)))$, $r \in (0,r(x)]$, are mutually disjoint, whence the set

$$R_2(x) := \big\{r \in (0,r(x)]\colon m\big(\partial(T(B(x,r)))\big) > 0\big\}$$

is countable. Thus, the set $R_1(x) \cup R_2(x)$ is countable; therefore, there exists $s(x) \in (0, r(x)]$ such that

$$m\big(\partial B(x, s(x))\big) = 0 \quad \text{and} \quad m\big(\partial(T(B(x, s(x))))\big) = 0. \tag{10.14}$$

Since $\{B(x, s(x)) \colon x \in V\}$ is an open cover of V and since V, as a separable metrizable space, is Lindelöf, there exists a countable set $\{x_n \colon n \geq 1\} \subseteq V$ such that

$$\bigcup_{n=1}^{\infty} B(x_n, s(x_n)) = V. \tag{10.15}$$

Define now inductively in the standard way a partition $(G_n)_{n=1}^{\infty}$ of V as follows.

$$G_1 := B(x_1, s(x_1)) \quad \text{and} \quad G_{n+1} := B\big(x_{n+1}, s(x_{n+1})\big) \setminus \bigcup_{k=1}^{n} B(x_k, s(x_k)) \quad \text{for all } n \geq 1.$$

Since, for every $n \geq 1$,

$$\partial G_n \subseteq \bigcup_{k=1}^{n} \partial B(x_k, s(x_k)),$$

it follows from (10.14) that

$$m(\partial G_n) = 0 \tag{10.16}$$

for every $n \geq 1$. Likewise, since

$$T(G_n) = T\big(B(x_n, s(x_n))\big) \setminus \bigcup_{k=1}^{n-1} T\big(B(x_k, s(x_k))\big),$$

we have that

$$\partial(T(G_n)) \subseteq \bigcup_{k=1}^{n} \partial\big(T\big(B(x_k, s(x_k))\big)\big);$$

it, therefore, follows from (10.14) that

$$m(\partial(T(G_n))) = 0 \tag{10.17}$$

for every $n \geq 1$. Since also, by (10.13) and by our construction, $\overline{G_n} \cap \Gamma = \emptyset$, we get from all the above, particularly (10.12), and from the hypotheses of our lemma that

$$m(T(A)) = m\left(\bigcup_{n=1}^{\infty} T(A\cap G_n)\right) \le \sum_{n=1}^{\infty} m(T(G_n)) = \sum_{n=1}^{\infty}\int_{G_n} g\,dm$$
$$= \int_V g\,dm < \int_A g\,dm + \varepsilon.$$

Since $\varepsilon > 0$ was arbitrary, it follows that

$$m(T(A)) \le \int_A g\,dm \tag{10.18}$$

for every special set $A \subseteq X$ disjoint from Γ. Using this fact for any special subset of X disjoint from Γ and also (10.12), we get that

$$m(T(A)) = m\left(\bigcup_{n=1}^{\infty} T(A\cap G_n)\right) = \sum_{n=1}^{\infty} m(T(A\cap G_n))$$
$$= \sum_{k=1}^{\infty}\big(m(T(G_n)) - m(T(G_n\backslash A))\big)$$
$$= \sum_{n=1}^{\infty}\left(\int_{G_n} g\,dm - m(T(G_n\backslash A))\right) \ge \sum_{n=1}^{\infty}\left(\int_{G_n} g\,dm - \int_{G_n\backslash A} g\,dm\right)$$
$$= \int_{\bigcup_{n=1}^{\infty} G_n} g\,dm - \int_{\bigcup_{n=1}^{\infty} G_n\backslash A} g\,dm = \int_V g\,dm - \int_{V\backslash A} g\,dm$$
$$\ge \int_A g\,dm - \int_{V\backslash A} g\,dm$$
$$\ge \int_A g\,dm - \varepsilon.$$

Letting $\varepsilon \searrow 0$, we, thus, get that $m(T(A)) \ge \int_A g\,dm$. Along with (10.18), this gives us that

$$m(T(A)) = \int_A g\,dm.$$

The proof of Lemma 10.1.7 is complete. ■

For our next result, we need the following small fact from topology, which is certainly well known but does not seem to be easily accessible in the literature. We, therefore, now provide its full proof.

Theorem 10.1.8 *Every locally compact metrizable topological space is completely metrizable.*

Proof Let X be such a space. Let ρ be a metric on X inducing its topology. Let $(\tilde{X}, \tilde{\rho})$ be the completion of the metric space (X, ρ). This means, we recall, that X is a dense subset of $\tilde{X}$, $\tilde{\rho}|_{X\times X} = \rho$, and the metric space $(\tilde{X}, \tilde{\rho})$ is complete. Since $\tilde{X}$ is completely metrizable (as $\tilde{\rho}$ is a complete metric on it), completely metrizable subspaces of $\tilde{X}$ coincide with its G_δ subsets. Therefore, it suffices to show that X is a G_δ subset of $\tilde{X}$. We shall show more; namely, that X is an open subset of $\tilde{X}$. To do this, we shall prove that every point $z \in X$ has an open neighborhood in $\tilde{X}$ which is contained in X. So, since the space X is locally compact, there exists radius $r > 0$ such that $\overline{B_\rho(z, 2r)}_X$, which is the closure of $B_\rho(z, 2r)$ in X, is a compact set. We shall show that

$$B_{\tilde{\rho}}(z, r) \subseteq B_\rho(z, 2r);$$

as $B_\rho(z, 2r) \subseteq X$, this will complete the proof. So, let $w \in B_{\tilde{\rho}}(z, r)$. Then, since X is a dense subset of $\tilde{X}$, there exists $(w_n)_{n=1}^\infty$, a sequence of points in X converging to w. So, by disregarding finitely many terms of this sequence, we may assume that all its elements belong to $B_{\tilde{\rho}}(z, r)$. Hence, these belong to $B_{\tilde{\rho}}(z, r) \cap X = B_\rho(z, r) \subseteq \overline{B_\rho(z, r)}_X$. But then, on the one hand, $(w_n)_{n=1}^\infty$ is a Cauchy sequence in (X, ρ) (because of converging in $(\tilde{X}, \tilde{\rho})$), and, on the other hand, it contains a subsequence converging to some point $w' \in \overline{B_\rho(z, r)}_X$ since this closure is compact. Thus, $w = w' \in \overline{B_\rho(z, r)}_X \subseteq B_\rho(z, 2r)$. ■

For the full picture, we would like to note that not every metric on a locally compact metrizable topological space is complete. The set $(0, 1) \subseteq \mathbb{R}$ with the standard Euclidean metric is an example.

We end this subsection with the following technical result, which will be substantially used when we deal with the Sullivan conformal measures for elliptic functions.

Lemma 10.1.9 *Let $T: X \to Y$ be a continuous map from a locally compact metrizable space X to a locally compact metrizable space Y. Let Γ be a compact subset of X containing $X_*(T)$. Suppose that, for all integers $n \geq 1$, there are respective uniformly bounded above continuous functions $g_n: X \longrightarrow [0, +\infty)$ and Borel probability measures m_n on X satisfying the following conditions.*

(a) (10.7) *holds for $g = g_n$ and for every special set $A \subseteq X$ with*

$$A \cap \Gamma = \emptyset.$$

(b)

$$m_n(T(A)) \geq \int_A g_n \, dm_n$$

for every special set $A \subseteq X$.

(c) *The sequence* $(g_n)_{n=1}^{\infty}$ *converges uniformly to a continuous function* $g\colon X \longrightarrow [0, +\infty)$.

Then, for every weak accumulation point m *of the sequence* $(m_n)_{n=1}^{\infty}$, *treated as measures on* Y, *we have that*

(1)

$$m(T(A)) = \int_A g\, dm$$

for all special sets $A \subseteq X$ *such that* $A \cap (\Gamma \cup \mathrm{Crit}(T)) = \emptyset$ *and*

(2)

$$m(T(A)) \geq \int_A g\, dm$$

for all special sets $A \subseteq X$ *such that* $A \cap \mathrm{Crit}(T) = \emptyset$.

Furthermore, if, in addition, $w \in X\backslash\Gamma$ *and*

(d) *there exist an integer* $q = q_w \geq 1$ *and special mutually disjoint sets* $W_1, W_2, \dots, W_q$ *(some of which can be empty), such that* $W_1 \cup W_2 \cup \cdots \cup W_q$ *is an open neighborhood of* w, *then*

(3)

$$m(T(w)) \leq g(w)m(\{w\}).$$

Consequently, if

(e) *the set* $\mathrm{Crit}(T)$ *is countable (possibly finite) and all functions* $g_n|_{\mathrm{Crit}(T)}$, $n \geq 1$, *are identically equal to* 0, *then*

(4)

$$m(T(A)) = \int_A g\, dm$$

for all special sets $A \subseteq X$ *such that* $A \cap \Gamma = \emptyset$.

Proof First notice that, according to Theorem 10.1.8, the topological space X is completely metrizable. This is needed to be sure that images of special sets are Borel. By passing to a subsequence, we may assume without loss of generality that the sequence $(m_n)_{n=1}^{\infty}$ converges to m in the weak topology on the space of Borel probability measures on Y.

First, aiming to prove (2), assume that A is special, compact, and $m(\partial A) = 0$. Since then $T(A)$ is also compact, thus closed, by using hypothesis (b), we get that

$$m(T(A)) \geq \limsup_{n\to\infty} m_n(T(A)) \geq \limsup_{n\to\infty} \int_A g_n \, dm_n = \int_A g \, dm.$$

Now drop the assumption $m(\partial A) = 0$ but keep A compact and assume in addition that there exists an open special set $\tilde{A}$ such that $A \subseteq \tilde{A}$. Since A is compact and X is locally compact and metrizable, there exists a descending sequence $(A_k)_{k=1}^{\infty}$ of compact subsets of $\tilde{A}$ whose intersection equals A and $m(\partial A_k) = 0$ for every $k \geq 1$. By what has already been proved, we have that

$$m(T(A)) = \lim_{k\to\infty} m(T(A_k)) \geq \lim_{k\to\infty} \int_{A_k} g \, dm = \int_A g \, dm.$$

Finally, keep A special assuming that $A \cap \mathrm{Crit}(T) = \emptyset$. Since $\mathrm{Crit}(T)$ is a closed set, for every $x \in A$ there exists $r_x > 0$ such that the set $B(x, r_x)$ is special. Since the space X, being completely metrizable (by Theorem 10.1.8) and separable, is Lindelöf, there exists a sequence $(x_n)_{n=1}^{\infty}$ of points in A such that

$$\bigcup_{n=1}^{\infty} B(x_n, r_{x_n}) \supseteq A.$$

Now form in the following standard way a countable partition $\{A_n\}_{n=1}^{\infty}$ of A out of cover $\{A \cap B(x_n, r_{x_n})\}_{n=1}^{\infty}$ of A:

$$A_1 := A \cap B(x_1, r_{x_1}), \quad A_{n+1} := A \cap B(x_{n+1}, r_{x_{n+1}}) \backslash \bigcup_{j=1}^{n} A_j, \quad n \geq 1.$$

Then, by the previous case,

$$\begin{aligned} m(T(A)) &= m\left(T\left(\bigcup_{n=1}^{\infty} A_n\right)\right) = m\left(\bigcup_{n=1}^{\infty} T(A_n)\right) = \sum_{n=1}^{\infty} m(T(A_n)) \\ &\geq \sum_{n=1}^{\infty} \int_{A_n} g \, dm = \int_{\bigcup_{n=1}^{\infty} A_n} g \, dm \\ &= \int_A g \, dm. \end{aligned}$$

In order to prove item (1), consider first any special set $A \subseteq X$ such that

$$A \cap (\Gamma \cup \mathrm{Crit}(T)) = \emptyset \quad \text{and} \quad m(\partial A) = m(\partial T(A)) = 0.$$

We then have that

$$m(T(A)) = \lim_{n\to\infty} m_n(T(A)) = \lim_{n\to\infty} \int_A g_n \, dm_n = \int_A g \, dm.$$

Therefore, since $\Gamma \cup \mathrm{Crit}(T)$ is a closed set containing $\mathrm{Sing}(T)$, item (1) holds, because of Lemma 10.1.7.

Now assuming that (d) also holds, we will prove item (3). By the same standard argument as the one giving countability of the set $R_1(x)$ in the proof of Lemma 10.1.7 (and since $w \notin X_*(T)$ and $X_*(T)$ is a closed subset of X), there exists a sequence $(r_j)_{j=1}^{\infty}$ in $(0,\infty)$ converging to 0 such that

$$\overline{B(w,r_j)} \subseteq W \backslash X_*(t) \quad \text{and} \quad m(\partial B(w,r_j)) = 0 \quad \forall\, j \in \mathbb{N}. \tag{10.19}$$

For every $i \in \{1,\dots,q\}$ and $j \in \mathbb{N}$, let

$$W_j(i) = W_i \cap B(w,r_j).$$

Then, for each $j \in \mathbb{N}$, all the sets $W_1(j), W_2(j), \dots, W_q(j)$ are special, mutually disjoint, and

$$B(w,r_j) := \bigcup_{i=1}^{q} W_i(j)$$

is an open neighborhood of w. By our hypotheses, we have, for every $j \geq 1$, that

$$\lim_{n\to\infty} \int_{B(w,r_j)} g_n \, dm_n = \int_{B(w,r_j)} g \, dm. \tag{10.20}$$

Fix $j \in \mathbb{N}$. As the sequence

$$\left(\int_{W_j(1)} g_n \, dm_n \right)_{n=1}^{\infty}$$

is bounded, there is an unbounded increasing sequence $\left(s_n^{(1)}\right)_{n=1}^{\infty}$ of positive integers, depending on j, such that the limit

$$\lim_{n\to\infty} \int_{W_j(1)} g_{s_n^{(1)}} \, dm_{s_n^{(1)}}$$

exists. Proceeding by induction, suppose that we have, for some $1 \leq l < q$, a subsequence $\left(s_n^{(l)}\right)_{n=1}^{\infty}$ of the sequence $(s_n)_{n=1}^{\infty}$ such that the limit

$$\lim_{n\to\infty} \int_{W_j(i)} g_{s_n^{(i)}} \, dm_{s_n^{(i)}}$$

exists for every $i = 1,\dots,l$. Since the sequence

$$\left(\int_{W_j(l)} g_{s_n^{(l)}} \, dm_{s_n^{(l)}} \right)_{n=1}^{\infty}$$

is bounded, there is a subsequence $\left(s_n^{(l+1)}\right)_{k=1}^{\infty}$ of the sequence $\left(s_n^{(l)}\right)_{n=1}^{\infty}$ such that the limit

$$\lim_{k\to\infty}\int_{W_j(l+1)} g_{s_n^{(l+1)}}\, dm_{s_n^{(l+1)}}$$

exists. Then the limit

$$\lim_{n\to\infty}\int_{W_j(l)} g_{s_n^{(q)}}\, dm_{s_n^{(q)}}$$

exists for all $= l,\dots,q$. By our hypotheses, for all $l = 1,\dots,q$, the limit

$$\lim_{n\to\infty} m_{s_n^{(q)}}(T(W_j(l)))$$

also exists and

$$\lim_{n\to\infty} m_{s_n^{(q)}}(T(W_j(l))) = \lim_{n\to\infty}\int_{W_j(l)} g_{s_n^{(q)}}\, dm_{s_n^{(q)}}.$$

Using this and (10.31), we deduce that

$$\begin{aligned}
m(\{T(w)\}) &\le m\big(T(B(w,r_j))\big) \le \lim_{n\to\infty} m_{s_n}^{(q)}\big(T(B(w,r_j))\big)\\
&= \lim_{n\to\infty} m_{s_n}^{(q)}\left(T\left(\bigcup_{l=1}^{q} W_j(l)\right)\right) = \lim_{n\to\infty} m_{s_n}^{(q)}\left(\bigcup_{l=1}^{q} T\left(W_j(l)\right)\right)\\
&\le \liminf_{n\to\infty}\sum_{l=1}^{q} m_{s_n}^{(q)}\big(T(W_j(l))\big)\\
&= \sum_{l=1}^{q}\lim_{n\to\infty} m_{s_n}^{(q)}\big(T(W_j(l))\big) = \sum_{l=1}^{q}\lim_{n\to\infty}\int_{W_j(l)} g_{s_n}^{(q)}\, dm_{s_n^{(q)}}\\
&= \lim_{n\to\infty}\sum_{l=1}^{q}\int_{W_j(l)} g_{s_n}^{(q)}\, dm_{s_n^{(q)}} = \lim_{n\to\infty}\int_{B(w,r_j)} g_{s_n}^{(q)}\, dm_{s_n^{(q)}}\\
&= \int_{B(w,r_j)} g\, dm.
\end{aligned}$$

Hence,

$$m(\{T(w)\}) \le \lim_{j\to\infty}\int_{B(w,r_j)} g\, dm = g(w)m(\{w\}).$$

The proof of item (3) is complete. ■

10.1.3 The Limit Construction

In this subsection, we shall provide a construction aiming to produce conformal measures. It will make use of the following simple fact. For a sequence $(a_n)_{n=1}^{\infty}$ of reals, the number

$$c := \limsup_{n\to\infty} \frac{a_n}{n} \tag{10.21}$$

will be called the transition parameter of the sequence $(a_n)_{n=1}^{\infty}$. It is uniquely determined by the property that

$$\sum_{n=1}^{\infty} \exp(a_n - ns)$$

converges for $s > c$ and diverges for $s < c$. For $s = c$, the sum may converge or diverge. By a simple argument, one obtains the following.

Lemma 10.1.10 *There exists a sequence $(b_n)_{n=1}^{\infty}$ of positive reals such that*

$$\sum_{n=1}^{\infty} b_n \exp(a_n - ns) \begin{cases} < \infty & \text{if } s > c \\ = \infty & \text{if } s \le c \end{cases}$$

and

$$\lim_{n\to\infty} \frac{b_n}{b_{n+1}} = 1.$$

Proof If

$$\sum_{n\ge 1} \exp(a_n - nc) = \infty,$$

put $b_n = 1$ for every $n \ge 1$. If

$$\sum_{n\ge 1} \exp(a_n - nc) < \infty,$$

choose a sequence $(n_k)_{k=1}^{\infty}$ of positive integers such that

$$\lim_{k\to\infty} \frac{n_k}{n_{k+1}} = 0 \quad \text{and} \quad \varepsilon_k := a_{n_k} n_k^{-1} - c \longrightarrow 0.$$

Setting then

$$b_n = \exp\left(n\left(\frac{n_k - n}{n_k - n_{k-1}}\varepsilon_{k-1} + \frac{n - n_{k-1}}{n_k - n_{k-1}}\varepsilon_k\right)\right) \quad \text{for } n_{k-1} \le n < n_k,$$

it is easy to check that the lemma follows. ∎

The actual setting of the current section is this.

(a) $T: X \to X$ is a continuous self-map of a completely metrizable topological space X.
(b) $(E_n)_{n=1}^{\infty}$ is a sequence of finite subsets of X such that

$$T^{-1}(E_n) \subset E_{n+1} \tag{10.22}$$

for every $n \geq 1$.
(c) $\phi: X \to (-\infty, +\infty]$ is a bounded below continuous function such that

$$\phi\left(\bigcup_{n=1}^{\infty} E_n\right) \subseteq (-\infty, +\infty). \tag{10.23}$$

We assume the standard convention that

$$e^{-\infty} = 0 \quad \text{and} \quad e^{+\infty} = +\infty.$$

With this convention, we have the following immediate observation.

Observation 10.1.11 The function $e^{-\phi}: X \to [0, +\infty)$ is bounded and continuous.

Let

$$a_n := \log\left(\sum_{x \in E_n} \exp(S_n\phi(x))\right), \tag{10.24}$$

where, we recall,

$$S_n\phi = \sum_{k=0}^{n-1} \phi \circ T^k.$$

Denote by c the transition parameter of this sequence. Choose a sequence $(b_n)_1^{\infty}$ of positive reals as in Lemma 10.1.10 for the sequence $(a_n)_1^{\infty}$. For every $s > c$, define

$$M_s := \sum_{n=1}^{\infty} b_n \exp(a_n - ns) \tag{10.25}$$

and the normalized measure

$$m_s = \frac{1}{M_s} \sum_{n=1}^{\infty} \sum_{x \in E_n} b_n \exp(S_n\phi(x) - ns)\delta_x, \tag{10.26}$$

where δ_x denotes the Dirac δ measure supported at the point $x \in X$. Let A be a special set. Using (10.22) and (10.26), it follows that

$$\begin{aligned} m_s(T(A)) &= \frac{1}{M_s}\sum_{n=1}^{\infty}\sum_{x\in E_n\cap T(A)} b_n \exp(S_n\phi(x) - ns) \\ &= \frac{1}{M_s}\sum_{n=1}^{\infty}\sum_{x\in A\cap T^{-1}(E_n)} b_n \exp(S_n\phi(T(x)) - ns) \\ &= \frac{1}{M_s}\sum_{n=1}^{\infty}\sum_{x\in A\cap E_{n+1}} b_n \exp[S_{n+1}\phi(x) - (n+1)s]\exp(s - \phi(x)) \\ &\quad - \frac{1}{M_s}\sum_{n=1}^{\infty}\sum_{x\in A\cap(E_{n+1}\setminus T^{-1}(E_n))} b_n \exp(S_n\phi(T(x)) - ns). \end{aligned} \tag{10.27}$$

Set

$$\begin{aligned} \Delta_A(s) := \Bigg| &\frac{1}{M_s}\sum_{n=1}^{\infty}\sum_{x\in A\cap E_{n+1}} b_n \exp[S_{n+1}\phi(x) - (n+1)s]\exp(s - \phi(x)) \\ &- \int_A \exp(c - \phi)\,dm_s \Bigg| \end{aligned}$$

and observe that

$$\begin{aligned} \Delta_A(s) = \frac{1}{M_s}\Bigg| &\sum_{n=1}^{\infty}\sum_{x\in A\cap E_{n+1}} \exp[S_{n+1}\phi(x) - (n+1)s] \\ &\times \exp(-\phi(x))\big[b_n e^s - b_{n+1}e^c\big] - b_1 \sum_{x\in A\cap E_1} e^{c-s} \Bigg| \\ \le \frac{1}{M_s}&\sum_{n=1}^{\infty}\sum_{x\in A\cap E_{n+1}} \left|\frac{b_n}{b_{n+1}} - e^{c-s}\right| b_{n+1}\exp(s - \phi(x)) \\ &\times \exp[S_{n+1}\phi(x) - (n+1)s] + \frac{1}{M_s} b_1 \exp(c - s)\,\sharp(A\cap E_1) \\ \le \frac{1}{M_s}&\sum_{n=1}^{\infty}\sum_{x\in E_{n+1}} \left|\frac{b_n}{b_{n+1}} - e^{c-s}\right| b_{n+1}\exp(s - \phi(x)) \\ &\times \exp[S_{n+1}\phi(x) - (n+1)s] + \frac{1}{M_s} b_1 \exp(c - s)\,\#E_1. \end{aligned}$$

By Lemma 10.1.10, we have that $\lim_{n\to\infty} b_{n+1}/b_n = 1$ and $\lim_{s\searrow c} M_s = \infty$. Therefore,

$$\lim_{s\searrow c} \Delta_A(s) = 0 \tag{10.28}$$

uniformly with respect to all special sets A.

Definition 10.1.12 Any weak accumulation point, when $s \searrow c$, of the measures $(m_s)_{s>c}$, defined by (10.26), will be called a PS limit measure (associated with the function ϕ and the sequence $(E_n)_1^\infty$) or just a PSl measure.

10.1.4 Conformality Properties of PS Limit Measures

In order to find conformal measures among the limit measures, it is necessary to examine (10.27) in greater detail. To begin with, for a Borel set $D \subseteq X$, consider the following condition:

$$\lim_{s\downarrow c} \frac{1}{M_s} \sum_{n=1}^{\infty} \sum_{x\in D\cap(E_{n+1}\setminus T^{-1}E_n)} b_n \exp\big(S_n\phi(T(x)) - ns\big) = 0. \tag{10.29}$$

As an immediate consequence of (10.27) and (10.28), we get the following.

Lemma 10.1.13 *Let $T: X \to X$ be a continuous self-map of a completely metrizable topological space X. Let $(E_n)_{n=1}^\infty$ be a sequence of finite subsets of X such that* (10.22) *holds and $\phi: X \longrightarrow (-\infty, +\infty]$ be a bounded below continuous function satisfying* (10.23)*. If a set $D \subseteq X$ satisfies condition* (10.29)*, then the sets of all accumulation points of the functions*

$$(c,\infty) \ni s \longmapsto m_s(T(D)) \quad \text{and} \quad (c,\infty) \ni s \longmapsto \int_D \exp(s-\varphi)dm_s,$$

as $s \searrow c$, coincide. More precisely, if $(s_n)_{n=1}^\infty$ is a sequence of real numbers in (c,∞) such that $\lim_{n\to\infty} s_n = c$, then $\lim_{n\to\infty} \int_D \exp(s_n - \varphi)dm_{s_n}$ exists if and only if $\lim_{n\to\infty} m_{s_n}(T(D))$ exists; if any of those exist, then they coincide.

Lemma 10.1.14 *Let $T: X \to X$ be a continuous self-map of a completely metrizable topological space X. Let $(E_n)_{n=1}^\infty$ be a sequence of finite subsets of X such that* (10.22) *holds and $\phi: X \longrightarrow (-\infty, +\infty]$ be a bounded below continuous function satisfying* (10.23)*. Let m be a limit measure of Definition 10.1.12 and Γ be a closed set containing* Sing(T)*. Assume that every special set $D \subseteq X$ with $m(\partial D) = m(\partial T(D)) = 0$ and $\bar{D} \cap \Gamma = \emptyset$ satisfies condition* (10.29)*. Then*

$$m(T(A)) = \int_A \exp(c-\phi)dm$$

for every special set A disjoint from Γ.

Proof It follows immediately from Lemma 10.1.13 and boundedness and continuity of the function $e^{-\phi}$ that

$$m(T(D)) = \int_D \exp(c - \phi)dm$$

for every special set $D \subseteq X$ for which $\bar{D} \cap \Gamma = \emptyset$ and $m(\partial D) = m(\partial T(D)) = 0$. Applying now Lemma 10.1.7 completes the proof. ■

The proof of the following lemma is very similar to the last part of the proof of Lemma 10.1.9. We provide this proof here since Lemma 10.1.15 is very important and for the sake of completeness and convenience for the reader.

Lemma 10.1.15 *Let $T: X \to X$ be a continuous self-map of a completely metrizable topological space X. Let $(E_n)_{n=1}^{\infty}$ be a sequence of finite subsets of X such that* (10.22) *holds and $\phi: X \longrightarrow (-\infty, +\infty]$ be a bounded below continuous function satisfying* (10.23). *Suppose that condition* (10.29) *holds for every special set D whose closure is disjoint from $X_*(T)$. Suppose also that $w \in X \backslash X_*(T)$ is a point for which there exist an integer $q \geq 1$ and special mutually disjoint sets $W_1, W_2, \ldots, W_q$ (some of which can be empty) such that $W_1 \cup W_2 \cup \cdots \cup W_q$ is an open neighborhood of w.*

If m is a limit measure of Definition 10.1.12, then

$$m(T(\{w\})) \leq \exp(c - \phi(w))m(\{w\}).$$

Proof By the same standard argument as the one giving countability of the set $R_1(x)$ in the proof of Lemma 10.1.7 (and since $w \notin X_*(T)$ and $X_*(T)$ is a closed subset of X), there exists a sequence $(r_n)_{n=1}^{\infty}$ in $(0, \infty)$ converging to 0 such that

$$\overline{B(w,r_n)} \subseteq W \backslash X_*(t) \quad \text{and} \quad m(\partial B(w,r_n)) = 0 \quad \forall n \in \mathbb{N}. \tag{10.30}$$

For every $i \in \{1, \ldots, q\}$ and $n \in \mathbb{N}$, let

$$W_n(i) = W_i \cap B(w,r_n).$$

Then, for each $n \in \mathbb{N}$, all the sets $W_1(n), W_2(n), \ldots, W_q(n)$ are special, mutually disjoint, and

$$B(w,r_n) := \bigcup_{i=1}^{q} W_i(n)$$

is an open neighborhood of w. From the definition of m, there exists a sequence $(s_k)_{k=1}^{\infty}$ in (c, ∞) such that

$$\lim_{k \to \infty} m_{s_k} = m.$$

Since the function $e^{-\varphi}\colon X \to [0,\infty)$ is bounded and continuous, by looking at the second assertion of (10.30), we infer, for every $n \in \mathbb{N}$, that

$$\lim_{k\to\infty} \int_{B(w,r_n)} e^{s_k-\varphi}\, dm_{s_k} = \int_{B(w,r_n)} e^{c-\varphi}\, dm. \tag{10.31}$$

Fix $n \in \mathbb{N}$. As the sequence

$$\left(\int_{W_n(1)} e^{s_k-\varphi}\, dm_{s_k}\right)_{k=1}^{\infty}$$

is bounded, there is a subsequence $\left(s_k^{(1)}\right)_{k=1}^{\infty}$, depending on n, of the sequence $(s_k)_{k=1}^{\infty}$ such that the limit

$$\lim_{k\to\infty} \int_{W_n(1)} e^{s_k^{(1)}-\varphi}\, dm_{s_k^{(1)}}$$

exists. Proceeding by induction, suppose that we have, for some $1 \le j < q$, a subsequence $\left(s_k^{(j)}\right)_{k=1}^{\infty}$ of the sequence $(s_k)_{k=1}^{\infty}$ such that the limit

$$\lim_{k\to\infty} \int_{W_n(i)} e^{s_k^{(i)}-\varphi}\, dm_{s_k^{(i)}}$$

exists for every $i = 1, \ldots, j$. Since the sequence

$$\left(\int_{W_n(j)} e^{s_k^{(j)}-\varphi}\, dm_{s_k^{(j)}}\right)_{k=1}^{\infty}$$

is bounded, there is a subsequence $\left(s_k^{(j+1)}\right)_{k=1}^{\infty}$ of the sequence $\left(s_k^{(j)}\right)_{k=1}^{\infty}$ such that the limit

$$\lim_{k\to\infty} \int_{W_n(j+1)} e^{s_k^{(j+1)}-\varphi}\, dm_{s_k^{(j+1)}}$$

exists. Then the limit

$$\lim_{k\to\infty} \int_{W_n(j)} e^{s_k^{(j)}-\varphi}\, dm^{(q)}_{s_k^{(j)}}$$

exists for all $j = 1, \ldots, q$. By virtue of the hypotheses and Lemma 10.1.13, for all $j = 1, \ldots, q$, the limit

$$\lim_{k\to\infty} m_{s_k^{(q)}}(T(W_n(j)))$$

also exists and

$$\lim_{k\to\infty} m_{s_k^{(q)}}(T(W_n(j))) = \lim_{k\to\infty} \int_{W_n(j)} e^{s_k^{(q)}-\varphi}\, dm_{s_k^{(q)}}.$$

Using this and (10.31), we deduce that

$$\begin{aligned}
m(\{T(w)\}) &\le m\big(T(B(w,r_n))\big) \le \lim_{k\to\infty} m_{s_k^{(q)}}\big(T(B(w,r_n))\big)\\
&= \lim_{k\to\infty} m_{s_k^{(q)}}\left(T\left(\bigcup_{j=1}^{q} W_n(j)\right)\right) = \lim_{k\to\infty} m_{s_k^{(q)}}\left(\bigcup_{j=1}^{q} T\big(W_n(j)\big)\right)\\
&\le \liminf_{k\to\infty}\sum_{j=1}^{q} m_{s_k^{(q)}}\big(T(W_n(j))\big)\\
&= \sum_{j=1}^{q}\lim_{k\to\infty} m_{s_k^{(q)}}\big(T(W_n(j))\big) = \sum_{j=1}^{q}\lim_{k\to\infty}\int_{W_n(j)} e^{s_k-\varphi}\,dm_{s_k^{(q)}}\\
&= \lim_{k\to\infty}\sum_{j=1}^{q}\int_{W_n(j)} e^{s_k^{(q)}-\varphi}\,dm_{s_k^{(q)}} = \lim_{k\to\infty}\int_{B(w,r_n)} e^{s_k^{(q)}-\varphi}\,dm_{s_k^{(q)}}\\
&= \int_{B(w,r_n)} e^{c-\varphi}\,dm.
\end{aligned}$$

Hence,

$$m(\{T(w)\}) \le \lim_{n\to\infty}\int_{B(w,r_n)} e^{c-\varphi}\,dm = e^{c-\varphi}m(\{w\}).$$

The proof of Lemma 10.1.16 is complete. ■

Lemma 10.1.16 *Let $T\colon X \to X$ be a continuous self-map of a separable locally compact metrizable space X. Let $(E_n)_{n=1}^{\infty}$ be a sequence of finite subsets of X such that* (10.22) *holds and $\phi\colon X \longrightarrow (-\infty, +\infty]$ be a bounded below continuous function satisfying* (10.23). *Let m be a limit measure of Definition 10.1.12. If condition* (10.29) *is satisfied for $D = X$, then*

$$m(T(A)) \ge \int_A \exp(c-\phi)dm$$

for every special set A disjoint from Crit(T).

Proof First notice that, according to Theorem 10.1.8, the topological space X is completely metrizable. This is, among others, needed to be sure that images of special sets are Borel. From the definition of the measure m, there exists a sequence $(s_k)_{k=1}^{\infty}$ in (c,∞) such that

$$\lim_{k\to\infty} m_{s_k} = m.$$

Suppose first that A is compact and $m(\partial A) = 0$. It follows from Lemma 10.1.13 that

$$\lim_{k\to\infty}\left|m_{s_k}(T(A)) - \int_A \exp(c-\phi)dm_{s_k}\right| = 0.$$

Since $T(A)$ is also compact, thus closed, this implies that

$$m(T(A)) \geq \limsup_{k\to\infty} m_{s_k}(T(A)) = \lim_{k\to\infty}\int_A \exp(c-\phi)dm_{s_k} = \int_A \exp(c-\phi)dm.$$

Now drop the assumption that $m(\partial A) = 0$ but keep A compact and assume in addition that there exists an open special set $\tilde{A}$ such that $A \subseteq \tilde{A}$. Since A is compact and X is locally compact and metrizable, there exists a descending sequence $(A_n)_{n=1}^{\infty}$ of compact subsets of $\tilde{A}$ whose intersection equals A and $m(\partial A_n) = 0$ for every $n \geq 1$. By what has already been proved, we have that

$$m(T(A)) = \lim_{n\to\infty} m(T(A_n)) \geq \lim_{n\to\infty}\int_{A_n} \exp(c-\phi)dm = \int_A \exp(c-\phi)dm.$$

Now drop the hypothesis that A is compact but keep the assumption that there exists a special open set $\tilde{A}$ containing A. If $K \subseteq A$ is a compact set, then, by the previous case,

$$m(T(A)) \geq m(T(K)) \geq \int_K \exp(c-\varphi)dm. \tag{10.32}$$

Since the space X is completely metrizable (by Theorem 10.1.8), the measure m is inner regular; therefore, taking in (10.32), the supremum over all compact subsets of A, we get that

$$m(T(A)) \geq \int_K \exp(c-\varphi)dm.$$

Finally, keep A special assuming that $A \cap \mathrm{Crit}(T) = \emptyset$. Since $\mathrm{Crit}(T)$ is a closed set, for every $x \in A$, there exists $r_x > 0$ such that the set $B(x, r_x)$ is special. Since the space X, being completely metrizable (by Theorem 10.1.8) and separable, is Lindelöf, there exists a sequence $(x_n)_{n=1}^{\infty}$ of points in A such that

$$\bigcup_{n=1}^{\infty} B(x_n, r_{x_n}) \supseteq A.$$

Now form in the following standard way a countable partition $\{A_n\}_{n=1}^{\infty}$ of A out of cover $\{A \cap B(x_n, r_{x_n})\}_{n=1}^{\infty}$ of A,

$$A_1 := A \cap B(x_1, r_{x_1}), \quad A_{n+1} := A \cap B\big(x_{n+1}, r_{x_{n+1}}\big) \setminus \bigcup_{j=1}^{n} A_j, \quad n \geq 1.$$

Then, by the previous case,

$$\begin{aligned}
m(T(A)) &= m\left(T\left(\bigcup_{n=1}^{\infty} A_n\right)\right) = m\left(\bigcup_{n=1}^{\infty} T(A_n)\right) = \sum_{n=1}^{\infty} m(T(A_n)) \\
&\geq \sum_{n=1}^{\infty} \int_{A_n} \exp(c-\varphi)dm = \int_{\bigcup_{n=1}^{\infty} A_n} \exp(c-\varphi)dm \\
&= \int_A \exp(c-\varphi)dm.
\end{aligned}$$

■

10.2 Sullivan Conformal Measures for Holomorphic Maps f in $\mathcal{A}(X)$ and in $\mathcal{A}^w(X)$

This section is devoted to a comprehensive study of a more special, though extremely important, kind of conformal measures called the Sullivan conformal measures. An extended historical and motivational discussion of these measures, and more general ones as well, was given at the beginning of this chapter. We, therefore, start immediately with actual mathematics.

In this section, X is a compact subset of a parabolic Riemann surface Y, i.e., either the complex plane or a complex torus and $f \in \mathcal{A}(X)$. Our goal here is to provide a construction aiming to establish the existence of measures that will be called (Sullivan) t-conformal measures for the function f. Indeed, we will prove the existence of measures with some slightly weaker properties. We will call them semi-t-conformal Sullivan measures. This will be the main result of the current section. It will be primarily used in Section 17.6 in Volume 2 to prove the existence of the Sullivan conformal measures for elliptic functions. Under the hypothesis that $X_*(f) = \emptyset$, i.e., if $f \in \mathcal{A}_{0,*}(X)$, this main result will give a genuine Sullivan t-conformal measure for the map $f \in \mathcal{A}(X)$ itself. Also, at the end of this section and in the next section, i.e., Section 10.3, we will introduce several dynamically significant subsets of X and several fractal characteristics of the set X, and we will establish some relations between them. In particular, in the case when $f \in \mathcal{A}_{0,*}(X)$, we will provide several fractal-type characterizations of the minimal exponent t for which such Sullivan t-conformal measures exist. In the second volume of the book, all these results will be applied to construct and to study the Sullivan conformal measures for elliptic functions.

To begin with, fix

$$\xi \in X \backslash \bigcup_{n=0}^{\infty} f^{-n}\left(\bigcup_{k=0}^{\infty} f^k(\mathrm{Crit}(f))\right) \tag{10.33}$$

and, for all, $n \geq 1$ set

$$E_n := f^{-(n-1)}(\xi).$$

Then

$$E_{n+1} = f^{-1}(E_n);$$

therefore, the sequence $(E_n)_{n=1}^{\infty}$ satisfies (10.22) and (10.29) with all sets $D \subseteq X$. Fix an arbitrary $t \geq 0$ and observe that the function

$$\phi := -t \log |f'|\colon X \longrightarrow (-\infty, +\infty]$$

is continuous, bounded below, satisfies (10.23), and the function $e^{-\phi} = |f'|^t$ is continuous. Let $c_\xi(t)$ be the transition parameter associated with the sequence $(E_n)_{n=1}^{\infty}$ and the function ϕ according to (10.21) and (10.24). As a fairly immediate consequence of Lemmas 10.1.14–10.1.16, we get the following.

Lemma 10.2.1 *If Y is a parabolic Riemann surface, $X \subseteq Y$ is compact, $f \in \mathcal{A}^w(X)$, and ξ comes from* (10.33)*, then, for every $t \geq 0$, there exists a Borel probability measure m_t on X such that*

$$m_t(f(A)) \geq \int_A e^{c_\xi(t)} |f'|^t \, dm_t$$

for every special set $A \subseteq X$ and

$$m_t(f(A)) = \int_A e^{c_\xi(t)} |f'|^t \, dm_t$$

if, in addition, $A \cap J(f)_* = \emptyset$.

Proof Let m_t be a limit measure of Definition 10.1.12. The first assertion (inequality $\geq$) holds for every special set $A \subseteq X$ disjoint from $\mathrm{Crit}(f)$ because of Lemma 10.1.16. Remembering that the set $\mathrm{Sing}(f)$ is closed, it follows from Lemma 10.1.14 that

$$m_t(f(A)) = \int_A e^{c_\xi(t)} |f'|^t \, dm_t$$

for every special set $A \subseteq X$ disjoint from $\mathrm{Sing}(f)$. But, since $f \in \mathcal{A}^w(X)$, the hypotheses of Lemma 10.1.15 are satisfied for every point $c \in \mathrm{Crit}(f)$. It, therefore, follows from this lemma for every point c that

$$0 \leq m(\{f(c)\}) \leq e^{c_\xi(t)} |f'(c)|^t m(\{c\}) = 0.$$

Hence, all inequalities in this formula become equalities and the proof of Lemma 10.2.1 is complete. ■

Assuming from now on that $f \in \mathcal{A}_0^w(X)$ (which, we recall, means that $f \in \mathcal{A}^w(X)$ and $\text{Crit}(f) = \emptyset$), we want to establish some useful properties of the function $c_\xi(t)$ and to get a parameter $t \geq 0$ such that $c_\xi(t) = 0$. Let

$$\text{P}(t) := \text{P}(f, -t \log |f'|)$$

be the topological pressure of the potential $-t \log |f'|$ with respect to the dynamical system $f : X \to X$. We shall prove the following.

Lemma 10.2.2 *If Y is a parabolic Riemann surface, $X \subseteq Y$ is compact, $f \in \mathcal{A}_0^w(X)$, and ξ comes from* (10.33)*, then, for every $t \geq 0$, we have that $c_\xi(t) \leq \text{P}(t)$.*

Proof Since the map $f : X \to X$ has no critical points it is locally one-to-one at all points of X. Since X is also compact, this implies that there exists $\delta > 0$ such that the map f restricted to any set with diameter $\leq \delta$ is one-to-one. Consequently, all the sets E_n are (n, ε)-separated for all $\varepsilon \in (0, \delta)$. Hence, the required inequality $c_\xi(t) \leq \text{P}(t)$ follows immediately from Theorem 7.2.8. ■

The standard straightforward arguments, such as those showing continuity of topological pressure, also prove the following.

Lemma 10.2.3 *If Y is a parabolic Riemann surface, $X \subseteq Y$ is compact, $f \in \mathcal{A}_0^w(X)$, and ξ comes from* (10.33)*, then the function* $[0, +\infty) \ni t \longmapsto c_\xi(t) \in \mathbb{R}$ *is continuous.*

Set

$$s_\xi(f) := \inf\{t \geq 0 : c_\xi(t) \leq 0\}. \tag{10.34}$$

We call the number

$$\text{DD}_\text{h}(X) := \sup\{\text{HD}(\mu)\} \tag{10.35}$$

the first dynamical dimension of X, where the supremum is taken over all Borel probability f-invariant ergodic measures μ on X with $\text{h}_\mu(f) > 0$.

Similarly, we call the number

$$\text{DD}_\chi(X) := \sup\{\text{HD}(\mu)\} \tag{10.36}$$

the second dynamical dimension of X, where the supremum is taken over all Borel probability f-invariant ergodic measures μ on X with $\chi_\mu(f) > 0$.

Because of Theorem 9.4.1 (Ruelle's Inequality), we have the following immediate inequality:

$$\mathrm{DD}_{\chi}(X) \leq \mathrm{DD}_{\mathrm{h}}(X). \tag{10.37}$$

We shall prove the following.

Lemma 10.2.4 *Let Y be a parabolic Riemann surface and X be a compact subset of Y. If $f \in \mathcal{A}_0^w(X)$, then, for every ξ coming from* (10.33), *we have that*

$$s_{\xi}(f) \leq \mathrm{DD}_{\mathrm{h}}(X) \leq 2.$$

Proof Suppose, on the contrary, that $\mathrm{DD}(X) < s_{\xi}(f)$ for some ξ coming from (10.33) and take $0 \leq \mathrm{DD}(X) < t < s_{\xi}(f)$. From this choice and by Lemma 10.2.2, we have that

$$0 < c_{\xi}(t) \leq \mathrm{P}(t),$$

and, by the Variational Principle, Theorem 7.5.1, there exists an ergodic f-invariant Borel probability measure μ on X such that

$$\mathrm{P}(t) \leq \mathrm{h}_{\mu}(f) - t\chi_{\mu}(f) + \frac{1}{2}c_{\xi}(t).$$

Therefore, by Corollary 9.1.3, we get that $\mathrm{h}_{\mu}(f) \geq c_{\xi}(t)/2 > 0$; applying, in addition, Ruelle's Inequality (Theorem 9.4.1), $\chi_{\mu}(f) > 0$. Hence, it follows from Theorem 9.5.1 that

$$t \leq \mathrm{HD}(\mu) - \frac{1}{2}\frac{c_{\xi}(t)}{\chi_{\mu}(f)} < \mathrm{HD}(\mu) \leq \mathrm{DD}_{\mathrm{h}}(X).$$

This contradiction finishes the proof. ■

Since $c_{\xi}(0) \geq 0$, as an immediate consequence of Lemmas 10.2.3 and 10.2.4, we get that

$$c_{\xi}(s_{\xi}(f)) = 0. \tag{10.38}$$

Inserting this into Lemma 10.2.1, we get the following main result of this section.

Lemma 10.2.5 *Let Y be a parabolic Riemann surface and X be a compact subset of Y. If $f \in \mathcal{A}_0^w(X)$, then, for every ξ coming from* (10.33), *there exists a Borel probability measure m_{ξ} on X such that*

$$m_{\xi}(f(A)) \geq \int_A |f'|^{s_{\xi}(f)}\, dm_{\xi}$$

for every special set $A \subseteq X$ and

$$m_\xi(f(A)) = \int_A |f'|^{s_\xi(f)}\, dm_\xi$$

if, in addition, $A \cap X_*(f) = \emptyset$.

Definition 10.2.6 Any measure with the former property above (with some parameter t in place of $s_\xi(f)$) will be called (Sullivan) semi-t-conformal for f. If, in addition, the latter property (with $s_\xi(f)$ replaced by t) holds for all special sets $A \subseteq X$, then the measure m is called (Sullivan) t-conformal for f.

As an immediate consequence of Lemma 10.2.5, we get the following.

Corollary 10.2.7 *Let Y be a parabolic Riemann surface and X be a compact subset of Y. If $f \in \mathcal{A}^w_{0,*}(X)$, then, for every ξ coming from* (10.33), *there exists a (Sullivan) $s_\xi(f)$-conformal measure for f.*

10.3 Radial Subsets of X, Dynamical Dimensions, and Sullivan Conformal Measures for Holomorphic Maps f in $\mathcal{A}(X)$ and in $\mathcal{A}^w(X)$

We want to say more about the structure of the Sullivan conformal measures, particularly about the exponents for which these exist, for maps in $\mathcal{A}_0(X)$, $\mathcal{A}_{0,*}(X)$, and, especially, in Volume 2, for elliptic functions. We need some new concepts and results to do this. These are also interesting on their own.

Definition 10.3.1 Let Y be a parabolic Riemann surface and X be a compact subset of Y. If $f \in \mathcal{A}^w(X)$, then we say that a point $z \in X$ is radial (or conical) if and only if there exists $\eta > 0$ such that, for infinitely many $n \geq 1$, the map f^n restricted to the connected component $\text{Comp}_z\big(f^{-n}(B(f^n(z),\eta))\big)$ of $f^{-n}(B(f^n(z),\eta))$ containing z is one-to-one. Denote the set of such ns by $N_z(f)$.

We call a radial point $z \in X$ expanding if and only if there exists a number $\lambda > 1$ such that

$$|(f^n)'(z)| \geq \lambda^n \tag{10.39}$$

for infinitely many ns in $N_z(f)$.

We, respectively, denote by $X_r(f)$ and $X_{er}(f)$ the sets of radial and expanding radial points in X.

Given $\eta > 0$, we define $X_r(f)(\eta)$ to be the set of all radial points of f that are witness to this definition to the number 2η. Then also

$$X_{er}(f)(\eta) := X_r(f)(\eta) \cap X_{er}(f).$$

The radial Julia sets for rational functions were indirectly introduced in [**Lj**] and independently, openly, in [**U2**] by analogy with radial/conical sets in the theory of Kleinian groups (see also [**DMNU**] and [**McM2**]). In the realm of transcendental meromorphic functions, these were for the first time introduced and dealt with in [**UZ1**] and then in many papers, especially those of Mayer, Rempe-Gilen, Zdunik, and the second named author of this book. In this book, radial Julia sets will ultimately serve as a good concept and tool to characterize the minimal exponent f for which the Sullivan conformal measures exist for elliptic functions. In this section, we used them for holomorphic maps $f \in \mathcal{A}(X)$ and the results obtained here will be used in Volume 2 to study elliptic functions. Among many other purposes, they have also served the same purpose in the theory of iteration of rational functions.

As an immediate consequence of Lemma 8.1.19, we get the following.

Proposition 10.3.2 *Let Y be a parabolic Riemann surface and X be a compact subset of Y. If $f \in \mathcal{A}(X)$, then*

$$\liminf_{N_z(f)\ni n\to\infty} |(f^n)'(z)| = +\infty.$$

Now we shall prove the first inequality we are really after in the study of exponents of the Sullivan conformal measures of holomorphic maps $f \in \mathcal{A}_0(X)$.

Proposition 10.3.3 *Let Y be a parabolic Riemann surface and X be a compact subset of Y. If $f \in \mathcal{A}^w(X)$, then*

$$\mathrm{DD_h}(X) \le \mathrm{DD}_\chi(X) \le \mathrm{HD}\big(X_{er}(f)\big) \le \mathrm{HD}\big(X_r(f)\big).$$

Proof Only the second inequality is nontrivial and we will prove it now. Our goal is to show that, if μ is a Borel probability f-invariant measure on X with positive Lyapunov exponent χ_μ, then μ-a.e. a point in X is exponentially radial. We will work in the Rokhlin natural extension $(\tilde f : \tilde X \longrightarrow \tilde X, \tilde\mu)$ and we will use Pesin's Theory. Indeed, fix $\varepsilon = 1/2$. Let $\tilde X(1/2) \subseteq \tilde X$ be the set produced in Theorem 9.3.5. Since, by this theorem $\tilde\mu\big(\tilde X(1/2)\big) > 1/2$, it follows from the ergodic part of the Birkhoff Ergodic Theorem (Theorem 2.3.2) and ergodicity of the measure $\tilde\mu$ that there exists a measurable set $Z \subseteq \tilde X$ such that $\tilde\mu(Z) = 1$ and the set

$$Q(z) := \big\{n \ge 0 \colon \tilde f^n(z) \in \tilde X(1/2)\big\}$$

is infinite for every $z \in Z$. For every $n \in Q(z)$, let

$$f^{-n}_{\tilde f^n(z)_n} : B\left(\tilde f^n(z)_0, r(\varepsilon)\right) \longrightarrow X$$

be the holomorphic inverse branch produced in Theorem 9.3.5. Since $\tilde{f}^n(z)_n = z_0$ and $\tilde{f}^n(z)_0 = f^n(z_0)$, and since $\pi_0(z) = z_0$, we conclude from the second to last formula of Theorem 9.3.5 that $\pi_0(Z) \subseteq J_{er}(f)$. But, since also $\mu(\pi_0(Z)) = \tilde{\mu}(\pi_0^{-1}(\pi_0(Z))) \geq 1$, we get that

$$\mathrm{HD}(J_{er}(f)) \geq \mathrm{HD}(\pi_0(Z)) \geq \mathrm{HD}(\mu).$$

So, taking the supremum over all such measures μ, we obtain that $\mathrm{HD}(J_{er}(f)) \geq \mathrm{DD}_X(X)$ and the proof of Proposition 10.3.3 is complete. ■

We say that if Z is a topological space, then a continuous map $T\colon Z \to Z$ is called topologically transitive if and only if there exists a point $z \in Z$ whose ω-limit set $\omega(z)$, i.e., the set of all limit points of the sequence $(T^n(z))_{n=1}^{\infty}$, is equal to z.

In order to provide all our characterizations of the least exponent of all the Sullivan conformal measures for holomorphic maps in $\mathcal{A}(X)$, we need the following lemma involving conformal measures and radial sets $X_r(f)$.

Lemma 10.3.4 *Let Y be a parabolic Riemann surface and X be a compact subset of Y. If $f \in \mathcal{A}_*(X)$, the map $f\colon X \to X$ is topologically transitive, $t \geq 0$, and ν is a t-conformal (probability) measure, then $t \geq \mathrm{HD}(X_r(f))$ and $\mathrm{H}^t|_{X_r(f)}$ is absolutely continuous with respect to ν, i.e., $\mathrm{H}^t|_{X_r(f)} \prec \nu$.*

Proof Since the map $f\colon X \to X$ is topologically transitive, t-conformality of ν immediately implies that $\mathrm{supp}(\nu) = X$. Fix $z \in X_r(f)$. Then there exists $\eta > 0$ such that $z \in X_r(f)(\eta)$, which, along with the infinite subset $N_z(f)$ of natural numbers, come from Definition 10.3.1. Fix $n \in N_z(f)$. It follows then from Lemma 8.3.13 that

$$f_z^{-n}\big(B(f^n(z),\eta)\big) \subseteq B\big(z, K\eta|(f^n)'(z)|^{-1}\big).$$

Applying Theorem 8.3.9, conformality of the measure m, and Theorem 1.2.12, we, thus, get that

$$\begin{aligned}
\nu\Big(B\big(z, K\eta|(f^n)'(z)|^{-1}\big)\Big) &\geq K^{-t}|(f^n)'(z)|^{-t}\nu\big(B(f^n(z),\eta)\big) \\
&\geq K^{-t}|(f^n)'(z)|^{-t}M(\nu,\eta) \\
&= M(\nu,\eta)K^{-2t}\eta^{-t}\big(K\eta|(f^n)'(z)|^{-1}\big)^t,
\end{aligned}$$

where $M(\nu,\eta) > 0$ comes from (1.19) of Theorem 1.2.12. Thus, letting $n \in N_z(f)$ tend to $+\infty$, and also using Proposition 10.3.2, we get that

$$\limsup_{r\to 0} \frac{\nu(B(z,r))}{r^t} \geq M(\nu,\eta)K^{-2t}\eta^{-t}.$$

Therefore, $\mathrm{H}^t|_{X_r(f)(\eta)}$ is absolutely continuous with respect to ν, i.e., $\mathrm{H}^t|_{X_r(f)(\eta)} \prec \nu$ because of Theorem 1.6.3(1). Since

$$X_r(f) = \bigcup_{k=1}^{\infty} X_r(f)(1/k),$$

the proof of Lemma 10.3.4 is complete. ■

As an immediate consequence of Corollary 10.2.7, Proposition 10.3.3, and Lemmas 10.3.4 and 10.2.4, we get the following.

Theorem 10.3.5 *Let Y be a parabolic Riemann surface and X be a compact subset of Y. If* $f \in \mathcal{A}_{0,*}(X)$, *then*

$$\mathrm{DD}_{\mathrm{h}}(X) = \mathrm{DD}_{\chi}(X) = \mathrm{HD}\big(X_{er}(f)\big) = \mathrm{HD}\big(X_r(f)\big);$$

denoting this common value by s_f, *there exists an* s_f*-conformal measure m on X for the map f.*

In addition, s_f *is the least exponent* $t \geq 0$ *for which there exists a t-conformal measure for f supported on X.*

10.4 Conformal Pairs of Measures

In this section, we examine in detail how the Sullivan conformal measures, or, rather, the Sullivan like conformal measures, behave under transformations, particularly those with critical points. We will eventually use them for one given conformal measure but, in order to see clearly what is going on and what matters, it is most natural to present these results in the setting of conformal and semi-conformal pairs of measures. The results of this section, which are quite technical, are all taken from Sections 2–4 of [**U3**]. We will apply them in Volume 2 of the book when dealing with conformal measures for elliptic functions. Motivated by the last definition in the previous section, we formulate the following.

Definition 10.4.1 Let Y be either an elliptic or parabolic Riemann surface. Fix $t \geq 0$. Let G and H be nonempty open subsets of Y. Let $f: G \to H$ be a holomorphic map. A pair (m_G, m_H) of Borel finite measures, respectively, on G and H is called semi-t-conformal for the map $f: G \longrightarrow H$ if and only if

$$m_H(f(A)) \geq \int_A |f'|^t \, dm_G \tag{10.40}$$

for every Borel set $A \subseteq G$ such that $f|_A$ is injective; this pair is said to be t-conformal for $f: G \to H$ if and only if

$$m_H(f(A)) = \int_A |f'|^t \, dm_G \tag{10.41}$$

for these sets A.

(a) If $Y = \mathbb{C}$, then, sometimes, to avoid any possible ambiguity, we will refer to the pair (m_G, m_H) as Euclidean (semi)-t-conformal for the map $f: G \to H$. Similarly, if $Y = \widehat{\mathbb{C}}$, we will sometimes refer to the pair (m_G, m_H) as spherical (semi)-t-conformal for the map $f: G \to H$. If G and H are contained in $\mathbb{C}$, but we want them to be treated as subsets of the Riemann surface $\widehat{\mathbb{C}}$ and we want to use spherical metrics and derivatives, we will write $|f'|_s$ in both (10.40) and (10.41).

(b) If both measures m_G and m_H are restrictions of the same Borel finite measure m defined on $G \cup H$, we refer to m as a (semi)-t-conformal measure for the map $f: G \to H$.

Definition 10.4.2 Let X be a metric space. Let ν be a Borel finite measure on X. Given $t > 0$, $r > 0$, and $L > 0$, a point $x \in X$ is said to be (r, L)-t-upper estimable with respect to a finite Borel measure ν if and only if

$$\nu(B(x,r)) \le Lr^t.$$

The point x is said to be (r, L)-t-lower estimable with respect to ν if and only if

$$\nu(B(x,r)) \ge Lr^t.$$

We will frequently abbreviate the notation, writing (r, L)-t-u.e. for (r, L)-t-upper estimable and (r, L)-t-l.e. for (r, L)-t-lower estimable.

We also say that the point x is t-upper estimable (t-lower estimable) if it is (r, L)-t-upper estimable ((r, L)-t-lower estimable) for some $L > 0$ and all $r > 0$ sufficiently small.

Definition 10.4.3 Let X be a metric space. Let ν be a Borel finite measure on X. Fix $t \ge 0$. Given $r > 0$, $\sigma > 0$, and $L > 0$, a point $x \in X$ is said to be (r, σ, L)-t-strongly lower estimable, or simply (r, σ, L)-t-s.l.e. with respect to the measure ν if and only if

$$\nu(B(y, \sigma r)) \ge Lr^t$$

for every $y \in B(x,r)$.

Lemma 10.4.4 *Let X be a metric space. Let ν be a Borel finite measure on X. Fix $t \ge 0$. Fix $r > 0$, $\sigma > 0$, and $L > 0$. If a point $z \in X$ is (r, σ, L)-t-s.l.e.*

with respect to the measure ν, *then every point* $x \in B(z, r/2)$ *is* $(r/2, 2\sigma, 2^t L)$-t-*s.l.e. with respect to this measure.*

Proof Let $y \in B(z, r/2)$. Then $y \in B(z, r)$; therefore,

$$\nu(B(y, 2\sigma(r/2))) = \nu(B(y, \sigma r)) \geq Lr^t = 2^t L(r/2)^t.$$

■

Lemma 10.4.5 *Let* X *be a metric space. Let* ν *be a Borel finite measure on* X. *Fix* $t \geq 0$. *Fix* $r > 0$, $\sigma > 0$, *and* $L > 0$. *If a point* $x \in X$ *is* (r, σ, L)-t-*s.l.e. with respect to the measure* ν, *then, for every* $0 < u \leq 1$, *the point* x *is* $(ur, \sigma/u, Lu^{-t})$-t-*s.l.e. with respect to this measure.*

Proof If $y \in B(x, ur)$, then $y \in B(x, r)$; therefore,

$$\nu(B(y, (\sigma/u)ur)) = \nu(B_e(y, \sigma r)) \geq Lr^t = Lu^{-t}(ur)^t.$$

■

We would like to finish this part, not involving mappings, with the following obvious fact.

Lemma 10.4.6 *Let* X *be a metric space. If* ν *is a Borel finite measure on* X *positive on nonempty open sets, i.e.,* $\text{supp}(\nu) = X$, *then, for every* $r > 0$, *there exists* $E(r) \geq 1$ *such that every point* $x \in X$ *is* $(r, E(r))$-t-*u.e. and* $(r, E(r)^{-1})$-t-*l.e. with respect to* ν.

The following lemma, the first one involving dynamics, is a straightforward consequence of Theorems 8.3.9 and 8.3.12.

Lemma 10.4.7 *Let* Y *be either an elliptic or parabolic Riemann surface. Fix* $t \geq 0$. *Let* G *and* H *be nonempty open subsets of* Y. *Let* (m_G, m_H) *be a semi-*t-*conformal pair of measures for a univalent holomorphic map* $f : G \to H$. *If* $z \in B(z, 2R) \subseteq G$, *then, for every* $0 \leq r \leq R$, *we have that*

$$m_G\big(B(z, K^{-1}r|f'(z)|^{-1})\big) \leq K^t|f'(z)|^{-t} m_H(B(f(z), r)).$$

If, in addition, the pair (m_G, m_H) *is* t-*conformal, then also*

$$m_G\big(B(z, Kr|f'(z)|^{-1})\big) \geq K^{-t}|f'(z)|^{-t} m_H(B(f(z), r)),$$

where the constant $K \geq 1$ *comes from either Theorem 8.3.9 or Theorem 8.3.12.*

Lemma 10.4.8 *Let* Y *be either an elliptic or parabolic Riemann surface. Fix* $t \geq 0$. *Let* G *and* H *be nonempty open subsets of* Y. *Let* (m_G, m_H) *be a semi-*t-*conformal pair of measures for a univalent holomorphic map* $f : G \to H$. *Let* $z \in B(z, 2R) \subseteq G$.

If the point $H(z)$ is (r,σ,L)-t-s.l.e. with respect to m_H, where $r \leq R/2$ and $\sigma \leq 1$, then the point z is $(K^{-1}|f'(z)|^{-1}r, K^2\sigma, L)$-$t$-s.l.e. with respect to m_G, where the constant $K \geq 1$ comes from either Theorem 8.3.9 or Theorem 8.3.12.

Proof In this proof, we apply Lemma 10.4.7 several times without indicating it. Consider

$$x \in B\big(z, K^{-1}r|f'(z)|^{-1}\big).$$

Then $f(x) \in B(f(z),r)$; therefore, $m_H(B(f(x),\sigma r)) \geq Lr^t$. Since

$$B(f(x),\sigma r) \subseteq B(f(z),2r) \subseteq B(f(z),R),$$

we have that

$$f_z^{-1}\big(B(f(x),\sigma r)\big) \subseteq B\big(x, K\sigma r|f'(z)|^{-1}\big) = B\big(x, K^2\sigma(K^{-1}|f'(z)|^{-1}r)\big).$$

Thus,

$$\nu\big(B\big(x, K^2\sigma\big(K^{-1}|f'(z)|^{-1}r\big)\big) \geq K^{-t}|f'(z)|^{-t}Lr^t = L\big(K^{-1}|f'(z)|^{-1}r\big)^t.$$

The proof is finished. ■

Lemma 10.4.9 *Let Y be either an elliptic or parabolic Riemann surface. Fix $t \geq 0$. Let G and H be nonempty open subsets of Y. Let (m_G, m_H) be a semi-t-conformal pair of measures for a univalent holomorphic map $f : G \to H$. If $0 < r \leq R(H,c)$ and $f(c)$ is (r,L)-t-l.e. with respect to m_H, then c is*

$$\big((A(c)r)^{1/p_c}, (A(c))^{-2t}L\big)\text{-}t\text{-l.e.}$$

with respect to m_G, where $A(c) := A(f,c)$ comes from (8.34) *and* (8.35) *in Section 8.4 and Remark 8.4.2.*

Proof By (8.34) along with Remark 8.4.2, we get that $B(f(c),r) = f(\mathrm{Comp}(c,f,r))$. If $x \in \mathrm{Comp}(c,f,r)$, then

$$A^{-1}(c)|x-c|^{p_c} \leq |f(x)-f(c)| < r,$$

which implies that $x \in B(c,(A(c)r)^{1/p_c})$. Thus, $B(f(c),r) \subseteq f\big(B(c,(A(c)r)^{1/p_c})\big)$; therefore,

$$\begin{aligned}
Lr^t &\le m_H(B(f(c),r)) \le m_H\big(f(B(c,(A(c)r)^{1/p_c}))\big) \\
&\le \int_{B(c,(A(c)r)^{1/p_c})} |f'(z)|^t \, dm_G(z) \\
&\le \int_{B(c,(A(c)r)^{1/p_c})} A^t(c)\big(|z-c|^{p_c-1}\big)^t \, dm_G(z) \\
&\le A^t(c)(Ar)^{\frac{p_c-1}{p_c}t} m_G\big(B(c,A(c)r)^{1/p_c}\big).
\end{aligned}$$

So, $m_G\big(B(c,(A(c)r)^{1/p_c})\big) \ge A^{-2t}(c)L\big((A(c)r)^{1/p_c}\big)^t$. ■

Lemma 10.4.10 *Let Y be either an elliptic or parabolic Riemann surface. Fix $t \ge 0$. Let G and H be nonempty open subsets of Y. Let (m_G, m_H) be a semi-t-conformal pair of measures for an analytic map $f: G \to H$. Let $c \in G$ be a critical point of f such that $m_G(c) = 0$. If $0 < r \le R(f,c)$ and $f(c)$ is (s,L)-t-u.e. with respect to m_H for all $0 < s \le r$, then c is*

$$\big((A^{-1}(c)r)^{1/p_c}, q(2(A(c))^2)^t(2^{t/p_c}-1)^{-1}L\big)\text{-}t\text{-}u.e.$$

with respect to m_G.

Proof Take any $0 < s \le r$. Then $f\big(B(c,(A^{-1}(c)s)^{1/p_c})\big) \subseteq B(f(c),r)$. Therefore, recalling that $A(c;a,b) = \{z\colon a \le |z-c| < b\}$, denoting that

$$A_n(c) := A\left(c; 2^{-1/p_c}(A^{-1}(c)s)^{1/p_c}, (A^{-1}(c)s)^{1/p_c}\right),$$

and using the decomposition of $B(c,A^{-1}(c)s)^{1/p_c})$ guaranteed by Observation 8.4.1 and Remark 8.4.2 along with (8.35) (and also Remark 8.4.2), we obtain that

$$\begin{aligned}
Ls^t &\ge m_H(B_e(f(c),s)) \ge m_H\Big(f\big(B(c,A^{-1}(c)s)^{1/p_c}\big)\Big) \\
&= p_c^{-1}\int_{B\left(c,(A^{-1}(c)s)\right)^{1/p_c}} |f'(z)|^t \, dm_G(z) \ge p_c^{-1}\int_{A_n(c)} |f'(z)|^t \, dm_G(z) \\
&\ge p_c^{-1}A^{-t}\big(2^{-1}A^{-1}(c)s\big)^{\frac{p_c-1}{p_c}t} m_G(R(c)).
\end{aligned}$$

So,

$$\begin{aligned}
&m_G\big(A(c;2^{-1/p_c}(A^{-1}(c)s)^{1/p_c},(A^{-1}(c)s)^{1/p_c})\big) \\
&\quad\le p_c 2^{t\left(1-\frac{1}{p_c}\right)}(A^{2t}(c))L\big((A^{-1}(c))^{1/p_c}\big)^t;
\end{aligned}$$

therefore,

$$\begin{aligned}
& m_G\big(B_e(c,(A^{-1}(c)r)^{1/p_c})\big) \\
&= m_G\left(\bigcup_{n=0}^{\infty} A\Big(c;2^{-\frac{n+1}{p_c}}(A^{-1}(c)r)^{1/p_c},2^{-\frac{n}{p_c}}((A(c))^{-1}r)^{1/p_c}\Big)\right) \\
&= \sum_{n=0}^{\infty} m_G\Big(A(c;2^{-\frac{1}{p_c}}(A^{-1}(c)2^{-n}r)^{1/p_c},(A^{-1}(c)2^{-n}r)^{1/p_c})\Big) \\
&\le p_c\big(2^{1-\frac{1}{p_c}}A^2(c)\big)^t L\sum_{n=0}^{\infty}\Big(A^{-1}(c)2^{-n}r\Big)^{t/p_c} \\
&= p_c\big(2^{1-\frac{1}{p_c}}A^2(c)\big)^t \frac{L}{1-2^{-\frac{t}{p_c}}}\big((A^{-1}(c)r)^{1/p_c}\big)^t \\
&= p_c\big(2A^2(c)\big)^t\big(2^{t/p_c}-1\big)^{-1}L\big((A^{-1}(c)r)^{1/p_c}\big)^t.
\end{aligned}$$

The proof is finished. ■

Lemma 10.4.11 *Let Y be either an elliptic or parabolic Riemann surface. Fix $t \ge 0$. Let G and H be nonempty open subsets of Y. Let (m_G, m_H) be a semi-t-conformal pair of measures for an analytic map $f: G \to H$. Let $c \in G$ be a critical point of the analytic map f. If $0 < r \le \frac{1}{3}R(H,c)$, $0 < \sigma \le 1$, and $f(c)$ is (r,σ,L)-t-s.l.e. with respect to m_H, then c is*

$$\big((A^{-1}(c)r)^{1/p_c},\tilde{\sigma},\tilde{L}\big)\text{-}t\text{-s.l.e.}$$

with respect to m_G, where $\tilde{\sigma} := (2^{p_c+1}K(A(c))^2\sigma)^{1/p_c}$ and $\tilde{L} := L\min\{K^{-t},((A(c))^2\sigma)^{\frac{1-p_c}{p_c}t}\}$.

Proof Let $x \in B\big(c,((A(c))^{-1}r)^{1/p_c}\big)$. If $\tilde{\sigma}((A(c))^{-1}r)^{1/p_c} \ge 2|x-c|$, then

$$\begin{aligned}
B(x,\tilde{\sigma}(A^{-1}(c)r)^{1/p_c}) &\supseteq B\big(c,\tilde{\sigma}(A^{-1}(c)r)^{1/p_c}/2\big) = B\big(c,(2K)^{1/p_c}(A(c)\sigma r)^{1/p_c}\big) \\
&\supseteq B\big(c,(A(c)\sigma r)^{1/p_c}\big).
\end{aligned}$$

It follows from our assumptions that $f(c)$ is $(\sigma r,\sigma^{-t}L)$-l.e. with respect to m_H; therefore, in view of Lemma 10.4.9, the critical point c is $((A(c)\sigma r)^{1/p_c},(A(c))^{-2t}\sigma^{-t}L)$-l.e. with respect to m_G. Thus,

$$\begin{aligned}
m_G\big(B(x,\tilde{\sigma}(A^{-1}(c)r)^{1/p_c})\big) &\ge A^{-2t}(c)\sigma^{-t}L(A(c)\sigma r)^{t/p_c} \\
&= (A^2(c)\sigma)^{\frac{1-p_c}{p_c}t}L\big((A^{-1}(c)r)^{1/p_c}\big)^t.
\end{aligned} \tag{10.42}$$

So, suppose that

$$\tilde{\sigma}\big((A(c))^{-1}r\big)^{1/p_c} < 2|x-c|. \tag{10.43}$$

Since c is a critical point, we have, by (8.35) and Remark 8.4.2, that

$$|f'(x)| \geq A^{-1}(c)|x-c|^{p_c-1} \geq A^{-1}(c)\tilde{\sigma}^{p_c-1}(A^{-1}(c)r)^{\frac{p_c-1}{p_c}} 2^{1-p_c},$$

which means that

$$\begin{aligned}\tilde{\sigma}(A^{-1}(c)r)^{1/p_c} &\geq A^{-1}(c)\tilde{\sigma}_c^{p} A^{-1}(c)r2^{1-p_c}|f'(x)|^{-1} \\ &= 4K\sigma r|f'(x)|^{-1} \\ &\geq K\sigma r|f'(x)|^{-1}.\end{aligned} \tag{10.44}$$

In view of (10.43),

$$|f(x)-f(c)| \geq A^{-1}(c)|x-c|^{p_c} \geq A^{-1}(c)2^{-p_c}\tilde{\sigma}^{p_c}A^{-1}(c)r = 2K\sigma r \geq 2\sigma r,$$

which implies that

$$f(c) \notin B_e(f(x), 2\sigma r). \tag{10.45}$$

Since $|f(x)-f(c)| \leq A(c)|x-c|^{p_c} \leq R/3$, we have that $B_e(f(x),2\sigma r) \subseteq B_e(f(c),R)$. So, (10.45) implies the existence of a holomorphic inverse branch

$$f_x^{-1}: B_e(f(x),2\sigma r) \longrightarrow G$$

of f, which sends $f(x)$ to x. Since, by our assumptions, $f(x)$ is $(\sigma r, \sigma^{-t}L)$-l.e. with respect to m_H, it follows from Lemma 10.4.8 that x is $(K\sigma r|f'(x)|^{-1}, (K^2\sigma)^{-t}L)$-l.e. with respect to m_G. Thus, using (10.44), we get that

$$\begin{aligned}m_G\big(B\big(x,\tilde{\sigma}((A(c))^{-1}r)^{1/p_c}\big)\big) &\geq m_G\big(B\big(x, K\sigma r|f'(x)|^{-1}\big)\big) \\ &\geq (K^2\sigma)^{-t}L\big(K\sigma r|f'(x)|^{-1}\big)^t \\ &\geq K^{-t}Lr^tA^{-t}(c)|x-c|^{(1-p_c)t} \\ &\geq K^{-t}L\big(A^{-1}(c)r\big)^t\big(A^{-1}(c)r\big)^{\frac{1-p_c}{p_c}t} \\ &= K^{-t}L\big((A^{-1}(c)r)^{1/p_c}\big)^t.\end{aligned}$$

In view of this and (10.42), the proof is complete. ■

11
Graph Directed Markov Systems

In this chapter, we describe a powerful method to construct and study geometric and dynamical properties of fractal sets. This method is given by the theory of countable alphabet conformal iterated function systems (IFSs), or more generally of countable alphabet conformal graph directed Markov systems (CGDMSs), as developed in the papers [**MU1**] and [**MU4**] and the book [**MU2**]. We present some elements of this theory now, primarily those related to conformal measures and a version of Bowen's Formula for the Hausdorff dimension of limit sets of such systems. In particular, we will get an almost cost-free, effective, lower estimate for the Hausdorff dimension of such limit sets. More about CGDMSs can be found in many papers and books, including [**MU3**]–[**MU5**], [**MPU**], [**MSzU**], [**U5**], [**U6**], [**Hen**], [**CTU**], [**CLU1**], [**CLU2**], and [**CU**].

Afterwards, in Volume 2, first in Chapter 17, we will apply the techniques developed here to get quite a good, very important explicit estimate from below of the Hausdorff dimensions of the Julia sets of all elliptic functions. Second, and this is a much more involved application, by means of the theory of nice sets presented in the next chapter, i.e., Chapter 12, we will apply in Chapter 22, the theory of CGDMSs to explore ergodic and stochastic properties of invariant versions of the Sullivan conformal measures for parabolic and subexpanding elliptic functions.

11.1 Subshifts of Finite Type over Infinite Alphabets: Topological Pressure

Let

$$\mathbb{N} := \{1, 2, \dots\}$$

be the set of all natural numbers, i.e., of all positive integers, and E be a countable, either finite or infinite, set, called in what follows an alphabet. Let

$$\sigma : E^{\mathbb{N}} \longrightarrow E^{\mathbb{N}}$$

be the shift map, i.e., the map cutting off the first coordinate. It is given by the formula

$$\sigma\left((\omega_n)_{n=1}^{\infty}\right) := \left((\omega_{n+1})_{n=1}^{\infty}\right).$$

We also set

$$E^* := \bigcup_{n=0}^{\infty} E^n.$$

For every $\omega \in E^*$, by $|\omega|$ we mean the only integer $n \geq 0$ such that $\omega \in E^n$. We call $|\omega|$ the length of ω. We make a convention that $E^0 = \emptyset$. If $\omega \in E^{\mathbb{N}}$ and $n \geq 1$, we put

$$\omega_{|n} := \omega_1, \ldots, \omega_n \in E^n.$$

Given $\omega, \tau \in E^{\infty}$, we define $\omega \wedge \tau \in E^{\infty} \cup E^*$ to be the longest initial block common to both ω and τ. For each $\alpha > 0$, we define a d_α metric d_α, on E^{∞}, by setting

$$d_\alpha(\omega, \tau) := \mathrm{e}^{-\alpha|\omega \wedge \tau|}. \tag{11.1}$$

These metrics are all mutually equivalent, meaning that they induce the same topology on E^{∞}. Of course, they then define the same σ-algebras of Borel sets.

A function defined on E^{∞} is uniformly continuous with respect to one of these metrics if and only if it is uniformly continuous with respect to all of them. Also, a function is Hölder continuous with respect to one of these metrics if and only if it is Hölder with respect to all of them; of course, the Hölder exponent depends on the metric. If no metric is specifically mentioned, we take it to be d_1.

Now consider a 0–1 matrix

$$A : E \times E \longrightarrow \{0, 1\}.$$

Any such matrix is called an incidence matrix. Set

$$E_A^{\infty} := \left\{\omega \in E^{\mathbb{N}} : A_{\omega_i \omega_{i+1}} = 1 \text{ for all } i \geq 1\right\}.$$

All elements of E_A^{∞} are called A-admissible. We also set

$$E_A^n := \left\{w \in E^{\mathbb{N}} : \ A_{\omega_i \omega_{i+1}} = 1 \text{ for all } 1 \leq i \leq n-1\right\}, \quad n \geq 1,$$

and

$$E_A^* := \bigcup_{n=0}^{\infty} E_A^n.$$

The elements of these sets are also called A-admissible. For every $\omega \in E_A^*$, we put

$$[\omega] := \left\{\tau \in E_A^{\infty}: \ \tau_{||\omega|} = \omega\right\}$$

and call this set the cylinder generated by ω. The following proposition is obvious.

Proposition 11.1.1 *If E is a countable set and $A\colon E \times E \longrightarrow \{0,1\}$ is an incidence matrix, then $E_A^{\mathbb{N}}$ is a closed subset of E^{∞} invariant under the shift map $\sigma: E^{\infty} \longrightarrow E^{\infty}$. The latter means that*

$$\sigma(E_A^{\infty}) \subseteq E_A^{\infty}.$$

The matrix $A\colon E \times E \longrightarrow \{0,1\}$ is said to be *finitely irreducible* if and only if there exists a finite set $\Lambda \subseteq E_A^*$ such that, for all $i, j \in E$, there exists a path $\omega \in \Lambda$ for which

$$i\omega j \in E_A^*.$$

Given a set $F \subseteq E$, we put

$$F^{\infty} := \{\omega \in E^{\mathbb{N}} : \omega_i \in F \ \text{ for all } \ i \geq 1\}.$$

Given $F \subseteq E$ and a function $f\colon F_A^{\infty} \longrightarrow \mathbb{R}$, we define the standard *nth partition function* by

$$Z_n(F, f) := \sum_{\omega \in F_A^n} \exp\left(\sup_{\tau \in [\omega]} \sum_{j=0}^{n-1} f(\sigma^j(\tau))\right).$$

Recall from Definition 7.1.20 that a sequence $\{a_n\}_{n=1}^{\infty}$ consisting of real numbers is said to be subadditive if and only if

$$a_{n+m} \leq a_n + a_n$$

for all $m, n \geq 1$. We will need the following.

Lemma 11.1.2 *The sequence* $\mathbb{N} \ni n \longmapsto \log Z_n(F, f)$ *is subadditive.*

Proof We need to show that the sequence $\mathbb{N} \ni n \longmapsto Z_n(F, f)$ is submultiplicative, i.e., that

$$Z_{m+n}(F, f) \leq Z_m(F, f) Z_n(F, f)$$

for all $m, n \geq 1$. And, indeed,

$$\begin{aligned}
Z_{m+n}(F, f) &= \sum_{\omega \in F_A^{m+n}} \exp\left(\sup_{\tau \in [\omega]} \sum_{j=0}^{mn-1} f(\sigma^j(\tau))\right) \\
&= \sum_{\omega \in F_A^{m+n}} \exp\left(\sup_{\tau \in [\omega]} \left\{\sum_{j=0}^{m-1} f(\sigma^j(\tau)) + \sum_{j=0}^{n-1} f(\sigma^j(\sigma^m(\tau)))\right\}\right) \\
&\leq \sum_{\omega \in F_A^{m+n}} \exp\left(\sup_{\tau \in [\omega]} \sum_{j=0}^{m-1} f(\sigma^j(\tau)) + \sup_{\tau \in [\omega]_F} \sum_{j=0}^{n-1} f(\sigma^j(\sigma^m(\tau)))\right) \\
&\leq \sum_{\omega \in F_A^m} \sum_{\rho \in F_A^n} \exp\left(\sup_{\tau \in [\omega]} \sum_{j=0}^{m-1} f(\sigma^j(\tau)) + \sup_{\gamma \in [\rho]} \sum_{j=0}^{n-1} f(\sigma^j(\gamma))\right) \\
&= \sum_{\omega \in F_A^m} \exp\left(\sup_{\tau \in [\omega]} \sum_{j=0}^{m-1} f(\sigma^j(\tau))\right) \cdot \sum_{\rho \in F_A^n} \exp\left(\sup_{\gamma \in [\rho]} \sum_{j=0}^{m-1} f(\sigma^j(\gamma))\right) \\
&= Z_m(F, f) Z_n(F, f).
\end{aligned}$$

■

We can now define the topological pressure of f with respect to the shift map $\sigma : F_A^{\mathbb{N}} \longrightarrow F_A^{\mathbb{N}}$. Indeed, combining Lemmas 11.1.2 and 7.1.22, we see that the following limit exists and the equality following it holds.

$$\mathrm{P}_F(f) := \lim_{n \to \infty} \frac{1}{n} \log Z_n(F, f) = \inf_{n \in \mathbb{N}} \left\{\frac{1}{n} \log Z_n(F, f)\right\}. \tag{11.2}$$

If $F = E$, we suppress the subscript F and write simply $\mathrm{P}(f)$ for $\mathrm{P}_E(f)$ and $Z_n(f)$ for $Z_n(E, f)$.

Definition 11.1.3 A function $f : E_A^\infty \longrightarrow \mathbb{R}$ is said to be acceptable provided it is uniformly continuous and

$$\mathrm{osc}(f) := \sup\left\{\sup(f|_{[e]}) - \inf(f|_{[e]}) : e \in E\right\} < +\infty.$$

Note that an acceptable function need not be bounded. In fact, in what follows, it usually will not. We shall prove the following.

Theorem 11.1.4 *If* $f\colon E_A^\infty \longrightarrow \mathbb{R}$ *is acceptable and* A *is finitely irreducible, then*

$$\mathrm{P}(f) = \sup\{\mathrm{P}_F(f)\},$$

where the supremum is taken over all finite subsets F *of* E.

Proof The inequality

$$\mathrm{P}(f) \geq \sup\{\mathrm{P}_F(f)\}$$

is obvious. In order to prove the converse, let $\Lambda \subseteq E_A^*$ witness finite irreducibilitly of the matrix A. We shall show first that

$$\mathrm{P}(f) < +\infty.$$

Put

$$q := \#\Lambda,\ p := \max\{|\omega|\colon \omega \in \Lambda\},\ \text{ and}$$
$$T := \min\left\{\inf\left\{\sum_{j=0}^{|\omega|-1} f\circ\sigma^j|_{[\omega]}\right\} : \omega \in \Lambda\right\}.$$

Fix $\varepsilon > 0$. By acceptability of the function $f_A^\infty \longrightarrow \mathbb{R}$, we have that $M := \mathrm{osc}(f) < +\infty$ and there exists $l \geq 1$ such that

$$|f(\omega) - f(\tau)| < \varepsilon$$

whenever $\omega|_l = \tau|_l$. Now fix $k \geq l$. By Lemma 11.1.2, $\frac{1}{k}\log Z_k(f) \geq \mathrm{P}(f)$. Therefore, there exists a finite set $F \subseteq I$ such that

$$\frac{1}{k}\log Z_k(F, f) > \mathrm{P}(f) - \varepsilon. \tag{11.3}$$

We may assume that F contains Λ. Put

$$\overline{f} := \sum_{j=0}^{k-1} f\circ\sigma^j.$$

Now, for every element $\tau = \tau_1, \tau_2, \ldots, \tau_n \in F_A^k \times \cdots \times F_A^k$ (n factors), one can choose elements $\alpha_1, \alpha_2, \ldots, \alpha_{n-1} \in \Lambda$ such that

$$\overline{\tau} := \tau_1\alpha_1\tau_2\alpha_2, \ldots, \tau_{n-1}\alpha_{n-1}\tau_n \in E^*.$$

Notice that the function $\tau \longmapsto \overline{\tau}$ defined in this way is at most u^{n-1}-to-one, where u is the number of lengths of words composing Λ. Then, for every $n \geq 1$,

$$
\begin{aligned}
q^{n-1} & \sum_{i=kn}^{kn+p(n-1)} Z_i(F,f) \\
& \geq \sum_{\tau \in (F_A^k)^n} \exp\left(\sup_{[\overline{\tau}]} \sum_{j=0}^{|\overline{\tau}|} f \circ \sigma^j\right) \geq \sum_{\tau \in (F_A^k)^n} \exp\left(\inf_{[\overline{\tau}]} \sum_{j=0}^{|\overline{\tau}|} f \circ \sigma^j\right) \\
& \geq \sum_{\tau \in (F_A^k)^n} \exp\left(\sum_{i=1}^{n} \inf \overline{f}|_{[\tau_i]} + T(n-1)\right) \\
& = \exp(T(n-1)) \sum_{\tau \in (F_A^k)^n} \exp \sum_{i=1}^{n} \inf \overline{f}|_{[\tau_i]} \\
& \geq \exp(T(n-1)) \sum_{\tau \in (F_A^k)^n} \exp\left(\sum_{i=1}^{n} (\sup \overline{f}|_{[\tau_i]} - (k-l)\varepsilon - Ml)\right) \\
& = \exp(T(n-1) - (k-l)\varepsilon n - Mln) \sum_{\tau \in (F_A^k)^n} \exp \sum_{i=1}^{n} \sup \overline{f}|_{[\tau_i]} \\
& = \mathrm{e}^{-T} \exp\big(n(T - (k-l)\varepsilon - Ml)\big) \left(\sum_{\tau \in F_A^k} \exp(\sup \overline{f}|_{[\tau]})\right)^n .
\end{aligned}
$$

Hence, there exists $kn \leq i_n \leq (k+p)n$ such that

$$
Z_{i_n}(F,f) \geq \frac{1}{pn} \mathrm{e}^{-T} \exp\big(n(T - (k-l)\varepsilon - Ml - \log q)\big) Z_k(F,f)^n;
$$

therefore, using (11.3), we obtain that

$$
\begin{aligned}
\mathrm{P}_F(f) &= \lim_{n \to \infty} \frac{1}{i_n} \log Z_{i_n}(F,f) \\
&\geq \frac{-|T|}{k} - \varepsilon + \frac{l\varepsilon}{k+p} - \frac{Ml + \log p}{k} + \mathrm{P}(f) - 2\varepsilon \geq \mathrm{P}(f) - 7\varepsilon,
\end{aligned}
$$

provided that k is large enough. Thus, letting $\varepsilon \searrow 0$, the theorem follows. The case $\mathrm{P}(f) = \infty$ can be treated similarly. ∎

11.2 Graph Directed Markov Systems

We start with the following definition of the main object of this chapter.

Definition 11.2.1 A graph directed Markov system (GDMS) consists of

(a) a directed multigraph (E, V) with a countable set of edges E and a finite set of vertices V,

(b) an incidence matrix $A: E \times E \longrightarrow \{0,1\}$,
(c) two functions $i,t: E \longrightarrow V$ such that $t(a) = i(b)$ whenever $A_{ab} = 1$,
(d) a family of nonempty compact metric spaces $\{X_v\}_{v\in V}$,
(e) a number $\beta \in (0,1)$, and
(f) a collection $\{\phi_e: X_{t(e)} \longrightarrow X_{i(e)}: e \in E\}$ of one-to-one contractions, all with a Lipschitz constant $\leq \beta$.

Briefly, the set

$$S := \{\phi_e: X_{t(e)} \longrightarrow X_{i(e)}\}_{e\in E}$$

is called a graph directed Markov system (GDMS). The corresponding number $\beta \in (0,1)$ is denoted by β_S.

A GDMS is called an iterated function system (IFS) if V, the set of vertices, is a singleton and $A(E \times E) = \{1\}$.

The main object of interest in this chapter will be the limit set of the system S and its geometric features. We now define the limit set. For each $\omega \in E_A^*$, say $\omega \in E_A^n$, we consider the map coded by ω:

$$\phi_\omega := \phi_{\omega_1} \circ \cdots \circ \phi_{\omega_n}: X_{t(\omega_n)} \longrightarrow X_{i(\omega_1)}.$$

For $\omega \in E_A^\infty$, the sets $\{\phi_{\omega|_n}(X_{t(\omega_n)})\}_{n\geq 1}$ form a descending sequence of nonempty compact sets; therefore, $\bigcap_{n\geq 1} \phi_{\omega|_n}(X_{t(\omega_n)}) \neq \emptyset$. Since, for every $n \geq 1$,

$$\operatorname{diam}\big(\phi_{\omega|_n}(X_{t(\omega_n)})\big) \leq \beta_S^n \operatorname{diam}\big(X_{t(\omega_n)}\big) \leq \beta_S^n \max\{\operatorname{diam}(X_v): v \in V\},$$

we conclude that the intersection

$$\bigcap_{n\geq 1} \phi_{\omega|_n}(X_{t(\omega_n)})$$

is a singleton and we denote its only element by $\pi(\omega)$. In this way, we have defined the "projection" map π

$$\pi: E_A^\infty \longrightarrow \bigoplus_{v\in V} X_v$$

from the coding space E_A^∞ to $\bigoplus_{v\in V} X_v$, the disjoint union of the compact sets X_v. The set

$$J = J_S := \pi(E_A^\infty)$$

will be called the *limit set* of the GDMS S.

Remark 11.2.2 As a matter of fact, we do not need all the maps $\phi_e: X_{t(e)} \longrightarrow X_{i(e)}$, $e \in E$, to be uniform contractions. It entirely suffices to know that there exist $\beta \in (0,1)$ and some integer $q \geq 1$ such that all the maps $\phi_\omega \colon X_{t(\omega)} \longrightarrow X_{i(\omega)}$, $\omega \in E_A^q$, are contractions with a Lipschitz constant $\leq \beta$.

11.3 Conformal Graph Directed Markov Systems

In order to be able to say anything meaningful about fractal properties of the limit set of a GDMS, the contracting maps ϕ_e, $e \in E$, need to have some additional smooth properties than merely continuity. There are basically two options: the contracting maps of the system S are commonly assumed to be either affine maps or conformal maps. In this book, we consider the latter case.

Definition 11.3.1 We call a GDMS a conformal graph directed Markov system (CGDMS) if the following conditions are satisfied.

(4a) For every vertex $v \in V$, X_v is a compact subset of a Euclidean space $\mathbb{R}^d$ (the dimension d common for all $v \in V$) and $X_v = \overline{\mathrm{Int}(X_v)}$.

(4b) (*Open Set Condition*)(*OSC*). For all $a, b \in E$ with $a \neq b$, we have that

$$\phi_a(\mathrm{Int}(X_{t(a)}) \cap \phi_b(\mathrm{Int}(X_{t(b)}) = \emptyset.$$

(4c) For every vertex $v \in V$, there exists an open connected set $W_v \supseteq X_v$ such that, for every $\omega \in E_A^*$, the map ϕ_ω extends to a C^1 conformal diffeomorphism from $W_{t(\omega)}$ into $\mathbb{R}^d$.

(4d) There are two constants $L \geq 1$ and $\alpha > 0$ such that

$$\left| \frac{|\phi_e'(y)|}{|\phi_e'(x)|} - 1 \right| \leq L \|y - x\|^\alpha$$

for every $e \in E$ and every pair of points $x, y \in W_{t(e)}$, where $|\phi_\omega'(x)|$ means the norm of the derivative; equivalently, this is the scaling factor of the similarity map $\phi_\omega'(x) \colon \mathbb{R}^d \longrightarrow \mathbb{R}^d$.

Remark 11.3.2 We would like to emphasize that, unlike [**MU1**] and [**MU2**], until Section 11.8, entitled The Strong Open Set Condition, we do NOT assume in this chapter any kind of cone condition or Strong Open Set Condition (SOSC).

Remark 11.3.3 The requirement of uniform contractions of the maps ϕ_e, $e \in E$, or, rather, its weaker form explained in Remark 11.2.2, is now replaced by a slightly stronger condition requiring that

$$\|\phi_\omega'\|_\infty := \sup\left\{|\phi_\omega'(x)| \colon x \in W_{t(\omega)}\right\} \leq \beta = \beta_S < 1 \qquad (11.4)$$

for some integer $q \geq 1$ and all $\omega \in E_A^q$. Passing to the (iterated) GDMS

$$\{\phi_\omega\colon X_t(\omega) \longrightarrow X_{i(\omega)} : \omega \in E_A^q\},$$

we will frequently assume without loss of generality that (11.4) holds with $q = 1$.

Remark 11.3.4 The condition *(4c)* follows from the following (stronger) condition.

(4c′) For every vertex $v \in V$, there exists an open connected set $W_v \supseteq X_v$ such that, for every $e \in E$ with $t(e) = v$, the map ϕ_ω extends to a C^1 conformal diffeomorphism from $W_{t(e)}$ into $W_{i(e)}$.

Remark 11.3.5 The entire theory of CGDMSs almost verbatim extends to the case when, instead of $\mathbb{R}^d$, the ambient space is a complex torus $\mathbb{T}_\Lambda = \mathbb{C}/\Lambda$, where Λ is a lattice in $\mathbb{C}$ and the diameters of all sets X_v, $v \in V$, do not exceed $u_{\mathbb{T}_\Lambda}$. So, it holds for all projective Riemann surfaces. This extension is what we will need, though not so substantially, in Chapter 12, and then, much more substantially, in Chapter 22 in Volume 2. It also easily extends to all tori $\mathbb{R}^d/\mathbb{Z}^d$ with appropriate restrictions on the diameters of the sets X_v. It even extends to all conformal manifolds of any dimension.

We start our investigations of CGDMSs by proving the following.

Lemma 11.3.6 *If $S = \{\phi_e\}_{e \in E}$ is a CGDMS, then, for every vertex $v \in V$, there exists an open, connected, and bounded set W'_v such that $X_v \subseteq W'_v \subseteq \overline{W'_v} \subseteq W_v$.*

Proof Fix $v \in V$. Let

$$\delta_v := \mathrm{dist}\big(X_v, \mathbb{R}^d \backslash W_v\big) > 0.$$

The neighborhood

$$B(X_v, \delta_v/2) = \bigcup_{x \in X_v} B(x, \delta_v/2)$$

of X_v is open and bounded but not necessarily connected. However, each open ball

$$B(x, \delta_v/2), \;\; x \in X_v,$$

is connected and, thus, every connected component of $B(X_v, \delta_v/2)$ contains at least one such ball. Since each set $B(X_v, \delta_v/2)$, $v \in V$, is bounded and its connected components are mutually disjoint and each contain at least one ball of radius $\delta_v/2$, the neighborhood $B(X_v, \delta_v/2)$ has only finitely many connected components. Denote these components by

$$B_v^{(1)}, B_v^{(2)}, \ldots, B_v^{(k_v)}.$$

For each $1 \le j \le k_v$, choose

$$x_v^{(j)} \in B_v^{(j)} \cap X_v.$$

Since the set W_v is path-connected, for every $1 \le j < k_v$, there is a piecewise C^1 curve $\gamma_v^{(j)}$ joining $x_v^{(j)}$ and $x_v^{(j+1)}$ in W_v. Since $\gamma_v^{(j)}$ is compact and W_v is open, there is $\delta_v^{(j)} > 0$ such that

$$B\left(\gamma_v^{(j)}, \delta_v^{(j)}\right) \subseteq W_v.$$

So, setting

$$W_v' := B(X_v, \delta_v/2) \cup \bigcup_{j=1}^{k-1} B\left(\gamma_v^{(j)}, \delta_v^{(j)}/2\right)$$

finishes the proof. ■

Proposition 11.3.7 *If $d \ge 2$ and a family $S = \{\phi_e\}_{e \in I}$ satisfies conditions* (4a) *and* (4c)*, then it also satisfies condition* (4d) *with $\alpha = 1$ after replacing all the sets W_v, $v \in V$, by the corresponding sets W_v', $v \in V$, produced in Lemma 11.3.6.*

Proof Since the set of vertices V is finite, this proposition is an immediate consequence of Theorem 8.3.7 in the case $d = 2$. In the case $d \ge 3$, which will not be considered in Parts 3 and 4 of this book, the proposition is proved in [**MU2**] and ultimately relays on the Liouville Classification Theorem of conformal homeomorphisms of $\overline{\mathbb{R}^d}$, which asserts that every conformal map in $\mathbb{R}^d$, $d \ge 3$, is a composition of the inversion with respect to a sphere with radius 1 (the center can be ∞) and a similarity map. ■

As a rather straightforward consequence of hypothesis (4d), we get the following.

Lemma 11.3.8 *If $S = \{\phi_e\}_{e \in E}$ is a CGDMS, then, for all $\omega \in E^*$ and all $x, y \in W_{t(\omega)}$, we have that*

$$\left|\log|\phi_\omega'(y)| - \log|\phi_\omega'(x)|\right| \le \frac{L}{1 - \beta_S^\alpha} \|y - x\|^\alpha.$$

Proof For every $\omega \in E^*$, say $\omega \in E^n$, and every $z \in W_{t(\omega)}$, put $z_k = \phi_{\omega_{n-k+1}} \circ \phi_{\omega_{n-k+2}} \circ \cdots \circ \phi_{\omega_n}(z)$; put also $z_0 = z$. In view of (4d), for any two points $x, y \in W_{t(\omega)}$, we have that

$$\begin{aligned}
\left|\log(|\phi'_\omega(y)|) - \log(|\phi'_\omega(x)|)\right| &= \left|\sum_{j=1}^{n} \log\left(1 + \frac{|\phi'_{\omega_j}(y_{n-j})| - |\phi'_{\omega_j}(x_{n-j})|}{|\phi'_{\omega_j}(x_{n-j})|}\right)\right| \\
&\le \sum_{j=1}^{n} \frac{\left||\phi'_{\omega_j}(y_{n-j})| - |\phi'_{\omega_j}(x_{n-j})|\right|}{|\phi'_{\omega_j}(x_{n-j})|} \\
&\le \sum_{j=1}^{n} L\|y_{n-j} - x_{n-j}\|^\alpha \\
&\le L\sum_{j=1}^{n} \beta_S^{\alpha(n-j)}\|y-x\|^\alpha \\
&\le \frac{L}{1-\beta_S^\alpha}\|y-x\|^\alpha. \qquad (11.5)
\end{aligned}$$

■

As a straightforward consequence of this lemma (and ultimately of (4d)), we get the following.

(4e) (Bounded Distortion Property). There exists $K \ge 1$ such that, for all $\omega \in E^*$ and all $x, y \in W_{t(\omega)}$,

$$|\phi'_\omega(y)| \le K|\phi'_\omega(x)|.$$

We shall now prove some basic geometric consequences of the properties (4a)–(4e). Because of the Mean Value Inequality, for all finite words $\omega \in E^*$, all convex subsets C of $W_{t(\omega)}$, all $x \in X_{t(\omega)}$, and all radii $0 < r \le \operatorname{dist}(X_{t(\omega)}, \partial W_{t(\omega)})$, we have that

$$\operatorname{diam}(\phi_\omega(C)) \le \|\phi'_\omega\|_\infty \operatorname{diam}(C), \quad \phi_\omega(B(x,r)) \subseteq B\big(\phi_\omega(x), \|\phi'_\omega\|r\big). \qquad (11.6)$$

We shall now prove the following.

Lemma 11.3.9 *Replacing, if necessary, all the sets W_v, $v \in V$, by the corresponding sets W'_v, $v \in V$, produced in Lemma 11.3.6, we will have that there exists a constant $D \ge 1$ such that*

$$\|\phi_\omega(y) - \phi_\omega(x)\| \le D\|\phi'_\omega\| \cdot \|y - x\| \qquad (11.7)$$

for all finite words $\omega \in E_A^*$ and all $x, y \in W_{t(\omega)}$.

Proof Fix

$$R := \min\{\operatorname{dist}(W'_v, \partial W_v) \colon v \in V\} > 0.$$

Since V, the set of all vertices, is finite and since each set $\overline{W'_v}$ is compact and connected, there exists an integer $D \ge 3$ such that each such set can be covered by finitely many balls $B(x_1, R/2), \dots, B(x_{D-1}, R/2)$ with centers $x_1, \dots, x_{D-1}$ in W'_v and with the property that

$$B(x_i, R/2) \cap B(x_{i+1}, R/2) \neq \emptyset$$

for all $i = 1, 2, \dots, D-2$. Therefore, for every vertex $v \in V$ and all points $x, y \in \overline{W'_v}$, there are $k \le D+1$ points $x = z_1, z_2, \dots, z_k = y$ in $B(W'_v, R)$ such that $\|z_i - z_{i+1}\| \le \|x - y\|$ for all $i = 1, 2, \dots, k-1$. Hence, using the Mean Value Inequality, we get that

$$\begin{aligned}\|\phi_\omega(x) - \phi_\omega(y)\| &\le \sum_{i=1}^{k-1} \|\phi_\omega(z_i) - \phi_\omega(z_{i+1})\| \le \sum_{i=1}^{k-1} \|\phi'_\omega\|_\infty \|z_i - z_{i+1}\| \\ &\le (k-1)\|\phi'_\omega\|_\infty \|x - y\| \\ &\le D\|\phi'_\omega\|_\infty \|x - y\|.\end{aligned}$$

Thus, the proof is complete. ■

Since the set V is finite, replacing the number D of Lemma 11.3.9 by

$$D \max\big\{1, \max\{\mathrm{diam}(W_v) : v \in V\}\big\},$$

as an immediate consequence of this lemma, we get the following.

Corollary 11.3.10 *For all finite words $\omega \in E_A^*$, we have that*

$$\mathrm{diam}\big(\phi_\omega(W_{t(\omega)})\big) \le D\|\phi'_\omega\|_\infty. \tag{11.8}$$

Now we shall prove the following.

Lemma 11.3.11 *For all finite words $\omega \in E_A^*$, all $x \in X_{t(\omega)}$, and all radii $0 < r < \mathrm{dist}(X_{t(\omega)}, \partial W_{t(\omega)})$, we have that*

$$\phi_\omega(B(x,r)) \supseteq B\big(\phi_\omega(x), K^{-1}\|\phi'_\omega\| r\big). \tag{11.9}$$

Proof First, notice that $B(x,r) \subseteq W_{t(\omega)}$. Take also any $\omega \in E^*$ and let $R \ge 0$ be the maximal radius such that

$$B(\phi_\omega(x), R) \subseteq \phi_\omega(B(x,r)). \tag{11.10}$$

Then

$$\partial\big(B(\phi_\omega(x), R)\big) \cap \partial\big(\phi_\omega(B(x,r))\big) \neq \emptyset; \tag{11.11}$$

and, in view of the Mean Value Inequality along with the Bounded Distortion Property (4e), we get that

$$\phi_\omega^{-1}\big(B(\phi_\omega(x), R)\big) \subseteq B\big(x, K\|\phi'_\omega\|_\infty^{-1} R\big).$$

If $K\|\phi'_\omega\|_\infty^{-1}R < r$, then the set $\phi_\omega\big(B(x, K\|\phi'_\omega\|_\infty^{-1}R)\big)$ is well defined and

$$B(\phi_\omega(x), R) \subseteq \phi_\omega\big(B(x, K\|\phi'_\omega\|_\infty^{-1}R)\big).$$

But then both (11.10) and (11.11) imply that $K\|\phi'_\omega\|_\infty^{-1}R \geq r$. This contradiction shows that

$$K\|\phi'_\omega\|_\infty^{-1}R \geq r.$$

So, using (11.10), we obtain (11.9), which completes the proof. ■

Let R_S be the radius of the largest open ball that can be inscribed in all the sets X_v, $v \in V$. Let $x_v \in \mathrm{Int}(X_v)$, $v \in V$, be the centers of the respective balls. As an immediate consequence of Lemma 11.3.11, we get, perhaps with a larger constant D, the following.

Lemma 11.3.12 *For all finite words $\omega \in E_A^*$, we have that*

$$\phi_\omega\big(\mathrm{Int}\big(X_{t(\omega)}\big)\big) \supseteq B\big(\phi_\omega(x_{t(\omega)}), K^{-1}\|\phi'_\omega\|R_S\big); \tag{11.12}$$

so,

$$\mathrm{diam}\big(\phi_\omega(X_{t(\omega)})\big) \geq D^{-1}\|\phi'_\omega\|_\infty. \tag{11.13}$$

11.4 Topological Pressure, θ-Number, and Bowen's Parameter

Let S be a finitely irreducible CGDMS. For every $t \geq 0$, let

$$Z_n(t) := \sum_{\omega \in E_A^n} \|\phi'_\omega\|_\infty^t.$$

Since

$$\|\phi'_{\omega\tau}\|_\infty \leq \|\phi'_\omega\|_\infty \cdot \|\phi'_\tau\|_\infty,$$

we see that $Z_{m+n}(t) \leq Z_m(t)Z_n(t)$ for all integers $m, n \geq 1$; consequently, the sequence

$$\mathbb{N} \ni n \longmapsto \log Z_n(t) \in \mathbb{R}$$

is subadditive. Thus, the limit

$$\lim_{n\to\infty} \frac{1}{n} \log Z_n(t)$$

exists and is equal to

$$\inf_{n\geq 1}\left\{\frac{1}{n}\log Z_n(t)\right\}.$$

This limit is denoted by $\mathrm{P}(t)$, or, if we want to be more precise, by $\mathrm{P}_E(t)$ or $\mathrm{P}_S(t)$. This is called the topological pressure of the system S evaluated at the parameter t. Let $\zeta : E_A^{\mathbb{N}} \longrightarrow \mathbb{R}$ be the function defined by the formula

$$\zeta(\omega) := t\log|\phi'_{\omega_1}(\pi(\sigma(\omega)))|.$$

As a straightforward consequence of Lemma 11.3.8, we get the following.

Lemma 11.4.1 *For every $t \geq 0$, the function $t\zeta : E_A^{\mathbb{N}} \longrightarrow \mathbb{R}$ is Hölder continuous and* $\mathrm{P}(\sigma, t\zeta) = \mathrm{P}(t)$.

Let

$$F(S) := \{t \geq 0 : \ \mathrm{P}(t) < \infty\}$$

and

$$\theta(S) := \inf(F(S)).$$

Having bounded distortion (4e) and uniform contraction of all generators of the systems S, the following facts are easy to prove.

Proposition 11.4.2 *If S is a finitely irreducible CGDMS, then*

(a) $\mathrm{P}(t) < +\infty \Leftrightarrow Z_1(t) < +\infty$.
(b) $\inf\{t \geq 0 : Z_1(t) < +\infty\} = \theta(S)$.
(c) *The topological pressure function* $\mathrm{P}(t)$ *is*

 (c1) *nonincreasing on* $[0, +\infty)$,
 (c2) *strictly decreasing on* $[0, +\infty)$ *with* $\lim_{t\to+\infty}\mathrm{P}(t) = -\infty$,
 (c3) *convex and continuous on* $F(S)$.

(d) $\mathrm{P}(0) = +\infty$ *if and only if E is infinite.*

All of these items are indeed easy and straightforward to prove. Proofs of most of them can be found in existing published sources, most notably in [**MU2**]. There is one exception, namely continuity of the pressure function $P|_{F(S)}$ at the point θ; we do not think that its proof has been published anywhere. But the proof is very easy, very short, and straightforward; since we have been frequently asked about it, we present this proof here. More precisely, we shall prove the following.

$$\lim_{t\searrow\theta}\mathrm{P}(t) = \mathrm{P}(\theta). \tag{11.14}$$

Indeed, the inequality

$$\limsup_{t \searrow \theta} \mathrm{P}(t) \le \mathrm{P}(\theta) \tag{11.15}$$

follows directly from item (c1) of Proposition 11.4.2. In order to prove that

$$\liminf_{t \searrow \theta} \mathrm{P}(t) \ge \mathrm{P}(\theta), \tag{11.16}$$

fix any number $s < \mathrm{P}(\theta)$. Then fix $\varepsilon > 0$. By virtue of Theorem 11.1.4, there exists a finite set $F \subseteq E$ containing all letters of a finite set witnessing finite irreducibility of the system S (more precisely, of the incidence matrix A) such that

$$\mathrm{P}_F(\theta) > s - \frac{\varepsilon}{2}.$$

Since, for finite systems, the pressure function is convex on the whole interval $(-\infty, +\infty)$ (this is true even without irreducibility), there exists $\delta > 0$ such that

$$|\mathrm{P}_F(t) - \mathrm{P}_F(\theta)| < \frac{\varepsilon}{2}$$

for all $t \in [\theta, \theta + \delta)$. But then

$$\mathrm{P}(t) \ge \mathrm{P}_F(t) > \mathrm{P}_F(\theta) - \frac{\varepsilon}{2} > s - \varepsilon.$$

Hence,

$$\liminf_{t \searrow \theta} \mathrm{P}(t) \ge s,$$

and (11.16) follows. Formula (11.14) is, thus, proved.

The number

$$h_S := h := \inf\{t \ge 0 \colon \mathrm{P}(t) \le 0\} \tag{11.17}$$

is called Bowen's parameter of the system S. It will turn out to be equal to the Hausdorff dimension of the limit set J_S. In view of Proposition 11.4.2(c1) we have that

$$\mathrm{P}(h_S) \le 0. \tag{11.18}$$

The following useful observation is now obvious.

Observation 11.4.3 If $\mathrm{P}(t) = 0$ for some $t \ge 0$, then $t = h$.

Definition 11.4.4 We say that the system S is

- regular if $\mathrm{P}(h_S) = 0$,
- strongly regular if there exists $t \ge 0$ such that $0 < \mathrm{P}(t) < +\infty$, and
- co-finitely regular if $\mathrm{P}(\theta_S) = +\infty$.

It is easy to see that each co-finitely regular system is strongly regular and each strongly regular one is regular. We shall prove the following.

Proposition 11.4.5 *If a finitely irreducible CGDMS S is co-finitely regular, then, for every co-finite finitely irreducible subset $F \subseteq E$, the system S_F is also co-finitely regular, thus regular.*

In addition, if S is a conformal IFS, and its every co-finite subsystem is regular, then S is co-finitely regular.

Proof First, notice that $\theta_F = \theta_S$ for every co-finite subset F of E. Suppose now that S is co-finitely regular. By virtue of Proposition 11.4.2(a), this means that $Z_1(\theta) = +\infty$. But then $Z_1(F,\theta) = +\infty$ for every co-finite subset F of E. This, however, by Proposition 11.4.2(a) again, means that each such finitely irreducible system is co-finitely regular, thus regular.

For the converse, suppose that S is an IFS and the system S is not co-finitely regular. By virtue of Proposition 11.4.2(a), used for the third time, this means that $Z_1(\theta) < +\infty$. But then there exists a co-finite subset F of E such that $Z_1(F,\theta) < 1$. Hence, by the definition of topological pressure, $\mathrm{P}_F(\theta) < 0$. But then F is not regular and we are done. ■

Let us record the following obvious fact.

Fact 11.4.6 If S is a finitely irreducible CGDMS, then $\theta(S) \leq h_S$. If S is strongly regular, in particular if S is co-finitely regular, then $\theta(S) < h_S$.

11.5 Hausdorff Dimensions and Bowen's Formula for GDMSs

In this section, we prove a formula for the Hausdorff dimension of the limit set of a finitely irreducible CGDMS. It is entirely expressed in dynamical terms and, because of its correspondence to the formula obtained in Bowen's breakthrough work [**Bow2**] dealing with quasi-Fuchsian groups, we refer to it as Bowen's Formula. In the case of finite alphabets E, S being an IFS, and the maps ϕ_e, $e \in E$, all being similarities, it was obtained in the seminal paper by Hutchinson [**Hut**], who also introduced the concept of the Open Set Condition (OSC). In the case when the alphabet E is still finite, S is an IFS, but the maps ϕ_e, $e \in E$, are all assumed only to be conformal, Bowen's Formula was obtained in [**Bed**]. In the case of arbitrary (countable alphabet) conformal IFSs, this formula was obtained in [**MU1**]. Finally, it was proved in its final form of arbitrary (countable alphabet) CGMMSs in [**MU2**]. This is its form we formulate and prove in the current section. It also has some extensions to conformal IFSs acting in Hilbert spaces; see [**MSzU**].

We start with the following simple geometrical observation following from the OSC.

Lemma 11.5.1 *If S is a CGDMS, then, for all $0 < \kappa_1 < \kappa_2 < +\infty$, for all $r > 0$, and for all $x \in X$, the cardinality of any collection of mutually incomparable words $\omega \in E_A^*$ that satisfy the condition*

$$B(x,r) \cap \phi_\omega(X_{t(\omega)}) \neq \emptyset$$

and

$$\kappa_1 r \le \operatorname{diam}(\phi_\omega(X_{t(\omega)})) \le \kappa_2 r$$

is bounded above by the number

$$\left((1+\kappa_2)KD(R_S\kappa_1)^{-1}\right)^d,$$

where, we recall, R_S is the radius of the largest open ball that can be inscribed in all the sets X_v, $v \in V$.

Proof Recall that λ_d is the Lebesgue measure on $\mathbb{R}^d$ and that V_d is the Lebesgue measure of the unit ball in $\mathbb{R}^d$. For every $v \in V$, let x_v be the center of a ball with radius R_S which is contained in $\operatorname{Int}(X_v)$. Let F be any collection of A-admissible words satisfying the hypotheses of our lemma. Then, for every $\omega \in F$, we have that

$$\phi_\omega(X_{t(\omega)}) \subseteq B\big(x, r + \operatorname{diam}(\phi_\omega(X_{t(\omega)}))\big) \subseteq B(x,(1+\kappa_2)r).$$

Since, by the OSC, all the sets $\{\phi_\omega(\operatorname{Int}X_{t(\omega)})\}_{\omega\in F}$ are mutually disjoint, applying Lemma 11.3.11, we, thus, get that

$$\begin{aligned}
V_d(1+\kappa_2)^d r^d &= \lambda_d(B(x,(1+\kappa_2)r)) \ge \lambda_d\left(\bigcup_{\omega\in F}\phi_\omega(X_{t(\omega)})\right)\\
&= \sum_{\omega\in F}\lambda_d(\phi_\omega(\operatorname{Int}X_{t(\omega)})) \ge \sum_{\omega\in F}\lambda_d\big(\phi_\omega(B(x_{t(\omega)},R_S))\big)\\
&\ge \sum_{\omega\in F}\lambda_d\big(B(\phi_\omega(x_{t(\omega)}), K^{-1}R_S\|\phi_\omega'\|_\infty)\big)\\
&\ge \sum_{\omega\in F}\lambda_d\big(B(\phi_\omega(x_{t(\omega)}), (KD)^{-1}R_S\operatorname{diam}(X_{i(\omega)}))\big)\\
&\ge \sum_{\omega\in F}\lambda_d\big(B(\phi_\omega(x_{t(\omega)}), (KD)^{-1}R_S\kappa_1 r)\big)\\
&= \sharp F((KD)^{-1}R_S\kappa_1)^d V_d r^d.
\end{aligned}$$

Hence,

$$\sharp F \le \left((1+\kappa_2)(KD)(R_S\kappa_1)^{-1}\right)^d.$$

We are done. ■

Now assume that the alphabet E is finite and keep the incidence matrix A (finitely) irreducible. Fix $t \ge 0$. Consider the operator $\mathcal{L}_t$ given by the formula

$$\mathcal{L}_t g(\omega) := \sum_{i:\, A_{i\omega_1}=1} g(i\omega)|\phi_i'(\pi(\omega))|^t, \quad \omega \in E_A^\infty, \tag{11.19}$$

where G is a real-valued function defined on $C(E_A^\infty)$. Note that

$$\mathcal{L}_t(C(E_A^\infty)) \subseteq C(E_A^\infty)$$

and the linear operator $\mathcal{L}_t$ acts continuously on $C(E_A^\infty)$, the Banach space of all real-valued continuous functions on E_A^∞ endowed with the supremum norm. In fact, the norm of $\mathcal{L}_t$ is bounded by $Z_1(t)$. A straightforward inductive calculation gives that, for all $n \ge 1$, we have that

$$\mathcal{L}_t^n g(\omega) = \sum_{\tau \in E_A^n,\, \tau\omega \in E_A^\infty} g(\tau\omega)|\phi_\tau'(\pi(\omega))|^t. \tag{11.20}$$

Let $\mathcal{L}_t^* \colon C^*(E_A^\infty) \longrightarrow C^*(E_A^\infty)$ be the dual operator of $\mathcal{L}_t$. Denote by M_A the set of all Borel probability measures on $E_A^{\mathbb{N}}$ considered to be a convex subset of the Banach space $C^*(E_A^\infty)$ via the canonical embedding

$$M_A \ni m \longmapsto \left(g \longmapsto \mu(g) = \int_{E_A^\infty} g\,dm\right) \in C^*(E_A^\infty).$$

Consider the map

$$M_A \ni m \longmapsto \frac{\mathcal{L}_t^* m}{\mathcal{L}_t^* m(1\!\!1)} \in M_A. \tag{11.21}$$

This map is well defined since

$$\mathcal{L}_t^* m(1\!\!1) = m(\mathcal{L}_t 1\!\!1) = \int_{E_A^{\mathbb{N}}} \sum_{i:\, A_{i\omega_1}=1} |\phi_i'(\pi(\omega))|^t\,dm,$$

and the integrand is everywhere positive, whence the integral is positive. Since this map is continuous in the weak* topology on M_A and since M_A is a compact (because E is finite) convex subset of the locally convex topological vector space $C^*(E_A^\infty)$ endowed with the weak* topology, it follows from the

Schauder–Tichonov Theorem that the map defined in (11.21) has a fixed point. Denote it by $\tilde{m}_t$ and put

$$\lambda_t = \mathcal{L}_t^* \tilde{m}_t(1\!\!1) > 0.$$

We then have that

$$\mathcal{L}_t^* \tilde{m}_t = \lambda_t \tilde{m}_t. \tag{11.22}$$

Iterating (11.22), we get, for every $n \geq 1$, that

$$\begin{aligned}
\lambda_t^n &= \lambda_t^n \tilde{m}_t(1\!\!1) = \mathcal{L}_t^{*n} \tilde{m}_t(1\!\!1) = \tilde{m}_t\big(\mathcal{L}_t^n 1\!\!1\big) = \int_{E_A^\infty} \sum_{\tau \in E_A^n : A_{\tau_n \omega_1} = 1} |\phi_\tau'(\pi(\omega))|^t d\tilde{m}_t \\
&\leq \int_{E_A^\infty} \sum_{\tau \in E_A^n} \|\phi_\tau\|_\infty^t d\,\tilde{m}_t = \int_{E_A^\infty} Z_n(t)\, d\tilde{m}_t \\
&= Z_n(t).
\end{aligned}$$

Therefore,

$$\mathrm{P}(t) = \lim_{n \to \infty} \frac{1}{n} \log Z_n(t) \geq \lim_{n \to \infty} \frac{1}{n} \log \lambda_t^n = \log \lambda_t. \tag{11.23}$$

Formulas (11.19) and (11.20) clearly extend to all Borel bounded functions $g \colon E_A^\infty \longrightarrow \mathbb{R}$. The standard approximation arguments show then that

$$\mathcal{L}_t^{*n} \tilde{m}_t(g) = \lambda_t^n \tilde{m}_t(g) \tag{11.24}$$

for all Borel bounded functions $g \colon E_A^\infty \longrightarrow \mathbb{R}$. In particular, taking $g := 1\!\!1_{\phi_{\omega(x)}} \circ \pi$ and $\omega \in E_A^*$, we get, with $n = |\omega|$, that

$$\begin{aligned}
\tilde{m}_t([\omega]) &= \lambda_t^{-n} \mathcal{L}_t^{*n} \tilde{m}_t(1\!\!1_{[\omega]}) = \lambda_t^{-n} \tilde{m}_t\big(\mathcal{L}_t^n(1\!\!1_{[\omega]})\big) \\
&= \lambda_t^{-n} \int_{E_A^{\mathbb{N}}} \sum_{\tau \in E_A^n,\ A_{\tau_n \rho_1} = 1} |\phi_\tau(\pi(\rho))|^t 1\!\!1_{[\omega]}(\tau\rho) d\tilde{m}_t(\rho) \\
&= \lambda_t^{-n} \int_{\rho \in E_A^{\mathbb{N}}} \sum_{A_{\omega_n \rho_1} = 1} |\phi_\omega'(\pi(\rho))|^t \, d\tilde{m}_t.
\end{aligned} \tag{11.25}$$

From this formula, we get that

$$\tilde{m}_t([\omega]) \leq \lambda_t^{-n} \|\phi_\omega'\|_\infty^t \tilde{m}_t\big(\{\rho \in E_A^\infty : A_{\omega_n \rho_1} = 1\}\big) \leq \lambda_t^{-n} \|\phi_\omega'\|_\infty^t \tag{11.26}$$

and

$$
\begin{aligned}
\tilde{m}_t([\omega]) &\ge K^{-t}\lambda_t^{-n}\|\phi'_\omega\|_\infty^t \tilde{m}_t\big(\{\rho \in E_A^\infty : A_{\omega_n\rho_1} = 1\}\big) \\
&= K^{-t}\lambda_t^{-n}\|\phi'_\omega\|_\infty^t \tilde{m}_t\left(\bigcup_{e\in E,\, A_{\omega_n e}=1} [e]\right) \\
&= K^{-t}\lambda_t^{-n}\|\phi'_\omega\|_\infty^t \sum_{e\in E,\, A_{\omega_n e}=1} \tilde{m}_t([e]).
\end{aligned}
\tag{11.27}
$$

Now, since $\tilde{m}_t\big(E_A^\infty\big) = 1$, there is $b \in E$ such that $\tilde{m}_t([b]) > 0$. Let Λ be the finite subset of E_A^* witnessing finite irreducibility of the matrix A. Put

$$
M := \min\{\|\phi'_\tau\|_\infty : \tau \in \Lambda\}.
$$

Then, for every $\omega \in E_A^*$, there exists a word $\tau \in \Lambda$ such that $\omega\tau b \in E_A^*$. We, therefore, get from (11.27) that

$$
\begin{aligned}
\tilde{m}_t([\omega]) &\ge \tilde{m}_t([\omega\tau]) \ge K^{-t}\lambda_t^{-n}\|\phi'_{\omega\tau}\|_\infty^t \tilde{m}_t([b]) \\
&\ge K^{-2t}\tilde{m}_t([b])\|\phi'_\tau\|_\infty^t \lambda_t^{-n}\|\phi'_\omega\|_\infty^t \\
&\ge K^{-2t}\tilde{m}_t([b])M^t\lambda_t^{-n}\|\phi'_\omega\|_\infty^t \\
&> 0.
\end{aligned}
\tag{11.28}
$$

Rearranging terms, we get that

$$
\|\phi'_\omega\|_\infty^t \le K^{2t}\big(\tilde{m}_t([b])M^t\big)^{-1}\lambda_t^n \tilde{m}_t([\omega]).
\tag{11.29}
$$

Summing over all A-admissible words of length n, this gives

$$
Z_n(t) \le K^{2t}\big(\tilde{m}_t([b])M^t\big)^{-1}\lambda_t^n \sum_{\omega\in E_A^n} \tilde{m}_t([\omega]) = K^{2t}\big(\tilde{m}_t([b])M^t\big)^{-1}\lambda_t^n.
$$

Therefore,

$$
\mathrm{P}(t) = \lim_{n\to\infty} \frac{1}{n}\log Z_n(t) \le \lim_{n\to\infty}\frac{1}{n}\log\lambda_t^n = \log\lambda_t.
$$

Along with (11.23), this yields

$$
\log\lambda_t = \mathrm{P}(t).
\tag{11.30}
$$

Since the system S is finite, with h defined in (11.17), we have that $\mathrm{P}(h) = 0$; therefore, (11.26) and (11.29), applied with $t = h$, yield

$$
M_h\|\phi'_\omega\|_\infty^h \le \tilde{m}_h([\omega]) \le \|\phi'_\omega\|_\infty^h
\tag{11.31}
$$

with some positive constant M_h. Let

$$
m_h := \tilde{m}_h \circ \pi^{-1}
\tag{11.32}
$$

and, more generally,

$$m_t := \tilde{m}_t \circ \pi^{-1} \tag{11.33}$$

for all $t \geq 0$. We call m_h the h-conformal measure for the system S. More generally, we call m_t the $\left(t, e^{\mathrm{P}(t)}\right)$-conformal measure for the system S. We shall prove the following.

Theorem 11.5.2 *Suppose that $S = \{\phi_e\}_{e \in E}$ is a finite irreducible CGDMS. Then*

(a) *there exists a constant $C \geq 1$ such that, for all $x \in J_S$ and all $0 < r \leq 1$, we have that*

$$C^{-1} \leq \frac{m_h(B(x,r))}{r^h} \leq C,$$

(b) $0 < \mathrm{H}_h(J_S), \Pi_h(J_S) < +\infty$,

(c) $\mathrm{HD}(J_S) = h$.

Proof Since the alphabet E is finite, we have that

$$\xi := \inf\{\|\phi_e'\|_\infty : e \in E\} > 0.$$

Fix $x \in J_S$ and $0 < r < \frac{1}{2}\min\{\mathrm{diam}(X_v) : v \in V\}$. Then $x = \pi(\tau)$ for some $\tau \in E_A^\infty$. Let $n = n(\tau) \geq 0$ be the least integer such that

$$\phi_{\tau|_n}\left(X_{t(\tau_n)}\right) \subseteq B(x,r).$$

Then by (11.31), we have the following:

$$m_h(B(x,r)) \geq m_h\left(\phi_{\tau|_n}(X_{t(\tau_n)}\right) \geq \tilde{m}_h([\tau|_n])) \geq M_h K^{-h} \|\phi_{\tau|_n}'\|_\infty^h.$$

By the definition of n and Corollary 11.3.10, we have that

$$r \leq \mathrm{diam}\left(\phi_{\tau|_{n-1}}(X_{t(\tau_{n-1})})\right) \leq D\|\phi_{\tau|_{n-1}}'\|_\infty.$$

Hence,

$$m_h(B(x,r)) \geq M_h(DK)^{-h} r^h. \tag{11.34}$$

In order to prove the opposite inequality, let Z be the family of all minimal (in the sense of length) words $\omega \in E_A^*$ such that

$$\phi_\omega(X_{t(\omega)}) \cap B(x,r) \neq \emptyset \quad \text{and} \quad \phi_\omega(X_{t(\omega)}) \subseteq B(x,2r). \tag{11.35}$$

Consider an arbitrary $\omega \in Z$. Then

$$\mathrm{diam}\left(\phi_\omega(X_{t(\omega)})\right) \leq 2r \tag{11.36}$$

and $\operatorname{diam}(\phi_{\omega|_{|\omega|-1}}(X_{t(\omega)})) \geq 2r$. Therefore, making use of Corollary 11.3.10 twice, we get the following:

$$\begin{aligned}\operatorname{diam}\big(\phi_\omega(X_{t(\omega)})\big) &\geq D^{-1}\|\phi'_\omega\|_\infty \geq (DK)^{-1}\|\phi_{\omega|_{|\omega|-1}}\|_\infty \cdot \|\phi_{\omega_{|\omega|}}\|_\infty \\ &\geq K^{-1}\xi(DK)^{-1}\operatorname{diam}\Big(\phi_{\omega|_{|\omega|-1}}\big(X_{t(\omega_{|\omega|-1})}\big)\Big) \\ &\geq 2(KD^2)^{-1}\xi r.\end{aligned} \tag{11.37}$$

Hence, by virtue of (11.12) of Lemma 11.3.12, we get that

$$\phi_\omega(X_{t(\omega)}) \supset B\big(\phi_\omega(x_{t(\omega)}), 2(K^2DL)^{-1}R_S\xi r\big). \tag{11.38}$$

Since, by its very definition, the family Z consists of mutually incomparable words, Lemma 11.5.1 along with (11.36) and (11.37) imply that

$$\sharp Z \leq \Gamma := \big((1+2(KD^2)^{-1})2KD(R_S)^{-1}\big)^Q. \tag{11.39}$$

Since

$$\pi^{-1}(B(x,r)) \subseteq \bigcup_{\omega\in Z}[\omega],$$

we get from (11.35), (11.31), (11.3.12), (11.36), and (11.39) that

$$\begin{aligned}m_h(B(x,r)) &= \tilde{m}_h \circ \pi^{-1}(B(x,r)) \leq \tilde{m}_h\left(\bigcup_{\omega\in Z}[\omega]\right) \\ &= \sum_{\omega\in Z}\tilde{m}_h([\omega]) \leq \sum_{\omega\in Z}\|\phi'_\omega\|_\infty^h \\ &\leq \sum_{\omega\in Z}\big(D\operatorname{diam}\big(\phi_\omega(X_{t(\omega)})\big)\big)^h \leq (4D)^h\sum_{\omega\in Z}r^h = (4D)^h\sharp Z r^h \\ &\leq (4D)^h\Gamma r^h.\end{aligned}$$

Along with (11.34), this proves item (a). Items (b) and (c) are now an immediate consequence of the Frostman Converse Lemmas, i.e., Theorems 1.6.1 and 1.6.2. The proof is complete. ■

We are now ready to provide a short simple proof of the following main theorem of this section proven in [**MU2**]; see also [**MU1**].

Theorem 11.5.3 (Bowen's Formula) *If S is a finitely irreducible CGDMS, then*

$$h_S = \inf\{t \geq 0\colon \mathrm{P}(t) \leq 0\} = \mathrm{HD}(J_S) = \sup\big\{\mathrm{HD}(J_F)\colon F \subseteq E \text{ and } F \text{ is finite}\big\}.$$

Proof The first equality is just the definition of h_S. Put

$$h_\infty := \sup\{\mathrm{HD}(J_F)\colon\ F \subseteq E \text{ and } F \text{ is finite}\}$$

and

$$H := \mathrm{HD}(J_S).$$

Fix $t > h$. Then $\mathrm{P}(t) < 0$. Therefore, for all $n \geq 0$ large enough, we have that

$$\sum_{\omega \in E_A^n} \|\phi'_\omega\|_\infty^t \leq \exp\left(\frac{1}{2}\mathrm{P}(t)n\right).$$

Hence,

$$\sum_{\omega \in E_A^n} \mathrm{diam}^t(\phi_\omega(X_{t(\omega)})) \leq D^t \sum_{\omega \in E_A^n} \|\phi'_\omega\|_\infty^t \leq D^t \exp\left(\frac{1}{2}\mathrm{P}(t)n\right).$$

Since the family $\{\phi_\omega(X_{t(\omega)})\}_{\omega \in E^n}$ forms a cover of J_S, letting $n \to \infty$, we, thus, get that $\mathrm{H}_t(J_S) = 0$. This implies that $t \geq H$; in consequence, $h_S \geq H$. Since, obviously, $h_\infty \leq H$, we, thus, have that

$$h_\infty \leq H \leq h_S.$$

We are left to show that $h_S \leq h_\infty$. If F is a finite subset of E, then $h_F \leq h_\infty$; by virtue of Theorem 11.5.2, $\mathrm{P}_F(h_\infty) \leq 0$. So, in view of Theorem 11.1.4 and Lemma 11.4.1, we have that

$$\mathrm{P}(h_\infty) = \sup\{\mathrm{P}_F(h_\infty) \colon F \subseteq E, \text{ and } F \text{ is finite}\} \leq 0.$$

Hence, $h_\infty \geq h_S$ and the proof is complete. ■

Combining this theorem and Fact 11.4.6, we get the following.

Theorem 11.5.4 *If S is a finitely irreducible CGDMS, then*

$$\mathrm{HD}(J_S) = h_S \geq \theta(\mathcal{S}).$$

If, in addition, S is strongly regular, in particular if S is co-finitely regular, then

$$\mathrm{HD}(J_S) = h_S > \theta(S).$$

11.6 Conformal and Invariant Measures for CGDMSs

Sticking to the previous section,

$$S = \{\phi_e \colon e \in E\}$$

is assumed to be a finitely irreducible CGDMS with a countable alphabet E. Obviously, for all

$$t \in F(S) = \{t \geq 0 \colon \mathrm{P}(t) < +\infty\} = \{t \geq 0 \colon Z_1(t) < +\infty\},$$

(11.19) defines a bounded linear operator from $C_b(E_A^\infty)$, into itself, where $C_b(E_A^\infty)$ is the Banach space of all bounded real (or complex)-valued continuous functions on E_A^∞ endowed with the supremum norm. This operator is here also denoted by $\mathcal{L}_t$. Our primary goal in this section is to prove, for every $t \in F(S)$, the existence of a Borel probability measure $\tilde{m}_t$ on E_A^∞ satisfying equation (11.22) (and with $\lambda_t = e^{\mathrm{P}(t)}$). We first need the following auxiliary result. We will provide its short proof for the sake of completeness and convenience of the reader. It is more natural and convenient to formulate it in the language of directed graphs. Let us recall that a directed graph is said to be strongly connected if and only if its incidence matrix is irreducible. In other words, it means that every two vertices can be joined by a path of admissible edges.

Lemma 11.6.1 *If $\Gamma = \langle E, V\rangle$ is a strongly connected directed graph, then there exists a sequence of strongly connected subgraphs $\langle E_n, V_n\rangle$ of Γ such that all the vertices $V_n \subseteq V$ and all the edges E_n are finite, $\{V_n\}_{n=1}^\infty$ is an ascending sequence of vertices, $\{E_n\}_{n=1}^\infty$ is an ascending sequence of edges, and*

$$\bigcup_{n=1}^\infty V_n = V \quad \text{and} \quad \bigcup_{n=1}^\infty E_n = E.$$

Proof Let

$$V = \{v_n : n \geq 1\}$$

be a sequence of all vertices of Γ and

$$E = \{e_n : n \geq 1\}$$

be a sequence of edges of Γ. We will proceed inductively to construct the sequences

$$\{V_n\}_{n=1}^\infty \quad \text{and} \quad \{E_n\}_{n=1}^\infty.$$

In order to construct $\langle E_1, V_1\rangle$, let α be a path joining v_1 and v_2 ($i(\alpha) = v_1$, $t(\alpha) = v_2$) and β be a path joining v_2 and v_1 ($i(\beta) = v_2$, $t(\beta) = v_1$). These paths exist since Γ is strongly connected. We define $V_1 \subseteq V$ to be the set of all vertices of paths α and β and $E_1 \subseteq E$ to be the set of all edges from α and β enlarged by e_1 if this edge is among all the edges joining the vertices of V_1. Obviously, $\langle E_1, V_1\rangle$ is strongly connected and the first step of the inductive procedure is complete.

Suppose now that a strongly connected graph $\langle E_n, V_n\rangle$ has been constructed.

If $v_{n+1} \in V_n$, we set $V_{n+1} = V_n$ and E_{n+1} is then defined to be the union of E_n and all the edges from $\{e_1, e_2, \dots, e_n, e_{n+1}\}$ that are among all the edges joining the vertices of V_n.

If $v_{n+1} \notin V_n$, let α_n be a path joining v_n and v_{n+1} and β_n be a path joining v_{n+1} and v_n. We define V_{n+1} to be the union of V_n and the set of all vertices of α_n and β_n. E_{n+1} is then defined to be the union of E_n, all the edges building the paths α_n and β_n and all the edges from $\{e_1, e_2, \dots, e_n, e_{n+1}\}$ that are among all the edges joining the vertices of V_{n+1}. Since $\langle E_n, V_n \rangle$ was strongly connected, so is $\langle E_{n+1}, V_{n+1} \rangle$. The inductive procedure is complete. It immediately follows from the construction that

$$V_n \subseteq V_{n+1}, \ E_n \subseteq E_{n+1}, \ \bigcup_{n=1}^{\infty} V_n = V, \ \text{ and } \ \bigcup_{n=1}^{\infty} E_n = E.$$

We are done. ■

The following theorem was proved in [**MU2**]; see also Corollary 6.32 and Theorem 7.4 in [**CTU**], in a more general setting of arbitrary Hölder continuous summable potentials rather than merely $t\zeta$. We provide here its proof for the convenience of the reader and the sake of completeness.

Theorem 11.6.2 *If S is a finitely irreducible GDMS, then, for all $t \in F(S)$, there exists a unique Borel probability measure $\tilde{m}_t$ on E_A^∞ such that* (11.22) *holds, i.e.,*

$$\mathcal{L}_t^* \tilde{m}_t = e^{\mathrm{P}(t)} \tilde{m}_t. \tag{11.40}$$

Proof Without loss of generality, we may assume that $E = \mathbb{N}$. Since the incidence matrix A is irreducible, it follows from Lemma 11.6.1 that we can reorder the set $\mathbb{N}$ such that there exists a sequence $(l_n)_{n \geq 1}$, increasing to infinity, and such that, for every $n \geq 1$, the matrix

$$A|_{\{1,\dots,l_n\} \times \{1,\dots,l_n\}}$$

is irreducible. Given an integer $q \geq 1$, denote

$$\mathbb{N}(q) := \{1, 2, \dots, q\} \subseteq E = \mathbb{N}.$$

In view of (11.40), holding for finite alphabets, there exists an eigenmeasure $\tilde{m}_n$ of the operator $\mathcal{L}_n^*$, conjugate to the Perron–Frobenius operator

$$\mathcal{L}_n \colon C\big(\mathbb{N}(l_n)_A^\infty\big) \longrightarrow C\big(\mathbb{N}(l_n)_A^\infty\big)$$

associated with the potential $t\zeta|_{\mathbb{N}(l_n)_A^\infty}$. More precisely, for every function $g \in C(\mathbb{N}(l_n)_A^\infty)$,

$$\mathcal{L}_n g(\omega) := \sum_{i\in\mathbb{N}(l_n):\, A_{i\omega_1}=1} g(i\omega)\left|\phi_i'(\pi(\omega))\right|^t, \quad \omega \in \mathbb{N}(l_n)_A^\infty.$$

A family of Borel probability measures $\mathcal{M}$ in a topological space X is called tight, or uniformly tight, if and only if, for every $\varepsilon > 0$, there exists a compact set $K_\varepsilon \subseteq X$ such that

$$\mu(K_\varepsilon) > 1 - \varepsilon$$

for all $\mu \in \mathcal{M}$. If X is a completely metrizable space and $\mathcal{M}$ is a tight family of Borel probability measures, then Prohorov's Theorem (see, for example, [**Bog**, Book II, Theorem 8.6.2]), asserts that every sequence in $\mathcal{M}$ contains a weakly convergent subsequence.

We shall prove the following.

Claim 1°. The sequence $\{\tilde{m}_n\}_{n\geq 1}$ is tight with all measures $\tilde{m}_n$, $n \geq 1$, treated as Borel probability measures on E_A^∞.

Proof Let

$$\mathrm{P}_n(t) := \mathrm{P}_{\mathbb{N}(l_n)}(t).$$

Obviously, $\mathrm{P}_n(t) \geq \mathrm{P}_1(t)$ for all $n \geq 1$. For every $k \geq 1$, let $\pi_k : E_A^\infty \longrightarrow \mathbb{N}$ be the projection onto the kth coordinate, i.e.,

$$\pi\left(\{(\tau_u)_{u\geq 1}\}\right) := \tau_k.$$

By (11.30), $e^{\mathrm{P}_n(t)}$ is the eigenvalue of $\mathcal{L}_n^*$ corresponding to the eigenmeasure $\tilde{m}_n$.

Therefore, applying (11.25), we obtain, for every $n \geq 1$, every $k \geq 1$, and every $s \in \mathbb{N}$, that

$$\begin{aligned}
\tilde{m}_n(\pi_k^{-1}(s)) &= \sum_{\omega\in\mathbb{N}(l_n)_A^k:\omega_k=s} \tilde{m}_n([\omega]) \leq \sum_{\omega\in\mathbb{N}(l_n)_A^k:\omega_k=s} \|\phi_\omega'\|_\infty^t e^{-\mathrm{P}_n(t)k} \\
&\leq e^{-\mathrm{P}_n(t)k} \sum_{\omega\in\mathbb{N}(l_n)_A^k:\omega_k=s} \|\phi_{\omega|_{k-1}}'\|_\infty^t \|\phi_s'\|_\infty^t \\
&\leq e^{-\mathrm{P}_1(t)k} \left(\sum_{i\in\mathbb{N}} \|\phi_i'\|_\infty^t\right)^{k-1} \|\phi_s'\|_\infty^t.
\end{aligned}$$

Therefore,

$$\tilde{m}_n\left(\pi_k^{-1}([s+1,\infty))\right) \le e^{-\mathrm{P}_1(t)k}\left(\sum_{i\in\mathbb{N}}\|\phi_i'\|_\infty^t\right)^{k-1}\sum_{j=s+1}^{\infty}\|\phi_s'\|_\infty^t.$$

Fix now $\varepsilon > 0$. Then for every $k \ge 1$, choose an integer $n_k \ge 1$ so large that

$$e^{-\mathrm{P}_1 k}\left(\sum_{i\in\mathbb{N}}\|\phi_i'\|_\infty^t\right)^{k-1}\sum_{j=n_k+1}^{\infty}\|\phi_s'\|_\infty^t \le \frac{\varepsilon}{2^k}.$$

Then, for every $n \ge 1$ and every $k \ge 1$,

$$\tilde{m}_n\left(\pi_k^{-1}([n_k+1,\infty))\right) \le \varepsilon/2^k.$$

Hence,

$$\begin{aligned}\tilde{m}_n\left(E_A^\infty \cap \prod_{k\ge 1}[1,n_k]\right) &\ge 1-\sum_{k\ge 1}\tilde{m}_n\left(\pi_k^{-1}([n_k+1,\infty))\right)\\ &\ge 1-\sum_{k\ge 1}\frac{\varepsilon}{2^k} = 1-\varepsilon.\end{aligned}$$

Since $E_A^\infty \cap \prod_{k\ge 1}[1,n_k]$ is a compact subset of $E_A^{\mathbb{N}}$, the tightness of the sequence $\{\tilde{m}_n\}_{n\ge 1}$ is, therefore, proved. ∎

Thus, in view of Prohorov's Theorem, there exists $\tilde{m}$, a Borel probability measure on $E_A^{\mathbb{N}}$, which is a weak* limit point of the sequence $\{\tilde{m}_n\}_{n\ge 1}$. Passing to a subsequence, we may assume that the sequence $\{\tilde{m}_n\}_{n\ge 1}$ itself converges weakly to the measure $\tilde{m}$. Let

$$\mathcal{L}_{0,n} = e^{-\mathrm{P}_n(t)}\mathcal{L}_n \quad\text{and}\quad \mathcal{L}_0 = e^{-\mathrm{P}(t)}\mathcal{L}_t$$

be the corresponding normalized operators. Fix $g \in C_b\left(E_A^\infty\right)$ and $\varepsilon > 0$. Let us now consider an integer $n \ge 1$ so large that the following requirements are satisfied:

$$\sum_{i=n+1}\|g\|_\infty\|\phi_i'\|_\infty^t e^{-\mathrm{P}(t)} \le \frac{\varepsilon}{6}, \tag{11.41}$$

$$\sum_{i\le n}\|g\|_\infty\|\phi_i'\|_\infty^t\left|e^{-\mathrm{P}(t)} - e^{-\mathrm{P}_n(t)}\right| \le \frac{\varepsilon}{6}, \tag{11.42}$$

$$|\tilde{m}_n(g) - \tilde{m}(g)| \leq \frac{\varepsilon}{3}, \tag{11.43}$$

and

$$\left| \int_{E_A^\infty} \mathcal{L}_0(g) d\tilde{m} - \int_{E_A^\infty} \mathcal{L}_0(g) d\tilde{m}_n \right| \leq \frac{\varepsilon}{3}. \tag{11.44}$$

It is possible to satisfy condition (11.42) since, owing to Theorem 11.1.4, $\lim_{n\to\infty} \mathrm{P}_n(t) = \mathrm{P}(t)$. Let

$$g_n := g|_{E_{I_n}^\infty}.$$

The first two observations are the following:

$$\begin{aligned}
\mathcal{L}_{0,n}^* \tilde{m}_n(g) &= \int_{E_A^\infty} \sum_{i \leq n: A_{i\omega_n}=1} g(i\omega) \left|\phi_i'(\pi(\sigma(\omega)))\right|^t e^{-\mathrm{P}_n(t)} \, d\tilde{m}_n(\omega) \\
&= \int_{\mathbb{N}(I_n)_A^\infty} \sum_{i \leq n: A_{i\omega_n}=1} g(i\omega) \left|\phi_i'(\pi(\sigma(\omega)))\right|^t e^{-\mathrm{P}_n(t)} \, d\tilde{m}_n(\omega) \\
&= \int_{\mathbb{N}(I_n)_A^\infty} \sum_{i \leq n: A_{i\omega_n}=1} g_n(i\omega) \left|\phi_i'(\pi(\sigma(\omega)))\right|^t e^{-\mathrm{P}_n(t)} \, d\tilde{m}_n(\omega) \\
&= \mathcal{L}_{0,n}^* \tilde{m}_n(g_n) = \tilde{m}_n(g_n)
\end{aligned} \tag{11.45}$$

and

$$\tilde{m}_n(g_n) - \tilde{m}_n(g) = \int_{\mathbb{N}(I_n)^\mathbb{N}} (g_n - g) d\tilde{m}_n = \int_{\mathbb{N}(I_n)^\mathbb{N}} 0 d\tilde{m}_n = 0. \tag{11.46}$$

Using the triangle inequality, we get the following:

$$\begin{aligned}
\left|\mathcal{L}_0^* \tilde{m}(g) - \tilde{m}(g)\right| \leq& \left|\mathcal{L}_0^* \tilde{m}(g) - \mathcal{L}_0^* \tilde{m}_n(g)\right| + \left|\mathcal{L}_0^* \tilde{m}_n(g) - \mathcal{L}_{0,n}^* \tilde{m}_n(g)\right| \\
&+ \left|\mathcal{L}_{0,n}^* \tilde{m}_n(g) - \tilde{m}_n(g_n)\right| + \left|\tilde{m}_n(g_n) - \tilde{m}_n(g)\right| \\
&+ \left|\tilde{m}_n(g) - \tilde{m}(g)\right|.
\end{aligned} \tag{11.47}$$

Let us look first at the second summand. Applying (11.42) and (11.41), we get that

$$\begin{aligned}
&\left|\mathcal{L}_0^*\tilde{m}_n(g) - \mathcal{L}_{0,n}^*\tilde{m}_n(g)\right| \\
&= \Bigg| \int_{E_A^\infty} \sum_{i \le n: A_{i\omega_n}=1} g(i\omega)\Big(\left|\phi_i'(\pi(\sigma(\omega)))\right|^t e^{-\mathrm{P}(t)} \\
&\quad - \left|\phi_i'(\pi(\sigma(\omega)))\right|^t e^{-\mathrm{P}_n(t)}\Big)\, d\tilde{m}_n(\omega) \\
&\quad + \int_{E_A^\infty} \sum_{i > n: A_{i\omega_n}=1} g(i\omega)\left|\phi_i'(\pi(\sigma(\omega)))\right|^t e^{-\mathrm{P}(t)}\, d\tilde{m}_n(\omega) \Bigg| \qquad (11.48)\\
&\le \sum_{i \le n} \|g\|_\infty \|\phi_i'\|_\infty \left|e^{-\mathrm{P}(t)} - e^{-\mathrm{P}_n}\right| + \sum_{i=n+1} \|g\|_\infty^t \|\phi_i'\|_\infty^t e^{-\mathrm{P}(t)} \\
&\le \frac{\varepsilon}{6} + \frac{\varepsilon}{6} = \frac{\varepsilon}{3}.
\end{aligned}$$

Combining now, in turn, (11.44), (11.48), (11.45), (11.46), and (11.43), we get from (11.47) that

$$\left|\mathcal{L}_0^*\tilde{m}(g) - \tilde{m}(g)\right| \le \frac{\varepsilon}{3} + \frac{\varepsilon}{3} + \frac{\varepsilon}{3} = \varepsilon.$$

Letting $\varepsilon \searrow 0$, we, therefore, get that $\mathcal{L}_0^*\tilde{m}(g) = \tilde{m}(g)$ or $\mathcal{L}_t^*\tilde{m}(g) = e^{\mathrm{P}(t)}\tilde{m}(g)$. Hence,

$$\mathcal{L}_t^*\tilde{m} = e^{\mathrm{P}(t)}\tilde{m}$$

and the proof is complete. ■

Remark 11.6.3 Note that once we have an eigenmeasure of $\mathcal{L}_t^*$, then all the formulas from (11.22) to (11.23) hold for any countable alphabet E regardless of whether it is finite or not.

We will now introduce the concept of Gibbs states and shift-invariant Gibbs states for a parameter $t \in F(S)$. In this book, these play only a very limited auxiliary role; they are used by us only to simplify notation, for the formulation of some results, and for proofs. This concept can, however be extended to the class of all Hölder continuous potentials on E_A^∞, leading to a rich, meaningful, and powerful theory whose systematic account can be found in [**MU2**]; see also [**CTU**] and [**MU4**].

If S is a finitely irreducible GDMS with an incidence matrix A and $t \in F(S)$, then a Borel probability measure $\tilde{m}$ on E_A^∞ is called a Gibbs state for t if there exist constants $Q_t \ge 1$ and $\mathrm{P}_{\tilde{m}} \in \mathbb{R}$ such that, for every $\omega \in E_A^*$ and every $\tau \in [\omega]$,

$$Q_t^{-1} \le \frac{\tilde{m}([\omega])}{\left|\phi'_\omega\left(\pi(\sigma^{|\omega|}(\tau))\right)\right|^t \exp\left(-\mathrm{P}_{\tilde{m}}|\omega|\right)} \le Q_t. \tag{11.49}$$

If, additionally, $\tilde{m}$ is shift invariant, then $\tilde{m}$ is called an invariant Gibbs state. Because of the Bounded Distortion Property (4e), (11.49) also takes on the following somewhat simpler form.

For every $\omega \in E_A^*$,

$$Q_t^{-1} \le \frac{\tilde{m}([\omega])}{\left\|\phi'_\omega\right\|_\infty^t \exp\left(-\mathrm{P}_{\tilde{m}}|\omega|\right)} \le Q_t. \tag{11.50}$$

Proposition 11.6.4 *If S is a finitely irreducible GDMS with an incidence matrix A and $t \in F(S)$, then the following hold.*

(a) *For every Gibbs state $\tilde{m}$ for t, $\mathrm{P}_{\tilde{m}} = \mathrm{P}(t)$.*
(b) *Any two Gibbs states for the function t are boundedly equivalent to Radon–Nikodym derivatives bounded away from zero and infinity.*

Proof We shall first prove (a). Towards this end, fix $n \ge 1$ and sum up (11.49) over all words $\omega \in E_A^n$. Since

$$\sum_{|\omega|=n} \tilde{m}([\omega]) = 1,$$

we then get that

$$Q_t^{-1} e^{-\mathrm{P}_{\tilde{m}} n} \sum_{|\omega|=n} \left\|\phi'_\omega\right\|_\infty^t \le 1 \le Q_t e^{-\mathrm{P}_{\tilde{m}} n} \sum_{|\omega|=n} \left\|\phi'_\omega\right\|_\infty^t.$$

Applying logarithms to all three terms of this formula, dividing all the terms by n, and taking the limit as $n \to \infty$, we obtain that

$$-\mathrm{P}_{\tilde{m}} + \mathrm{P}(t) \le 0 \le -\mathrm{P}_{\tilde{m}} + \mathrm{P}(t),$$

which means that

$$\mathrm{P}_{\tilde{m}} = \mathrm{P}(t).$$

The proof of item (a) is, thus, complete.

In order to prove part (b), suppose that m and ν are two Gibbs states of t. Notice now that part (a) implies the existence of a constant $T \ge 1$ such that

$$T^{-1} \le \frac{\nu([\omega])}{m([\omega])} \le T$$

for all words $\omega \in E_A^*$. Straightforward reasoning now gives that ν and m are equivalent and

$$T^{-1} \leq \frac{d\nu}{dm} \leq T.$$

The proof of Proposition 11.6.4 is complete. ■

Theorem 11.6.5 *If S is a finitely irreducible GDMS with an incidence matrix A and $t \in F(S)$, then any eigenmeasure $\tilde{m}$ of the dual operator $\mathcal{L}_t^*: C^*(E_A^\infty) \longrightarrow C^*(E_A^\infty)$ is a Gibbs state for t. In addition, its corresponding eigenvalue λ is equal to $e^{\mathrm{P}(t)}$.*

Proof Let Λ be a minimal set which witnesses the finite irreducibility of A. Let

$$T := \min\left\{ \inf(|\phi_\alpha'|)e^{-\mathrm{P}(t)|\alpha|} : \alpha \in \Lambda \right\}.$$

Fix $\omega \in E_A^n$. Put $n := |\omega|$. For every $\alpha \in \Lambda$, let

$$E_\alpha(\omega) := \left\{ \tau \in E_A^\infty : \omega\alpha\tau \in E_A^\infty \right\}.$$

By the definition of Λ, $\bigcup_{\alpha\in\Lambda} \alpha(\omega) = E_A^\infty$. Hence, there exists $\gamma \in \Lambda$ such that

$$\tilde{m}(E_\gamma) \geq (\#\Lambda)^{-1}.$$

Writing $p = |\gamma|$ and using the Bounded Distortion Property (4e), we, therefore, get that

$$\begin{aligned}
\tilde{m}([\omega]) \geq \tilde{m}([\omega\gamma]) &= \lambda^{-(n+p)} \int_{\rho\in E_A^\infty : A_{\gamma_p\rho_1}=1} \left|\phi_{\omega\gamma}'(\pi(\rho))\right|^t d\tilde{m}(\rho) \\
&= \lambda^{-(n+p)} \int_{\rho\in E^\infty : A_{\gamma_p\rho_1}=1} \left|\phi_\omega'(\pi(\gamma\rho))\right|^t \cdot \left|\phi_\gamma'(\pi(\rho))\right|^t d\tilde{m}(\rho) \\
&\geq \lambda^{-n} T \lambda^{-p} \int_{\rho\in E_A^\infty : A_{\gamma_p\rho_1}=1} \left|\phi_\omega'(\pi(\gamma\rho))\right|^t d\tilde{m}(\rho) \\
&= \lambda^{-n} T \lambda^{-p} \int_{E_\gamma(\omega)} \left|\phi_\omega'(\pi(\gamma\rho))\right|^t d\tilde{m}(\rho) \\
&\geq T\lambda^{-p} K^{-t} \tilde{m}(E_\gamma(\omega)) \lambda^{-n} \left\|\phi_\omega'\right\|_\infty^t \\
&\geq T\lambda^{-p} K^{-t} (\#\Lambda)^{-1} \lambda^{-n} \left\|\phi_\omega'\right\|_\infty^t.
\end{aligned} \tag{11.51}$$

Along with (11.26), which holds because of Remark 11.6.3, this shows that $\tilde{m}$ is a Gibbs state for t. The equality $\lambda = e^{\mathrm{P}(t)}$ follows now immediately from Proposition 11.6.4. The proof of Theorem 11.6.5 is complete. ■

Theorem 11.6.6 *If S is a finitely irreducible GDMS with an incidence matrix A, then, for every $t \in F(S)$, there exists a unique shift-invariant Gibbs state $\tilde{\mu}_t$ of t. The shift-invariant Gibbs state $\tilde{\mu}_t$ is ergodic. In addition, if the incidence matrix A is finitely primitive, then the Gibbs state $\tilde{\mu}_t$ is completely ergodic.*

Proof Let $\tilde{m}$ be a Gibbs state for $t \in F(S)$. Fixing $\omega \in E_A^*$ and using (11.49), and Proposition 11.6.4(a), we get, for every $n \geq 1$, that

$$\begin{aligned}
\tilde{m}(\sigma^{-n}([\omega])) &= \sum_{\tau \in E_A^n : A_{\tau_n \omega_1}=1} \tilde{m}([\tau\omega]) \\
&\leq \sum_{\tau \in E_A^n : A_{\tau_n \omega_1}=1} Q_t \|\phi_{\tau\omega}\|_\infty \exp(-\mathrm{P}(t)(n+|\omega|)) \\
&\leq \sum_{\tau \in E_A^n : A_{\tau_n \omega_1}=1} Q_t \|\phi'_\tau\|_\infty e^{-\mathrm{P}(t)n} \|\phi'_\omega\|_\infty e^{-\mathrm{P}(t)|\omega|} \qquad (11.52) \\
&\leq \sum_{\tau \in E_n^\omega} Q_t Q_t \tilde{m}([\tau]) Q_t \tilde{m}([\omega]) \\
&\leq Q_t^3 \tilde{m}([\omega]).
\end{aligned}$$

Let the finite set of words Λ witness the finite irreducibility of the incidence matrix A and p be the maximal length of a word in Λ. Let

$$T := \min\left\{ \inf(|\phi'_\alpha|)e^{-\mathrm{P}(t)|\alpha|} : \alpha \in \Lambda \right\} \in (0, +\infty).$$

For each $\tau, \omega \in E_A^*$, let $\alpha = \alpha(\tau,\omega) \in \Lambda$ be such that $\tau\alpha\omega \in E_A^*$. Then, we have, for all $\omega \in E_A^*$ and all integers $n \geq 1$, that

$$\begin{aligned}
&\sum_{i=n}^{n+p} \tilde{m}(\sigma^{-i}([\omega])) \\
&\quad = \sum_{i=n}^{n+p} \sum_{\tau \in E_A^i : A_{\tau_i \omega_1}=1} \tilde{m}([\tau\omega]) \geq \sum_{\tau \in E_A^n} \tilde{m}([\tau\alpha(\tau,\omega)\omega]) \\
&\quad \geq \sum_{\tau \in E_A^n} Q_t^{-1} \inf\left(\left|\phi'_{\tau\alpha(\tau,\omega)\omega}\right|^t\right) \exp\left(-\mathrm{P}(t)(|\tau| + |\alpha(\tau,\omega)| + |\omega|)\right) \\
&\quad \geq Q_t^{-1} \sum_{\tau \in E_A^n} \inf\left(\left|\phi'_\tau\right|^t\right) \exp\left(-\mathrm{P}(t)(|\tau|)\right) \\
&\qquad \times \inf\left(\left|\phi'_{\alpha(\tau,\omega)}\right|^t\right) \exp\left(-\mathrm{P}(t)|\alpha(\tau,\omega)|\right) \inf\left(\left|\phi'_\omega\right|^t\right) \exp\left(-\mathrm{P}(t)(|\omega|)\right)
\end{aligned}$$

$$
\begin{aligned}
&\geq Q_t^{-1} T \inf\big(|\phi'_\omega|^t\big) \sum_{\tau \in E_A^n} \exp\big(-\mathrm{P}(t)|\omega|\big) \inf\big(|\phi'_\tau|^t\big) \exp\big(-\mathrm{P}(t)(|\tau|)\big) \\
&\geq Q_t^{-2} T \tilde{m}([\omega]) \sum_{\tau \in E_A^n} \exp\big(-\mathrm{P}(t)|\omega|\big) \inf\big(|\phi'_\tau|^t\big) \exp\big(-\mathrm{P}(t)(|\tau|)\big) \\
&\geq Q_t^{-2} T \tilde{m}([\omega]) Q_t^{-1} \sum_{\tau \in E_A^n} \tilde{m}([\tau]) \\
&= Q_t^{-3} T \tilde{m}([\omega]).
\end{aligned}
\tag{11.53}
$$

Let

$$
l_B : l_\infty \longrightarrow \mathbb{R}
$$

be a Banach limit; see (2.36) and properties (a)–(g) following it. It is then not difficult to check that

$$
\tilde{\mu}(B) = l_B\big((\tilde{m}(\sigma^{-n}(B)))_{n \geq 0}\big)
$$

defines a shift-invariant, finitely additive probability measure on Borel sets of E_A^∞ satisfying

$$
\frac{Q_t^{-3} T}{p} \tilde{m}(B) \leq \tilde{\mu}(B) \leq Q_t^3 \tilde{m}(B) \tag{11.54}
$$

for every Borel set $B \subset E_A^\infty$. Since $\tilde{m}$ is a countably additive measure, we easily deduce that $\tilde{\mu}$ is also countably additive. It then immediately follows from (11.54) that $\tilde{\mu}$ is a Gibbs state for t.

Let us prove the ergodicity of $\tilde{\mu}$. For each $\tau \in E_A^*$, as in (11.53), we get that

$$
\sum_{i=n}^{n+p} \tilde{\mu}(\sigma^{-i}([\tau]) \cap [\omega]) \geq \tilde{\mu}([\omega\alpha(\omega,\tau)\tau]) \geq Q_t^{-3} T \tilde{\mu}([\tau]) \tilde{\mu}([\omega]). \tag{11.55}
$$

Take now an arbitrary Borel set $B \subseteq E_A^\infty$ and fix $\varepsilon > 0$. Since the nested family of sets $\{[\tau] : \tau \in E_A^*\}$ generates the Borel σ-algebra on E_A^∞, for every $n \geq 0$ and every $\omega \in E_A^n$, we can find a subfamily Z of E_A^* consisting of mutually incomparable words such that

$$
B \subseteq \bigcup_{\tau \in Z} [\tau]
$$

and, for all $n \leq i \leq n + p$,

$$
\sum_{\tau \in Z} \tilde{\mu}(\sigma^{-i}([\tau]) \cap [\omega]) \leq \tilde{\mu}\big([\omega] \cap \sigma^{-i}(B)\big) + \varepsilon/p.
$$

Then, using (11.55), we get that

$$\begin{aligned}\varepsilon + \sum_{i=n}^{n+p} \tilde{\mu}\big([\omega] \cap \sigma^{-I}(B)\big) &\geq \sum_{i=n}^{n+p} \sum_{\tau \in Z} \tilde{\mu}\big([\omega] \cap \sigma^{-i}([\tau])\big) \\ &\geq \sum_{\tau \in Z} Q_t^{-3} T \tilde{\mu}([\tau]) \tilde{\mu}([\omega]) \\ &\geq Q_t^{-3} T \tilde{\mu}(B) \tilde{\mu}([\omega]).\end{aligned} \tag{11.56}$$

Hence, letting $\varepsilon \searrow 0$, we get that

$$\sum_{i=n}^{n+p} \tilde{\mu}\big([\omega] \cap \sigma^{-(i)}(B)\big) \geq Q_t^{-3} e^T \tilde{\mu}(B) \tilde{\mu}([\omega]).$$

From this inequality, we get that

$$\begin{aligned}\sum_{i=n}^{n+p} \tilde{\mu}\big(\sigma^{-i}(E_A^{\mathbb{N}} \backslash B) \cap [\omega]\big) &= \sum_{i=n}^{n+p} \tilde{\mu}\big([\omega] \backslash \sigma^{-i}(B) \cap [\omega]\big) \\ &= \sum_{i=n}^{n+p} \tilde{\mu}([\omega]) - \tilde{\mu}\big(\sigma^{-i}(B) \cap [\omega]\big) \\ &\leq (p - Q_t^{-3} T \tilde{\mu}(B)) \tilde{\mu}([\omega]).\end{aligned}$$

Thus, for every Borel set $B \subseteq E_A^{\infty}$, for every $n \geq 0$, and for every $\omega \in E_A^n$, we have that

$$\sum_{i=n}^{n+p} \tilde{\mu}\big(\sigma^{-i}(B) \cap [\omega]\big) \leq (p - Q_t^{-3} T (1 - \tilde{\mu}(B))) \tilde{\mu}([\omega]). \tag{11.57}$$

In order to conclude the proof of the ergodicity of σ, suppose that

$$\sigma^{-1}(B) = B \quad \text{with} \quad 0 < \tilde{\mu}(B) < 1.$$

Put

$$\gamma := 1 - Q_g^{-3} e^T (1 - \tilde{\mu}(B))/p.$$

Note that (11.54) implies that $Q_g^{-3} T p^{-1} \leq 1$; hence, $0 < \gamma < 1$. In view of (11.57), for every $\omega \in E_A^*$, we get that

$$\tilde{\mu}(B \cap [\omega]) = \tilde{\mu}\big(\sigma^{-i}(B) \cap [\omega]\big) \leq \gamma \tilde{\mu}([\omega]).$$

Take now $\eta > 1$ so small that $\gamma\eta < 1$ and choose a subfamily R of E_A^* consisting of mutually incomparable words such that

$$B \subseteq \bigcup_{\omega \in R} [\omega] \quad \text{and} \quad \tilde{\mu}\left(\bigcup_{\omega \in R} [\omega]\right) \leq \eta \tilde{\mu}(B).$$

Then

$$\tilde{\mu}(B) \leq \sum_{\omega \in R} \tilde{\mu}(B \cap [\omega]) \leq \sum_{\omega \in R} \gamma \tilde{\mu}([\omega]) = \gamma \tilde{\mu}\left(\bigcup_{\omega \in R} [\omega]\right) \leq \gamma \eta \tilde{\mu}(B < \tilde{\mu}(B)).$$

This contradiction finishes the proof of the existence part.

It follows from ergodicity of $\tilde{\mu}$ and Proposition 11.6.4(b) that any two Gibbs states are ergodic. So, the uniqueness of invariant Gibbs states follows immediately from Theorem 3.1.1 and Proposition 11.6.4(b) invoked again.

Finally, let us prove the complete ergodicity of $\tilde{\mu}$ in the case when A is finitely primitive. Essentially, we repeat the argument just given. Let Λ be a finite set of words all of length q that witness the finite primitivity of A. Fix $r \in \mathbb{N}$. Let $\omega \in E_A^n$. For each $\tau \in E_A^*$, we get the following improvement of (11.53):

$$\begin{aligned} \tilde{\mu}(\sigma^{-(n+qr)}([\tau]) \cap [\omega]) &\geq \sum_{\alpha \in \Lambda^r \cap E^{qr} : A_{\omega_n \alpha_1} = A_{\alpha_{qr} \tau_1} = 1} \tilde{\mu}([\omega \alpha \tau]) \\ &\geq Q_t^{-3} T^r \tilde{\mu}([\tau]) \tilde{\mu}([\omega]). \end{aligned} \tag{11.58}$$

Take now an arbitrary Borel set $B \subseteq E_A^\infty$. Fix $\varepsilon > 0$. Since the nested family of sets $\{[\tau] : \tau \in E_A^*\}$ generates the Borel σ-algebra on E_A^∞, for every $n \geq 0$ and every $\omega \in E_A^n$, we can find a subfamily Z of E_A^* consisting of mutually incomparable words such that

$$B \subseteq \bigcup_{\tau \in Z} [\tau]$$

and

$$\sum_{\tau \in Z} \tilde{\mu}\big(\sigma^{-(n+qr)}([\tau]) \cap [\omega]\big) \leq \tilde{\mu}\big([\omega] \cap \sigma^{-(n+qr)}(B)\big) + \varepsilon.$$

Then, using (11.58), we get that

$$\varepsilon + \tilde{\mu}\big([\omega] \cap \sigma^{-(n+qr)}(B)\big) \geq \sum_{\tau \in Z} Q_t^{-3} T^r \tilde{\mu}([\tau]) \tilde{\mu}([\omega]) \geq Q_t^{-3} T^r \tilde{\mu}(B) \tilde{\mu}([\omega]).$$

Hence, letting $\varepsilon \searrow 0$, we get that

$$\tilde{\mu}\big([\omega] \cap \sigma^{-(n+qr)}(B)\big) \geq \tilde{Q}(r) \tilde{\mu}(B) \tilde{\mu}([\omega]),$$

where $\tilde{Q}(r) := Q_t^{-3} \exp(rT)$. Note that it follows from this last inequality that $\tilde{Q}(r) \leq 1$. Also, from this inequality, we find that

$$\begin{aligned} \tilde{\mu}\big(\sigma^{-(n+qr)}(E_A^\infty \backslash B) \cap [\omega]\big) &= \tilde{\mu}\big([\omega] \backslash \sigma^{-n}(B) \cap [\omega]\big) \\ &= \tilde{\mu}([\omega]) - \tilde{\mu}\big(\sigma^{-(n+qr)}(B) \cap [\omega]\big) \\ &\leq (1 - \tilde{Q}(r)\, \tilde{\mu}(B)) \tilde{\mu}([\omega]). \end{aligned}$$

Thus, for every Borel set $B \subseteq E_A^\infty$, for every $n \geq 0$, and for every $\omega \in E_A^n$, we have that

$$\tilde{\mu}\big(\sigma^{-(n+qr)}(B) \cap [\omega]\big) \leq \big(1 - \tilde{Q}(r)\,(1 - \tilde{\mu}(B))\big)\tilde{\mu}([\omega]). \tag{11.59}$$

In order to conclude the proof of the complete ergodicity of σ, suppose that

$$\sigma^{-r}(B) = B \quad \text{with} \quad 0 < \tilde{\mu}(B) < 1.$$

Let

$$(E_A^r)^* := \bigcup_{k \in \mathbb{N}} (E_A^r)^k.$$

Put

$$\gamma := 1 - \tilde{Q}(r)(1 - \tilde{\mu}(B)).$$

Note that $0 < \gamma < 1$. In view of (11.59), for every $\omega \in (E_A^r)^*$, we get that

$$\tilde{\mu}(B \cap [\omega]) = \tilde{\mu}\big(\sigma^{-(|\omega|+qr)}(B) \cap [\omega]\big) \leq \gamma\tilde{\mu}([\omega]).$$

Take now $\eta > 1$ so small that $\gamma\eta < 1$ and choose a subfamily R of $(E_A^r)^*$ consisting of mutually incomparable words such that

$$B \subseteq \bigcup_{\omega \in R} [\omega]$$

and

$$\tilde{\mu}\left(\bigcup_{\omega \in R} [\omega]: \right) \leq \eta\tilde{\mu}(B).$$

Then

$$\tilde{\mu}(B) \leq \sum_{\omega \in R} \tilde{\mu}(B \cap [\omega]) \leq \sum_{\omega \in R} \gamma\tilde{\mu}([\omega]) = \gamma\tilde{\mu}\left(\bigcup_{\omega \in R} [\omega]\right) \leq \gamma\eta\tilde{\mu}(B) < \tilde{\mu}(B).$$

This contradiction finishes the proof of the complete ergodicity of $\tilde{\mu}$. The proof of Theorem 11.6.6 is complete. ■

Theorem 11.6.7 *If S is a finitely irreducible GDMS with an incidence matrix A and $t \in F(S)$, then the conjugate operator*

$$e^{-\mathrm{P}(t)}\mathcal{L}_t^*: C_b\big(E_A^\infty\big) \longrightarrow C_b\big(E_A^\infty\big)$$

fixes at most one Borel probability measure on E_A^∞.

Proof Suppose that $\tilde{m}$ and $\tilde{m}_1$ are two such fixed points. In view of Proposition 11.6.4(b) and Theorem 11.6.5, the measures $\tilde{m}$ and $\tilde{m}_1$ are equivalent. Consider the Radon–Nikodym derivative

$$\rho := \frac{d\tilde{m}_1}{d\tilde{m}}.$$

Temporarily fix $\omega \in E_A^*$, say $\omega \in E_A^n$. Denote

$$Z(\sigma(\omega)) := \{\tau \in E_A^\infty : A_{\omega_n \tau_1} = 1\}.$$

Note that if $n \geq 2$, which we assume from now on, then

$$Z(\sigma(\omega)) = Z(\sigma(\omega)).$$

It then follows from (11.25) and Theorem 11.6.5 that

$$\begin{aligned}
\tilde{m}([\omega]) &= \int_{Z(\omega)} |\phi'_\omega|^t \exp(-\mathrm{P}(t)n) d\tilde{m} \\
&= \int_{Z(\omega)} |\phi'_{\sigma(\omega)}(\pi(\tau))|^t \exp\big(-\mathrm{P}(t)(n-1)\big) |\phi'_{\omega_1}(\pi(\sigma(\omega)\tau))|^t \\
&\quad \times \exp(-\mathrm{P}(t)) d\tilde{m}(\tau) \\
&= \int_{Z(\sigma(\omega))} |\phi'_{\sigma(\omega)}(\pi(\tau))|^t \exp\big(-\mathrm{P}(t)(n-1)\big) |\phi'_{\omega_1}(\pi(\sigma(\omega)\tau))|^t \\
&\quad \times \exp(-\mathrm{P}(t)) d\tilde{m}(\tau).
\end{aligned}$$

Since, by the same token, also

$$\tilde{m}([\sigma(\omega)]) = \int_{Z(\sigma(\omega))} |\phi'_{\sigma(\omega)}(\pi(\tau))|^t \exp\big(-\mathrm{P}(t)(n-1)\big) d\tilde{m}(\tau),$$

we, thus, conclude that

$$\begin{aligned}
&\inf\Big\{ |\phi'_{\omega_1}(\pi(\sigma(\omega)\tau))|^t \exp(-\mathrm{P}(t)) : \tau \in Z(\sigma(\omega)) \Big\} \tilde{m}([\sigma(\omega)]) \\
&\leq \tilde{m}([\omega]) \\
&\leq \sup\Big\{ |\phi'_{\omega_1}(\pi(\sigma(\omega)\tau))|^t \times \exp(-\mathrm{P}(t)) : \tau \in Z(\sigma(\omega)) \Big\} \tilde{m}([\sigma(\omega)]).
\end{aligned}$$

We, therefore, conclude that, for every $\omega \in E_A^\infty$,

$$\lim_{n\to\infty} \frac{\tilde{m}\big([\omega|_n]\big)}{\tilde{m}\big([\sigma(\omega)|_{n-1}]\big)} = |\phi'_{\omega_1}(\pi(\sigma(\omega)))|^t \exp(-\mathrm{P}(t)). \tag{11.60}$$

Of course, the same formula is true with $\tilde{m}$ replaced by $\tilde{m}_1$. Since the measures $\tilde{m}$ and $\tilde{m}_1$ are equivalent, by applying Theorems 11.6.5 and 11.6.6 along with Proposition 11.6.4 (ergodicity), we deduce that there exists a measurable σ-invariant set $\Gamma \subseteq E_A^\infty$ such that the Radon–Nikodym derivative $\rho(\omega)$ is defined for every $\omega \in \Gamma$. So, if $\omega \in \Gamma$, then the Radon–Nikodym derivatives $\rho(\omega)$ and $\rho(\sigma(\omega))$ are both defined. Then from (11.60) and its version for $\tilde{m}_1$, we obtain that

$$
\begin{aligned}
\rho(\omega) &= \lim_{n\to\infty}\left(\frac{\tilde{m}_1([\omega|_n])}{\tilde{m}([\omega|_n])}\right) \\
&= \lim_{n\to\infty}\left(\frac{\tilde{m}_1([\omega|_n])}{\tilde{m}_1([\sigma(\omega)|_{n-1}])}\cdot\frac{\tilde{m}_1([\sigma(\omega)|_{n-1}])}{\tilde{m}([\sigma(\omega)|_{n-1}])}\cdot\frac{\tilde{m}([\sigma(\omega)|_{n-1}])}{\tilde{m}([\omega|_n])}\right) \\
&= \left|\phi'_{\omega_1}(\pi(\sigma(\omega)))\right|^t \exp(-\mathrm{P}(t))\rho(\sigma(\omega))\left|\phi'_{\omega_1}(\pi(\sigma(\omega)))\right|^{-t}\exp(\mathrm{P}(t)) \\
&= \rho(\sigma(\omega)).
\end{aligned}
$$

But, since, according to Theorem 11.6.6 again, the shift map $\sigma\colon E_A^\infty \longrightarrow E_A^\infty$ is ergodic with respect to a shift-invariant measure equivalent to $\tilde{m}$, we conclude that the Radon–Nikodym derivative ρ is $\tilde{m}$-a.e. constant. Since $\tilde{m}_1$ and $\tilde{m}$ are both probability measures, we, thus, have that $\tilde{m}_1 = \tilde{m}$. The proof of Theorem 11.6.7 is complete. ■

As an immediate consequence of Theorem 11.6.2 and Theorems 11.6.5–11.6.7, we get the following result summarizing what we did about the thermodynamic formalism.

Corollary 11.6.8 *If S is a finitely irreducible GDMS with an incidence matrix A and $t \in F(S)$, then*

(1) *There exists a unique eigenmeasure $\tilde{m}_t$ of the conjugate Perron–Frobenius operator $\mathcal{L}_t^*\colon C_b^*(E_A^\infty) \longrightarrow C_b^*(E_A^\infty)$ and the corresponding eigenvalue is equal to $e^{\mathrm{P}(t)}$.*
(2) *The eigenmeasure $\tilde{m}_t$ is a Gibbs state for t.*
(3) *There exists a unique shift-invariant Gibbs state $\tilde{\mu}_t$ for t.*
(4) *The Gibbs state $\tilde{\mu}_t$ is ergodic and equivalent to $\tilde{m}_t$ and $\log(d\tilde{\mu}_t/d\tilde{m}_t)$ is uniformly bounded.*
(5) *If the incidence matrix A is finitely primitive, then the Gibbs state $\tilde{\mu}_t$ is completely ergodic.*

For $t = h$, this corollary gives the following.

Proposition 11.6.9 *If S is a finitely irreducible CGDMS, then S is regular if and only if there exists a Borel probability measure ν on E_A^∞ such that*

$$\mathcal{L}_h^* \nu = \nu. \tag{11.61}$$

Furthermore, if such a measure exists, then it is unique.

In addition, if $\mathcal{L}_t^* \nu = \nu$, for some $t \geq 0$ and some Borel probability measure ν, then $t = h$ and $\nu = \tilde{m}_h$. In particular, (11.31) *holds.*

11.7 Finer Geometrical Properties of CGDMSs

In this section, we continue our investigations of CGDMSs. We deal with the pressure function, conformal measures, and strongly regular systems, and we prove the finiteness of the Hausdorff measure along with positivity of the packing measure.

Throughout the whole section, S is a finitely irreducible GDMS with an incidence matrix A and $t \in F(S)$.

With the measures $\tilde{m}_t$ and $\tilde{\mu}_t$ coming from Corollary 11.6.8, we denote

$$m_t := \tilde{m}_t \circ \pi^{-1}$$

and

$$\mu_t := \tilde{\mu}_t \circ \pi^{-1}.$$

In Volume 2, we will need the following three results, out of which the last two concern strongly regular conformal systems.

Proposition 11.7.1 *If $S_E = \{\phi_e \colon e \in E\}$ is a finitely irreducible CGDMS and E' is a proper subset of E such that the system $S_{E'} = \{\phi_e \colon e \in E'\}$ is also finitely irreducible, then*

$$\mathrm{P}_{E'}(t) < \mathrm{P}_E(t)$$

for all $t \in F(S)$.

Proof Fix $b \in E \backslash E'$. Suppose, for a contradiction, that $\mathrm{P}_{E'}(t) = \mathrm{P}_E(t)$ for some $t \in F(S)$. Let $\tilde{m}_t'$ be the measure produced in Theorem 11.6.2 for the system $S_{E'}$. It then follows from Remark 11.6.3, (11.23), (11.26), (11.29), and the third assertion of Corollary 11.6.8(4) that

$$\tilde{\mu}_t'([\omega]) \leq C \tilde{\mu}_t([\omega])$$

for some constant $C > 0$ and all $\omega \in {E'_{A'}}^*$, where $A' = A|_{E' \times E'}$. Furthermore, this inequality holds for all $\omega \in E_A^*$ since if $\omega \notin {E'_{A'}}^*$, then

$\tilde{\mu}'_t([\omega]) = 0$. Thus, the measure $\tilde{\mu}'_t$, considered as a Borel probability measure on E_A^∞, is absolutely continuous with respect to the measure $\tilde{\mu}_t$. Since, by the first assertion of Corollary 11.6.8(4), both measures $\tilde{\mu}'_t$ and $\tilde{\mu}_t$ are ergodic, we conclude (see Theorem 3.1.1) that these two measures are equal. This, however, is a contradiction as $\tilde{\mu}'_t([b]) = 0$ while $\tilde{\mu}_t([b]) > 0$. We are done. ■

Corollary 11.7.2 *If $S_E = \{\phi_e : e \in E\}$ is a strongly regular finitely irreducible CGDMS and E' is a proper subset of E such that the system $S_{E'} = \{\phi_e : e \in E'\}$ is also finitely irreducible, then*

$$h_{S_{E'}} < h_{S_E}.$$

Proof Since the system S_E is strongly regular, there exists $u < h_E$ such that $0 < \mathrm{P}_E(u) < +\infty$. Hence,

$$\mathrm{P}_{E'}(u) \le \mathrm{P}_E(u) < +\infty;$$

therefore, the function $\mathrm{P}_{E'}$ restricted to $[u, h_E]$ is continuous. Since also, by Proposition 11.7.1, $\mathrm{P}_{E'}(h_E) < \mathrm{P}_E(h_E) = 0$, we, therefore, see that there exists $t \in [u, h_E)$ such that $\mathrm{P}_{E'}(t) < 0$. Hence, it follows from Theorem 11.5.3 that $h_{E'} \le t < h_E$. The proof is complete. ■

Theorem 11.7.3 *If S is a finitely irreducible strongly regular CGDMS, then the metric entropy $\mathrm{h}_{\tilde{\mu}_{h_S}}(\sigma)$ of the dynamical system $\left(\sigma : E_A^\infty \longrightarrow E_A^\infty\right)$ with respect to the σ-invariant measure $\tilde{\mu}_{h_S}$ is finite.*

Proof Let α be the partition of $E_A^{\mathbb{N}}$ into initial cylinders of length 1, i.e.,

$$\alpha = \{[e]\}_{e \in E}.$$

Since the system S is strongly regular, by virtue of Proposition 11.4.2, there exists $\eta > 0$ such that $Z_1(h - \eta) < \infty$. This means that

$$\sum_{e \in E} \|\phi'_e\|_\infty^{h_S - \eta} < +\infty.$$

Since $\|\phi'_e\|_\infty^{-\eta} \ge -h_S \log \|\phi'_e\|_\infty$ for all but, perhaps, fintely many $e \in E$, the series

$$\sum_{e \in E} -h_S \log\left(\|\phi'_e\|_\infty\right) \|\phi'_e\|_\infty^{h_S}$$

converges too. Hence, by (11.31),

$$\mathrm{H}_{\tilde{\mu}_{h_S}}(\alpha) = \sum_{i \in I} -\log(\tilde{\mu}_{h_S}([e]))\tilde{\mu}_{h_S}([e]) < +\infty.$$

Since the partition α is a metric (even topological) generator of the dynamical system $\sigma : E_A^\infty \longrightarrow E_A^\infty$, we have that

$$\mathrm{h}_{\tilde{\mu}_{h_S}}(\sigma) = \mathrm{H}_{\tilde{\mu}_{h_S}}(\alpha) < +\infty$$

and the proof is complete. ■

Since we always have that $\mathrm{P}(h_S) \leq 0$, (11.29) and (11.8) immediately give the following.

Proposition 11.7.4 *If S is a finitely irreducible CGDMS, then the Hausdorff measure $\mathrm{H}_{h_S}(J_S) < +\infty$.*

We need, however, a slightly stronger statement.

Theorem 11.7.5 *If S is finitely irreducible regular CGDMS, then the Hausdorff measure H_{h_S} restricted to the limit set J_S is absolutely continuous with respect to the conformal measure m_h and*

$$\|d\mathrm{H}_h/dm_h\|_\infty < +\infty.$$

In particular, we get again that the Hausdorff measure $\mathrm{H}_{h_S}(J_S)$ is finite.

Proof Let A be an arbitrary closed subset of J_S. For every integer $n \geq 1$, put

$$A_n := \left\{\omega \in E_A^n : \phi_\omega(J_S) \cap A \neq \emptyset\right\}.$$

Then the sequence of sets

$$\left(\bigcup_{\omega \in A_n} \phi_\omega(X_{t(\omega)})\right)_{n=1}^\infty$$

is descending and

$$\bigcap_{n \geq 1} \left(\bigcup_{\omega \in A_n} \phi_\omega(X_{t(\omega)})\right) = A.$$

Therefore, using (11.8) in Corollary 11.3.10 and (11.31),

$$\begin{aligned}\mathrm{H}_h(A) &\leq \varliminf_{n\to\infty} \sum_{\omega \in A_n} \left(\mathrm{diam}(\phi_\omega(X_{t(\omega)}))\right)^h \leq D^h \varliminf_{n\to\infty} \sum_{\omega \in A_n} \|\phi'_\omega\|^h \\ &\leq D^h M_h^{-1} \varliminf_{n\to\infty} \sum_{\omega \in A_n} \tilde{m}_h([\omega])\end{aligned}$$

$$\le D^h M_h^{-1} \varliminf_{n\to\infty} \left(m \left(\bigcup_{\omega\in A_n} \phi_\omega(X_{t(\omega)}) \right) \right)$$
$$= D^h M_h^{-1} m_h(A).$$

Since J_S is a metric separable space, the measure m_h is regular; therefore, the inequality

$$\mathrm{H}_t(A) \le D^h M_h^{-1} m_h(A)$$

extends to all Borel subsets of J_S. The proof is finished. ■

A dual statement holds for packing measures. It, however, requires an additional mild hypothesis common in the theory of CGDMSs. Since this hypothesis has many more significant consequences, we formulate it in the next section, where the first result is about packing measures and other results then follow.

11.8 The Strong Open Set Condition

As indicated at the end of the previous section, in this one we formulate the SOSC and derive several of its remarkable consequences. We want to emphasize now that this is a mild natural condition and it is satisfied for "most" GDMSs. Here it is.

Definition 11.8.1 A CGDMS $S = \{\phi_e\}_{e\in E}$ (satisfying the OSC) is said to satisfy the SOSC if

$$J_S \cap \mathrm{Int} X \neq \emptyset.$$

Proposition 11.8.2 *If S is a finitely irreducible regular CGDMS satisfying the SOSC, then the packing measure $\Pi_h(J_S) > 0$.*

Proof By our (SOSC) hypothesis, there exists $\tau \in E_A^\infty$ such that $x := \pi(\tau) \in \mathrm{Int}(X)$. There then exists $R > 0$ such that $B(x, R) \subseteq \mathrm{Int}(X)$. Hence, there exists $q \ge 1$ so large that

$$\pi([\tau|_q]) \subseteq B(x, R/2). \qquad (11.62)$$

Since $\tilde{\mu}_h([\tau|_q]) > 0$ and the σ-invariant mesure $\tilde{\mu}_h$ is ergodic, it follows from Theorem 2.3.9 that $\tilde{\mu}_h(E_\tau^\infty) = 1$, where

$$E_\tau^\infty := \{\omega \in E_A^\infty : \sigma^n(\omega) \in [\tau|_q] \text{ for infinitely many } ns\}.$$

So, fix $\omega \in E_\tau^\infty$ and $n \geq 1$ such that $\sigma^n(\omega) \in [\tau|_q]$. Then

$$B(\pi(\sigma^n(\omega)), R/2) \subseteq B(x, R) \subseteq \text{Int}(X).$$

It also follows from Lemma 11.3.11 that

$$\phi_{\omega|_n}\big(B(\pi(\sigma^n(\omega)), R/2)\big) \supseteq B\big(\pi(\omega), (2K)^{-1}R\|\phi'_{\omega|_n}\|_\infty\big).$$

Therefore, also using the last assertion of Proposition 11.6.9, we get that

$$\begin{aligned} m_h\big(B\big(\pi(\omega), (2K)^{-1}R\|\phi'_{\omega|_n}\|_\infty\big)\big) &\leq m_h\big(\phi_{\omega|_n}(B(\pi(\sigma^n(\omega)), R/2))\big) \\ &\leq m_h\big(\phi_{\omega|_n}(\text{Int}(X))\big) \\ &= \tilde{m}_h\big([\omega|_n]\big) \leq \|\phi'_{\omega|_n}\|_\infty^h \\ &= (2K/R)^h\big((2K)^{-1}R\|\phi'_{\omega|_n}\|_\infty\big)^h. \end{aligned}$$

Hence, denoting by $\mathbb{N}(\omega)$ the infinite set of integers $n \geq 1$ such that $\sigma^n(\omega) \in [\tau|_q]$,

$$\varliminf_{r \to 0} \frac{m_h\big(B\big(\pi(\omega), r\big)\big)}{r^h} \leq \varliminf_{\substack{n \to \infty \\ n \in \mathbb{N}(\omega)}} \frac{m_h\big(B\big(\pi(\omega), (2K)^{-1}R\|\phi'_{\omega|_n}\|_\infty\big)\big)}{\big((2K)^{-1}R\|\phi'_{\omega|_n}\|_\infty\big)^h} \leq (2K/R)^h.$$

So, applying the Frostman Converse Theorem for packing measures, i.e., Theorem 1.6.4(1), we, therefore, get that

$$\Pi_h\big(J_S\big) \geq \Pi_h\big(\pi(E_\tau^\infty)\big) > 0.$$

The proof is complete. ■

A straightforward observation is that if

$$\phi_\omega\big(X_{t(\omega)}\big) \subseteq \text{Int}\big(X_{i(\omega)}\big)$$

for some $\omega \in E_A^*$, then (SOSC) holds. Before we provide further significant consequences of (SOSC), we prove several technical ones. First, it directly follows from (OSC) that

$$\phi_\omega\big(X_{t(\omega)}\big) \cap \phi_\tau\big(X_{t(\tau)}\big) = \phi_\omega\big(\partial X_{t(\omega)}\big) \cap \phi_\tau\big(\partial X_{t(\tau)}\big) \tag{11.63}$$

for all incomparable words $\omega, \tau \in E_A^*$. Second:

Lemma 11.8.3 *If $S = \{\phi_e\}_{e \in E}$ is a CGDMS, then*

$$\sigma(\pi^{-1}(\partial X)) \subseteq \pi^{-1}(\partial X).$$

Proof Since $\phi_e(X_{t(e)}) \subseteq \text{Int}(X_{i(e)})$ and $\partial X_v \cap \text{Int}X_v = \emptyset$ for every $e \in E$ and every $v \in V$, we conclude that

$$\phi_e^{-1}(\partial X_{i(e)}) \subseteq \partial X_{t(e)}. \tag{11.64}$$

Now if $\omega \in \pi^{-1}(\partial X)$, then $\pi(\omega) \in \partial X_{i(\omega_1)}$. Since also $\pi(\omega) = \phi_{\omega_1}(\pi(\sigma(\omega)))$, using (11.64), we conclude that

$$\pi(\sigma(\omega)) \in \phi_{\omega_1}^{-1}(\partial X_{i(\omega_1)}) \subseteq \partial X_{t(\omega_1)}.$$

Hence, $\sigma(\omega) \in \pi^{-1}(\partial X_{i(\omega_1)}) \subseteq \pi^{-1}(\partial X)$ and the proof is complete. ■

Lemma 11.8.4 *If $S = \{\phi_e\}_{e\in E}$ is a CGDMS satisfying (SOSC), then, for every $\omega \in E_A^*$, we have that*

$$\sigma^{|\omega|}\big(\pi^{-1}(\phi_\omega(\partial X_{t(\omega)}))\big) \subseteq \pi^{-1}(\partial X).$$

Proof Put $n = |\omega|$. Let $\tau \in \sigma^{|\omega|}\big(\pi^{-1}(\phi_\omega(\partial X_{t(\omega)}))\big)$. Then $\tau = \sigma^n(\gamma)$ for some $\gamma \in \pi^{-1}(\phi_\omega(\partial X_{t(\omega)}))$. So,

$$\phi_{\gamma|_n}(\pi(\sigma^n(\gamma))) = \pi(\gamma) \in \phi_\omega(\partial X_{t(\omega)}).$$

It, thus, follows from (OSC) that $\pi(\tau) = \pi(\sigma^n(\gamma)) \in \partial X_{t(\gamma_n)} \subseteq \partial X$. Hence, $\tau \in \pi^{-1}(\partial X)$ and the proof is complete. ■

Now we are in position to prove the following first main result of this section.

Theorem 11.8.5 *If $S = \{\phi_e\}_{e\in E}$ is a CGDMS satisfying (SOSC) and μ is a Borel probability σ-invariant ergodic measure on E_A^∞ with full topological support, then*

(a) $\mu \circ \pi^{-1}(\partial X) = 0$.

(b) $\mu \circ \pi^{-1}(\phi_\omega(\partial X_{t(\omega)})) = 0$ *for each* $\omega \in E_A^*$.

(c) *If $\omega, \tau \in E_A^*$ are incomparable, then*

$$\mu \circ \pi^{-1}\big(\phi_\omega(\partial X_{t(\omega)}) \cap \phi_\tau(\partial X_{t(\tau)})\big) = 0.$$

Proof Since S satisfies (SOSC), there exists $\omega \in E_A^\infty$ such that $\pi(\omega) \in \text{Int}X_{i(\omega_1)}$. But then there exists an integer $n \geq 1$ such that

$$\phi_{\omega|_n}(X_{t(\omega_n)}) \subseteq \text{Int}X_{i(\omega_1)}.$$

Hence, using also the fact that $\text{supp}(\mu) = E_A^\infty$, we get that

$$\begin{aligned}\mu \circ \pi^{-1}(\text{Int}X) &\geq \mu \circ \pi^{-1}(\text{Int}X_{i(\omega_1)}) \geq \mu \circ \pi^{-1}(\phi_{\omega|_n}(X_{t(\omega_n)})) \\ &\geq \mu \circ \pi^{-1}([\omega|_n]) > 0.\end{aligned}$$

Therefore, $\mu \circ \pi^{-1}(\partial X) < 1$. So, it follows from ergodicity of the measure μ and Lemma 11.8.3 that $\mu \circ \pi^{-1}(\partial X) = 0$. The proof of item (a) is complete.

Proving item (b), it follows from item (a) of Lemma 11.8.4 and σ-invariance of the measure μ that

$$0 = \mu \circ \pi^{-1}(\partial X) \geq \mu\big(\sigma^{|\omega|}\big(\pi^{-1}\big(\phi_\omega\big(\partial X_{t(\omega)}\big)\big)\big)\big) \geq \mu\big(\pi^{-1}\big(\phi_\omega\big(\partial X_{t(\omega)}\big)\big)\big).$$

The proof of item (b) is complete.

Now we shall prove item (c). Since ω and τ are incomparable, $\omega = \gamma a \alpha$ and $\tau = \gamma b \beta$ with some $\gamma \in E_A^*$, $\alpha, \beta \in E_A^\infty$ and $a, b \in E$ with $a \neq b$. Then, by (OSC),

$$\begin{aligned}\phi_\omega\big(X_{t(\omega)}\big) \cap \phi_\tau\big(X_{t(\tau)}\big) &\subseteq \phi_\gamma\big(\phi_a\big(X_{t(a)}\big) \cap \phi_b\big(X_{t(b)}\big)\big) \\ &= \phi_\gamma\big(\phi_a\big(\partial X_{t(a)}\big) \cap \phi_b\big(\partial X_{t(b)}\big)\big) \\ &\subseteq \phi_{\gamma a}\big(\partial X_{t(a)}\big).\end{aligned}$$

So, by item (b),

$$\mu \circ \pi^{-1}\big(\phi_\omega\big(\partial X_{t(\omega)}\big) \cap \phi_\tau\big(\partial X_{t(\tau)}\big)\big) \leq \mu \circ \pi^{-1}\big(\phi_{\gamma a}\big(\partial X_{t(a)}\big)\big) = 0.$$

Thus, the proof of item (c) is complete and we are done. ■

Since, for every $t \in F(S)$, the invariant measure μ_t is of full topological support and since the measures μ_t and m_t are equivalent, as an immediate consequence of Theorem 11.8.5, we get the following.

Theorem 11.8.6 *If $S = \{\phi_e\}_{e \in E}$ is a finitely irreducible CGDMS satisfying (SOSC) and $t \in F(S)$, then*

(a) $\mu_t(\partial X) = 0$.
(b) $\mu_t\big(\phi_\omega\big(\partial X_{t(\omega)}\big)\big) = 0$ *for each* $\omega \in E_A^*$.
(c) *If* $\omega, \tau \in E_A^*$ *are incomparable, then*

$$\mu_t\big(\phi_\omega\big(\partial X_{t(\omega)}\big) \cap \phi_\tau\big(\partial X_{t(\tau)}\big)\big) = 0.$$

Also, the same holds with μ_t replaced by m_t.

Now, we are in position to prove the conformality property of the measure m_t. For every $e \in E$, let

$$E_A^+(e) := \big\{\omega \in E_A^\infty : A_{e\omega_1} = 1\big\}.$$

Theorem 11.8.7 *If $S = \{\varphi_e\}_{e \in E}$ is a CGDMS satisfying (SOSC) and $t \in F(S)$, then*

$$m_t(\varphi_\omega(F)) = e^{-\mathrm{P}(t)|\omega|} \int_F |\varphi_\omega'|^t \, dm_t$$

for every $\omega \in E_A^*$ and every Borel set $F \subseteq \pi\big(E_A^+(\omega_{|\omega|})\big)$.

Proof Put $n := |\omega|$. Denote

$$Z_\omega(F) := \{\gamma \in \pi^{-1}(F) : A_{\omega_n \gamma_1} = 1\}.$$

Note that if $x \in F$, then $x = \pi(\gamma)$, where $\gamma \in E^+(\omega_n)$. So, $\gamma \in \pi^{-1}(F)$ and $A_{\omega_n\gamma_1} = 1$. Whence $\gamma \in Z_\omega(F)$. We have, thus, proved that

$$\pi(Z_\omega(F)) = F. \tag{11.65}$$

Now if $\tau \in \pi^{-1}(F)\backslash Z_\omega(F)$, then $A_{\omega_n\tau_1} = 0$. Hence,

$$\pi^{-1}(F)\backslash Z_\omega(F) \subseteq \bigcup_{e\in E: A_{\omega_n e}=0} [e].$$

So, using (11.65), we get that

$$\begin{aligned}
\pi(\pi^{-1}(F)\backslash Z_\omega(F)) &\subseteq F \cap \pi\left(\bigcup_{b\in E: A\omega_n b=0} [e]\right) \\
&\subseteq \bigcup_{a\in E: A_{\omega_n a}=1} \varphi_a\big(X_{t(a)}\big) \cap \bigcup_{b\in E: A_{\omega_n b}=1} \varphi_b\big(X_{t(b)}\big) \\
&= \bigcup_{a\in E: A_{\omega_n} a=1} \ \bigcup_{b\in E: A_{\omega_n} b=1} \varphi_a\big(X_{t(a)}\big) \cap \varphi_b\big(X_{t(b)}\big).
\end{aligned}$$

Hence, it follows from Theorem 11.8.6(c) that $m_t\big(\pi(\pi^{-1}(F)\backslash Z_\omega(F))\big) = 0$. Therefore,

$$\tilde{m}_t(\pi^{-1}(F)\backslash Z_\omega(F)) = 0. \tag{11.66}$$

Now $\tau \in \pi^{-1}(\varphi_\omega(F))$ if and only if $\pi(\tau) \in \varphi_\omega(F)$, i.e., $\varphi_{\tau|n}(\pi(\sigma^n(\tau))) \in \varphi_\omega(F)$. Hence,

$$\omega Z_\omega(F) \in \pi^{-1}(\varphi_\omega(F)) \subseteq Z_\omega(F) \cup \pi^{-1}\left(\bigcup_{\tau\in E_A^n\backslash\{\omega\}} \varphi_\tau\big(X_{t(\tau)}\big) \cap \varphi_\omega\big(X_{t(\omega)}\big)\right).$$

It, therefore, follows from Theorem 11.8.6(c) that

$$\tilde{m}_t(\omega Z_\omega(F)) = \tilde{m}_t(\pi^{-1}(\varphi(F))).$$

So, remembering that $\lambda_t = e^{\mathrm{P}(t)}$ and using (11.24), (11.65), and (11.66), we get that

$$\begin{aligned}
m_t(\varphi_\omega(F)) &= \tilde{m}_t \circ \pi^{-1}(\varphi_\omega(F)) = \tilde{m}_t(\omega Z_\omega(F)) = e^{-\mathrm{P}(t)n}\mathcal{L}_t^{*n}\tilde{m}_t\big(1\!\!1_{[\omega Z_\omega(F)]}\big) \\
&= e^{-\mathrm{P}(t)}\tilde{m}_t\big(\mathcal{L}_t^n 1\!\!1_{[\omega Z_\omega(F)]}\big) \\
&= e^{-\mathrm{P}(t)n}\int_{E_A^{\mathbb{N}}}\sum_{\tau \in E_A^n} 1\!\!1_{[\omega Z_\omega(F)]}(\tau\gamma)|\varphi_\tau'(\pi(\gamma))|^t \, d\tilde{m}_t(\gamma) \\
&= e^{-\mathrm{P}(t)n}\int_{Z_\omega(F)} |\varphi_\omega'(\pi(\gamma))|^t \, d\tilde{m}_t(\gamma) \\
&= e^{-\mathrm{P}(t)n}\int_{\pi^{-1}(F)} |\varphi_\omega'(\pi(\gamma))|^t \, d\tilde{m}_t(\gamma) \\
&= e^{-\mathrm{P}(t)n}\int_F |\varphi_\omega'|^t \, dm_t.
\end{aligned}$$

The proof is complete. ■

11.9 Conformal Maximal Graph Directed Systems

We already know the significance of the measure m_h defined by (11.32). In this section, we want to give an intrinsic characterization of this measure, i.e., one that does not invoke any symbol dynamics. For this, we, however, need to restrict our attention to a narrower class of GDMSs. In fact, we will do it more generally for all measures m_t, $t \in F(\mathcal{S})$.

Definition 11.9.1 A GDMS $\mathcal{S} = \{\phi_e : e \in E\}$ is called maximal if $A_{ab} = 1$ whenever $i(b) = t(a)$. Equivalently, $A_{ab} = 1$ if and only if $i(b) = t(a)$.

In particular, every IFS is a maximal GDMS. Now, let $\mathcal{S} = \{\phi_e : e \in E\}$ be a conformal maximal GDMS. For every $t \in F(\mathcal{S})$, the (intrinsic) Perron–Frobenius operator $L_t : C(X) \to C(X)$ is defined as follows. If $v \in V$ and $x \in X_v$, then

$$L_t g(x) := \sum_{e \in E : t(e) = v} g(\phi_e(x))|\phi_e'(x)|^t. \tag{11.67}$$

The following theorem of this section was partly proved in [**MU1**] and [**MU2**].

Theorem 11.9.2 *Let $\mathcal{S} = \{\phi_e : e \in E\}$ be a finitely irreducible conformal maximal GDMS satisfying (SOSC) and $t \in F(\mathcal{S})$. If m is a Borel probability measure on X, then the following conditions are equivalent.*

(a) $m = m_t$.
(b) $L_t^*(m) = e^{\mathrm{P}(t)} m$.
(c)

$$m(\phi_e(F)) = e^{-\mathrm{P}(t)} \int_F |\phi_e'|^t \, dm$$

for every $e \in E$ *and every Borel set* $F \subseteq X_{t(e)}$.
(d) *Item (c) holds and*

$$m\big(\phi_a(X_{t(a)}) \cap \phi_b(X_{t(b)})\big) = 0$$

whenever $a, b \in E$ *and* $a \neq b$.
(e) $L_t^*(m) = \gamma m$ *for some* $\gamma > 0$.
(f) $m(J_S) = 1$ *and*

$$m(\phi_e(F)) = \gamma^{-1} \int_F |\phi_e'|^t \, dm$$

for some $\gamma > 0$, *every* $e \in E$, *and evry Borel set* $F \subseteq X_{t(e)}$.
(g) *Item (f) holds and*

$$m\big(\phi_a(X_{t(a)}) \cap \phi_a(X_{t(a)})\big) = 0$$

whenever $a, b \in E$ *and* $a \neq b$.

Proof We will first establish a close relation between the operators $\mathcal{L}_t$ and L_t, where $t \in F(S)$.

Claim 1°. Let $t \in F(S)$. If $n \geq 0$, then, for every $g \in C(X)$, we have that

$$\mathcal{L}_t^n(g \circ \pi) = L_t^n(g) \circ \pi,$$

where, we recall, $\pi : E_A^\infty \longrightarrow J_S$ is the canonical Hölder continuous projection from the symbol space onto the limit set of the GDMS S. ■

Proof The proof goes through direct calculation. For every $\omega \in E_A^\infty$, we have that

$$\begin{aligned}
\mathcal{L}_t^n(g \circ \pi)(\omega) &= \sum_{\tau \in E_A^n : A_{\tau_n \omega_1} = 1} g \circ \pi(\tau\omega) \big|\varphi_\tau'(\pi(\omega))\big|^t \\
&= \sum_{\tau \in E_A^n : t(\tau) = i(\omega)} g(\varphi_\tau(\pi(\omega))) \big|\varphi_\tau'(\pi(\omega))\big|^t \\
&= L_t^n(g)(\pi(\omega)).
\end{aligned}$$

The proof of Claim 1° is complete. ■

The standard approximation argument, based on Theorem 2.4.6 in [**MU2**], which involves Hölder continuous functions only, gives the following.

Fact 11.9.3 For every $k \in C_b(E_A^\infty)$, the sequence $\left(e^{-\mathrm{P}(t)n}\mathcal{L}_t^n(k)\right)_{n=0}^\infty$ converges uniformly to $\tilde{m}_t(k)\tilde{\rho}_t$, where $\tilde{\rho}_t = \frac{d\tilde{\mu}_t}{d\tilde{m}_t}$.

Claim 2°. There exists a unique continuous function $\rho_t : J_S \longrightarrow (0, +\infty)$ such that

$$\tilde{\rho}_t = \rho_t \circ \pi,$$

and, for every $g \in C_b(J_S)$, the sequence $\left(e^{-\mathrm{P}(t)n}L_t^n g\right)_{n=0}^\infty$ converges uniformly to the function $m_t(g)\rho_t$.

Proof Applying Fact 11.9.3 to the function $k = 1\!\!1$, along with Claim 1°, applied to the function $g = 1\!\!1$, we get, for every $x \in J_S$ and every $\omega \in \pi^{-1}(x)$, that

$$\begin{aligned}\tilde{\rho}_t(\omega) &= \lim_{n\to\infty}\left(e^{-\mathrm{P}(t)n}\mathcal{L}_t^n(1\!\!1 \circ \Pi)\right)(\omega) = \lim_{n\to\infty}\left(e^{-\mathrm{P}(t)n}L_t^n 1\!\!1\right)(\pi(\omega)) \\ &= \lim_{n\to\infty}\left(e^{-\mathrm{P}(t)n}L_t^n 1\!\!1\right)(x)\end{aligned}$$

and the convergence is uniform with respect to $\omega \in E_A^\infty$ and $x \in J_S$. In particular, $\tilde{\rho}_t$ is constant on $\pi^{-1}(x)$. Since also $\tilde{\rho}_t : E_A^\infty \longrightarrow (0, +\infty)$ is continuous, we, thus, conclude that there exists a continuous function $\rho_t : J_S \longrightarrow (0, 1)$ such that

$$\tilde{\rho}_t = \rho_t \circ \pi.$$

Now the uniqueness of ρ_t follows from surjectivity of the projection $\pi : E_A^\infty \longrightarrow J_S$. Applying again Claim 1° with the function g and Fact 11.9.3 with the function $k = g \circ \Pi$, we get, for every $x \in J_S$ and every $x \in \pi^{-1}(\omega)$, that

$$\begin{aligned}\lim_{n\to\infty}\left(e^{-\mathrm{P}(t)n}L_t^n 1\!\!1\right)(x) &= \lim_{n\to\infty}\left(e^{-\mathrm{P}(t)n}L_t^n g\right)(\pi(\omega)) = \lim_{n\to\infty}\left(e^{-\mathrm{P}(t)n}\mathcal{L}_t^n(g\circ\pi)\right)(\omega) \\ &= \tilde{m}_t(g\circ\pi)\tilde{\rho}_t(\omega) = m_t(g)\rho_t(\pi(\omega)) \\ &= m_t(g)\rho_t(x)\end{aligned}$$

and the convergence is uniform with respect to $\omega \in E_A^\infty$ and $x \in J_S$. The proof of Claim 2° is complete. ■

Now we shall prove that (a) ⇒ (b). Indeed, by Claim 1°, we have, for every $g \in C(X)$, that

$$\begin{aligned}L_t^* m_t(g) &= m_t(L_t g) = \tilde{m}_t \circ \pi^{-1}(L_t g) = \tilde{m}_t((L_t g)\circ\pi) \\ &= \tilde{m}_t(\mathcal{L}_t(g\circ\pi)) = \mathcal{L}_t^*\tilde{m}_t(g\circ\pi) \\ &= e^{\mathrm{P}(t)}\tilde{m}_t(g\circ\pi) = e^{\mathrm{P}(t)}\tilde{m}_t\circ\pi^{-1}(g) \\ &= e^{\mathrm{P}(t)}m_t(g).\end{aligned}$$

Therefore, $L^* m_t = m_t$ and the implication (a) $\Rightarrow$ (b) is established.

Of course, (b) $\Rightarrow$ (e). Now we shall prove that (e) $\Rightarrow$ (a). Because of Claim 2°, we get, for every $g \in C(X)$, that

$$\lim_{n\to\infty} e^{-\mathrm{P}(t)n} L_t^{*n} m(g) = \lim_{n\to\infty} m\big(e^{-\mathrm{P}(t)n} L_t^n(g)\big)$$
$$= \lim_{n\to\infty} \int_{J_S} e^{-\mathrm{P}(t)n} L_t^n g(x) dm(x)$$
$$= m_t(g) m(\rho_t).$$

Invoking (e), we then obtain that

$$\lim_{n\to\infty} \big((\gamma e^{-\mathrm{P}(t)})^n m(g)\big) = m_t(g) m(\rho_t).$$

Taking $g = 1\!\!1$ and noting that $m(\rho_t) > 0$ (as ρ_t is everywhere positive on E_A^∞), we, thus, obtain that $\gamma = e^{\mathrm{P}(t)}$ and that

$$m(g) = m_t(g) m(\rho_t). \tag{11.68}$$

Taking again $g = 1\!\!1$, we, thus, get that $m(\rho_t) = 1\!\!1$. Then (11.68) becomes $m(g) = m_t(g)$. This means that $m = m_t$ and the proof of the implication (e) $\Rightarrow$ (a) is complete.

Now we shall prove that (a) $\Rightarrow$ (d). The second part of (d) holds because of (a) and Theorem 11.8.6. Because of Theorem 11.8.7, in order to prove the first part of (d), i.e., item (c), it suffices to show that

$$F \cap J_S \subseteq \pi\big(E_A^+(e)\big) \tag{11.69}$$

and

$$m_t\big((J_S \cap \varphi_e(F)) \backslash \varphi_e(F \cap J_S)\big) = 0. \tag{11.70}$$

Indeed, proving (11.69), if $x \in F \cap J_S$, then there exists $\omega \in E_A^\infty$ such that $x = \Pi(\omega)$ and $i(\omega_1) = t(e)$. But, since our system S is maximal, this means that $A_{e\omega_1} = 1$. Hence, $\omega \in E_A^+(e)$. Thus, $x = \pi(\omega) \in \pi(E_A^+(e))$. Thus, (11.69) holds.

Proving (11.70), let

$$x \in (J_S \cap \varphi_e(F)) \backslash \varphi_e(F \cap J_S).$$

Then $x = \varphi_e(z)$ for some $z \in F \backslash J_S$ and $x = \pi(\omega)$ for some $\omega \in E_A^\infty$. So, $x = \varphi_{\omega_1}\big(\pi(\sigma(\omega))\big)$. Since the map φ_e is one-to-one and $\pi(\sigma(\omega)) \in J_S$, we, thus, infer that $\omega_1 \neq e$. Since also $\in \varphi_e\big(X_{t(e)}\big) \cap \varphi_{\omega_1}\big(X_{t(\omega_1)}\big)$, we, thus, conclude that

$$(J_S \cap \varphi_e(F)) \backslash \varphi_e(F \cap J_S) \subseteq \bigcup_{a \neq b} \phi_a(X_{t(a)}) \cap \phi_b(X_{t(b)}).$$

In conjunction with Theorem 11.8.6, this finishes the proof of (11.70) and (d) is proved.

Obviously, (g) $\Rightarrow$ (f).

We will show now that (f) $\Rightarrow$ (c). This means that we are to show that $\gamma = e^{\mathrm{P}(t)}$. We have, for every $n \geq 1$, that

$$\begin{aligned}
1 = m(J_S) &\leq m\left(\bigcup_{\omega \in E_A^n} \varphi_\omega(X_{t(\omega)})\right) \leq \sum_{\omega \in E_A^n} m(\varphi_\omega(X_{t(\omega)})) \\
&= \sum_{\omega \in E_A^n} \gamma^{-n} \int_{X_t(\omega)} |\varphi_\omega'|^t \, dm \\
&\leq \gamma^{-n} \sum_{\omega \in E_A^n} \|\varphi_\omega'\|^t m(X_{t(\omega)}) \\
&\leq \gamma^{-n} \sum_{\omega \in E_A^n} \|\varphi_\omega'\|^t \\
&= \gamma^{-n} Z_n(t).
\end{aligned}$$

Hence, $\gamma^n \leq Z_n(t)$; therefore,

$$\log \gamma = \lim_{n \to \infty} \frac{1}{n} \log \gamma^n \leq \liminf_{n \to \infty} \frac{1}{n} \log Z_n(t) = \mathrm{P}(t). \tag{11.71}$$

On the other hand, because we have already proved that (a) $\Rightarrow$ (d) (so (c) holds with m replaced by m_t), we get, for every $\omega \in E_A^*$, that

$$m_t\big(\varphi_\omega(\mathrm{Int}X_{t(\omega)})\big) = e^{-\mathrm{P}(t)|\omega|} \int_{\mathrm{Int}X_{t(\omega)}} |\varphi_\omega'|^t \, dm_t \leq e^{-\mathrm{P}(t)|\omega|} \|\varphi_\omega'\|^t. \tag{11.72}$$

Since the system S is irreducible and satisfies (SOSC), it immediately follows from (f) that $m(\mathrm{Int}X_v) > 0$ for all $v \in V$. Therefore,

$$M := \min\{m(\mathrm{Int}X_v) \colon v \in V\} > 0. \tag{11.73}$$

Therefore, we get, for every $\omega \in E_A^*$, that

$$\begin{aligned}
m(\varphi_\omega(\mathrm{Int}X_{t(\omega)})) = \gamma^{-|\omega|} \int_{\mathrm{Int}X_{t(\omega)}} |\varphi_\omega'|^t \, dm &\geq \gamma^{-|\omega|} K^{-t} \|\varphi_\omega'\|^t m(\mathrm{Int}X_{t(\omega)}) \\
&\geq M K^{-t} \gamma^{-|\omega|} \|\varphi_\omega'\|^t.
\end{aligned}$$

Combining this with (11.72), we get that

$$m_t\big(\varphi_\omega(\mathrm{Int}X_{t(\omega)})\big) \leq M^{-1} K^t \gamma^{|\omega|} e^{-\mathrm{P}(t)|\omega|} m\big(\varphi_\omega\big(\mathrm{Int}X_{t(\omega)}\big)\big). \tag{11.74}$$

Hence, we get, for every integer $n \geq 1$, that

$$1 = \sum_{\omega\in E_A^n} m_t\big(\varphi_\omega(\mathrm{Int}X_{t(\omega)})\big) \leq M^{-1}K^t\left(\frac{\gamma}{e^{\mathrm{P}(t)}}\right)^n \sum_{\omega\in E_A^n} m\big(\varphi_\omega(\mathrm{Int}X_{t(\omega)})\big)$$
$$\leq M^{-1}K^t\left(\frac{\gamma}{e^{P(t)}}\right)^n.$$

Letting $n \to \infty$, we, thus, conclude that $\gamma \geq e^{\mathrm{P}(t)}$. Along with (11.71), this finishes the proof of equality $\gamma = e^{\mathrm{P}(t)}$, thus establishing (c).

So, in order to conclude the proof of our theorem, it suffices to show that (c) $\Rightarrow$ (a). We will do it now. We start with the following.

Claim 4°. *If* (c) *holds, then m_t is absolutely continuous with respect to m.*

Proof Let $G \subseteq X$ be an open set. Thus, for every $x \in J_S \cap G$, there exists at least one $\omega \in E_A^*$ such that $x \in \varphi_\omega(X_{t(\omega)}) \subseteq G$. Therefore, there exists $\mathfrak{F} \subseteq E_A^*$, a family of mutually incomparable finite words such that

$$J_S \cap G \subseteq \bigcup_{\omega\in\mathfrak{F}} \varphi_\omega(X_{t(\omega)}) \subseteq G.$$

Hence, using also (11.74) with $\gamma = e^{\mathrm{P}(t)}$, Theorem 11.8.6(b), and OSC, we obtain that

$$\begin{aligned} m_t(G) = m_t(J_S \cap G) &\leq \sum_{\omega\in\mathfrak{F}} m_t(\varphi_\omega(X_{t(\omega)})) = \sum_{\omega\in\mathfrak{F}} m_t(\varphi_\omega(\mathrm{Int}X_{t(\omega)}))\\ &\leq M^{-1}K^t\sum_{\omega\in\mathfrak{F}} m(\varphi_\omega(\mathrm{Int}X_{t(\omega)}))\\ &= M^{-1}K^t m\left(\bigcup_{\omega\in\mathfrak{F}} \varphi_\omega(\mathrm{Int}X_{t(\omega)})\right)\\ &\leq M^{-1}K^t m(G). \end{aligned} \tag{11.75}$$

Now if $\Gamma \subseteq X$ is a G_δ set, then $\Gamma = \bigcap_{n=1}^\infty G_n$, where $(G_n)_{n=1}^\infty$ is a descending sequence of open sets. So, using (11.75), we obtain that

$$m_t(\Gamma) = \lim_{n\to\infty} m_t(G_n) \leq M^{-1}K^t \lim_{n\to\infty} m(G_n) = M^{-1}K^t m(G). \tag{11.76}$$

Finally, if Y is an arbitrary Borel subset of X, then there exists $\hat{Y}$, a G_δ subset of X, such that $Y \subseteq \hat{Y}$, $m_t(\hat{Y}) = m_t(Y)$, and $m(\hat{Y}) = m(Y)$. So, using (11.76), we obtain that

$$m_t(Y) = m_t(\hat{Y}) \leq M^{-1}K^t m(\hat{Y}) = M^{-1}K^t m(Y).$$

The proof of Claim 4° is complete. ∎

We assume that (c) holds. Let

$$\rho := \frac{dm_t}{dm}.$$

We then have, for every $\omega \in E_A^*$ and every Borel set $F \in X_{t(\omega)}$, that

$$m_t(\varphi_\omega(F)) = \int_{\varphi_\omega(F)} \rho dm = e^{-\mathrm{P}(t)|\omega|} \int_F \rho \circ \varphi_\omega |\varphi_\omega'|^t dm.$$

On the other hand, because we have already proved that (a) $\Rightarrow$ (d) (so (c) holds with m replaced by m_t), we get that

$$m_t(\varphi_\omega(F)) = e^{-\mathrm{P}(t)|\omega|} \int_F |\varphi_\omega'|^t dm_t = e^{-\mathrm{P}(t)|\omega|} \int_F \rho |\varphi_\omega'|^t dm.$$

Therefore,

$$\rho \circ \varphi_\omega(x) = \rho(x)$$

for m_t-a.e. $x \in X_{t(\omega)}$. So, for $\tilde{m}_t$-a.e. $\tau \in E_A^\infty$, we get that π

$$\rho \circ \pi(\sigma(\tau)) = \rho \circ \varphi_{\tau_1}(\pi(\sigma(\tau))) = \rho \circ \pi(\tau).$$

This means that the function $\rho \circ \pi : E_A^\infty \longrightarrow \mathbb{R}$ is σ-invariant and ergodicity of the measure $\tilde{\mu}_t$ with respect to σ yields that the function $\rho \circ \pi$ is $\tilde{m}_t$-a.e.constant. Thus, we also have the following.

Claim 5°. *The function $\rho: J_S \longrightarrow \mathbb{R}$ is m_t-a.e. constant.*

The next step in the proof of (a) is to show the following.

Claim 6°. *$\rho = 1$ m_t-a.e.; consequently, $m = m_t$.*

Proof Seeking contradiction, assume that $\rho > 1$. Then $1 = m_t(J_S) = \rho m(J_S)$. So, $m(J_S) = 1/\rho < 1$. Then $m(X \backslash J_S) = 1 - \frac{1}{\rho} > 0$, whence the formula

$$\hat{m}(A) := \frac{\rho}{\rho - 1} m(A \backslash J_S)$$

defines a Borel probability measure on X. For every $e \in E$ and every Borel set $F \subseteq X_{t(e)}$, we, thus, have that

$$\begin{aligned}
m(\varphi_e(F)\backslash J_S) &= m(\varphi_e(F)) - m(\varphi_e(F)\cap J_S)\\
&= e^{-\mathrm{P}(t)}\int_F |\varphi_e'|^t dm - \frac{1}{\rho}m_t(\varphi_e(F)\cap J_S)\\
&= e^{-\mathrm{P}(t)}\int_F |\varphi_e'|^t dm - \frac{1}{\rho}m_t(\varphi_e(F))\\
&= e^{-\mathrm{P}(t)}\int_F |\varphi_e'|^t dm - \frac{1}{\rho}e^{-\mathrm{P}(t)}\int_F |\varphi_e'|^t dm_t\\
&= e^{-\mathrm{P}(t)}\int_F |\varphi_e'|^t dm - e^{-\mathrm{P}(t)}\int_{F\cap J_S} |\varphi_e'|^t dm\\
&= e^{-\mathrm{P}(t)}\int_{F\backslash J_S} |\varphi_e'|^t dm.
\end{aligned}$$

Therefore,

$$\hat{m}(\varphi_e(F)) = e^{-\mathrm{P}(t)}\int_F |\varphi_e'|^t d\hat{m}.$$

Hence, $\hat{m}$ is a Borel probability measure on X satisfying condition (c). So, by the already proven Claim 4°, m_t is absolutely continuous with respect to $\hat{m}$. Consequently, $m_t(J_S) = 0$ as $\hat{m}(J_S) = 0$. This contradiction shows that $\rho = 1$ and $m = m_t$. The implication (c) $\Rightarrow$ (a), i.e., Claim 6°, is, thus, established. ■

By the just established implication (c) $\Rightarrow$ (a), we conclude that if (c) holds, then $m(J_S) = 1$. Having this, the implication (d) $\Rightarrow$ (g) becomes obvious and the proof of Theorem 11.9.2 is complete. ■

11.10 Conjugacies of Conformal Graph Directed Systems

Definition 11.10.1 We say that two CGDMSs $S_1 = \{\phi_e\colon X_1 \to X_1\colon e \in E\}$ and $S_2 = \{\psi_e\colon X_2 \to X_2\colon e \in E\}$ with the same alphabet E, the same set of vertices, and the same incidence matrix $A\colon E \times E \to \{0,1\}$ are topologically conjugate if there exists a homeomorphism $H\colon J_{S_1} \longrightarrow J_{S_2}$ such that

$$H \circ \phi_e = \psi_e \circ H \tag{11.77}$$

for every $e \in E$.

We say that S_1 and S_2 are bi-Lipschitz conjugate if the map H is bi-Lipschitz continuous. Then H uniquely extends to a bi-Lipschitz map from $\overline{J}_{S_1}$ to $\overline{J}_{S_1}$ and (11.77) continues to hold for this extension.

We say that S_1 and S_2 are conformally conjugate if the map $H: J_{S_1} \longrightarrow J_{S_2}$ has a conformal extension to a map from X_1 to X_2.

We record the following two immediate observations.

Observation 11.10.2 Any two conformally conjugate CGDMSs are bi-Lipschitz conjugate and any two bi-Lipschitz conjugate systems are topologically conjugate. Also, then (11.77) continues to hold for these extensions.

Observation 11.10.3 If two CGDMSs $S_1 = \{\phi_e : X_1 \to X_1 : e \in E\}$ and $S_2 = \{\psi_e : X_2 \to X_2 : e \in E\}$ are bi-Lipschitz conjugate then the following hold:

(1) $\mathrm{P}_{S_1}(t) = \mathrm{P}_{S_2}(t)$ for all $t \geq 0$.
(2) $\theta(S_1) = \theta(S_2)$.
(3) The system S_2 is regular, strongly regular, or hereditarily regular, respectively, if and only if S_1 is regular, strongly regular, or hereditarily regular.

12

Nice Sets for Analytic Maps

In this chapter, we deal with a powerful tool of nice sets. Indeed, in this chapter, we introduce the concept of pre-nice sets and nice sets and prove their existence. Nice sets naturally emerge in dynamical systems in the context of self-maps of an interval, where their existence was sort of obvious. They were adapted, although in a slightly obscure way, to holomorphic endomorphisms of the Riemann sphere, by Rivera-Letelier in [**Ri**]. A clearer proof of the existence of nice sets and in a more general context of maps that are meromorphic, transcendental, or rational alike, from the complex plane to the Riemann sphere, was provided by Dobbs in [**Do**]. Nice sets are a powerful tool indeed. They formed a central theme in the fairly complete treatment of Collet–Eckmann rational functions given in [**PR**], as well as in later papers by Przytycki and Rivera-Letelier.

In this chapter, we define pre-nice sets and nice sets for holomorphic maps between elliptic and parabolic Riemann surfaces (one of which is an open subset of the other). We then prove their existence. We need such generality in order to deal, in Volume 2 of the book, with projected maps:

$$\hat{f}: \mathbb{T}_f \backslash \Pi_f(f^{-1}(\infty)) \longrightarrow \mathbb{T}_f,$$

where $f: \mathbb{C} \to \widehat{\mathbb{C}}$ is a nonconstant elliptic function, Λ_f is the lattice generated by f, $\mathbb{T}_f = \mathbb{C}/\Lambda_f$ is the corresponding torus, and $\Pi_f: \mathbb{C} \longrightarrow \mathbb{T}_f$ is the corresponding projection.

Nice sets naturally give rise to countable alphabet conformal iterated function systems, which were systematically explored in Chapter 11. In this way, many problems of transcendental and rational holomorphic dynamics can be successfully treated (or even reduced to) by means of the theory of conformal iterated function systems, or their generalization formed by graph directed Markov systems (GDMSs), presented, as mentioned above, in Chapter 11.

In this book, we apply the techniques of nice sets, via the means of iterated function systems, in Chapter 22 (in Volume 2), to demonstrate the finiteness of Krengel's Entropy for the dynamics generated by a subexpanding or parabolic elliptic function (and invariant measure equivalent to the conformal one), and, most notably, to prove the refined stochastic laws, such as the exponential decay of correlations, the Central Limit Theorem, or the Law of the Iterated Logarithm, for subexpanding elliptic functions and the parabolic elliptic function for which the above-mentioned invariant measure is finite.

Throughout this chapter, unless otherwise stated, Y denotes either a parabolic or an elliptic Riemann surface, i.e., either the complex plane $\mathbb{C}$, a complex torus $\mathbb{T}_\Lambda = \mathbb{C}/\Lambda$, where Λ is a lattice on $\mathbb{C}$, or the Riemann sphere $\widehat{\mathbb{C}}$.

12.1 Pre-Nice and Nice Sets and the Resulting CGMDS: Preliminaries

In this section, we define pre-nice and nice sets, and we prove several of their basic properties. We also define the maximal GDMS resulting from nice sets.

We start with somewhat long and involved topological preparations.

Proposition 12.1.1 *Let W be a subset of a Hausdorff topological space Y. Let V be a subset of W such that $W \backslash V$ is connected. Let G be an open subset of a Hausdorff topological space X such that $\overline{G} \subseteq X$ is compact. If $f\colon X \to Y$ is an open continuous map such that*

$$f(\partial G) \subseteq V \quad \text{and} \quad f(G) \not\supseteq W \backslash V,$$

then

$$f(G) \subseteq V.$$

Proof Since G is open, $\overline{G}$ is compact, and f is continuous and open, we have that $\partial f(G) \subseteq f(\partial G) \subseteq V$. Therefore,

$$\begin{aligned}\partial_{W\backslash V}\big(f(G) \cap (W\backslash V)\big) &\subseteq (W\backslash V) \cap \partial f(G) \subseteq (W\backslash V) \cap f(\partial G) \\ &\subseteq (W\backslash V) \cap V = \emptyset.\end{aligned}$$

Thus,

$$\partial_{W\backslash V}\big(f(G) \cap (W\backslash V)\big) = \emptyset.$$

Since the set $W \backslash V$ is connected, this implies that either

$$f(G) \cap (W\backslash V) = W\backslash V \quad \text{or} \quad f(G) \cap (W\backslash V) = \emptyset.$$

If the first part of this alternative holds, then $f(G) \supset W\backslash V$, contrary to our hypothesis. So, the other part of this alternative holds, meaning that $f(G) \subseteq V$. The proof is complete. ■

Since every nonconstant holomorphic mapping between Riemann surfaces is open, as an immediate consequence of this proposition, we get the following.

Corollary 12.1.2 *Let W be a subset of a Riemann surface Y. Let V be a subset of W such that $W\backslash V$ is connected. Let G be an open subset of a Riemann surface X such that $\overline{G} \subseteq X$ is compact. If $f: X \to Y$ is a holomorphic map, $f(\partial G) \subseteq V$ and $f(G) \not\supseteq W\backslash V$, then*

$$f(G) \subseteq V.$$

Proposition 12.1.3 *Let X and Z be two Riemann surfaces.*

1. *Let $D \subseteq X$ be an open set conformally equivalent to the unit disk $\mathbb{D}$.*
2. *Let W be a subset of Z and V be a subset of W such that $W\backslash V$ is connected.*
3. *Let $G \subseteq D$ be an open connected set such that $\bar{G} \subseteq \mathbb{D}$ is compact.*
4. *Let Γ be the only connected component of $D\backslash G$ which is not compact.*
5. *Let $\hat{G} := D\backslash\Gamma \supset G$.*

Then the set $\hat{G}$ is open connected, simply connected and $\partial\hat{G} \subseteq \partial G$. Furthermore, if $f: D \to Z$ is a holomorphic map, $f(\partial G) \subseteq V$, and $f(D) \not\supseteq W\backslash V$, then

$$f(\hat{G}) \subseteq V.$$

Proof The set $\hat{G}$ is open because D is open and Γ is closed in D. As an immediate consequence of the (very general) Theorem 5, page 140, in [**Kur**], we conclude that $\hat{G}$ is connected. Simple connectivity of $\hat{G}$ follows from Theorem 4.4 in [**Ne**].

By the definition of Γ, we have that $G \subseteq \bar{G} \subseteq \overline{\hat{G}} \subseteq D$. Since both sets G and $\hat{G}$ are also open, we get that

$$\partial G = \partial_D G \quad \text{and} \quad \partial\hat{G} = \partial_D\hat{G}. \tag{12.1}$$

Because of Theorem 3, page 238, in [**Kur**] (which applies since X is locally connected), we have that

$$\partial_D\hat{G} = \partial_D(D\backslash\Gamma) = \partial_D\Gamma \subseteq \partial_D(D\backslash G) = \partial_D G.$$

Along with (12.1), this gives that $\partial\hat{G} \subseteq \partial G$ and the proof of the first part of our proposition is complete. Therefore,

$$f(\partial\hat{G}) \subseteq f(\partial G) \subseteq V.$$

Since $\hat{G} \subseteq D$ and $f(D) \not\supseteq W \setminus V$, we also have that $f(\hat{G}) \not\supseteq W \setminus V$. Thus, a direct application of Corollary 12.1.2 gives that $f(\hat{G}) \subseteq V$. The proof of Proposition 12.1.3 is complete. ■

Definition 12.1.4 Let Y be a Riemann surface and X be a nonempty open subset of Y. A nonempty set $V \subseteq X$ is said to be a pre-nice set for the analytic map $f\colon X \to Y$ if the following conditions are satisfied:

(a) $\overline{V}$ is compact.
(b) V has finitely many connected components.
(c) Every connected component of V is simply connected.
(d) $V \cap \bigcup_{n=0}^{\infty} f^n(\partial V) = \emptyset$.

Definition 12.1.5 Let Y be a Riemann surface. A nonempty open set $V \subseteq Y$ is said to be a nice set for the analytic map $f\colon X \to Y$ if the following conditions are satisfied:

(a) $\overline{V}$ is compact.
(b) V has finitely many connected components.
(c) If Γ is a connected component of V, then Γ is simply connected and there exists $\hat{\Gamma}$, an open connected, simply connected subset of Y containing $\overline{\Gamma}$, such that

$$\overline{\hat{\Gamma}} \cap \overline{\mathrm{PS(f)}} = \emptyset.$$

(d) $V \cap \bigcup_{n=0}^{\infty} f^n(\partial V) = \emptyset$.
(e) There exists $\lambda > 1$ such that

$$|(f^n)'(z)| \geq \lambda$$

for every $n \geq 1$ and all $z \in V \cap f^{-n}(V)$.

Of course, every nice set is pre-nice. Before establishing the existence of nice sets and pre-nice sets we shall establish the most characteristic and particularly useful, for us, properties of such sets. We start with the following.

Proposition 12.1.6 *Let Y be a Riemann surface. Let X be an open subset of Y. Suppose that V is a pre-nice set for a holomorphic map $f\colon X \to Y$. Let U and W be two distinct components of V. If $j,k \geq 0$ are two integers and A and B are connected components, respectively, of $f^{-j}(U)$ and $f^{-k}(W)$, then either*

(a) $A \cap B = \emptyset$,
(b) $A \subseteq B$, *or*
(c) $B \subseteq A$.

Proof Assume without loss of generality that $j \leq k$. Seeking contradiction, assume that $A \cap B \neq \emptyset$ but neither $A \subseteq B$ nor $B \subseteq A$. Then the first and the third of these properties yield $B \cap \partial A \neq \emptyset$. As $f^j(\partial A) = \partial U$, we, thus, get that

$$W \cap f^{k-j}(\partial U) = f^k(B) \cap f^k(\partial A) \supset f^k(B \cap \partial A) \neq \emptyset.$$

Since $\partial U \subseteq \partial V$ and $W \subseteq V$, this gives $V \cap f^{k-j}(\partial V) \neq \emptyset$, contrary to condition (d) of Definition 12.1.4. The proof is finished. ■

Now, given a nice set V, let $\mathcal{C}_1^\infty(V)$ be the family of all connected components of the set $V \cap \bigcup_{n=1}^\infty f^{-n}(V)$. Proposition 12.1.6 entails the following.

Proposition 12.1.7 *Let Y be a Riemann surface. Let X be an open subset of Y. Suppose that V is a pre-nice set for a holomorphic map $f: X \to Y$. Then the following hold.*

(a) *If $\Gamma \in \mathcal{C}_1^\infty(V)$, then there exists a unique connected component W of V and a unique integer $n(\Gamma) \geq 1$ such that Γ is a connected component of $f^{-n(\Gamma)}(W)$.*
(b) *If W is a connected component of V, $n \geq 1$, and Γ is a connected component of $f^{-n}(W)$ such that*

$$\Gamma \cap V \neq \emptyset \quad \text{and} \quad f^k(\Gamma) \cap V = \emptyset$$

for all $1 \leq k < n$, then $\Gamma \in \mathcal{C}_1^\infty(V)$ and $n(\Gamma) = n$.

This proposition implies that if V is a nice set, then, for every $U \in \mathcal{C}_1^\infty(V)$, there exists a unique connected component Γ^* of V and a unique holomorphic inverse branch $f_{\hat{\Gamma}^*}^{-n(\Gamma)}: \hat{\Gamma}^* \to X$ of $f^{n(\Gamma)}$ such that

$$f_\Gamma^{-n(\Gamma)}(\Gamma^*) = \Gamma.$$

We, therefore, obtain the following fundamental consequence of the existence of nice sets.

Theorem 12.1.8 *Let Y be either a parabolic or an elliptic ($\widehat{\mathbb{C}}$) Riemann surface. Let X be an open subset of Y, which, moreover, is a subset of $\mathbb{C}$ if $Y = \widehat{\mathbb{C}}$. If $f: X \to Y$ is an analytic map and V is a nice set for f, then*

$$\mathcal{S}_V = \left\{f_\Gamma^{-n(\Gamma)}: \hat{\Gamma}^* \longrightarrow X\right\}_{\Gamma \in \mathcal{C}_1^\infty(V)}$$

is a maximal conformal graph directed Markov system (CGDMS) in the sense of Chapter 11 and Remark 11.3.5, the maximality of a GDMS being defined in Definition 11.9.1.

12.2 The Existence of Pre-Nice Sets

The main and actually the sole result of this section concerns the existence of pre-nice sets.

Theorem 12.2.1 *Let Y be either a parabolic or an elliptic ($\widehat{\mathbb{C}}$) Riemann surface. Let X be an open subset of Y, which, moreover, is a subset of $\mathbb{C}$ if $Y = \widehat{\mathbb{C}}$. Let $f : X \to Y$ be an analytic map.*

Fix

- *a radius $R \in (0, u_Y)$,*
- *$\kappa \in (1, 2]$,*
- *F, a finite subset of $J(f)$,*
- *a collection $\{U_0(b)\}_{b \in F}$ of open subsets of Y,*
- *a vector $\mathbf{r} = \big(r_b : b \in F\big)$.*

Assume that the following nine properties are satisfied.

(1)

$$U_0(b) \subseteq B(b, r_b)$$

for all $b \in F$.

(2) *F can be represented as a disjoint union:*

$$F = F_0 \cup F_1$$

such that

(a) *$f(b) = b$ for every $b \in F_0$.*
(b) *For each $b \in F_0$, the map $f|_{B(b,6r_b)}$ is one-to-one and the holomorphic inverse branch $f_b^{-1} : B(b, 6r_b) \longrightarrow X$, sending b to b, is well defined.*
(c) *$f_b^{-1}\big(B(b, 3r_b)\big) \subseteq B(b, 6r_b)$.*
(d) *$b \in \partial U_0(b)$ for all $b \in F_0$.*
(e) *For each $b \in F_0$, there exists an integer $p_b \geq 1$ such that $U_0(b)$ has exactly p_b (open) connected components $U_0(b, j)$, $j = 1, 2, \ldots, p_b$, each of which is simply connected and its closure is a closed topological disk.*
(f) *For each $b \in F_0$ and each $j = 1, 2, \ldots, p_b$,*

$$f_b^{-1}(U_0(b, j)) \subseteq U_0\big(b, j\big)$$

and

(g) *The sequence*

$$\left(f_b|_{U_0(b,j)}^{-n} : U_0(b, j) \longrightarrow X\right)_{n=0}^{\infty}$$

converges uniformly to the constant function, which assigns to each point in $U_0(b, j)$ *the value* b.

(h) $B(b, r_b/2) \subseteq U_0(b)$ *for all* $b \in F_1$.

(i) *For each* $b \in F_1$, *the set* $U_0(b)$ *is connected, simply connected, and its closure is a closed topological disk. We then put* $p_b := 1$ *and denote also* $U_0(b)$ *by* $U_0(b, 1)$.

(j) *For each* $b \in F_0$ *and* $s > 0$, *there exists* $s_b^- \in (0, r_b]$ *such that if*

$$z \in B(b, s_b^-) \backslash \bigcup_{j=1}^{p_b} U_0(b, j),$$

then

$$f^n(z) \in B(b, s)$$

for all $n \geq 0$.

(3)

$$r_b \in \left(0, \frac{1}{6} \min\left\{R, \min\left\{|b - c| : c \in F \backslash \{b\}\right\}\right\}\right)$$

for all $b \in F_0$.

(4)

$$F_1 \cap \overline{\mathrm{PS}(f)} = \emptyset.$$

Moreover,

$$r_b \in \left(0, \frac{1}{6} \min\left\{R, \min\left\{|b - c| : c \in F \backslash \{b\}\right\}, \mathrm{dist}\left(b, \overline{\mathrm{PS}(f)}\right)\right\}\right)$$

for all $b \in F_1$.

(5) *Suppose that* $a, b \in F$ *and* $n \geq 1$ *is an integer. Assume that either* $b \in F_1$ *or there exists a point*

$$w \in f^{-n}(U_0(b)) \tag{12.2}$$

such that if

$$f^{n-1}(w) \notin U_0(b), \tag{12.3}$$

then

(a) *The holomorphic inverse branch* $f_w^{-n} \colon B(b, 6r_b) \to X$ *of* f^n, *sending* $f^n(w)$ *to* w, *is well defined (this is only an extra hypothesis if* $b \in F_0$; *if* $b \in F_1$, *this follows from (4)).*

(b) *If, in addition, the connected component of* $f^{-n}(B(b,2r_b))$ *containing* w *intersects* $B(a,2r_a)$ *and*

$$f^j(w) \notin \bigcup_{c\in F} U_0(c) \tag{12.4}$$

for all $j = 1,2,\dots,n-1$, *then*

$$|(f^n)'(z)| \geq \frac{8\kappa}{\kappa - 1} \cdot \frac{r_b}{r_a} \tag{12.5}$$

for all $z \in f_w^{-n}(B(b,2r_b))$.

(6) *For every* $b \in F_0$ *and every* $j \in \{1,2,\dots,p_b\}$, *there exists* $V_j(b)$, *an open neighborhood of* b *in* Y *such that*

(a) *The set* $\overline{V_j(b)} \cap \overline{U_0(b,j)}$ *is a closed topological disk.*
(b) *The set* $V_j(b)\backslash\overline{U_0(b,j)}$ *is connected (and, in consequence, the set* $\overline{V_j(b)\backslash U_0(b,j)}$ *is connected too).*
(c) $\overline{V_j(b)} \subseteq B(b,r_b'/4)$ *with some* $r_b' \in (0,(r_b)_b^-)$.

(7) $|f'(b)| = 1$ *for every* $b \in F_0$.
(8)

$$\mathrm{Crit}(f) \cap \bigcup_{w\in F_0} B(w,6r_w) = \emptyset.$$

(9) $\forall b \in F_0 \quad \omega(\mathrm{Crit}(f)) \cap B(b,6r_b) \subseteq \{b\}$.

Then there exists a pre-nice set $U = U_{\mathbf{r}}$ *(although we do not indicate it here, this set does depend also at least on* F *and the sets* $U_0(\xi)$, $\xi \in F_1$, *and* $U_0(\xi,i)$, $\xi \in F_0$, $j = 1,2,\dots,p_\xi$*) with the following additional properties:*

(A) *For all* $b \in F$,

$$U_0(b) \subseteq U \subseteq \bigcup_{b\in F} B(b,\kappa r_b) \subseteq \bigcup_{b\in F} B(b,2\kappa r_b).$$

(B) *For all* $b \in F_1$,

$$U_0(b) \subseteq U \subseteq \bigcup_{b\in F} B(b,\kappa r_b) \subseteq \bigcup_{b\in F} B(b,2\kappa r_b)$$
$$\subseteq \bigcup_{b\in F} B(b,3\kappa r_b) \subseteq Y\backslash\overline{\mathrm{PS}(f)}.$$

(C) *If* W *is a connected component of* U, *then there exists a unique* $b \in F$ *and a unique* $j \in \{1,2,\dots,p_b\}$ *such that*

$$U_0(b,j) = W.$$

We then denote W *by* $U(b,j)$.

(D)

$$U(b,j) \cap B(b,r_b') = U_0(b,j) \cap B(b,r_b')$$

for all $b \in F_0$ *and all* $j \in \{1,2,\ldots,p_b\}$.

(E)

$$f_b^{-1}(U(b,j)) \subseteq U(b,j) \subseteq B(b,r_b)$$

for all $b \in F_0$ *and all* $j \in \{1,2,\ldots,p_b\}$.

(F) *For all* $b \in F_0$ *and all* $j \in \{1,2,\ldots,p_b\}$, *there exists an open connected set* $W(b,j)$ *such that*

$$\overline{U(b,j)}\backslash\{b\} \subseteq W(b,j) \tag{12.6}$$

and, for every integer $n \geq 0$ *and every* $z \in f^{-n}(b)$, *the holomorphic branch*

$$f_z^{-n}\colon W(b,j) \longrightarrow X$$

of f^{-n}, *sending* $f^n(z)$ *to* z, *is well defined.*

(G) *For every* $b \in F_0$, *every* $j \in \{1,2,\ldots,p_b\}$, *and every* $u \in (0,r_b'/4)$ *small enough, there exist* $V_j(b,u) \subseteq V_j(b)$, *an open neighborhood of* b, *and an open connected set* $W(b,j;u) \subseteq W(b,j)$ *such that*

$$\overline{U(b,j)}\backslash V_j(b,u) \subseteq W(b,j;u) \tag{12.7}$$

and the maps

$$f_b^{-n}\colon W(b,j;u) \longrightarrow X,\ \ n \geq 0,$$

converge uniformly to the constant function whose range is equal to $\{b\}$.

Proof For all $w \in F$ and all $1 \leq i \leq p_w$, define $(U_n(w,i))_0^\infty$, an ascending sequence of open connected sets, as follows. For all integers $n \geq 1$, the set $U_n(w,i)$ is defined to be the connected component of the union

$$U_0(w,i) \cup \bigcup_{\xi\in F} \bigcup_{k=0}^{n} \bigcup_{V\in \mathrm{Comp}_k(U_0(\xi,i))} V$$

that contains w. Set

$$U_0(w) := \bigcup_{i=1}^{p_w} U_0(w,i).$$

We will prove by induction that

$$U_n(w) \subseteq B(w,\kappa r_w) \tag{12.8}$$

for all $w \in F$ and all $n \geq 0$. For $n = 0$, this is immediate as, by hypothesis (1),

$$U_0(w) \subseteq B(w, r_w) \subseteq B(w, \kappa r_w).$$

For the inductive step, suppose that (12.8) is true for all $0 \leq j < n$ with some $n \geq 1$. Fix $w \in F$ and $1 \leq i \leq p_w$. Let C be a connected component of

$$U_n(w,i) \backslash \overline{B(w, r_w)}.$$

Then

$$\overline{C} \cap \overline{B(w, r_w)} \neq \emptyset \tag{12.9}$$

and there exist $\xi^* \in F$, $s \in \{1, 2, \ldots, p_{\xi^*}\}$, a minimal integer $1 \leq k \leq n$, and a connected component $W \in \operatorname{Comp}_k\big(U_0(\xi^*, s)\big)$ such that

$$W \cap C \neq \emptyset. \tag{12.10}$$

By the definitions of C, W, and k, we have that

$$C \subseteq \bigcup_{d \in F} \bigcup_{l=1}^{p_d} \bigcup_{j=k}^{n} \bigcup_{V \in \operatorname{Comp}_j(U_0(d,l))} V;$$

therefore,

$$f^k(C) \subseteq f^k \left(\bigcup_{d \in F} \bigcup_{l=1}^{p_d} \bigcup_{j=k}^{n} \bigcup_{V \in \operatorname{Comp}_j(U_0(d,l))} V \right) \subseteq \bigcup_{d \in F} \bigcup_{l=1}^{p_d} \bigcup_{j=0}^{n-k} \bigcup_{V \in \operatorname{Comp}_j(U_0(d,l))} V.$$

Since the set $f^k(C)$ is connected and since, by (12.10), $f^k(C) \cap U_0(\xi^*, s) \neq \emptyset$, we, therefore, conclude that

$$f^k(C) \subseteq U_{n-k}(\xi^*, s). \tag{12.11}$$

Therefore, since $0 \leq n - k < n$, the inductive assumption yields

$$f^k(C) \subseteq U_{n-k}(\xi^*, s) \subseteq B(\xi^*, \kappa r_{\xi^*}) \subseteq B(\xi^*, 2r_{\xi^*}). \tag{12.12}$$

But, then, $W \cup C$ is a connected set (by (12.10)) with

$$\begin{aligned} f^k(W \cup C) &\subseteq U_0(\xi^*, s) \cup U_{n-k}(\xi^*, s) \\ &= U_{n-k}(\xi^*, s) \subseteq B\big(\xi^*, \kappa r_{\xi^*}\big) \subseteq B(\xi^*, 2r_{\xi^*}). \end{aligned} \tag{12.13}$$

Consider two cases. First, assume that either

- $\xi^* \in F_1$, or
- $k \geq 2$, or
- $k = 1$, $\xi^* \in F_0$, and $w \neq \xi^*$.

Then as, by minimality of k, (12.4) holds for all $z \in C$ (so (12.3) also holds), it follows from hypothesis (5) that the holomorphic inverse branch

$$f_W^{-k} \colon B(\xi^*, 6r_{\xi^*}) \longrightarrow X,$$

satisfying $f_W^{-k}(B(\xi^*, 6r_{\xi^*})) \supseteq W$, is well defined. Consequently, using also the fact that $W \cup C$ is connected, we get that

$$W \cup C \subseteq f_W^{-k}\big(B(\xi^*, 2r_{\xi^*})\big). \tag{12.14}$$

Also, with the help of (12.9), we have that

$$B(w, 2r_w) \cap f_W^{-k}\big(B(\xi^*, 2r_{\xi^*})\big) \neq \emptyset.$$

Therefore, we get from hypothesis (5b) that

$$\left|\left(f_W^{-k}\right)'(x)\right| \leq \frac{\kappa - 1}{8\kappa} \cdot \frac{r_w}{r_{\xi^*}} \leq \frac{\kappa - 1}{2\kappa} \cdot \frac{r_w}{r_{\xi^*}}$$

for all $x \in B\big(\xi^*, \kappa r_{\xi^*}\big)$. Consequently, recalling also (12.14), we get that

$$\begin{aligned}
\operatorname{diam}(C) &\leq \operatorname{diam}\big(f_W^{-k}\big(B(\xi^*, \kappa r_{\xi^*})\big)\big) < \frac{\kappa - 1}{8\kappa} \cdot \frac{r_w}{r_{\xi^*}} \operatorname{diam}\big(B(\xi^*, \kappa r_{\xi^*})\big) \\
&= \frac{\kappa - 1}{2\kappa} \cdot \frac{r_w}{r_{\xi^*}} \cdot 2\kappa r_{\xi^*} = \frac{\kappa - 1}{4} r_w.
\end{aligned}$$

Because of (12.9), we, thus, conclude that

$$C \subseteq \overline{B}\left(w, r_w + \frac{\kappa - 1}{4} r_w\right) = \overline{B}\left(w, \frac{\kappa + 3}{4} r_w\right) \subseteq B(w, \kappa r_w). \tag{12.15}$$

Suppose now that the remining case holds, meaning that $\xi^* \in F_0$, $w = \xi^*$, and $k = 1$. It then follows from (12.13) that

$$f(W \cup C) \subseteq B(w, 2r_w). \tag{12.16}$$

So,

$$f(\overline{W} \cup \overline{C}) \subseteq B(w, 3r_w). \tag{12.17}$$

Since, by (12.10), the set $\overline{W} \cup \overline{C}$ is connected, since, by (12.9), $\overline{W} \cup \overline{C}$ intersects $\overline{B}(w, r_w)$, since, by (2b), the map $f|_{B(w, 6r_w)}$ is one-to-one, and since, by (2c), $f_w^{-1}(B(w, 3r_w)) \subseteq B(w, 6r_w)$, we conclude, with the help of (12.17), that

$$W \subseteq \overline{W} \cup \overline{C} \subseteq f_w^{-1}(B(w, 3r_w)).$$

But, since $W \in \mathrm{Comp}_1(U_0(w, s))$ and $U_0(w, s) \subseteq B(w, r_w) \subseteq B(w, 3r_w)$, we, thus, conclude that

$$W = f_w^{-1}(U_0(w, s)).$$

So, by (2f) and by (1),

$$W \subseteq U_0(w,s) \subseteq B(w,r_w).$$

This, however, contradicts (12.10) and the very definition of C. Hence, the considered case is ruled out and (12.15) holds.

Thus, taking the union over all $i \in \{1,2,\ldots,p_w\}$ and over all connected components of $U_n(w,i)\backslash\overline{B(w,r_w)}$ along with $U_0(w) \subseteq B(w,r_w)$, we, thus, see that the inductive proof of (12.8) is complete.

Now for every $w \in F$ and all $1 \le i \le p_w$, define

$$U'(w,i) := \bigcup_{n=0}^{\infty} U_n(w,i).$$

From our construction and from (12.8), $U'(w)$ is, thus, an open connected set such that

$$B(w,r_w) \subseteq U'(w,i) \subseteq B(w,\kappa r_w) \tag{12.18}$$

for all $w \in F$ and all $1 \le i \le p_w$.

Let $\Gamma(w,j)$ be the connected component of $\overline{B}(w,4r_w)\backslash U'(w,i)$ containing $\partial B(w,4r_w)$. It immediately follows from (12.18) that

$$\Gamma(w,j) \supseteq \overline{B}(w,4r_w)\backslash B(w,\kappa r_w). \tag{12.19}$$

Let

$$U(w,i) := \overline{B}(w,4r_w)\backslash\Gamma(w,j). \tag{12.20}$$

Then

$$U_0(w,i) \subseteq U'(w,i) \subseteq U(w,i) \subseteq B(w,\kappa r_w) \tag{12.21}$$

and we have that following.

Observation 12.2.2 For every $w \in F$ and all $1 \le i \le p_w$, the following hold.

(1*) For every $\gamma \in [\kappa r_w, 6r_w]$,

$$U(w,i) = \overline{B}(w,\gamma)\backslash\Gamma(w,j) = B(w,\gamma)\backslash\Gamma(w,j).$$

(2*) $U(w,i)$ is an open subset of $B(w,\kappa\gamma)$.
(3*) The set $U(w,i)$ is connected.
(4*) The set $U(w,i)$ is simply connected.
(5*) $\partial U(w,i) \subseteq \partial U'(w,i)$.

Proof Item (1*) is obvious. Item (2*) follows immediately from (1*) and the fact that the set $\Gamma(w,j)$ is closed. Item (3*) is an immediate consequence of the (very general) Theorem 5, page 140, in [**Kur**]. Item (4*) follows from Theorem 4.4 in [**Ne**]. Because of (1*) and Theorem 3, page 238, in [**Kur**] (which applies since Y is locally connected), we have that

$$\begin{aligned}\partial U(w,i) &\subseteq \overline{B}(w,2r_w) \cap \big(\partial\Gamma(w,j) \cup \partial B(w,6r_w)\big) = \overline{B}(w,2r_w) \cap \partial\Gamma(w,j)\\ &\subseteq \overline{B}(w,2r_w) \cap \partial\big(B(w,4r_w)\backslash U'(w,j)\big)\\ &\subseteq \overline{B}(w,2r_w) \cap \big(\partial B(w,4r_w) \cup \partial U'(w,j)\big)\\ &= \overline{B}(w,2r_w) \cap \partial U'(w,j)\\ &= \partial U'(w,j).\end{aligned}$$

This means that item (5*) is proved and, simultaneously, the entire Observation 12.2.2 is. ■

For every element $b \in F$, set

$$U(b) := \bigcup_{i=1}^{p_b} U(p,i)$$

and

$$U := \bigcup_{w\in F} U(w) = \bigcup_{w\in F}\bigcup_{i=1}^{p_w} U(w,i).$$

Our goal now is to show that, for each $b \in F_0$, the sets

(*) $\qquad U(b,i),\ 1 \le i \le p_b$, are pairwise disjoint.

Having this done, Property (c) of Definition 12.1.4 follows from (4*), while its property (a) follows from the inclusion $\overline{U(w,i)} \subseteq \overline{B(w,\kappa r)}$ and since the latter set is compact. Properties (A) and (B) of Theorem 12.2.1 directly follow from (12.21) and hypothesis (4) of this theorem. Property (C) of Theorem 12.2.1 also holds since all the sets $U(w,i)$, $w \in F$, $1 \le i \le p_w$, are connected and simply connected since

$$U(w) \subseteq B(w,\kappa r_w) \subseteq B(w,2r_w)$$

and since the sets $B(w,2r_w)$, $w \in F$, are pairwise disjoint.
So, in order to have all the claims of the preceding paragraph established, we now focus on proving (*). Toward this end, we shall first prove the following.

Claim 1°. For every $b \in F_0$ and every $i \in \{1,2,\dots,p_b\}$, all holomorphic iterates

$$f_b^{-n} : U'(b,i) \longrightarrow X,$$

$n \geq 0$, are well defined and

$$f_b^{-n}(U'(b,i)) \subseteq U'(b,i) \subseteq B(b,\kappa r_b).$$

Proof We proceed by induction. For $n = 0$, the claim follows immediately from (12.18). Suppose that it holds for some $n \geq 0$. Then, by (2c) (and as $\kappa \leq 2$),

$$f_b^{-(n+1)} := f_b^{-1} \circ f_b^{-n} \colon U'(b,i) \longrightarrow X$$

is well defined. Since $U_0(b,i) \subseteq U'(b,i)$ and since $f_b^{-1}(U_0(b,i)) \subseteq U_0(b,i)$, we conclude from the construction of the set $U'(b,i)$ that $f_b^{-(n+1)}(U'(b,i)) \subseteq U'(b,i)$. Since also, by (2c), $U'(b,i) \subseteq B(b,\kappa r_b)$, the proof of Claim 1° is complete. ■

As an immediate consequence of this claim, the connectedness of the sets $U'(b,i)$, and hypothesis (2g), we get the following.

Claim 2°. For every $b \in F_0$ and every $i \in \{1,2,\dots,p_b\}$, the sequence

$$\left(f_b^{-n}|_{U'(b,i)} \colon U'(b,i) \longrightarrow U'(b,i)\right)_{n=0}^{\infty}$$

converges uniformly on compact subsets of $U'(b,i)$ to the constant function whose range is equal to $\{b\}$.

Now we are in a position to prove that the property (*) holds. Indeed, because of Claim 1°, we have that

$$f_b^{-1}(U'(b,i)) \subseteq U'(b,i) \subseteq U(b,i).$$

So, since $U(b,i)$ is conformally equivalent to the unit disk $\mathbb{D}$, it follows from Proposition 12.1.3 and (12.21) that

$$f_b^{-1}(U(b,i)) \subseteq U(b,i) \subseteq B(b,r_b). \tag{12.22}$$

So, item (E) is proved. Furthermore, all iterates

$$f_b^{-n} \colon U(b,i) \longrightarrow U(b,i), \quad n \geq 0,$$

are well defined and form a normal family. Then, from hypothesis (2g), we get the following.

Claim 3°. For every $b \in F_0$ and every $i \in \{1,2,\dots,p_b\}$, the sequence

$$\left(f_b^{-n}|_{U(b,i)} \colon U(b,i) \longrightarrow U(b,i)\right)_{n=0}^{\infty}$$

converges uniformly on compact subsets of $U(b,i)$ to the constant function whose range is equal to $\{b\}$.

We conclude from this claim and property (2h) that, for every $z \in U(b,i)$, there exists $n \geq 0$ such that

$$f_b^{-n}(z) \in \bigcup_{j=1}^{p_b} U_0(b,j).$$

So, if, for every $j \in \{1,2,\dots,p_b\}$,

$$U_\infty(b,i;j) := \left\{z \in U(b,i) \colon \exists n \geq 0 \ f_b^{-n}(z) \in U_0(b,j)\right\},$$

then

$$\bigcup_{j=1}^{p_b} U_\infty(b,i;j) := U(b,i). \tag{12.23}$$

Obviously, all the sets $U_\infty(b,i;j)$, $j = 1,\dots,p_b$, are open and, because of (2e) and (2f), they are pairwise disjoint. Since the set $U(b,i)$ is connected, we conclude that only one term in the union of (12.23) is nonempty. Since $U_\infty(b,i;i) \supset U_0(b,i) \neq \emptyset$, we further conclude that

$$U(b,i) = U_\infty(b,i;i).$$

Since, by (2e), applied again, the sets $\{U_\infty(b,i,;i)\}$, $i = 1,\dots,p_b$, are pairwise disjoint, the property (*) follows.

We, thus, conclude that the sets $U(w,i)$, $w \in F$, $i = 1,2,\dots,p_b$, coincide with the collection of all connected components of U. Hence, the number of connected components of U is finite, which means that condition (b) of Definition 12.1.4 holds.

We shall now show that condition (d) of Definition 12.1.4 holds. Indeed, seeking contradiction, suppose that

$$U \cap f^n(\partial U) \neq \emptyset$$

with some $n \geq 1$. Consider a minimal $n \geq 1$ with this property. Then there exists

$$x \in \partial U \quad \text{such that} \quad f^n(x) \in U.$$

Assume first that

$$f^n(x) \in U',$$

where

$$U' := \bigcup_{w \in F} \bigcup_{i=1}^{p_w} U'(w,i),$$

and assume only that $x \in \partial U'$ (this is a weaker requirement than $x \in \partial U$ as $\partial U \subseteq \partial U'$). Therefore, there exist $\xi \in F$, $i \in \{1, 2, \ldots, p_\xi\}$, an integer $k \geq 0$, and $W \in \mathrm{Comp}_k\big(U_0(\xi, i)\big)$ such that $f^n(x) \in W$. But then $x \in \tilde{W}$, the connected component of $f^{-n}(W)$ containing x. We also immediately see that

$$\tilde{W} \in \mathrm{Comp}_{n+k}\big(U_0(\xi, i)\big).$$

Furthermore, $\tilde{W} \cap U' \neq \emptyset$ as $x \in \partial U'$. By our construction of the set U', this implies that $\tilde{W} \subseteq U'$. Hence, $x \in U'$ and this contradiction rules out the considered case.

Now, continuing the general case, there exist $w \in F$ and $1 \leq i \leq p_w$ such that

$$f^n(x) \in U(w, i).$$

It, therefore, follows from (12.21), the formula $\kappa \leq 2$, and the minimality of n that (12.3) holds. Hence, by hypothesis (5a), there exists $f_x^{-n} \colon B(w, 6r_w) \longrightarrow X$, a unique holomorphic inverse branch of f^n defined on $B(w, 6r_w)$ and sending $f^n(x)$ back to x. If

$$f_x^{-n}(U'(w, i)) \cap U'(y, j) \neq \emptyset$$

for some $y \in F$ and $j \in \{1, 2, \ldots, p_y\}$, then, by the previous case,

$$f_x^{-n}(U'(w, i)) \subseteq U'(y, j) \subseteq U(y, j).$$

By virtue of Proposition 12.1.3, we, thus, then get $f_x^{-n}(U(w, i)) \subseteq U(y, j)$. In particular, $x = f_x^{-n}(f^n(x)) \in U(y, j) \subseteq U$. This contradiction yields

$$U' \cap f_x^{-n}(U'(w, i)) = \emptyset. \tag{12.24}$$

But

$$x \in \partial U(\xi, k) \cap f_x^{-n}(U(w, i)) \subseteq \partial U'(\xi, k) \cap f_x^{-n}(U(w, i)) \tag{12.25}$$

for some $\xi \in F$ and $k \in \{1, 2, \ldots, p_\xi\}$, where the inclusion part of this formula holds because of Observation 12.2.2(5*). Hence,

$$U'(\xi, k) \cap f_x^{-n}(U(w, i)) \neq \emptyset.$$

Because of Observation 12.2.2(5*), applied this time to $U(w, i)$, it follows from this and (12.24) that

$$U'(\xi, k) \subseteq f_x^{-n}(U(w, i)). \tag{12.26}$$

But then, using (12.21), we get that

$$f_x^{-n}(B(w, \kappa r_w)) \cap B(\xi, \kappa r_\xi) \neq \emptyset.$$

Therefore, remembering also about the minimality of n, we see that hypothesis (5) of our theorem is satisfied; so, it follows from (12.5), with the help of hypothesis (1) and (12.21), that

$$\begin{aligned}\operatorname{diam}\big(f_x^{-n}(U(w,i))\big) &\le \frac{\kappa-1}{8\kappa}\cdot\frac{r_\xi}{r_w}\operatorname{diam}(U(w,i))\\ &\le \frac{\kappa-1}{8\kappa}\cdot\frac{r_\xi}{r_w}2\kappa r_w = \frac{\kappa-1}{4}r_\xi \le r_\xi/4. \qquad (12.27)\end{aligned}$$

On the other hand, if $\xi \in F_1$, then it follows from (12.26) and hypothesis (2h) that

$$\operatorname{diam}\big(f_x^{-n}(U(w,i))\big) \ge \operatorname{diam}(U'(\xi,k)) \ge r_\xi/2.$$

This, however, contradicts (12.27) and finishes the proof of condition (d) of Definition 12.1.4 in the case when $\xi \in F_1$.

So, suppose that $\xi \in F_0$. It then follows from (12.26) and (2d) that

$$\xi \in \overline{U'(\xi,k)} \subseteq f_x^{-n}\big(\overline{U(w,i)}\big).$$

Then, by virtue of (2a), we get that

$$\xi = f^n(\xi) \in f^n(f_x^{-n}(\overline{U(w,i)})) = \overline{U(w,i)}.$$

It, thus, follows from hypothesis (3) and (12.21) (remember also that $\kappa \le 2$) that $\xi = w$. So, $w \in f_x^{-n}(\overline{U(w,i)})$ and, as $f^n(w) = w$ and $w \in \overline{U(w,i)} \subseteq B(w,2r_w)$, we conclude that $f_x^{-n}(w) = w$ and $f_x^{-n} = f_w^{-n}$. We deduce from this, (12.26), the first inclusion of (12.22), and the property (*) that $k = i$. It eventually follows from (12.25) and (12.22) that $x \in \partial U(w,i) \cap U(w,i)$. Since the set $U(w,i)$ is open, this is a contradiction and the proof of condition (d) of Definition 12.1.4 is complete.

We conclude the proof of this theorem by proving item (D). We start with the following.

Claim 4°. For all $b \in F_0$ and all $j \in \{1,2,\dots,p_b\}$, we have that

$$U'(b,j) \cap B(b,r'_b) = U_0(b,j) \cap B(b,r'_b).$$

Proof Of course,

$$U_0(b,j) \cap B(b,r'_b) \subseteq U'(b,j) \cap B(b,r'_b).$$

In order to prove the opposite inclusion, take any point

$$z \in U'(b,j) \cap B(b,r'_b).$$

If $z \notin U_0(b,j)$, then $f^n(z) \in B(b,r_b)$ for all $n \ge 0$ by virtue of condition (*), hypothesis (2j), and since $r'_b \le (r_b)_b^-$. Since also $z \in U'(b,j)$, by using property (*) and Claim 1°, we, thus, conclude that $f^k(z) \in U_0(B, j)$ for some $k \geq 0$. It then follows from Clam 1°, used again, that

$$z = f_b^{-k}\big(f^k(z)\big) \in f_b^{-k}(U_0(b, j)) \subseteq U_0(b, j).$$

This contradiction shows that $z \in U_0(b, j)$, finishing the proof of Claim 4°. ■

Fix $b \in F_0$ and $j \in \{1, 2, \dots, p_b\}$. By our hypothesis (6), the set $V_j(b) \backslash \overline{U_0(b, j)}$ is connected.

Seeking contradiction, suppose that

$$G := U(b, j) \cap (V_j(b) \backslash \overline{U_0(b, j)}) \neq \emptyset.$$

Then

$$\partial G \subseteq \overline{V_j(b)} \cap (\partial U(b, j) \cup \partial U_0(b, j)). \tag{12.28}$$

So, using item (5*) of Observation 12.2.2 and Claim 4°, we get that

$$\begin{aligned} \partial G &\subseteq \overline{V_j(b)} \cap (\partial U'(b, j) \cup \partial U_0(b, j)) \\ &\subseteq (\overline{V_j(b)} \cap \overline{U'(b, j)}) \cup \overline{U_0(b, j)} = \overline{U_0(b, j)}. \end{aligned}$$

Hence,

$$\partial_{V_j(b) \backslash \overline{U_0(b, j)}} G \subseteq (V_j(b) \backslash \overline{U_0(b, j)}) \cap \partial G = \emptyset.$$

Thus, because of (6b),

$$\text{either} \quad G = \emptyset \quad \text{or} \quad G = V_j(b) \backslash \overline{U_0(b, j)}. \tag{12.29}$$

If $G = V_j(b) \backslash \overline{U_0(b, j)}$, then $V_j(b) \backslash \overline{U_0(b, j)} \subseteq U(b, j)$; therefore, $V_j(b) \subseteq \overline{U(b, j)}$. So, if $s_1 > 0$ is so small that $B(b, s_1) \subseteq V_j(b)$, then

$$\overline{B(b, s_1)} \subseteq \mathrm{Int}(\overline{U(b, j)}) = U(b, j).$$

It, therefore, follows from Claim 3° that

$$f_b^{-q}(\overline{B(b, s_1)}) \subseteq B(b, s_1/2)$$

for all (just one suffices) $q \geq 1$ large enough. But then the Schwarz Lemma yields $|(f_b^{-q})'(b)| < 1$. Therefore, $|f'(b)| > 1$. This, however, contradicts our hypothesis (7) and proves that $G = \emptyset$. Therefore, keeping also in mind that the set $U_0(b, j)$ is open and $\overline{U_0(b, j)}$ is a closed topological disk, we get that

$$\begin{aligned} U(b, j) \cap (V_j(b) \backslash U_0(b, j)) &= U(b, j) \cap (V_j(b) \cap \partial U_0(b, j)) \\ &= U(b, j) \cap (V_j(b) \cap \partial U_0(b, j)) \\ &\subseteq U(b, j) \cap (V_j(b) \cap \overline{(Y \backslash \overline{U_0(b, j)})}). \end{aligned}$$

Seeking contradiction, suppose that this set is not empty. Since $U(b,j)\cap V_j(b)$ is open, we would then have that

$$U(b,j)\cap V_j(b)\cap(Y\backslash\overline{U_0(b,j)})\neq\emptyset.$$

Equivalently,

$$U(b,j)\cap(V_j(b)\backslash\overline{U_0(b,j)})\neq\emptyset,$$

meaning that $G\neq\emptyset$. This contradiction shows that

$$U(b,j)\cap(V_j(b)\backslash U_0(b,j))=\emptyset,$$

yielding

$$U(b,j)\cap V_j(b)=U_0(b,j)\cap V_j(b). \tag{12.30}$$

Since $V_j(b)$ is an open set containing b, there exists $\beta>0$ such that $B(b,\beta)\subseteq V_j(b)$. Consequently, for all $\alpha\in(0,\beta]$, we have that

$$U(b,j)\cap B(b,\alpha)=U_0(b,j)\cap B(b,\alpha) \tag{12.31}$$

and the proof of item D of Theorem 12.2.1 is complete.

Now, assuming conditions (8) and (9), we shall prove (F) and (G). Let $\xi\in\partial U(b,j)\backslash\{b\}$ for some $b\in F_0$ and $j\in\{1,2,\dots,p_b\}$. Fix some $s>0$ so small that $B(\xi,s)\subseteq B(b,3r_b)$. Then, because of condition (9), there exists an integer $N_\xi\geq 0$ such that

$$\left(\bigcup_{n=N_\xi}^{\infty}f^n(\mathrm{Crit}(f))\right)\cap B(\xi,s)=\emptyset.$$

Equivalently,

$$\mathrm{Crit}(f)\cap\bigcup_{n=N_\xi}^{\infty}f^{-n}(B(\xi,s))=\emptyset. \tag{12.32}$$

It immediately follows from (8), (12.22), and the continuity of f_b^{-1} that, for every $n\geq 0$, there exists $r_n(\xi)\in(0,+\infty)$ such that

$$\mathrm{Crit}(f)\cap\mathrm{Comp}(f_b^{-n}(\xi),k,8r_n(\xi))=\emptyset \tag{12.33}$$

for every $k\in\{0,1,\dots,n\}$. Put

$$r_\xi:=\min\{r_{N_\xi},s/8\}.$$

It then follows from (12.32) and (12.33) that

$$\mathrm{Crit}(f) \cap \bigcup_{n=0}^{\infty} \mathrm{Comp}\big(f_b^{-n}(\xi), n, 8r_\xi\big) = \emptyset.$$

Hence, for all $n \geq 0$, we have that

$$\mathrm{Crit}(f^n) \cap \mathrm{Comp}\big(f_b^{-n}(\xi), n, 8r_\xi\big) = \emptyset;$$

therefore, there exists a unique holomorphic branch $f_{b,\xi}^{-n}: B(\xi, 8r_\xi) \longrightarrow X$ of f^{-n} of sending ξ to $f_b^{-n}(\xi)$. So, taking

$$W(b,j) := U(b,j) \cup \bigcup_{\xi \in \partial U(b,j) \backslash \{b\}} B(\xi, 8r_\xi)$$

and also invoking condition (5a), we conclude that item (F) holds.

We shall now prove the following.

Claim 5°. For every $\xi \in \partial U(b,j) \backslash \{b\}$, the sequence $f_{b,\xi}^{-n}|_{B(\xi,4r_\xi)}: B(\xi, 4r_\xi) \longrightarrow X$ converges uniformly to the constant function whose range is equal to $\{b\}$.

Proof Since $\xi \in \partial U(b,j)$, we have that $B(\xi, r_\xi) \cap U(b,j) \neq \emptyset$. Fix a point y in this intersection. Take $\rho > 0$ so small that

$$B(y, 8\rho) \subseteq B(\xi, r_\xi) \cap U(b,j).$$

By virtue of Claim 3°, the sequence $f_{b,\xi}^{-n}|_{B(y,4\rho)}: B(y, 4\rho) \longrightarrow X$ converges uniformly to the constant function whose range is equal to $\{b\}$. Consequently,

$$\lim_{n\to\infty} \left\| \left(f_{b,\xi}^{-n}|_{B(y,4\rho)}\right)' \right\|_\infty = 0.$$

It follows from this and Theorem 8.3.9 that

$$\lim_{n\to\infty} \left\| \left(f_{b,\xi}^{-n}|_{B(\xi,4r_\xi)}\right)' \right\|_\infty = 0.$$

Hence,

$$\lim_{n\to\infty} \mathrm{diam}\left(f_{b,\xi}^{-n}\left(B\left(\xi, 4r_\xi\right)\right)\right) = 0$$

and the claim follows. ■

By condition (6), for every $u \in (0, r_b')$ small enough, there exists an open set $V_j(b,u) \subseteq B(b,u) \cap V_j(b)$, containing b, with the following properties.

(1*) The set $\overline{V_j(b,u)} \cap \overline{U_0(b,j)}$ is a closed topological disk.
(2*) The set $V_j(b,u) \backslash \overline{U_0(b,j)}$ is connected.
(3*) $V_j(b,u) \subseteq B\big(b, r_b'/4\big)$.
(4*) $\partial\Big(\overline{V_j(b,u)} \cap \overline{U_0(b,j)}\Big) = \Big(\overline{V_j(b,u)} \cap \partial U_0(b,j)\Big) \cup \Big(\overline{U_0(b,u)} \cap \partial V_j(b,j)\Big)$.

(5*) Both sets $\overline{V_j(b,u)} \cap \partial U_0(b,j)$ and $\overline{U_0(b,j)} \cap \partial V_j(b,u)$ are compact topological arcs.

(6*) The intersection

$$\left(\overline{V_j(b,u)} \cap \partial U_0(b,j)\right) \cap \left(\overline{U_0(b,u)} \cap \partial V_j(b,j)\right)$$

consists of two distinct points, which we denote by $x_j(b,u)$ and $y_j(b,u)$.

We shall prove the following.

Claim 6°. The set $\overline{U(b,j)} \backslash V_j(b,u)$ is connected.

Proof Fix two points $w, z \in U(b,j) \backslash V_j(b,u)$. Since $U(b,j)$ is an open connected subset of the Riemann surface Y, it is arcwise connected. Therefore, there exists a homeomorphic embedding

$$\gamma\colon [0,1] \longrightarrow U(b,j)$$

such that $\gamma(0) = w$ and $\gamma(1) = z$. If $\gamma([0,1]) \subseteq U(b,j) \backslash V_j(b,u)$, we are done. So, suppose that

$$\gamma([0,1]) \cap V_j(b,u) \neq \emptyset.$$

But, because of (12.30),

$$\gamma([0,1]) \cap V_j(b,u) \subseteq U_0(b,j). \tag{12.34}$$

Denote

$$t_w := \inf\{t \in [0,1]\colon\ \gamma(t) \in V_j(b,u)\}$$

and

$$t_z := \sup\{t \in [0,1]\colon\ \gamma(t) \in V_j(b,u)\}.$$

Then, also keeping in mind that the set $V_j(b,u)$ is open, we have that

$$\gamma([0,t_w]) \cap V_j(b,u) = \emptyset \quad \text{and} \quad \gamma([t_z,1]) \cap V_j(b,u) = \emptyset;$$

because of (12.34),

$$\gamma(t_w), \gamma(t_z) \in \overline{U_0(b,j)}.$$

We also have that

$$\gamma(t_w), \gamma(t_z) \in \partial V_j(b,u).$$

In conclusion,

$$\gamma(t_w), \gamma(t_z) \in \overline{U_0(b,j)} \cap \partial V_j(b,u).$$

Therefore, by (5*), there exists a compact topological arc $\Delta \subseteq \overline{U_0(b,j)} \cap \partial V_j(b,u)$, with $\gamma(t_w)$ and $\gamma(t_z)$ being its endpoints, which is entirely contained in $U_0(b,j) \cap \partial V_j(b,u)$ except, perhaps, the endpoints $\gamma(t_w)$ and $\gamma(t_z)$. But both $\gamma(t_w)$ and $\gamma(t_z)$ belong to $U(b,j)$. Thus,

$$\Delta \subseteq U(b,j)\backslash V_j(b,u).$$

Hence, $[\gamma(0), \gamma(t_w)] \cup \Delta \cup [\gamma(t_z), \gamma(1)]$ is a compact topological arc joining $\gamma(0)$ and $\gamma(1)$, entirely contained in $U(b,j)\backslash V_j(b,u)$. So, any two points in $U(b,j)\backslash V_j(b,u)$ belong to some connected subset of $U(b,j)\backslash V_j(b,u)$. Therefore, the set $U(b,j)\backslash V_j(b,u)$ is also connected. Since also

$$\overline{U(b,j)}\backslash V_j(b,u) \subseteq \overline{U(b,j)\backslash V_j(b,u)},$$

we, thus, conclude that the set $\overline{U(b,j)}\backslash V_j(b,u)$ is also connected. The proof of Claim 6° is complete. ■

Obviously,

$$\left(\overline{U(b,j)}\backslash V_j(b,u)\right) \backslash \bigcup_{\xi \in \partial U(b,j)\backslash V_j(b,u)} B(\xi, 4r_\xi) \subseteq U(b,j)\backslash V_j(b,u)$$
$$= U_0(b,j)\backslash V_j(b,u), \quad (12.35)$$

where the equality sign was written because of (12.31) and the choice of u. Since $U_0(b,j)$ is an open set, for every $\xi \in U_0(b,j)\backslash V_j(b,u)$, there exists $r_\xi > 0$ so small that

$$B(\xi, 8r_\xi) \subseteq U_0(b,j).$$

Since, by (12.35), the collection

$$\left\{B(\xi, 4r_\xi) : \xi \in \left(\partial U(b,j)\backslash V_j(b,u)\right) \cup \left(U_0(b,j)\backslash V_j(b,u)\right)\right\}$$

is an open cover of the compact set $\overline{U(b,j)}\backslash V_j(b,u)$, there exists a finite set

$$Z \subseteq \left(\partial U(b,j)\backslash V_j(b,u)\right) \cup \left(U_0(b,j)\backslash V_j(b,u)\right)$$

such that

$$\overline{U(b,j)}\backslash V_j(b,u) \subseteq W(b,j;u) := \bigcup_{\xi \in Z} B(\xi, 4r_\xi). \quad (12.36)$$

Obviously, the set $W(b,j;u)$ is open. It is connected because, by Claim 6°, the set $\overline{U(b,j)}\backslash V_j(b,u)$ is connected, each ball $B(\xi, 4r_\xi)$, $\xi \in Z$, is connected, and each such ball intersects $\overline{U(b,j)}\backslash V_j(b,u)$. Since $W(b,j;u) \subseteq W(b,j)$, for every $n \geq 0$ and every $z \in f^{-n}(b)$, the holomorphic branch $f_z^{-n}\colon W(b,j;u) \longrightarrow X$ of f^{-n} is well defined. Also, the maps

$f_b^{-n}: W(b, j; u) \longrightarrow X$ converge uniformly to the constant function whose range is equal to $\{b\}$ because of Claim 5°, hypothesis (2g), and since the set Z is finite. The proof of item (G) is complete.
Thus, the proof of Theorem 12.2.1 is complete. ■

We remark that the hypotheses about the behavior of f around points $b \in F_0$ are modeled on those when b is a rationally indifferent fixed point of f. These will be needed and addressed in depth and length in Chapter 22 in Volume 2 of the book.

12.3 The Existence of Nice Sets

In this section, we derive several consequences of Theorem 12.2.1. All of them are about the existence of nice sets under various assumptions. We start with the following.

Corollary 12.3.1 *Let Y be either a parabolic or an elliptic ($\widehat{\mathbb{C}}$) Riemann surface. Let X be an open subset of Y, which, moreover, is a subset of $\mathbb{C}$ if $Y = \widehat{\mathbb{C}}$. Let $f: X \to Y$ be an analytic map.*

Fix

- *a radius $R \in (0, u_Y)$,*
- *$\lambda > 1$ and $\kappa \in (1, 2]$,*
- *a finite subset F of $J(f) \backslash \overline{\mathrm{PS}(f)}$,*
- *a vector $\mathbf{r} = \big(r_b : b \in F\big)$.*

Assume that the following two conditions are satisfied.

(1)

$$r_b \in \left(0, \frac{1}{6} \min\left\{R, \min\left\{|b - c| : c \in F \backslash \{b\}\right\}, \mathrm{dist}\big(b, \overline{\mathrm{PS}(f)}\big)\right\}\right)$$

for all $b \in F$.

(2) *If $a, b \in F$, $n \geq 1$ is an integer, $w \in f^{-n}(B(b, 2r_b))$ is such that the connected component of $f^{-n}(B(b, 2r_b))$ containing w intersects $B(a, 2r_a)$, and*

$$f^j(w) \notin \bigcup_{c \in F} B(c, r_c) \tag{12.37}$$

for all $j = 1, 2, \ldots, n-1$, *then*

$$|(f^n)'(z)| \geq \max\left\{\frac{8\kappa}{\kappa - 1} \cdot \frac{r_b}{r_a}, \lambda\right\} \tag{12.38}$$

for all $z \in f_w^{-n}(B(b, 2r_b))$.

Then there exists a nice set $U = U_{\mathbf{r}}$ *with the following properties.*

(a) *For all* $b \in F$,

$$B(b, r_b) \subseteq U \subseteq \bigcup_{b \in F} B(b, \kappa r_b) \subseteq B(b, 3\kappa r_b) \subseteq Y \backslash \overline{\mathrm{PS}(f)}.$$

(b) *If* W *is a connected component of* U, *then* $W \cap F$ *is a singleton.*

Proof Setting $F := F_1$ and $U_0(b) := B(b, r_b)$ for all $b \in F$, this corollary is an immediate consequence of Theorem 12.2.1 except perhaps condition (e) of Definition 12.1.5. But this one follows immediately from (12.38) of condition (2) of our corollary. ∎

As an immediate consequence of Corollary 12.3.1, we get the following.

Corollary 12.3.2 *Let* Y *be either a parabolic or an elliptic* ($\widehat{\mathbb{C}}$) *Riemann surface. Let* X *be an open subset of* Y, *which, moreover, is a subset of* $\mathbb{C}$ *if* $Y = \widehat{\mathbb{C}}$. *Let* $f : X \to Y$ *be an analytic map.*

Fix

- *a radius* $R \in (0, u_Y)$,
- $\lambda > 1$ *and* $\kappa \in (1, 2]$,
- *a finite subset* F *of* $J(f) \backslash \overline{\mathrm{PS}(f)}$,
- *a radius*

$$r \in \left(0, \frac{1}{6} \min\left\{R, \min\left\{|a - b| : a, b \in F, a \neq b\right\}, \mathrm{dist}\left(F, \overline{\mathrm{PS}(f)}\right)\right\}\right).$$

Assume that if $a, b \in F$, $n \geq 1$ *is an integer,* $w \in f^{-n}(B(b, 2r))$ *is such that the connected component of* $f^{-n}(B(b, 2r))$ *containing* w *intersects* $B(a, 2r)$, *and*

$$f^j(w) \notin \bigcup_{c \in F} B(c, r) \tag{12.39}$$

for all $j = 1, 2, \ldots, n-1$, *then*

$$|(f^n)'(z)| \geq \max\left\{\frac{8\kappa}{\kappa - 1}, \lambda\right\} \tag{12.40}$$

for all $z \in f_w^{-n}(B(b, 2r))$.

Then there exists a nice set $U = U_r$ with the following properties.

(a) *For all $b \in F$,*

$$B(b,r) \subseteq U \subseteq \bigcup_{b\in F} B(b,\kappa r) \subseteq B(b,3\kappa r) \subseteq Y\backslash\overline{\mathrm{PS}(f)}.$$

(b) *If W is a connected component of U, then $W \cap F$ is a singleton.*

We shall now provide some sufficient conditions for the hypotheses of Corollary 12.3.2 to be satisfied. We first shall prove the following.

Theorem 12.3.3 *Let Y be either a parabolic or an elliptic ($\widehat{\mathbb{C}}$) Riemann surface. Let X be an open subset of Y, which, moreover, is a subset of $\mathbb{C}$ if $Y = \widehat{\mathbb{C}}$. Let*

$$f\colon X \longrightarrow Y$$

be an analytic map with the Standard Property. Let $Q \subseteq Y$ be a set witnessing this property.

Fix a finite subset F of

$$\begin{aligned} J(f)\backslash\left(Q \cup \overline{\mathrm{PS}(f)} \cup \bigcup_{n=0}^{\infty} f^{-n}(Y\backslash X)\right) &= J(f) \cap \bigcap_{n=0}^{\infty} f^{-n}(X)\backslash\big(Q \cup \overline{\mathrm{PS}(f)}\big) \\ &= J(f) \cap \bigcap_{n=0}^{\infty} f^{-n}(Y)\backslash\big(Q \cup \overline{\mathrm{PS}(f)}\big) \end{aligned}$$

such that

$$F \cap \bigcup_{n=1}^{\infty} f^{n}(F) = \emptyset. \tag{12.41}$$

Fix also $\kappa \in (1,2]$ and a radius $R \in (0,u_Y)$.

Then, for every

$$r \in \left(0, \frac{1}{6}\min\left\{R, \min\big\{|a-b|: a,b \in F, a \neq b\big\}, \mathrm{dist}\big(F, Q \cup \overline{\mathrm{PS}(f)}\big)\right\}\right)$$

small enough, there exists a nice set $U = U_r$ with the following properties.

(a) *$B(F,r) \subseteq U \subseteq B(F,\kappa r) \subseteq B(F,2\kappa r) \subseteq J(f)\backslash\overline{\mathrm{PS}(f)}$.*
(b) *If W is a connected component of U, then $W \cap F$ is a singleton. Denoting this singleton by b, we then set $\hat{W} := B(b,\kappa r)$.*

Proof We shall prove that, for every

$$r \in \left(0, \Delta := \frac{1}{6}\min\left\{R, \min\big\{|a-b|: a,b \in F, a \neq b\big\}, \mathrm{dist}\big(F, Q \cup \overline{\mathrm{PS}(f)}\big)\right\}\right)$$

small enough, the hypotheses of Corollary 12.3.2 are satisfied. This, in fact, means that we are supposed to check that (12.40) of this corollary holds for all such radii r. Indeed, by Lemma 8.1.19, there exists $N \geq 1$ so large that

$$|(f^n)'(z)| \geq \max\left\{\frac{8\kappa}{\kappa - 1}, 2\right\} \tag{12.42}$$

for all $n \geq N$ and all $z \in f^{-n}(B(F,\Delta))$ whenever the connected component of $f^{-n}(B(F,\Delta))$ containing z intersects $B(F,\Delta)$. On the other hand, since the set F is finite, it follows from (12.41) that there exists $\Delta_1 \in (0,\Delta/4)$ so small that

$$B(F,2\Delta_1) \cap \bigcup_{n=1}^{N} f^n\big(B(F,2\Delta_1)\big) = \emptyset. \tag{12.43}$$

So, taking any $r \in (0,\Delta_1)$ finishes the proof. ■

12.4 The Maximal GDMSs Induced by Nice Sets

As an immediate consequence of Corollary 12.3.1 and Theorem 12.1.8, we get the following.

Theorem 12.4.1 *The system $\mathcal{S}_{U_r}$ resulting from Corollary 12.3.1 and Theorem 12.1.8 is a maximal CGDMS in the sense of Chapter 11.*

As an immediate consequence of Corollary 12.3.2 and Theorem 12.1.8, we get the following.

Theorem 12.4.2 *The system $\mathcal{S}_{U_r}$ resulting from Corollary 12.3.2 and Theorem 12.1.8 is a maximal CGDMS in the sense of Chapter 11.*

As an immediate consequence of Theorems 12.3.3 and 12.1.8, we get the following.

Theorem 12.4.3 *For all $r > 0$ small enough, the systems $\mathcal{S}_{U_r}$ resulting from Theorems 12.3.3 and* 12.1.8 *are maximal CGDMSs in the sense of Chapter 11.*

We end this section with the theorem formulated below whose proof requires the full power of Theorem 12.2.1.

With the notation of Theorem 12.2.1 and its proof, for every $b \in F_1$, set

$$X_b := X(b,1;u) := \overline{U(b,1)} \quad \text{and} \quad W(b,1;u) := B(b,4r_b). \tag{12.44}$$

For every $b \in F_0$ and $j \in \{1,2,\ldots,p_b\}$, set

$$X^*(b,j) := \overline{U(b,j)} \backslash f_b^{-1}(U(b,j)). \tag{12.45}$$

For every $u \in (0, r_b'/4)$ as small as required in Theorem 12.2.1(G), set

$$X(b,j;u) := X^*(b,j) \backslash V_j(b,u). \tag{12.46}$$

For every $A = X(a,j;u)$, $a \in F$, $j \in \{1,\ldots,p_a\}$, put

$$U_A := U(a,j;u) \quad \text{and} \quad W_A := W(a,j;u).$$

We form a system $\mathcal{S}_U = \mathcal{S}_{U_{\mathrm{r}}}$ as follows. The pairs

$$\big\{\big(X(b,j;u), W(b,j;u)\big),\, b \in F,\ j \in \{1,2,\ldots,p_b\}\big\} \tag{12.47}$$

form the domains of $\mathcal{S}_U$. In order to define the maps of $\mathcal{S}_U$, look at all sets

$$A, B \in \big\{X(b,j;u) \colon b \in F \ \text{ and } j \in \{1,2,\ldots,p_b\}\big\} \tag{12.48}$$

and all integers $n \geq 1$ such that

$$\mathrm{Int}(A) \cap f^{-n}(\mathrm{Int}(B)) \neq \emptyset. \tag{12.49}$$

Because of hypothesis (5) and item (F) of Theorem 12.2.1, for every $\xi \in \mathrm{Int}(A) \cap f^{-n}(\mathrm{Int}(B))$ there exists

$$f_\xi^{-n} \colon W_B \longrightarrow X, \tag{12.50}$$

a unique holomorphic branch of f^{-n}, defined on W_B and sending $f^n(\xi) \in \mathrm{Int}(B) \subseteq W_B$ to ξ. We declare that the map $f_\xi^{-n} \colon W_B \longrightarrow X$ belongs to $\mathcal{S}_U$ if

$$f^s\big(f_\xi^{-n}(\mathrm{Int}(B))\big) \cap \bigcup_{w \in F} \bigcup_{j=1}^{p_w} \mathrm{Int}(X(w,j;r)) = \emptyset \tag{12.51}$$

for all integers $0 < s < n$. It follows from (12.49) that

$$U_A \cap f_\xi^{-n}(U_B) \neq \emptyset. \tag{12.52}$$

Theorem 12.4.4 *Let Y be either a parabolic or an elliptic ($\widehat{\mathbb{C}}$) Riemann surface. Let X be an open subset of Y, which, moreover, is a subset of $\mathbb{C}$ if $Y = \widehat{\mathbb{C}}$. Let*

$$f \colon X \longrightarrow Y$$

be an analytic map with the Standard Property. Let $Q \subseteq Y$ be a set witnessing this property.

Fix

- $\kappa \in (1,2]$,
- F, *a finite subset of* $J(f)\backslash O_+(Q)$,
- *a radius* $R \in (0, u_Y)$,
- *a collection* $\{U_0(b)\}_{b\in F}$ *of open subsets of* Y, *and*
- *a vector* $\mathbf{r} = \big(r_b : b \in F\big)$ *with the following properties.*

(1)

$$U_0(b) \subseteq B(b, r_b)$$

for all $b \in F$.

(2) F *can be represented as a disjoint union:*

$$F = F_0 \cup F_1$$

such that

(a) $f(b) = b$ *for every* $b \in F_0$.

(b) *For each* $b \in F_0$, *the map* $f|_{B(b,6r_b)}$ *is one-to-one and the holomorphic inverse branch* $f_b^{-1} : B(b, 6r_b) \longrightarrow X$, *sending* b *to* b, *is well defined.*

(c) $f_b^{-1}\big(B(b, 3r_b)\big) \subseteq B(b, 6r_b)$.

(d) $b \in \partial U_0(b)$ *for all* $b \in F_0$.

(e) *For each* $b \in F_0$, *there exists an integer* $p_b \geq 1$ *such that* $U_0(b)$ *has exactly* p_b *(open) connected components* $U_0(b, j)$, $j = 1, 2, \ldots, p_b$, *each of which is simply connected and its closure is a closed topological disk.*

(f) *For each* $b \in F_0$ *and each* $j = 1, 2, \ldots, p_b$,

$$f_b^{-1}(U_0(b, j)) \subseteq U_0\big(b, j\big).$$

(g) *The sequence*

$$\Big(f_b|_{U_0(b,j)}^{-n} : U_0(b, j) \longrightarrow X\Big)_{n=0}^{\infty}$$

converges uniformly to the constant function, which assignes to each point in $U_0(b, j)$ *the value* b.

(h) $B(b, r_b/2) \subseteq U_0(b)$ *for all* $b \in F_1$.

(i) *For each* $b \in F_1$, *the set* $U_0(b)$ *is connected, simply connected, and its closure is a closed topological disk. We then put* $p_b := 1$ *and denote also* $U_0(b)$ *by* $U_0(b, 1)$.

(j) *For each* $b \in F_0$ *and* $s > 0$, *there exists* $s_b^- \in (0, r_b]$ *such that if*

$$z \in B\big(b, s_b^-\big) \setminus \bigcup_{j=1}^{p_b} U_0(b, j),$$

then

$$f^n(z) \in B(b, s)$$

for all $n \geq 0$.

(3)

$$r_b \in \left(0, \frac{1}{6}\min\Big\{R, \min\big\{|b - c| : c \in F\backslash\{b\}\big\}\Big\}\right)$$

for all $b \in F_0$.

(4)

$$F_1 \cap \overline{\mathrm{PS}(f)} = \emptyset.$$

Moreover,

$$r_b \in \left(0, \frac{1}{6}\min\Big\{R, \min\big\{|b - c| : c \in F\backslash\{b\}\big\}, \mathrm{dist}\big(b, \overline{\mathrm{PS}}(f)\big)\Big\}\right)$$

for all $b \in F_1$.

(5) *Suppose that* $a, b \in F$ *and* $n \geq 1$ *is an integer. Assume that either* $b \in F_1$ *or there exists a point*

$$w \in f^{-n}(U_0(b)) \tag{12.53}$$

such that if

$$f^{n-1}(w) \notin U_0(b), \tag{12.54}$$

then

(a) *the holomorphic inverse branch* $f_w^{-n} : B(b, 6r_b) \longrightarrow X$ *of* f^n, *sending* $f^n(w)$ *to* w, *is well defined (this is only an extra hypothesis if* $b \in F_0$; *if* $b \in F_1$, *this follows from (4)).*

(b) *If, in addition, the connected component of* $f^{-n}(B(b, 2r_b))$ *containing* w *intersects* $B(a, 2r_a)$ *and*

$$f^j(w) \notin \bigcup_{c \in F} U_0(c) \tag{12.55}$$

for all $j = 1, 2, \ldots, n-1$, *then*

$$|(f^n)'(z)| \geq \frac{8\kappa}{\kappa - 1} \cdot \frac{r_b}{r_a} \tag{12.56}$$

for all $z \in f_w^{-n}(B(b, 2r_b))$.

(6) *For every* $b \in F_0$ *and every* $j \in \{1, 2, \ldots, p_b\}$, *there exists* $V_j(b)$, *an open neighborhood of* b *in* Y *such that*

   (a) *The set* $\overline{V_j(b)} \cap \overline{U_0(b, j)}$ *is a closed topological disk.*
   (b) *The set* $V_j(b) \backslash \overline{U_0(b, j)}$ *is connected (and, in consequence, the set* $\overline{V_j(b) \backslash U_0(b, j)}$ *is connected too).*
   (c) $\overline{V_j(b)} \subseteq B(b, r_b'/4)$ *with some* $r_b' \in (0, (r_b)_b^-)$.

(7) $|f'(b)| = 1$ *for every* $b \in F_0$.
(8)

$$\mathrm{Crit}(f) \cap \bigcup_{w \in F_0} B(w, 6r_w) = \emptyset.$$

(9) $\forall b \in F_0 \quad \omega(\mathrm{Crit}(f)) \cap B(b, 6r_b) \subseteq \{b\}$.

Then the system $\mathcal{S}_U$ *generated by the domains of* (12.47) *and by the maps of* (12.50) *is a maximal CGDMS in the sense of Definitions 11.3.1 and* 11.9.1.

Proof The number $u \in (0, r_b'/4)$ will be further required to be sufficiently small in the course of the proof. For every integer $n \geq 1$, every $b \in F_1$, and every $j \in \{1, \ldots, b\}$, we have that

$$\begin{aligned} f_b^{-n}(X(b, j; u)) &\subseteq f_b^{-n}(X^*(b, j)) \subseteq f_b^{-n}\left(\overline{U(b, j)} \backslash f_b^{-1}(U(b, j))\right) \\ &\subseteq f_b^{-n}(\overline{U(b, j)}) \subseteq f_b^{-1}(\overline{U(b, j)}). \end{aligned}$$

Hence,

$$\begin{aligned} &\mathrm{Int}\big((X(b, j; u)) \cap f_b^{-n}(X(b, j; u))\big) \\ &\subseteq \mathrm{Int}\left(\overline{U(b, j)} \backslash f_b^{-1}(U(b, j))\right) \cap f_b^{-1}(\overline{U(b, j)}) \\ &= \left(\mathrm{Int}(\overline{U(b, j)}) \backslash \overline{f_b^{-1}(U(b, j))}\right) \cap f_b^{-1}(\overline{U(b, j)}) \\ &\subseteq \left(\mathrm{Int}(\overline{U(b, j)}) \backslash \overline{f_b^{-1}(U(b, j))}\right) \cap \overline{f_b^{-1}(U(b, j))} \\ &= \emptyset. \end{aligned} \tag{12.57}$$

Fix two sets A, B as in (12.48). Suppose that (12.49) holds and fix

$$\xi \in \mathrm{Int}(A) \cap f^{-n}(\mathrm{Int}(B)).$$

We consider several cases.

Case 1°. $A = X_a$ and $B = X_b$ for some $a, b \in F_1$. It then follows from Theorem 12.2.1, Proposition 12.1.6, and (12.52) that

$$\begin{aligned} f_\xi^{-n}(X(b,j;u)) &= f_\xi^{-n}(\overline{U(b,1)}) \\ &\subseteq \overline{f_\xi^{-n}(U(b,1))} \subseteq \overline{U(a,1),} = X_a = X(a,1;u) \end{aligned} \tag{12.58}$$

and we are done in this case.

Case 2°. $A = X_a$ for some $a \in F_1$ and $B = X(b,j;u)$ for some $b \in F_0$ and $j \in \{1,2,\dots,p_b\}$. Then, similar to the previous case, it follows from Theorem 12.2.1, Proposition 12.1.6, and (12.52) that

$$\begin{aligned} f_\xi^{-n}(X(b,j;u)) &\subseteq f_\xi^{-n}(\overline{U(b,j)}) \\ &\subseteq \overline{f_\xi^{-n}(U(b,j))} \subseteq \overline{U(a,1)} = X_a = X(a,1;u) \end{aligned} \tag{12.59}$$

and we are done in this case too.

Case 3°. $A = X(a,j;r)$ for some $a \in F_0$ and $j \in \{1,2,\dots,p_a\}$ and $B = X_b$ for some $b \in F_1$. Seeking contradiction, suppose that

$$f_\xi^{-n}(U(b,1)) \not\subseteq (U(a,j)) \backslash f_a^{-1}(\overline{U(a,j)}).$$

But by Theorem 12.2.1, Proposition 12.1.6, and (12.52),

$$f_\xi^{-n}(U(b,1)) \subset (U(a,j)).$$

So,

$$f_\xi^{-n}(U(b,1)) \cap f_a^{-1}(\overline{U(a,j)}) \neq \emptyset.$$

Hence, $f_\xi^{-n}(U(b,1)) \cap f_a^{-1}(U(a,j)) \neq \emptyset$. Thus, $f \circ f_\xi^{-n}(U(b,1)) \cap U(a,j) \neq \emptyset$. Let then $k \geq 1$ be the largest integer such that

$$f^s \circ f_\xi^{-n}(U(b,1)) \cap U(a,j) \neq \emptyset \tag{12.60}$$

for all $1 \leq s \leq k$. Such an integer k exists and $k < n$ because

$$U(b,1) \cap U(a,j) = \emptyset \quad (\text{as } a \neq b). \tag{12.61}$$

It also follows from (12.60) and (12.51) that

$$\begin{aligned} \emptyset \neq f^k \circ f_\xi^{-n}(U(b,1)) \cap U(a,j) &\subseteq U(a,j) \backslash \left(\mathrm{Int}(\overline{U(a,j)}) \backslash \overline{f_a^{-1}(U(a,j))} \right) \\ &\subseteq U(a,j) \cap \overline{f_a^{-1}(U(a,j))}. \end{aligned}$$

Since the set $f^k \circ f_\xi^{-n}(U(b,1))$ is open, this implies that

$$f^k \circ f_\xi^{-n}(U(b,1)) \cap f_a^{-1}(U(a,j)) \neq \emptyset.$$

Hence,

$$f^{k+1} \circ f_\xi^{-n}(U(b,1)) \cap U(a,j) \neq \emptyset,$$

contrary to the definition of k. We have, thus, proved that

$$f_\xi^{-n}(U(b,1)) \subseteq U(a,j) \backslash f_a^{-1}(\overline{U(a,j)}). \tag{12.62}$$

Since the sets $\{U(c,k);\ c \in F,\ k \in \{1,\dots,p_c\}\}$ are pairwise disjoint, it follows from (12.62) and item (E) of Theorem 12.2.1 that

$$f_\xi^{-n}(U(b,1)) \cap \bigcup_{c\in F}\bigcup_{k=1}^{p_c} f_c^{-1}(U(c,k)) = \emptyset.$$

Hence,

$$f \circ f_\xi^{-n}(U(b,1)) \cap \bigcup_{c\in F}\bigcup_{k=1}^{p_c} U(c,k) = \emptyset. \tag{12.63}$$

Now, seeking contradiction, suppose that

$$f_\xi^{-n}(U(b,1)) \nsubseteq X(a,j;u).$$

It then follows from (12.45), (12.46), and (12.62) that

$$f_\xi^{-n}(U(b,1)) \cap V_j(a,u) \neq \emptyset. \tag{12.64}$$

Now it is the moment to fix $u \in (0, r_b')$ sufficiently small. We assume, in addition to all other assumptions, that the number r is so small that

$$V_i(w,u) \cap f(V_i(w,u)) \cup f^2(V_i(w,u)) \subseteq B(w, r_w')$$

for all $w \in F_0$ and $i \in \{1,\dots,p_w\}$. It then follows from (12.64) that

$$f \circ f_\xi^{-n}(U(b,1)) \cap B(a, r_a') \neq \emptyset. \tag{12.65}$$

This, along with (12.63) and hypothesis (2j) (and since $r_b' \leq (r_b)_b^-$ and $U_0(c,k) \subseteq U(c,k)$ for all $c \in F$ and $k \in \{1,\dots,p_c\}$), yields

$$\emptyset \neq f^l\left(f \circ f_\xi^{-n}(U(b,1)) \cap B(a,r_a')\right) \subseteq B(a,r_a) \tag{12.66}$$

for all $l \geq 0$. In particular, taking $l = n-1 \geq 0$, we get that $U(b,1) \cap B(a,r_a) \neq \emptyset$. Since $a \neq b$, this contradicts hypotheses (3) and (4). The inclusion

$$f^k \circ f_\xi^{-n}(U(b,1)) \subseteq X(a,j;u)$$

is, thus, proved. This, in turn, by minimality of n, implies that $k = 0$, whence

$$f_\xi^{-n}(U(b,1)) \subseteq X(a,j;u). \tag{12.67}$$

Therefore, as the set $X(b,j;u)$ is closed,

$$f_\xi^{-n}(B) = f_\xi^{-n}(X_b) = f_\xi^{-n}(\overline{U(b,1)}) \subseteq \overline{f_\xi^{-n}(U(b,1))} \subseteq X(a,j;u) = A. \tag{12.68}$$

Case 4°. $A = X(a,i;u)$ and $B = X(b,j;u)$ for some $a \in F_0$, $i \in \{1,\dots,p_a\}$, $b \in F_0 \setminus \{a\}$, and $j \in \{1,\dots,p_b\}$.

Since $U(b,j) \subseteq W(b,j;u)$ (so f_ξ^{-n} is well defined on $U(b,j)$, we can proceed in this case in entirely the same way as in Case 3°, with only $U(b,1)$ replaced by $U(b,j)$. We then end up with the formula $f_\xi^{-n}(U(b,j)) \subseteq X(a,i;u)$, corresponding to (12.67) of Case 3°. Then

$$f_\xi^{-n}(X(b,j;u)) \subseteq f_\xi^{-n}(\overline{U(b,j)}) \subseteq \overline{f_\xi^{-n}(U(b,j))} \subseteq X(a,i;u). \tag{12.69}$$

Consider the following in turn.

Case 5°. $A = X(c,i;u)$ and $B = X(c,j;r)$ for some $c \in F_0$ and $i \neq j \in \{1,\dots,p_c\}$.

We proceed exactly as in Case 3° with obvious replacements $X(a,j;u)$ by $X(c,i;u)$ and $U(b,1)$ by $U(b,j)$. Also, now the formula

$$U(a,i) \cap U(a,j) = \emptyset$$

replaces (12.61) and holds not because $a \neq b$ but because $i \neq j$. We then obtain the formula

$$f_\xi^{-n}(U(c,j)) \subseteq X(c,i;u),$$

corresponding to (12.67) of Case 3°. Hence,

$$f_\xi^{-n}(X(c,j;u)) \subseteq f_\xi^{-n}(\overline{U(c,j)}) \subseteq \overline{f_\xi^{-n}(U(c,j))} \subseteq X(c,i;u). \tag{12.70}$$

We now consider the following.

Case 6°. Eventually, assume that $A = X(a,j;u)$ and $B = X(a,j;u)$ with some $c \in F_0$ and $j \in \{1,\dots,p_c\}$. If

$$f^u(f_\xi^{-n}(U(c,j)) \cap B(c,6r_c)) \neq \emptyset$$

for all integers $u = 0,1,\dots,n$, then $f_\xi^{-n} = f_c^{-n}$; so, this is ruled out by (12.57). Hence,

$$f^u(f_\xi^{-n}(U(c,j)) \cap B(c,6r_c)) = \emptyset \tag{12.71}$$

for some integer $u \in \{0, 1, \ldots, n\}$. We now proceed again exactly as in Case 3° with obvious replacements $X(a, j; u)$ by $X(c, j; u)$ and $U(b, 1)$ by $U(c, j)$. But now the fact that the integer $k \geq 1$, involved in (12.60), is strictly smaller than n follows not from $a \neq b$ but from (12.71). As in the previous case (Case 5°), we then obtain the formula

$$f_\xi^{-n}(U(c, j)) \subseteq X(c, j; u),$$

corresponding to (12.67) of Case 3°. Hence,

$$f_\xi^{-n}(X(c, j; u)) \subseteq f_\xi^{-n}(\overline{U(c, j)}) \subseteq \overline{f_\xi^{-n}(U(c, j))} \subseteq X(c, j; u). \quad (12.72)$$

Because of this, Lemma 8.1.19, and Remarks 11.2.2 and 11.3.3, condition (f) of Definition 11.2.1 of GDMSs is, thus, satisfied.

Let us now prove condition (4b) of Definition 11.3.1 of CGDMSs, i.e., the Open Set Condition. So, suppose that A, B, C, and D are in the set of (12.48),

$$\xi \in \mathrm{Int}(A) \cap f^{-m}(\mathrm{Int}(B)), \quad \zeta \in \mathrm{Int}(C) \cap f^{-n}(\mathrm{Int}(D)),$$

(12.51) holds for ξ and B and for ζ and D, and

$$f_\xi^{-m}(\mathrm{Int}(B)) \cap f_\zeta^{-n}(\mathrm{Int}(D)) \neq \emptyset \quad (12.73)$$

for some integers $m, n \geq 1$. Because of the, already proven, condition (f) of Definition 11.2.1, we conclude from (12.73) that $A \cap C \neq \emptyset$. Therefore, $A = C$. Suppose without loss of generality that $m \leq n$. It also follows from (12.73) that

$$\mathrm{Int}(B) \cap f^m\left(f_\zeta^{-n}(\mathrm{Int}(D))\right) \neq \emptyset.$$

Along with (12.51), this implies that $m = n$. But then it follows from (12.73) that $B = D$. Applying then (12.73) once more and remembering that $U_A = U_B$ is an open connected, simply connected set, we conclude that $f_\xi^{-m} = f_\zeta^{-n}$. The Open Set Condition, i.e., condition (4b) of Definition 11.3.1, is satisfied.

Since condition (4c) of Definition 11.3.1 of CGDMSs is satisfied because of (12.50), the proof of Theorem 12.4.4 is complete. ■

Remark 12.4.5 Instead of assuming in Theorem 12.4.4 that f satisfies the Standard Property and $F \cap O_+(Q) = \emptyset$, we could have assumed that $r_a = r_b$ for all $a, b \in F$ and used this assumption rather than Lemma 8.1.19 to prove condition (f) of Definition 11.2.1.

12.5 An Auxiliary Technical Result

We will prove in this very short section the following result, which is somewhat close to the subject of the current chapter, is interesting on its own, and which will be used in Volume 2 of this book.

Lemma 12.5.1 *Let Y be either a parabolic or an elliptic ($\widehat{\mathbb{C}}$) Riemann surface. Let X be an open subset of Y, which, moreover, is a subset of $\mathbb{C}$ if $Y = \widehat{\mathbb{C}}$. Let*

$$f : X \longrightarrow Y$$

be an analytic map.

- *Fix a radius $R \in (0, u_Y)$.*
- *Assume that the Euclidean area of X is finite.*
- *Let $w \in X$. Let $r \in (0, R]$ be so small that $B(w, 2r) \subseteq X$.*

Then, for every $n \geq 1$,

$$\sup\left\{\left|\left(f_V^{-n}\right)'(z)\right| : V \in \bigcup_{k=1}^{n} \mathrm{Comp}_k^*(w,r),\ z \in B(w,r)\right\} < +\infty$$

and, for every $\varepsilon > 0$,

$$\#\left\{V \in \bigcup_{k=1}^{n} \mathrm{Comp}_k^*(w,r) : \left|\left(f_V^{-n}\right)'(z)\right| \geq \varepsilon \text{ for some } z \in B(w,r)\right\} < +\infty.$$

Proof Obviously, it is enough to prove that

$$\sup\left\{\left|\left(f_V^{-n}\right)'(z)\right| : V \in \mathrm{Comp}_n^*(w,r),\ z \in B(w,r)\right\} < +\infty \tag{12.74}$$

for all $n \geq 1$ and

$$\#\left\{V \in \mathrm{Comp}_n^*(w,r) : \left|\left(f_V^{-n}\right)'(z)\right| \geq \varepsilon \text{ for some } z \in B(w,r)\right\} < +\infty \tag{12.75}$$

for all $n \geq 1$ and all $\varepsilon > 0$. Since the components $V \in \mathrm{Comp}_n^*(w,r)$ are mutually disjoint, by applying Theorem 8.3.9, we get that

$$+\infty > \mathrm{Area}(X) \geq \sum_{V \in \mathrm{Comp}_n^*(w,r)} \mathrm{Area}(V) \geq K^{-2}\pi r^2 \sum_{V \in \mathrm{Comp}_n^*(w,r)} \left|\left(f_V^{-n}\right)'(w)\right|^2. \tag{12.76}$$

Therefore, using Theorem 8.3.9 again, we get that

$$\begin{aligned}\sup\{\left|\left(f_V^{-n}\right)'(z)\right| &: V \in \mathrm{Comp}_n^*(w,r), z \in B(w,r)\} \\ &\leq K \sup\left\{\left|\left(f_V^{-n}\right)'(w)\right| : V \in \mathrm{Comp}_n^*(w,r)\right\} < +\infty.\end{aligned}$$

This means that (12.74) is proved.

Denote the set of components V involved in (12.75) by $C_n(\varepsilon)$ and the corresponding points $z \in B(w,r)$ by z_V. Using (12.76) and employing Theorem 8.3.9, we get that

$$
\begin{aligned}
+\infty > \text{Area}(X) &\geq r^2 \sum_{V \in C_n(\varepsilon)} |(f_V^{-n})'(w)|^2 \\
&\geq K^{-2}\pi K^{-2} r^2 \sum_{V \in C_n(\varepsilon)} |(f_V^{-n})'(z_V)\|^2 \\
&\geq K^{-4}\pi r^2 \varepsilon^2 \# C_n(\varepsilon).
\end{aligned}
$$

Hence, $\#C_n(\varepsilon) \leq K^4\pi^{-2}r^{-2}\varepsilon^{-2}\text{Area}(X) < +\infty$. This means that (12.75) is proved and we are done. ■

References

[Aa] J. Aaronson, *An Introduction to Infinite Ergodic Theory*, Mathematical Surveys and Monographs, vol. 50, American Mathematical Society (1997). 2.3, 5.1, 5.2, 5.3, 5.3

[ADU] J. Aaronson, M. Denker, M. Urbański, *Ergodic theory for Markov fibered systems and parabolic rational maps*, Trans. Am. Math. Soc. 337 (1993), 495–548. (Preface)

[Ab] L.M. Abramov, *On the entropy of a flow*, Dokl. Akad. Nauk. SSSR. 128 (1959), 873–875. 6.6

[Ah] L.V. Ahlfors, *Zur Theorie der Überlagerungsflächen*, Acta Math. 65 (1935), 157–194. 8.2

[Al] D.S. Alexander, *A History of Complex Dynamics from Schröder to Fatou and Julia*, Springer Fachmedien Wiesbaden (1994). Originally published by Friedr. Vieweg & Sohn Verlagsgesellschaft (1994). Softcover reprint of the hardcover first edition (1994). (Preface), 8

[AIR] D.S. Alexander, F. Iavernaro, A. Rosa, *Early Days in Complex Dynamics: A History of Complex Dynamics in One Variable During 1906–1942*, History of Mathematics, vol. 38, American Mathematical Society, London Mathematical Society (2012). (Preface), 8

[AIM] K. Astala, T. Iwaniec, G. Martin, *Elliptic Partial Differential Equations and Quasiconformal Mappings in the Plane*, Princeton Mathematical Series, vol. 48, Princeton University Press (2009). 8.2

[Ba1] I.N. Baker, *Multiply connected domain of normality in iteration theory*, Math. Z. 81 (1963), 206–214. (Preface)

[BKL1] I.N. Baker, J. Kotus, Y. Lü, *Iterates of meromorphic functions I*, Ergodic Theory Dyn. Syst. 11 (1991), 241–248. (Preface)

[BKL2] I.N. Baker, J. Kotus, Y. Lü, *Iterates of meromorphic functions II: Examples of wandering domains*, J. Lond. Math. Soc. 42 (1990), 267–278.

[BKL3] I.N. Baker, J. Kotus, Y. Lü, *Iterates of meromorphic functions III: Preperiodic domains*, Ergodic Theory Dyn. Syst. 11 (1991), 603–618.

[BKL4] I.N. Baker, J. Kotus, Y. Lü, *Iterates of meromorphic functions IV: Critically finite functions*, Results Math. 22 (1992), 651–656. (Preface)

[Ba] K. Barański, *Hausdorff dimension and measures on Julia sets of some meromorphic maps*, Funda. Math. 147 (1995), 239–260. (Preface)

[BKZ1] K. Barański, B. Karpińska, A. Zdunik, *Bowen's formula for meromorphic functions*, Ergodic Theory Dyn. Syst. 32:4 (2012), 1165–1189. 10

[BKZ2] K. Barański, B. Karpińska, A. Zdunik, *Conformal measures for meromorphic maps*, Ann. Acad. Sci. Fenn. Math. 43:1 (2018), 247–266. 10

[Bea] A.F. Beardon, *Iteration of Rational Maps*, Graduate Texts in Mathematics, vol. 132, Springer-Verlag (1991). (Introduction), 8.6

[Bed] T. Bedford, *Hausdorff dimension and box dimension in self-similar sets*, Proc. Conf. Topology and Measure V, Ernst–Moritz–Arndt Universität Greisfwald (1988). 11.5

[Ber1] W. Bergweiler, *Iteration of meromorphic functions*, Bull. New Ser. Am. Math. Soc. 29:2 (1993), 151–188. (Preface)

[Bie] L. Bieberbach, *Über die Koeffizienten derjenigen Potenzreihen, welche eine schlichte Abbildung des Einheitskreises vermitteln*, Preuss. Akad. Wiss., Phys.-Math. Kl., 138 (1916), 940–955. 8.3

[Bil1] P. Billingsley, *Ergodic Theory and Information*, R. E. Krieger Pub. Co. (1978). 9.5

[Bil2] P. Billingsley, *Probability and Measure*, 3rd ed., John Wiley & Sons (1995). 1.4, 1.4, 1.4, 3.3

[Bir] G.D. Birkhoff, *Proof of the ergodic theorem*, Proc. Natl. Acad. Sci. USA 17 (1931), 656–660. 2, 2.3

[Bog] V.I. Bogachev, *Measure Theory*, vols. I and II, Springer-Verlag (2007). 11.6

[Bow1] R. Bowen, *Equilibrium states and the ergodic theory for Anosov diffeomorphisms*, Lecture Notes in Mathematics, vol. 470, Springer-Verlag (1975). 7, 7.5

[Bow2] R. Bowen, *Hausdorff dimension of quasi-circles*, Publ. Math. IHES 50 (1980), 11–25. 9.5, 11.5

[BF] B. Branner, N. Fagella, *Quasiconformal Surgery in Holomorphic Dynamics*, Cambridge Studies in Advanced Mathematics, vol. 141, Cambridge University Press (2014). 8.2

[CLU1] V. Chousionis, D. Leykekhman, M. Urbański, *Dimension spectrum of conformal graph directed Markov systems*, Sel. Math. New Ser. 25 (2019), 40. (Introduction), 11

[CLU2] V. Chousionis, D. Leykekhman, M. Urbański, *On the dimension spectrum of infinite subsystems of continued fractions*, Trans. Am. Math. Soc. 373 (2020), 1009–1042. (Introduction), 11

[CU] V. Chousionis, M. Urbański, *Porosity in conformal dynamical systems*, Math. Proc. Camb. Philos. Soc. 172 (2022), 303–371. (Introduction), 11

[CTU] V. Chousionis, J. Tyson, M. Urbański, *Conformal Graph Directed Markov Systems on Carnot Groups*, Memoirs of the American Mathematical Society, vol. 266, American Mathematical Society (2020). (Introduction), 11, 11.6, 11.6

[Coh] D. Cohn, *Measure Theory*, 2nd ed., Birkhäuser Advanced Texts/Basler Lehrbücher, Birkhäuser (2013). 1.1, 1.2

[DMNU] M. Denker, R.D. Mauldin, Z. Nitecki, M. Urbański, *Conformal measures for rational functions revisited*, Funda. Math. 157 (1998), 161–173. 10.3

[DU1] M. Denker, M. Urbański, *On the existence of conformal measures*, Trans. Am. Math. Soc. 328 (1991), 563–587. (Introduction), 10, 10.1

[DU2] M. Denker, M. Urbański, *Hausdorff and conformal measures on Julia sets with a rationally indifferent periodic point*, J. Lond. Math. Soc. 43 (1991), 107–118. 1.3, 10

[DU3] M. Denker, M. Urbański, *On Sullivan's conformal measures for rational maps of the Riemann sphere*, Nonlinearity 4 (1991), 365–384. 10

[DU4] M. Denker, M. Urbański, *On Hausdorff measures on Julia sets of subexpanding rational maps*, Israel J. Math. 76 (1991), 193–214. (Preface)

[DU5] M. Denker, M. Urbański, *Geometric measures for parabolic rational maps*, Ergodic Theory Dyn. Syst. 12 (1992), 53–66. (Preface), 10

[DU6] M. Denker, M. Urbański, *The capacity of parabolic Julia sets*, Math. Z. 211 (1992), 73–86. 10

[Do] N. Dobbs, *Nice sets and invariant densities in complex dynamics*, Math. Proc. Camb. Philos. Soc. 150 (2011), 157–165. 12

[DS] N. Dunford, J. Schwartz, *Linear Operators, Part 1: General Theory*, Wiley-Interscience (1988). 3.2

[Eg] H. Eggleston, *Sets of fractional dimensions which occur in some problems of number theory*, Proc. Lond. Math. Soc. 54 (1952), 42–93. 9.5

[EL] A. Eremenko, M. Lyubich, *Dynamical properties of some classes of entire functions*, Ann. Inst. Fourier, 42 (1992), 989–1020. (Preface)

[Fal1] K. Falconer, *Fractal Geometry: Mathematical Foundations and Applications*, 2nd ed., Wiley (2003). 1.5

[Fal12] K. Falconer, *Techniques in Fractal Geometry*, Wiley (1997). 1.5

[Fal13] K. Falconer, *The Geometry of Fractal Sets*, Cambridge University Press (1985). 1.5

[FdM] E. de Faria, W. de Melo, *Mathematical Tools for One-Dimensional Dynamics*, Cambridge Studies in Advanced Mathematics, vol. 115, Cambridge University Press (2008). 8.2

[Fat1] P. Fatou, *Sur les équations fonctionelles*, Bull. Soc. Math. France 47 (1919), 161–271. (Preface)

[Fat2] P. Fatou, *Sur les équations fonctionelles transcendantes*, Bull. Soc. Math. France 48 (1920), 208–314. (Preface)

[Fat3] P. Fatou, *Sur l'itération des fonctions transcendantes entèries*, Acta Math. 47 (1926), 337–370. (Preface)

[FM] A. Fletcher, V. Markovic, *Quasiconformal Maps and Teichmüller Theory*, Oxford Graduate Texts in Mathematics, Oxford University Press (2006). 8.2

[For] O. Forster, *Lectures on Riemann Surfaces*, Springer-Verlag (1981). 8.1.1, 8.5

[Fr] G. Folland, *Real Analysis: Modern Techniques and Their Applications*, 2nd ed., Wiley (2007). 1.1, 1.2

[Gar] F. Gardiner, N. Lakic, *Quasiconformal Teichmuller Theory*, Mathematical Surveys & Monographs, vol. 76, American Mathematical Society (1999). 8.2

[Go1] S. Gouëzel, *Central limit theorem and stable laws for intermittent maps*, Probab. Theory Relat. Fields 128 (2004), 82–122. (Introduction), 4, 4.2, 4.2

[Go2] S. Gouëzel, *Almost sure invariance principle for dynamical systems by spectral methods*, Ann. Probab. 38 (2010), 1639–1671. 4.2

[Gr] H. Grötzsch, *Über die Verzerrung bei schlichten nicht-konformen Abbildungen und über eine damit zusammenhängende Erweiterung des Picardschen Satzes*, Sitzungsberichte sächs. Akad. Wiss., Math.–Phys. Klasse, 80 (1928), 503–507. 8.2

[Gu] M. Guzmán, *Differentiation of Integrals in* $\mathbb{R}^n$, Lecture Notes in Mathematics, vol. 541, Springer Verlag (1976). 1.3

[Ha] A. Hatcher, *Algebraic Topology*, Cambridge University Press (2002). 8.1.1, 8.2

[H] F. Hausdorff, *Dimension und äußeres Maß*, Math. Ann. 79 (1919), 157–179. 1.5, 1.5

[HK1] J. Hawkins, L. Koss, *Ergodic properties and Julia sets of Weierstrass elliptic functions*, Monatsh. Math. 137 (2002), 273–301. (Preface)

[Heino] J. Heinonen, *Lectures on Analysis on Metric Spaces*, Universitext, Springer (2001). 1.3

[Hen] D. Hensley, *Continued Fractions*, World Scientific (2006). 11

[Hi] E. Hille, *Analytic Function Theory*, vols. I and II, Ginn and Company (1962). 8.3

[Hub] J. Hubbard, *Teichmuller Theory and Applications to Geometry, Topology, and Dynamics*, Matrix Editions (2006). 8.2

[Hut] J.E. Hutchinson, *Fractals and self-similarity*, Indiana Univ. Math. J. 30 (1981), 713–747. 11.5

[Ju] G. Julia, *Mémoire sur l'iteration des fonctions rationnelles*, J. Math. Pures Appl. 8 (1918), 47–245. (Preface)

[KH] A. Katok, B. Hasselblatt, *Introduction to the Modern Theory of Dynamical Systems*, revised ed., Encyclopedia of Mathematics and its Applications, vol. 54, Cambridge University Press, (1996). 2.3

[Kod] K. Kodaira, *Complex Analysis*, Cambridge University Press (2007). 8.1.1

[Kot] J. Kotus, *On the Hausdorff dimension of Julia sets of meromorphic functions. II*, Bull. Soc. Math. Fr. 128 (1995), 33–46. (Preface)

[KU1] J. Kotus, M. Urbański, *Conformal, geometric and invariant measures for transcendental expanding functions*, Math. Ann. 324 (2002), 619–656. (Preface)

[KU2] J. Kotus, M. Urbański, *Existence of invariant measures for transcendental subexpanding functions*, Math. Z. 243 (2003) 25–36. 2.4, 10

[KU3] J. Kotus, M. Urbański, *Hausdorff dimension and Hausdorff measures of elliptic functions*, Bull. Lond. Math. Soc. 35 (2003), 269–275. (Preface)

[KU4] J. Kotus, M. Urbański, *Geometry and ergodic theory of non-recurrent elliptic functions*, J. Anal. Math. 93 (2004) 35–102. (Preface)

[Kur] K. Kuratowski, *Topology II*, Academic Press and PWN, (1968). 12.1, 12.1, 12.2

[Le] O. Lehto, *Univalent Functions and Teichmüller Spaces*, Springer (2011). 8.2

[LV] O. Lehto, K.I. Virtanen, *Quasiconformal Mappings in the Plane*, Springer-Verlag (1973). 8.2, 8.2

[Lj] M. Ljubich, *Dynamics of rational transforms: topological picture*, Russ. Math. Surv. 41:4 (1986), 43–117. 10.3

[Ly] M. Lyubich, *The measurable dynamics of the exponential map*, Siberian J. Math. 28 (1987), 111–127. (Preface)

[M2] R. Mañé, *The Hausdorff dimension of invariant probabilities of rational maps*, Lecture Notes in Mathematics, vol. 1331, (1988), 86–117. 9.5

[M3] R. Mañé, *Ergodic Theory and Differentiable Dynamics*, Springer (1987). 9.2, 9.2.3

[MM] A. Manning, H. McCluskey, *Hausdorff dimension for horseshoes*, Ergodic Theory Dyn. Syst. 3 (1983), 251–260. 9.5

[Mar] M. Martens, *The existence of σ-finite invariant measures, Applications to real one-dimensional dynamics*. https://arxiv.org/abs/math/9201300. (Introduction), 2.4

[Mas] W. Massey, *A Basic Course in Algebraic Topology*, Springer (1991). 8.2

[Mat] P. Mattila, *Geometry of Sets and Measures in Euclidean Spaces, Fractals and Rectifiability*, Cambridge Studies in Advanced Mathematics, Cambridge University Press (1995). 1.3, 1.5

[MPU] R.D. Mauldin, F. Przytycki, M. Urbański, *Rigidity of conformal iterated function systems*, Compos. Math. 129 (2001), 273–299. (Introduction), 11

[MSzU] R.D. Mauldin, T. Szarek, M. Urbański, *Graph directed Markov systems on Hilbert spaces*, Math. Proc. Camb. Philos. Soc. 147 (2009), 455–488. (Introduction), 1.3, 11, 11.5

[MU1] R.D. Mauldin, M. Urbański, *Dimensions and measures in infinite iterated function systems*, Proc. Lond. Math. Soc. 73:3 (1996), 105–154. (Preface), (Introduction), 10, 11, 11.3.2, 11.5, 11.5, 11.9

[MU2] R.D. Mauldin, M. Urbański, *Graph Directed Markov Systems: Geometry and Dynamics of Limit Sets*, Cambridge Tracts in Mathematics, Cambridge University Press (2003). (Preface), (Introduction), 4.2, 10, 11, 11.3.2, 11.3, 11.4, 11.5, 11.5, 11.6, 11.6, 11.9, 11.9

[MU3] R.D. Mauldin, M. Urbański, *Conformal iterated function systems with applications to the geometry of continued fractions*, Trans. Am. Math. Soc. 351 (1999), 4995–5025. (Introduction), 11

[MU4] R.D. Mauldin, M. Urbański, *Gibbs states on the symbolic space over an infinite alphabet*, Israel J. Math. 125 (2001), 93–130. (Preface), (Introduction), 11, 11.6

[MU5] R.D. Mauldin, M. Urbański, *Fractal measures for parabolic IFS*, Advances in Mathematics, vol. 168 (2002), 225–253. (Introduction), 11

[MSU] V. Mayer, B. Skorulski, M. Urbański, *Random Distance Expanding Mappings, Thermodynamic Formalism, Gibbs Measures, and Fractal Geometry*, Lecture Notes in Mathematics, vol. 2036, Springer (2011). 10

[MyU1] V. Mayer, M. Urbański, *Gibbs and equilibrium measures for elliptic functions*, Math. Z. 250 (2005), 657–683. (Preface)

[MyU2] V. Mayer, M. Urbański, *Geometric Thermodynamical Formalism and Real Analyticity for Meromorphic Functions of Finite Order*, Ergodic Theory Dyn. Syst. 28 (2008), 915–946. (Preface), 10

[MyU3] V. Mayer, M. Urbański, *Thermodynamical Formalism and Multifractal Analysis for Meromorphic Functions of Finite Order*, Memoirs of the American Mathematical Society, vol. 203, no. 954, American Mathematical Society (2010). (Preface), 10

[MyU4] V. Mayer, M. Urbański, *Thermodynamical formalism for entire functions and integral means spectrum of asymptotic tracts*, Trans. Am. Math. Soc. 373 (2020), 7669–7711. (Preface)

[MyU5] V. Mayer, M. Urbański, *Thermodynamic formalism and geometric applications for transcendental meromorphic and entire functions*, in M. Pollicott, S. Vaienti (eds.), Thermodynamic Formalism: CIRM Jean-Morlet Chair, Fall 2019, Springer (2021), 99–139. (Preface)

[McM1] C. McMullen, *Area and Hausdorff dimension of Julia sets of entire functions*, Trans. Am. Math. Soc. 300 (1987), 329–342. 1.7, 1.7

[McM2] C. McMullen, *Hausdorff dimension and conformal dynamics II: Geometrically finite rational maps*, Comment. Math. Helv. 75 (2000), 535–593. 10.3

[MN] I. Melbourne, M. Nicol, *Almost sure invariance principle for nonuniformly hyperbolic systems*, Commun. Math. Phys. 260 (2005), 131–146. 4.2

[MT] I. Melbourne, D. Terhesiu, *Decay of correlations for nonuniformly expanding systems with general return times*, Ergodic Theory Dyn. Syst. 34 (2014), 893–918. 5.3

[Mis] M. Misiurewicz, *A short proof of the variational principle for a $\mathbb{Z}_+^{\mathbb{N}}$ action on a compact space*, Bull. Acad. Pol. Math. 24 (1976), 1069–1075. 7, 7.5

[Ne] Z. Nehari, *Conformal Mapping*, Dover Books on Advanced Mathematics, Dover Publications (1975). 8.2

[vN] J. von Neumann, *Proof of the quasi-ergodic hypothesis*, Proc. Natl. Acad. Sci. USA 18 (1932), 70–82. 2.3

[Ne] M.H.A. Newman, *Elements of the Topology of Plane sets of Points*, Greenswood Press (1985). 12.1, 12.2

[Par] K.R. Parthasarathy, *Introduction to Probability and Measure*, Hindustan Book Agency (2005). 4.3

[Pat1] S.J. Patterson, *The limit set of a Fuchsian group*, Acta Math. 136 (1976), 241–273. (Introduction), 10, 10.1.1

[Pat2] S.J. Patterson, *Lectures on measures on limit sets of Kleinian groups*, in D.B.A. Epstein (ed.), Analytical and Geometric Aspects of Hyperbolic Space, London Mathematical Society, Lecture Note Series, vol. 111, Cambridge University Press (1987). (Introduction), 10, 10.1.1

[Pe1] Ya. Pesin, *Families of invariant manifolds corresponding to nonzero characteristic exponents*, Math. USSR Izv. 10 (1976), 1261–1305. 9.3

[Pe2] Ya. Pesin, *Characteristic exponents and smooth ergodic theory*, Russ. Math. Surv. 32 (1977), 55–114. 9.3

[Ph] R. Phelps, *Lectures on Choquet's Theorem*, Lecture Notes in Mathematics, vol. 1757, Springer (2001). 7.5.1

[P1] F. Przytycki, *Hausdorff dimension of harmonic measure on the boundary of an attractive basin for a holomorphic map*, Invent. Math. 80 (1985), 169–171. 9.5

[P2] F. Przytycki, *Lyapunov characteristic exponents are nonnegative*, Proc. AMS 119 (1993), 309–317. 9.1

[PR] F. Przytycki, J. Rivera-Letelier, *Statistical properties of topological Collet-Eckmann maps*, Ann. Sci. Éc. Norm. Supér. 40 (2007), 135–178. (Preface), 12

[PU2] F. Przytycki, M. Urbański, *Conformal Fractals: Ergodic Theory Methods*, London Mathematical Society Lecture Notes Series, vol. 371, Cambridge University Press (2010). 1.5, 7, 9.2, 9.3

[PUZI] F. Przytycki, M. Urbański, A. Zdunik, *Harmonic, Gibbs and Hausdorff measures on repellers for holomorphic maps I*, Ann. Mathe. 130 (1989), 1–40. 9.5

[Re] M. Rees, *The exponential map is not recurrent*, Math. Z. 191 (1986), 593–598. (Preface)

[Ri] J. Rivera-Letelier, *A connecting lemma for rational maps satisfying a no-growth condition*, Ergodic Theory Dyn. Syst. 27 (2007), 595–636. 12

[Ro] C.A. Rogers, *Hausdorff Measures*, Cambridge University Press (1998). 1.5

[RF] H. Royden, P. Fitzpatrick, *Real Analysis*, Classic Version, 4th ed., Pearson Modern Classics for Advanced Mathematics Series, Pearson (2017). 1.1, 1.2

[Ru1] D. Ruelle, *Thermodynamic Formalism*, Encyclopedia of Mathematics and its Applications, vol. 5, Addison-Wesley, Reading, MA. (1976). 7

[Ru2] D. Ruelle, *An inequality for the entropy of differentiable maps*, Bol. Soc. Brasil. Mat. 9 (1978), 83–87. 9.4

[Sc] J.L. Schiff, *Normal Families*, Universitext, Springer (1993). 8.1.2

[Sh1] C. Shannon, *A mathematical theory of communication*, Bell Syst. Techn. J., 27 (1948), 379–423, 623–656. 6

[Sh2] C. Shannon, *The Mathematical Theory of Communication*, University of Illinois Press (1949). 6

[Sin1] Ya.G. Sinai, *On the notion of entropy of a dynamical system*, Dokl. Akad. Nauk SSSR 124 (1959), 768–771. 6

[Sin2] Ya.G. Sinai, *Gibbs measures in ergodic theory*, Russ. Math. Surv. 27 (1972), 21–70. 7

[SkU1] B. Skorulski, M. Urbański, *The Law of Iterated Logarithm and equilibrium measures versus Hausdorff measures for dynamically semi-regular meromorphic functions*, in J. Barral, S. Seuret (eds.), Further Developments in Fractals and Related Fields, Trends in Mathematics, Birkhäuser (2013), 213–234. 4.1

[Sr] S.M. Srivastava, *A Course on Borel Sets,* Graduate Texts in Mathematics, vol. 180, Springer-Verlag (1998). 10.1.1, 10.1.1

[Sta1] G. Stallard, *Entire functions with Julia sets of zero measure*, Math. Proc. Camb. Philos. Soc. 108 (1990), 551–557. (Preface)

[Sta2] G. Stallard, *The Hausdorff dimension of Julia sets of entire functions*, Ergodic Theory Dyn. Syst. 11 (1991), 769–777. (Preface)

[Su2] D. Sullivan, *Seminar on conformal and hyperbolic geometry*, Institut des Hautes Études Scientifiques, preprint (1982). (Introduction), 10, 10.1.1

[Su3] D. Sullivan, *Conformal dynamical systems*. in J. Palis (ed.), Geometric Dynamics, Lecture Notes in Mathematics, vol. 1007, Springer Verlag (1981), 725–752. 10.1.1

[Su4] D. Sullivan, *Quasiconformal homeomorphisms in dynamics, topology, and geometry*, Proceedings of the International Congress of Mathematicians, Berkeley, American Mathematical Society, (1986), 1216–1228. 10

[Su5] D. Sullivan, *The density at infinity of a discrete group*, Inst. Hautes Études Sci. Pub. Math. 50 (1979), 259–277. 10, 10.1.1

[Su6] D. Sullivan, *Entropy, Hausdorff measures old and new, and the limit set of a geometrically finite Kleinian groups*, Acta Math. 153 (1984), 259–277. 1.5, 1.5

[Su7] D. Sullivan, *Disjoint spheres, approximation by imaginary quadratic numbers and the logarithmic law for geodesics*, Acta Math. 149 (1982), 215–237. (Introduction), 10, 10.1.1

[SU] H. Sumi, M. Urbański, *Measures and dimensions of Julia sets of semi-hyperbolic rational semigroups*, Discrete Contin. Dyn. Syst. 30 (2011), 313–363. 2.4

[SUZ] M. Szostakiewicz, M. Urbański, A. Zdunik, *Fine inducing and equilibrium measures for rational functions of the Riemann sphere*, Israel J. Math. 210 (2015), 399–465. 4.1, 4.2

[TT] S.J. Taylor, C. Tricot, *Packing measure, and its evaluation for a Brownian path*, Trans. Am. Math. Soc. 288 (1985), 679–699. 1.5, 1.5

[TZ] M. Thaler, R. Zweimüller, *Distributional limit theorems in infinite ergodic theory*, Probab. Theory Relat. Fields 135, (2006), 15–52. 5.3, 5.3, 5.3

[Tr] C. Tricot, *Two definitions of fractional dimension*, Math. Proc. Camb. Philos. Soc. 91 (1982), 57–74. 1.5, 1.5

[U1] M. Urbański, *The Hausdorff dimension of the set of points with non-dense orbit under a hyperbolic dynamical system*, Nonlinearity 4 (1991), 385–397. 1.7, 1.7

[U2] M. Urbański, *On some aspects of fractal dimensions in higher dimensional dynamics*, Proc. of the Göttingen Workshop Problems on Higher Dimensional Complex Dynamics, Mathematica Gottingensis 3 (1995), 18–25. 10.3

[U3] M. Urbański, *Rational functions with no recurrent critical points*, Ergodic Theory Dyn. Syst. 14 (1994), 391–414. (Preface), 8.4, 10.4

[U4] M. Urbański, *Geometry and ergodic theory of conformal non-recurrent dynamics*, Ergodic Theory Dyn. Syst. 17 (1997), 1449–1476. (Preface), 5.4

[U5] M. Urbański, *Diophantine approximation of self-conformal measures*, J. Number Theory 110 (2005) 219–235. (Introduction), 11

[U6] M. Urbański, *Porosity in conformal infinite iterated function systems*, J. Number Theory 88 (2001), 283–312. (Introduction), 11

[UZ1] M. Urbański, A. Zdunik, *The finer geometry and dynamics of exponential family*, Mich. Math. J. 51 (2003), 227–250. (Preface), 10, 10.3

[UZ2] M. Urbański, A. Zdunik, *Real analyticity of Hausdorff dimension of finer Julia sets of exponential family*, Ergodic Theory Dyn. Syst. 24 (2004), 279–315. (Preface), 10

[Wa1] P. Walters, *A variational principle for the pressure of continuous transformations*, Am. J. Math. 97 (1975), 937–971. 7, 7.5

[Wa2] P. Walters, *An Introduction to Ergodic Theory*, Graduate Text in Mathematics, Springer-Verlag (2000). 7

[LSY1] L.-S. Young, *Dimension, entropy and Lyapunov exponents*, Ergodic Theory Dyn. Syst. 2 (1982), 109–124. 9.5

[LSY2] L.-S. Young, *Statistical properties of dynamical systems with some hyperbolicity*, Ann. Math. 147 (1998) 585–650. (Introduction), 4, 4.2

[LSY3] L.-S. Young, *Recurrence times and rates of mixing*, Israel J. Math. 110 (1999), 153–188. (Preface), (Introduction), 4, 4.2, 4.2, 4.2

Index of Symbols

Subject Index